Toxic Metals in
Soil–Plant Systems

Toxic Metals in Soil–Plant Systems

Edited by
SHEILA M. ROSS
Department of Geography,
University of Bristol, UK

JOHN WILEY & SONS
Chichester · New York · Brisbane · Toronto · Singapore

Other Wiley Editorial Offices

John Wiley & Sons, Inc., 605 Third Avenue,
New York, NY 10158-0012, USA

Jacaranda Wiley Ltd, 33 Park Road, Milton,
Queensland 4064, Australia

John Wiley & Sons (Canada) Ltd, 22 Worcester Road,
Rexdale, Ontario M9W 1L1, Canada

John Wiley & Sons (SEA) Pte Ltd, 37 Jalan Pemimpin #05-04,
Block B, Union Industrial Building, Singapore 2057

Typeset in 10/12 Times by Mathematical Composition Setters Ltd., Salisbury, Wiltshire.
Printed and bound in Great Britain by Biddles Ltd., Guildford, Surrey.

Contents

List of Contributors

A. J. M. Baker
Department of Animal and Plant Sciences, University of Sheffield, PO Box 601, Sheffield S10 2UQ, UK

G. W. Barrett
Department of Zoology, Miami University, Oxford, Ohio 45056, USA

M. Benninger-Truax
Department of Biology, Hiram College, Hiram, Ohio 44234, USA

S. R. Brewer
Department of Zoology, Miami University, Oxford, Ohio 45056, USA

R. J. Bullock
Department of Botany, University of Bristol, University Road, Bristol BS8 1SS, UK

G. Clint
Institute of Terrestrial Ecology, Merlewood Experimental Station, Grange-over-Sands, Cumbria LA11 6LF, UK

N. M. Dickinson
School of Biological and Earth Sciences, Liverpool John Moores University, Byrom Street, Liverpool L3 3AF, UK

C. M. Finlayson
Office of the Supervising Scientist, Alligator Rivers Region Research Institute, Private Mail Bag 2, Jabiru, NT 0886, Australia

S. P. Hopkin
School of Animal and Microbial Sciences, University of Reading, PO Box 228, Reading, Berkshire RG6 2AJ, UK

A. D. Horrill
Institute of Terrestrial Ecology, Merlewood Experimental Station, Grange-over-Sands, Cumbria LA11 6LF, UK

K. J. Kaye
School of Geography, Oxford University, Mansfield Road, Oxford, OX1 3TB, UK

N. W. Lepp
School of Biological and Earth Sciences, Liverpool John Moores University, Byrom Street, Liverpool L3 3AF, UK

M. H. Martin
Department of Botany, University of Bristol, University Road, Bristol, BS8 1SS, UK

S. P. McGrath
Soil Science Department, AFRC Institute of Arable Crops Research, Rothamsted Experimental Station, Harpenden, Hertfordshire AL5 2JQ, UK

J. Proctor
Department of Biological and Molecular Sciences, University of Stirling, Stirling FK9 4LA, UK

S. M. Ross
Department of Geography, University of Bristol, University Road, Bristol BS8 1SS, UK

A. P. Turner
Department of Biological Sciences, Wye College, University of London, Wye, near Ashford, Kent TN25 5AH, UK

Preface

The aim of this text is to describe and discuss the processes of trace metal cycling in different contaminated ecosystems under conditions where their concentrations become toxic, treating the entire soil–plant system as a whole. The text is divided into an initial section in which key compartments of the soil–plant system are reviewed in detail, and a second section containing case studies from different environments and ecosystems. Some explanation is required of the terminology used throughout the text for the metals studied. Generally, two terms will be used in the text: either "trace metals" or "toxic metals". These terms are not synonymous. The intention will be to discuss the processes associated with those metals which are normally present in soils in trace amounts, defined, for example, by Mattigod et $al.$ (1981) as less than 1 mol m^{-3}. Concentrations of trace metals in soil can build up, due either to local variations in geochemistry, or to anthropogenic activities. When they occur in ecosystems in high quantities they can become toxic to plants, organisms and animals, including humans.

The term "heavy metal", although *generally* taken to mean metals of the periodic table with atomic numbers greater than 20, but excluding the alkali metals and the alkali earths (Tiller, 1989), or metals having specific gravities $> 5 \text{ g cm}^{-2}$, has not been *uniformly* defined in terms of atomic weight or specific gravity and will not be used here. Discussion will concentrate on metals which generally occur in trace amounts in environmental and ecological materials, and will not consider in detail the so-called metalloids (Sposito, 1986) such as arsenic, selenium or antimony. Further chemical definition of metals considered to be potentially toxic in soil–plant systems are discussed in Chapter 3. In this text, we are generally concerned with ecosystem pollution in which metal concentrations have been elevated to toxic levels through anthropogenic influences such as mining spoil or effluent, the application of sewage sludges, aerial inputs from exhausts, smelters or incinerator wastes, or metal mobilisation through the influence of acid precipitation. The metals considered most problematic in terms of environmental pollution and toxicity are lead, cadmium, zinc, copper and nickel but there are now many empirical studies showing evidence of the toxicity in soil–plant systems of other metals including manganese, mercury, tin and chromium. The toxicity of metals that are ubiquitous in soil, including aluminium and iron, may be mentioned in passing, but will not be discussed in detail.

Two central issues in discussing the effects that metals have on soil–plant systems are (i) factors influencing metal toxicity and (ii) factors influencing metal bioavailability. Important secondary issues include, on the biological side, development of adaptability and tolerance in affected individuals and populations and, on the soil side, the chemical speciation and long-term fate of trace metals under the influence of prevailing soil physical and chemical conditions.

Some concepts from classical toxicology will be used to discuss the specific toxicity of trace metals in soil organisms and in plants. Important determinants of metal toxicity in the environment and in different ecosystems are the mobility and bioavailability of various inorganic and organic chemical species formed in soils and waters. Mobility and bioavailability themselves are determined by a range of environmental factors including soil pH, redox, adsorption and exchange capacities, organic chelation, soil hydrology and microbial activity.

Section I of the text introduces concepts and processes which provide the background to understanding the effect of metals on soils and plants. The principal metal sources, both natural and anthropogenic, and their chemical forms, are introduced in Chapter 1. The complicated issue of the meaning of metal toxicity for soil organisms and for plants is discussed in Chapter 2. The effect of soil conditions and processes on metal speciation and chemical transformations, together with techniques which purport to assess "availability" of soil metals, is reviewed and discussed in Chapter 3. Chapter 4 reviews the ways in which plants cope with potentially toxic soil metal conditions. The first section concludes with Chapter 5 discussing some of the problems of studying and comparing the fate and distribution of metals in different impacted ecosystems, including forestry and agriculture. These five contributions provide the background for a range of ecosystem case studies in Section II of the text.

Our aim in examining a range of different case studies in the second section of the book is to compare and contrast the relative mobilities, retentions and cycling of metals under a range of soil and vegetation conditions. The spatial scale of approach in the case studies varies from field scale to whole catchments, while the temporal scale varies from one year to more than ten. In some cases, notably (Chapter 8 (Hopkin) and Chapter 6 (McGrath), detail of specific operative parts of ecosystem processes are discussed. In other studies, such as Chapter 7 (Barrett) and Chapter 9 (Martin and Bullock), whole integrated ecosystems are examined. Overall, the objective of the second section in the book is to present both new and continuing ecosystem studies, and to bring together published data on quite different environments and habitats. A subsequent aim is to identify both common and system-specific toxic metal processes, retention mechanisms or compartmentalisation. The habitats discussed in the example case studies range from agricultural land to forest and from salt marsh to upland heather moorland. The emphasis is predominantly on temperate ecosystems since this is where the wealth of experience in toxic metal research lies. The need for more detailed quantification and interpretation of toxic metal effects in tropical environments is emphasised.

Despite the fact that there are now numerous studies of the effects of unusual loadings of metals on different ecosystems (illustrated in many of the case studies and in the selective review given in Chapter 5, this volume), there are serious gaps in current knowledge of the effects of metals on soil physico-chemical processes, microbial processes and the growth and health of higher plants. Other serious gaps exist in knowledge of metal uptake and transport processes through biological cells and membranes. Fluxes between many ecosystem compartments are extremely difficult to measure and, if they are estimated at all, it is from difference calculations from compartment audits. In addition to these points, an understanding of the sources and forms of metals input to soils and their influence on living organisms requires clarification of the definition of toxicity and the meaning of critical metal concentrations for biological processes.

Many metal transformation, mobility, bioavailability and toxicology issues remain unresolved, but they provide technical, theoretical and ethical challenges for future environmental scientists to resolve.

REFERENCES

Mattigod, S. V., Sposito, G. and Page, A. L. (1981) Factors affecting the solubilities of metals in soils. In: Dowdy, R. H., Ryan, J. A., Volk, V. V. and Baker, D. E. (Eds) *Chemistry in the Soil Environment*, pp. 203–221. Americal Society of Agronomy and Soil Science Society of America, Special Publication No. 40. Americal Society of Agronomy, Madison, USA.

Sposito, G. (1986) Distribution of potentially hazardous trace metals. In: Sigel, H. (Ed.) *Metal Ions in Biological Systems*. Marcel Dekker, New York.

Tiller, K. G. (1989) Heavy metals in soils and their environmental significance. *Advances in Soil Science*, **9**, 113–142.

Acknowledgements

The idea for this book arose from a thematic conference session organised under the auspices of the Institute of British Geographers, in which a series of field case studies were presented, discussing the cycling of metals in ecosystems and the adaptation of plants to metal contamination. The final format and content of this text, however, has evolved considerably during the preparation stages. During discussions with various colleagues, the idea developed to include a mixture of forefront soil and plant process theory as well as a range of case studies. Compiling this mixture of subject matter has proved to be an extremely interesting and rewarding challenge. Many people have been particularly patient in discussing with me concepts of metal toxicity and soil restoration. In particular, I am indebted to Dr Mike Martin, Dr Denis Brown and Dr Richard Campbell of the Botany Department, Bristol University, Dr David Watson and Dr Bob Bell of the Environmental Advisory Unit, Cardiff and Liverpool Offices respectively, Dr Richard Bullock of Bristol Ecological Consultants and Dr Graham Nickless and Dr David Roberts of the Department of Chemistry, Bristol University. My postgraduate students in the Geography Department, Bristol University, have been especially supportive.

All chapters in the book have been refereed and my sincere thanks are due to a large number of international referees for giving so freely of their time. Thanks are due to Simon Godden who drew the majority of the diagrams in Section 1 of the book and to Liz Humphries and Anna Paszkowicz who typed many of the tables in this section. I am grateful to Krystina Brown who helped with copy-editing and to Anne-Marie Bremner for superb indexing. The highly professional support of the Wiley editorial team is most greatly appreciated.

S.M. Ross
Bristol, February 1994

Abbreviations, Acronyms and Symbols Used in the Text

[ ]	concentrations
( )	thermodynamic activities
AAF	Air accumulation factor (total lead concentration in foliage/lead concentration in air)
ASV	Anodic stripping voltametry
BTC	Breakthrough curve (soil solutes)
CEC	Cation exchange capacity
CF	Concentration factor (total metal concentration in plant/total metal concentration in soil)
DCPD	Dicalcium phosphate dihydrate
DTPA	Diethylenetriaminepentaacetic acid
EDTA	Ethylenediaminetetraacetic acid
Eh	Redox potential
EPA	Environmental Protection Agency (USA)
ESR	Electron spin resonance (spectrometry)
FA	Fulvic acid (organic molecules, usually of molecular weight 500–2000)
FYM	Farmyard manure
G_r°	Standard Gibbs free energy of a reaction
HA	Humic acid (organic molecules, usually of molecular weight 2000–10 000)
ICRCL	Inter-departmental Committee on the Redevelopment of Contaminated Land (UK).
ICP–AES	Inductively coupled plasma–atomic emission spectrometry
ICP–MS	Inductively coupled plasma–mass spectrometry
IR	Infrared (spectrophotometry)
ISE	Specific ion electrode
°K	degrees Kelvin (temperature)
K_r°	Standard reaction activity constant
LC	Lethal concentration
M	Molar solution concentration
NMR	Nuclear magnetic resonance (spectrometry)
NOEC	No observed effect concentration
NRA	National Rivers Authority (UK)
P_{CO_2}	Partial pressure of carbon dioxide
pe	Negative log of the electron activity
pH	Negative log of the hydrogen ion activity

r	ionic radius
RTE	Relative topsoil enrichment ratio: (total metal concentration in 0–15 cm zone of soil/total metal concentration in 30–45 cm zone of soil)
SOM	Soil organic matter
TCP	Tricalcium phosphate
TI	Tolerance index
TP	Technophility index
VAM	Vesicular-arbuscular mycorrhiza
Z	ionic charge

Section I

THEORY AND PROCESSES

1 Sources and Forms of Potentially Toxic Metals in Soil–Plant Systems

SHEILA M. ROSS
University of Bristol, UK

ABSTRACT

Mineral rock weathering and anthropogenic sources provide two of the main types of metal inputs to soil–plant systems. Around 100 times more lead can be input to ecosystems from anthropogenic sources than by natural processes. Mn, Ni and Cr are the trace metals generally present in highest concentrations in soil, although these are not necessarily the most toxic. Agricultural amendments, including inorganic fertilisers and organic manures and sludges, are very important sources of metals in soils. In many countries, their rates of application are controlled to limit metal loadings. Atmospheric inputs of metals from mining, industrial processes and combustion of fuels and waste materials, supply metals predominantly in the form of oxides and halides. The fate and transformation of these metal species in soil–plant systems depends very much on soil type and prevailing soil conditions.

1.1 INTRODUCTION

Trace metals occur naturally in rocks and soils, but chiefly in forms that are not available to living organisms, such as in the form of constituent and replacement elements in rock and soil minerals. Increasingly higher quantities of certain trace metals are being released into the environment by anthropogenic activities, primarily associated with industrial processes, manufacturing and the disposal of industrial and domestic refuse and waste materials. This chapter describes the principal sources of trace metals in soils and their geochemical origins. Comparisons are made between the quantities of potentially toxic metals supplied via natural and anthropogenic processes. The chapter ends with a brief introduction to the chemical forms of metals likely to be deposited in soils from various anthropogenic emissions.

Toxic Metals in Soil–Plant Systems. Edited by S. M. Ross
© 1994 John Wiley & Sons Ltd

Table 1.1. Typical trace metal concentrations in major rock types ($\mu g\,g^{-1}$)

Element	Igneous rocks			Sedimentary rocks		
	Ultramafic (ultrabasic, e.g. serpentine)	Mafic (basic, e.g. basalt)	Granite	Limestone	Sandstone	Shale
Cr	2000–2980	200	4	10–11	35	90–100
Mn	1040–1300	1500–2200	400–500	620–1100	4–60	850
Co	110–150	35–50	1	0.1–4	0.3	19–20
Ni	2000	150	0.5	7–12	2–9	68–70
Cu	10–42	90–100	10–13	5.5–15	30	39–50
Zn	50–58	100	40–52	20–25	16–30	100–120
Cd	0.12	0.13–0.2	0.09–0.2	0.028–0.1	0.05	0.2
Sn	0.5	1–1.5	3–3.5	0.5–4	0.5	4–6
Hg	0.004	0.01–0.08	0.08	0.05–0.16	0.03–0.29	0.18–0.5
Pb	0.1–14	3–5	20–24	5.7–7	8–10	20–23

Sources: Levinson (1974) and Alloway (1990a).

1.2 SOURCES, QUANTITIES AND FORMS OF TOXIC METALS INPUT TO SOILS

1.2.1 TRACE METAL INPUT FROM MINERAL WEATHERING

Trace metals accumulate locally in soils due to *in situ* weathering of rock minerals. There are generally higher quantities of metals in igneous than in sedimentary rocks, with Mn, Cr, Co, Ni, Cu and Zn being present in the highest quantities in most rock types (Table 1.1). Igneous and metamorphic rocks are the commonest natural source of trace metals in soil. They account for 95% of the earth's crust, with sedimentary rocks making up the remaining 5%. Of the sedimentary rocks, 80% are shales, 15% sandstones and 5% limestones (Mitchell, 1964). Sedimentary rocks are much more important soil parent materials since they overlie most igneous formations, to account for 75% of the outcrops at the earth's surface. The availability of trace metals to plants and ecosystem cycling depends on the ease of weathering of rocks. Sandstones are composed of minerals that weather with difficulty and contribute the smallest amounts of trace metals to soils. Several easily weathered minerals from igneous and metamorphic rocks, including olivine, hornblende and augite, contribute significant quantities of Mn, Co, Ni, Cu and Zn to soils (Table 1.2). Many trace metals are most commonly found in sulphide ores. Examples are galena (PbS), cinnabar (HgS), chalcopyrite ($CuFeS_2$), sphalerite (ZnS) and pentlandite ($(NiFe)_9S_8$). Cd and Zn are geochemically closely associated. They have similar ionic structures and electronegativities and tend to be found together in complex sulphide and carbonate minerals. Trace metals substitute isomorphically in silicate and other mineral lattices for other metal cations of similar ionic radius. Examples are the substitution of Pb^{2+} for K^+ in silicates, substitution of Mn^{2+} for Fe^{2+} in octahedral sites of ferromagnesian minerals, substitution of Ni^{2+} for Fe^{2+} in pyrites, substitution of Ni^{2+} and Co^{2+} for Mg^{2+} in ultrabasic minerals, and the substitution of Cr^{3+} for Fe^{3+} and Cr^{6+} for Al^{3+} in minerals of igneous rocks.

1.2.2 ANTHROPOGENIC SOURCES OF TRACE METALS IN SOILS

Apart from metals originating in parent geological materials and entering the soil through chemical weathering processes, there are many anthropogenic sources of toxic metals. First indications of increasing contamination caused by man have been identified by calculating indices of relative pollution potential. Nikiforova and Smirnova (1975) calculated "technophility indices" which are the ratio of annual mining activity to the mean concentration of trace metal in the earth's crust. Their data indicate that the highest degree of change is associated with Cd, Pb and Hg. These calculations place emphasis

Table 1.2. Trace metal constituents of common rocks and soil minerals

Ease of weathering	Mineral	Occurrence	Trace metals constituents
Easily weathered	Olivine	Igneous rocks	Mn, Co, Ni, Cu, Zn
	Anorthite		Mn, Cu, Sr
	Augite	Ultrabasic and basic volcanic rocks	Mn, Co, Ni, Cu, Zn, Pb
	Hornblende	Widespread in igneous and metamorphic rocks	Mn, Co, Ni, Cu, Zn
	Albite	Coarse, intermediate igneous rocks	Cu
	Biotite	Igneous and metamorphic rocks	Mn, Co, Ni, Cu, Zn
	Orthoclase	Acid igneous rocks	Cu, Sr
	Muscovite	Granites, schists, glass	Cu, Sr
Increasing mineral stability	Magnetite	Igneous and metamorphic rocks	Cr, Co, Ni, Zn

After Mitchell (1964).

on mining as the main source of transportable contaminants, but we can identify at least five main groups of anthropogenic sources of trace metal contamination in soil–plant systems (Table 1.3).

Campbell *et al.* (1983) compare natural and anthropogenic quantities of trace metals emitted to the atmosphere and show that around 15 times more

Table 1.3. Sources of toxic metals in the environment

1. *Metalliferous mining and smelting*:
   (a) Spoil heaps and tailings—contamination through weathering, wind erosion (As, Cd, Hg, Pb)
   (b) Fluvially dispersed tailings—deposited on soil during flooding, river dredging, etc. (As, Cd, Hg, Pb)
   (c) Transported ore separates—blown from conveyance onto soil (As, Cd, Hg, Pb)
   (d) Smelting—contamination due to wind-blown dust, aerosols from stack (As, Cd, Hg, Pb, Sb, Se)
   (e) Iron and steel industry (Cu, Ni, Pb)
   (f) Metal finishing (Zn, Cu, Ni, Cr, Cd)

2. *Industry*
   (a) Plastics (Co, Cr, Cd, Hg)
   (b) Textiles (Zn, Al, Z, Ti, Sn)
   (c) Microelectronics (Cu, Ni, Cd, Zn, Sb)
   (d) Wood preserving (Cu, Cr, As)
   (e) Refineries (Pb, Ni, Cr)

3. *Atmospheric deposition*:
   (a) Urban/industrial sources, including incineration plants, refuse disposal (Cd, Cu, Pb, Sn, Hg, V)
   (b) Pyrometallurgical industries (As, Cd, Cr, Cu, Mn, Ni, Pb, Sb, Tl, Zn)
   (c) Automobile exhausts (Mo, Pb (with Br and Cl), V)
   (d) Fossil fuel combustion (including power stations) (As, Pb, Sb, Se, U, V, Zn, Cd)

4. *Agriculture*:
   (a) Fertilisers (e.g. As, Cd, Mn, U, V, and Zn in some phosphatic fertilisers)
   (b) Manures (e.g. As, and Cu in some pig and poultry manures, Mn and Zn in farmyard manure)
   (c) Lime (As, Pb)
   (d) Pesticides (Cu, Mn, and Zn in fungicides, As and Pb used in orchards)
   (e) Irrigation waters (Cd, Pb, Se)
   (f) Corrosion of metals (e.g. galvanised and metal objects (fencing, troughs, etc.) Fe, Pb, Zn)

5. *Waste disposal on land*:
   (a) Sewage sludge (Cd, Cr, Cu, Hg, Mn, Mo, Ni, Pb, V, Zn)
   (b) Leachate from landfill (As, Cd, Fe, Pb)
   (c) Scrapheaps (Cd, Cr, Cu, Pb, Zn)
   (d) Bonfires, coal ash, etc. (Cu, Pb)

Cd, 100 times more Pb, 13 times more Cu, and 21 times more Zn are emitted by man's activities than by natural processes. Inevitably this plethora of sources of metals in the environment results in a vast number of different metal ions, species and complexes to contaminate soils and ecosystems. Whether such contamination becomes environmentally toxic or not depends very much on: (i) prevailing soil physical and chemical conditions, such as acidity, waterlogging and the presence of clays, Fe/Mn oxides and soil organic matter, which provide adsorption and reaction surfaces; (ii) soil and site hydrology, which cannot only dilute the contaminant *in situ* effect, but can also transport the pollutant effect from its source to cause contamination in new, unaffected sites; and (iii) the plant and soil microbial component of ecosystems which uptake and recycle metals.

1.2.3   QUANTITIES OF TRACE METALS IN SOILS AND PLANTS

The rather generalised data presented in Table 1.4. are an attempt to summarise what is known about "normal" trace metal concentrations in uncontaminated soils and plants and soil and plant concentrations generally thought to represent metal toxic conditions. It can be seen that "trace" metal total concentrations can vary widely in different soils derived from different parent materials. Table 1.1 indicates that the more basic igneous parent materials potentially contribute highest quantities of Cr, Mn, Co and Ni to soils, while amongst the sedimentary parent materials, shales potentially contribute the highest quantities of Cr, Co, Ni, Zn and Pb. Mineral weathering rates determine the release of these elements into soil, initially in the form of simple or complex inorganic ligands, depending on mineral solubilities under prevailing soil pH and Eh conditions.

Of all the trace metals listed in Table 1.4, Mn, Ni and Cr are present in highest quantities in soils, while Cd and Hg are present in smallest amounts. Likely soil and soil solution toxicity levels follow the same pattern, with lowest critical concentrations for Hg and Cd, as well as Pb, Cr and Zn. The fourth column in Table 1.4 represents the maximum concentration of each element in the soil solution, assuming that all of the element at its average level in the soil (column 3) were to dissolve in the water present at 10% of the dry weight of the soil (Lindsay, 1979). This is not a realistic assumption, since many of the mineral forms of trace metals are quite insoluble in normal soil conditions. These soil solution concentrations therefore represent *maximum* possible solubilities and molar concentrations in the soil solution. Since Pb and Zn are present in soils in higher concentrations than Hg, Cd or Cr, one might expect their toxicity concentrations to be exceeded more readily than Hg, Cd or Cr. However, mineral solubilities dictate how readily these levels are exceeded. These issues are discussed in more detail in Chapter 3.

Studies of downprofile soil trace metals show that elements such as Mn, Ni and Cr, whose main source in soil is the parent material, can accumulate to

Table 1.4. Typical concentrations of some trace metals in soils, the soil solution and in plants

	Soils					Plants		
			Average value in soil and maximum dissolved in the soil solution[c]					
Element	Normal range in soil (total) (μg g^{-1} dry wt)[a]	Concentration in soil considered toxic (total) (μg g^{-1} dry wt)[b]	Average (μg g^{-1} dry wt)	Molar concentration at 10% soil moisture	Soil solution concentration considered as toxic (mg litre^{-1})[d]	Normal range in plant material (μg g^{-1} fresh wt)[e]	Concentration in contaminated plants (μg g^{-1})[b]	Annual plant uptake (kg ha^{-1} year^{-1})[d]
Cr	5–1000	75–100	100	$10^{-1.72}$	0.001	0.03–15	5–30	nd
Mn	200–2000	1500–3000	600	$10^{-0.96}$	0.1–10	15–1000	300–500	1.0
Co	1–70	25–50	8	$10^{-2.87}$	0.01	0.05–0.5	15–50	0.0006
Ni	10–1000	100	40	$10^{-2.17}$	0.05	0.02–5	10–100	nd
Cu	2–100	60–125	30	$10^{-2.33}$	0.03–0.3	4–15	20–100	0.006
Zn	10–300	70–400	50	$10^{-2.12}$	<0.005	8–400	100–400	0.01
Cd	0.01–7	3–8	0.06	$10^{-5.57}$	0.001	0.2–0.8	5–30	nd
Sn	<5	50	10	$10^{-3.07}$	nd	0.2–6.8	60	0.001
Hg	0.02–0.2	0.3–5	0.03	$10^{-5.83}$	0.001	0.005–0.5	1–3	nd
Pb	2–200	100–400	10	$10^{-3.32}$	0.001	0.1–10	30–300	nd

Sources: [a]Swaine (1955); [b]Kabata-Pendias and Pendias (1984); [c]Lindsay (1979); [d]Bohn et al. (1985); [e]Allaway (1968) and Bowen (1979).
nd = Not determined.

high levels in subsoils. For example, concentrations of Ni in soils developed on serpentine can be as high as $100–7000 \, \mu g$ Ni g^{-1} (Brooks, 1987), but with only small concentrations in the surface organic litter (Berrow and Reaves, 1986). Brooks *et al.* (1990) even describe serpentine soils in Brazil whose Ni contents range from $2500–15\,000 \, \mu g$ Ni g^{-1} at the soil surface to $15\,000–30\,000 \, \mu g$ Ni g^{-1} at depth. Since many minerals containing trace metals are insoluble and resistant to weathering, higher concentrations can be found in tropical soils where longer and more intensive weathering regimes prevail. For other trace metals whose main input to soil is at the soil surface, through aerial deposition, as in the case of lead from combustion and exhaust emissions into the atmosphere, or cadmium, where large quantities are applied to agricultural soils in phosphatic fertilisers, the highest concentrations occur in topsoils. Topsoil concentration is maintained by vegetation recycling and, at best, these elements show very slow movement downprofile (e.g. Biddappa *et al.*, 1982; also see Chapter 9, this volume). There are many examples of Pb accumulation in surface soils. Colbourn and Thornton (1978) calculated the relative topsoil enhancement (RTE) as the ratio of Pb concentration in topsoils (< 15 cm) to the Pb concentration in subsoil (> 15 cm). They reported RTE values of $1.2–2.0$ in remote agricultural areas and values of $4–20$ in areas contaminated by mining.

Although all trace metals are released in varying quantities into soil from parent materials, significant amounts of certain soil metals are derived from other sources, particularly atmospheric deposition (mainly Pb and Zn) and agricultural amendments (mainly Cd). The relative proportions of trace metal inputs to soils worldwide are shown in Table 1.5 in data calculated by Nriagu and Pacyna (1988). The two main anthropogenic sources of soil trace metals are: (i) fly ash residues from coal combustion, providing 34.2%, 33.3%, 62.86% and 51.3% of the annual inputs of Cd, Hg, Mn and Ni respectively; and (ii) corrosion of commercial waste products, which provides 47%, 55.8%, 42% and 36.17% of all Cr, Cu, Pb and Zn inputs respectively. Other sources are individually important, such as agricultural and animal manures which provide 22.9% of Zn inputs, and urban refuse which provides 20.5% of Cd inputs to soils. Nriagu and Pacyna (1988) calculated that if total metal inputs were dispersed uniformly over the cultivated land area of $16 \times 10^{12} \, m^2$, the annual rates of metal application would vary from 1.0 g ha^{-1} for Cd to about 50 g ha^{-1} for Pb, Cu and Cr, to over 65 g ha^{-1} for Zn and Mn. Evidence from industrialised areas of Japan, the USA and Europe show that soils have already become overloaded with toxic metals, at a time when technology for decontaminating such conditions is yet to be developed.

1.2.4   TRACE METAL INPUT TO SOILS FROM AGRICULTURAL AMENDMENTS

Many agricultural amendments, including organic manures, inorganic

Table 1.5. Worldwide emissions of trace metals input into soils (10^6 kg year^{-1}) and the proportional input of each element (in parentheses as a percentage) from different sources

Source category	Metal								Annual discharge (10^{12} kg)
	Cd	Cr	Cu	Hg	Mn	Ni	Pb	Zn	
Agriculture and animal wastes (includes fertilisers)	0.23–4.45 (11.7)	14.5–150 (11.4)	17–119 (8.7)	0–1.7 (11.3)	65–253 (9.6)	9.2–82 (15)	5–49 (4.4)	162–471 (22.9)	183
Logging and wood wastes	0–2.2 (5.8)	2–18 (1.4)	3–52 (3.8)	0–2.2 (14.7)	18–104 (3.9)	2.2–23 (4.2)	6.6–8.2 (0.7)	13–65 (3.16)	11
Urban refuse and sewage/sludge	0.9–7.8 (20.5)	8–44 (3.3)	18–62 (4.5)	0–1.0 (6.7)	115–54 (2.05)	7.3–35 (6.4)	20.8–73 (6.6)	40–156 (7.6)	670
Fly ash	1.5–13 (34.2)	149–446 (34)	93–335 (24.5)	0.4–5 (33.3)	498–1655 (62.86)	56–279 (51.3)	45–242 (21.7)	112–484 (23.56)	3720
Atmospheric fallout	2.2–8.4 (22.1)	5.1–38 (2.9)	14–36 (2.6)	0.63–4.3 (28.6)	7.4–46 (1.75)	11–37 (6.8)	202–263 (26.6)	49–135 (6.6)	
Other[a]	0.77–2.15 (5.4)	305–616 (47)	396–763 (55.8)	0.57–0.8 (5.3)	106–521 (19.8)	20.3–88 (16.2)	199–478 (42.9)	313–743 (36.17)	
Total annual input to soil	5.6–38	484–1309	541–1367	1.6–15	706–2633	106–544	479–1113	689–2054	

Source: Calculated from the data of Nriagu and Pacyna (1988).
[a]This category includes wastage of commercial products, and assumes that 1–15% of the total annual production of metals is discarded.

fertilisers, pesticides and even irrigation waters, add trace metals, often in toxic amounts, to agricultural soils. Even if original additions are not at concentrations high enough to be initially toxic, critical levels can build up with repeated applications. Typical concentrations of trace metals in several types of agricultural amendments are listed in Table 1.6. Sewage sludges and composted refuse provide the biggest trace metal inputs to soils. Zn, Cd and Pb are the three main metals concentrated in sewage sludges, although significant quantities of Cr, Cu and Hg may also be added. Compositions of sewage sludges vary greatly depending on origin so it is difficult to generalise about their soil polluting potential. In order to minimise accumulation of metals in sludge-amended soils, recommended and mandatory levels of sludge and metals additions have been introduced in many countries. In the USA, sludge application levels are based on both the trace metal content and the cation exchange capacity (CEC) of the soil. For cadmium, for example, the maximum cumulative Cd loadings are: 5.5 kg Cd ha^{-1} for CEC <5 meq 100 g^{-1}; 11 kg Cd ha^{-1} for CEC $= 5$–15 meq 100 g^{-1}; and 22 kg Cd ha^{-1} for CEC > 15 meq 100 g^{-1} (US Environmental Protection Agency, 1979). Since sewage treatment removes only small proportions of trace metals, around 25% Cd, 23% Cu and only 7% Pb (Baker, 1990), sludges still contain appreciably higher trace metal concentrations than exist in soils or plants. European Union (EU) recommended sludge applications to agricultural land limit trace metal concentrations to 4000 μg Zn g^{-1} (mandatory) and 2500 μg Pb g^{-1} (recommended). These represent land loadings of 30 kg Zn ha^{-1} year^{-1} and 10–15 kg Pb ha^{-1} year^{-1}.

Inorganic phosphate fertilisers are important sources of Cd and other trace metals such as Cr and Pb to agricultural soils. Alloway (1990b) even suggests that all soils used for commercial agriculture will have had their Cd levels raised to some extent by phosphatic fertiliser application. Alloway's (1990b) data indicate that phosphatic rocks from Senegal and Togo contain the highest Cd contents at 255 and 160 g Cd t^{-1} P$_2$O$_5$ respectively. Worldwide, Nriagu (1980) estimates that phosphatic fertilisers, with an average Cd content of 7 μg g^{-1}, contribute 660 t Cd year^{-1} into the environment. Significant quantities of Cr are also added to soil in phosphatic fertilisers, but since the Cr is in the form Cr$^{(III)}$, it is not likely to be toxic (McGrath and Smith, 1990).

1.2.5 ATMOSPHERIC INPUTS OF TRACE METALS TO SOILS

The history of atmospheric metal pollution in north-west Europe and North America has been evaluated from geochemical studies of peat bog and lake sediment cores. More widespread atmospheric metal pollution is evident in ice cores from polar regions. These approaches have been reviewed by Levitt (1988). The metal pollution impact of smelting industries has been identified in peat deposits in the Gordano Valley, south-west England, dating from as early as 2000 years ago (Martin *et al.*, 1979), and probably relating to localised

Table 1.6. Typical ranges of trace metals in agricultural amendments (all values are $\mu g\,g^{-1}$)

Element	Sewage sludge	Composted refuse	Farmyard manure	Phosphate fertilisers	Nitrate fertilisers	Lime	Pesticides	Irrigation waters
Cr	8–40 600	1.8–410	1.1–55	66–245	3.2–19	10–15	—	—
Mn	60–3900	—	30–969	40–2000	—	40–1200	—	—
Co	1–260	—	0.3–24	1–12	5.4–12	0.4–3	—	—
Ni	6–5300	0.9–279	2.1–30	7–38	7–34	10–20	—	—
Cu	50–8000	13–3580	2–172	1–300	—	2–125	—	—
Zn	91–49 000	82–5894	15–566	50–1450	1–42	10–450	—	—
Cd	<1–3410	0.01–100	0.1–0.8	0.1–190	0.05–8.5	0.04–0.1	—	<0.05
Hg	0.1–55	0.09–21	0.01–0.36	0.01–2.0	0.3–2.9	0.05	0.6–6	—
Pb	2–7000	1.3–2240	0.4–27	4–1000	2–120	20–1250	11–26	<20

Source: Data reviewed from a range of sources by Alloway (1990*a*) and Fergusson (1990).

Roman metal smelting. In Gordano and elsewhere in Europe, steep increases in metal accumulation occurred around 200 years ago (e.g. Aaby and Jacobsen (1979) for Dravel Mose, Denmark and Levitt *et al.* (1979) for Grassington Moss, West Yorkshire, England). In North America, evidence of atmospheric influx of metals is more recent, dating from the last 80–100 years (e.g. Norton, 1986).

Influx of of metals from the atmosphere occurs mainly in particulate form, as dry deposition, wet deposition (as washout by precipitation) or as occult deposition (from fog and mist). Much evidence of long distance pollutant transport, particularly ionic species contributing to acid precipitation, has been presented in the last 10 years. Studies of atmospheric transport of metals indicate some long distance transport (Pacyna *et al.*, 1984) but more commonly, rather shorter distance transport, locally concentrated near industrial sources. Such a pattern was illustrated by Steinnes (1987) for relatively high levels of Pb, Cd, Zn, As, Sb and Se in southern Norway, but much lower concentrations in central and northern regions. Metal contamination around primary smelters can cover large areas, as seen around Sudbury, in Ontario, Canada, for example, where as much as 500 km^2 is metal-affected (Tiller, 1989). At Port Pirie in South Australia, vegetation foliar metals levels were higher than background levels as much as 12 km away from the smelter (Merry and Tiller, 1978) and liver tissues from sheep showed elevated Cu and Zn level over 40 km away from the smelter (Koh and Judson, 1986).

Natural inputs of metals to the atmosphere by volcanic activity are relatively high, especially for Hg, Pb and Ni (Table 1.7). Much higher quantities of metals transported in the atmosphere are of anthropogenic origin, from combustion, incineration, exhausts, and from the mining and smelting of metal ores. The high proportion (22.1%) of the Cd entering soils through atmospheric deposition (Table 1.5) is mainly derived from mining activities

Table 1.7. Typical concentration ranges of trace metals in the air at various locations (ng m^{-3})

Element	Europe	North America	Volcanoes (Hawaii/Etna)
Cr	1–140	1–300	45–67
Mn	9–210	6–900	55–1300
Co	0.2–37	0.13–23	4.5–27
Ni	4–120	<1–120	330
Cu	8–4900	5–1100	200–300
Zn	13–16 000	<10–1700	1000
Cd	0.5–620	<1–41	8–92
Hg	<0.009–2.8	0.007–38	18–250
Pb	55–340	45–13 000	27–1200

Source: Bowen (1979).

(Nriagu and Pacyna, 1988). The high proportion of Hg entering soils fro $\lrcorner$ atmospheric sources (28.6%) is mainly of natural origin such as vulcanism and through biochemical processes which occur in water and soils. Lead is the only other metal added to soils in significant quantities from the atmosphere (6.6% of all Pb additions to soils) (Table 1.5). It is also the most publicised atmospheric pollutant since the prime source, apart from the mining industry, is vehicular exhausts. Lindberg and Harriss (1989) report that total atmospheric Pb deposition ranges from 3.1–31 $mg\,m^{-2}\,year^{-1}$ in remote and rural areas to 27–140 $mg\,m^{-2}\,year^{-1}$ in urban and industrial areas. Higher than average Pb concentrations in roadside soils are due to the use of leaded petrols, with global exhaust Pb emissions estimated by Pacyna (1986) at around $176 \times 10^9\,mg\,year^{-1}$, or 45% of all atmospheric Pb sources. Pb deposition becomes concentrated in a zone adjacent to roads. Davies (1990) suggests that there is a zone of around 15 m on either side of a road in which the Pb concentration exceeds 1 $\mu g\,m^{-3}$ for every 1000 vehicles (averaged over a 24-h period). The distance away from the road which is affected by lead fallout depends on particle size, with the majority of particles in the smallest range, $<2\,\mu m$, dispersed some distance from the road. Wheeler and Rolfe (1979) noted a significant difference between soil lead concentrations, which declined at around 10–15 m from the road, and vegetation lead concentrations which remained high up to 40 m from the road. They concluded that soil Pb may be more determined by coarse particulate fallout, while plant leaves have large surface areas which are able to trap fine aerosols from the atmosphere.

Smaller amounts of other metals arrive in soil from atmospheric deposition, although there are many reports of local concentrations due to fallout in the vicinity of point sources such as metal smelters (e.g. Little and Martin, 1972; Brabec et al., 1983), or mining activities (e.g. Hemphill et al., 1983). Direct soil contamination concentrates metals in the vicinity of the point source while atmospheric pollution disperses the metal downwind. Soil metal concentrations can rise to exceptionally high concentrations close to metal smelting complexes. A large number of publications report metal pollution in the vicinity of metal smelters and smelting complexes. A range of soil metal concentrations measured close to different types of smelters is given in Table 1.8. Extremely high metal concentrations have been observed in these circumstances for Pb, Zn, Cd and Cu, but pollution can extend significant distances away from the source. Rutherford and Bray (1979), for example, found significantly elevated total soil Ni, Cu and Zn up to 3.5 km from a nickel smelter at Coniston, Ontario. Studying metals in vegetation close to a Pb–Zn smelter at Avonmouth, England, Burkitt et al. (1972) found Pb, Zn and Cd concentrations in perennial rye grass to be 225 $\mu g\,Pb\,g^{-1}$, 1600 $\mu g\,Zn\,g^{-1}$ and 50 $\mu g\,Cd\,g^{-1}$ immediately adjacent to the smelter, decreasing in concentration with distance away from the smelter to 25 $\mu g\,Pb\,g^{-1}$, 200 $\mu g\,Zn\,g^{-1}$ and 5 $\mu g\,Cd\,g^{-1}$ at 13 km distant. They found a sharper decline in vegetation concentrations of soil Cd and Zn with distance than in soil Pb

Table 1.8. Selected examples of total soil metal concentrations in the vicinity of metal smelters

Location	Smelter type	Metal	Concentration range ($\mu g\,g^{-1}$)	Source
South Australia: Port Pirie	Pb–Zn	Pb	4–2100	Tiller and de Vries (1977)
		Cd	0–14	
USA: Kellog, Idaho	Pb	Pb	1700–7600	Ragaini *et al.* (1977)
		Cd	71–140	
		Zn	up to 29 000	
USA: Palmerton, Montana	Zn	Zn	50 000–80 000	Buchauer (1973)
		Pb	200–1100	
		Cd	900–1500	
		Cu	600–1200	
USA: Missouri	Pb	Pb	22 740	Bolter *et al.* (1972)
		Zn	4170	
Canada: Sudbury, Ontario	Cu–Ni	Ni	3309	Hutchinson and Whitby (1973)
		Cu	2071	
		Co	154	
Canada: Trail, British Columbia	Pb	Pb	589–9293	John *et al.* (1975)

concentrations. A similar pattern was observed around the Cu–Zn smelter at Sudbury, Ontario, where soil Cu concentrations within 7.5 km of the complex are often higher than 1000 μg Cu g^{-1} (Hutchinson, 1979). Although it is difficult to generalise about these patterns, Davies (1990) suggests that within 1–3 km of a well-established Pb smelter, soil Pb concentrations are likely to be enhanced by around 15 times the background soil concentrations (i.e. concentrations of around 1500 μg Pb g^{-1}). A more detailed pattern of down-profile soil metal concentrations near to smelters was suggested by Martin and Coughtrey (1982) who reviewed a range of published soil metal data from different smelter sites. When they recalculated the data from different smelter sites to show the ratio of metal in the topsoil compared to that in the subsoil, they found that surface enrichment of Cd, Ni, Pb, Zn and Cu occurred not only close to smelters, but also at distances of up to 8 km from the source.

1.3   FORMS OF TOXIC METALS INPUT TO SOIL–PLANT SYSTEMS

The chemical forms of metals produced anthropogenically and emitted into the atmosphere or discharged into water bodies, and their stability in the environment, determine how mobile they are and whether they become toxic in receiving ecosystems. A wide range of possible forms can be emitted from different processes, even at a single point source. From a metal smelter, for example, sources emitting different metal forms include: (a) atmospheric particulate from furnace stack emissions, ore-crushing plants and windblow from stockpiled ore and from temporary and permanent waste heaps; or (b) direct soil contamination from conveyor belts, spillage from trucks, or sediment erosion from stockpiles and waste heaps (Martin and Coughtrey, 1982).

   In industrial processes which involve high temperatures, including coal and oil combustion for power generation, domestic refuse incineration, or smelting of ores and metal refinery processes, the temperature achieved during the process is one of the main criteria that determine the amount, form, particle size and fate of emitted metals. A second important criterion is the presence in the process of other elements, such as chlorides, which are also volatilised and can react with metals. In an early report, Bertine and Goldberg (1971) noted the relative volatility of elemental forms of metals as: Hg > As > Cd > Zn > Sb ≫ Mn > Ag, Sn, Cu; the order of volatility of oxides, sulphates, carbonates, silicates and phosphates as: As, Hg > Cd > Pb > Zn > Cu > Sn; and the volatility of sulphides as: As, Hg > Sn > Cd > Sb, Pb > Zn > Cu > Fe, Co, Ni, Mn, Ag. These sequences indicate that Hg, Cd, Zn and Pb in almost any inorganic form are more likely to be volatilised in an industrial process than are Cu, Fe or Mn. Klein *et al.* (1975) have classified metals on the basis of their boiling points and volatilisation potentials (Table 1.9).

Table 1.9. Classification of potentially toxic metals according to their boiling points and volatility during incineration

Class	Metals	
Class I	Al, Ca, Co, Fe, K, Mg, Mn	High boiling point, metals not volatilised in the combustion area. Make up the matrix of fly ash.
Class II	As, Cd, Cu, Pb, Sn, Zn	Volatilised during combustion. On cooling, these metals condense onto surfaces of fly ash particles.
Class III	Hg	Remains in gas phase throughout incineration. Does not condense.

Source: After Klein *et al.* (1975).

Table 1.10. Major metal chemical species evolved during various industrial and anthropogenically derived processes and input to soils

Process	Metal form					Source
	Cd	Pb	Hg	As	Cr	
Coal combustion	Cd(O), CdO CdS	$PbCl_2$, PbO PbS, Pb	$Hg^\circ{}_{(g)}$	As(O), As_2O_3 As_2S_3	—	Pacyna (1987)
Oil combustion	Cd(O), CdO	PbO	$Hg^\circ{}_{(g)}$	As(O), As_2O_3 organic arsenics	—	Pacyna (1987)
Non-ferrous metal production	CdO, CdS	PbO, $PbSO_4$ PbO. $PbSO_4$	—	As_2O_3	—	Pacyna (1987)
Iron and steel manufacture	CdO	PbO	—	—	—	Pacyna (1987)
Refuse incineration	Cd(O), CdO $CdCl_2$	Pb(O), PbO Pb, Cl_2	$Hg^\circ{}_{(g)}$	As(O), As_2O_3	—	Pacyna (1987)
Vehicular exhausts	—	$PbBr_2$, PbBrCl Pb(OH)Br, $(PbO)_2PbBr$ $(PbO)_2PbBrCl$ Change to carbonates and oxides in soil	—	—	—	Ter Haar and Bayard (1971); Post and Buseck (1985)
Ferrochrome smelters	—	—	—	—	$Cr^{(VI)}$	McGrath and Smith (1990)
Industrial emissions	CdO	PbO	—	—	Aerosols of $Cr^{(III)}$ and $Cr^{(IV)}$	Nriagu et al. (1988)
Biochemical industries	—	—	$Hg^\circ{}_{(g)}$ (gaseous volatile), $(CH_3)_2Hg$ produces soluble $Hg(OH)_2$ in rain	—	—	Lindquist et al. (1983)

Table 1.11. Forms of potentically toxic metals in waters and soils

Element	Soil[a]			Water[b]		
	Species	Adsorption	Processes	Species	Size (nm)	Phase state
Cd	CdO, CdCO$_3$, Cd(PO$_4$)$_2$ oxic CdS (reducing) CdCl$_2$	Al/Fe oxides clays SOM	— Precipitation of CdCO$_3$ at high pH	Cd^{2+}: free metal ion Cd fulvates: low molecular weight organic chelates Cd—clay: absorbed	<1 1–10 >1000	Dissolved
Pb	PbS, PbSO$_4$, Pb(OH)$_2$, PbCO$_3$ PbO, Pb(PO$_4$)$_2$ PbO(PO$_4$)$_2$ PbCl$^+$	Mn/Fe oxides clays SOM	— Absorbed on Fe oxides — Exchanged on clays and SOM — Biomethylated into volatile and toxic forms: (CH$_3$)$_4$ Pb and (CH$_3$)$_{4-n}$ Pb^{n+}	Pb^{2+}: free metal ion Pb—fulvates: low molecular weight organic chelates Pb–FeOOH: chemisorbed 2 PbCO$_3$.Pb(OH)$_2$: precipitates	<1 1–10 1–10 100–1000 >1000	Dissolved Dissolved Dissolved Colloidal Particular
Hg	Hg(OH)$_2$, HgCO$_3$ (oxic) HgS (reducing) Organo mercury compounds	Clays SOM Fe/Mn oxides	— Redox processes produce free Hg in soil — Biomethylation into toxic forms	Hg(OH)$_2$ Hg fulvate chelates Hg humic chelates	<1 1–10 10–100	Dissolved Dissolved Colloidal
As	AsO$_4^{3-}$ (oxic) AlAsO$_4$, FeAsO$_4$ Ca$_3$(AsO$_4$)$_2$	Strongly absorbed clays SOM Fe/Mn oxides	— Biomethylation	HAsO$_4^{2-}$ As fulvate chelates	<1 1–10	Dissolved Dissolved

Source: [a]Fergusson (1990); [b]Harrison (1987).

According to metal affinities for oxides, sulphides and halides, and according to the concentrations of these ligands in the air during the incineration process, various volatile, aerosol and particulate metal species are produced. Some of the chemical forms of metals emitted from a number of different anthropogenic processes are shown in Table 1.10. During the incineration of domestic refuse, the burning of different plastics, including polyvinyl chloride, provides halides for metal complexation. Metal halides, particularly chlorides, are an extremely important chemical form arriving in soil from atmospheric deposition since most are soluble at low pH and at high chloride concentrations in the soil solution. Fernandez *et al.* (1992) used thermodynamic data to classify metals according to their likely stabilities in incinerator fly ash. The classification depends on whether the thermodynamic stability of the metal oxide is greater than or less than its chloride. If the oxide is more stable than the chloride, the metal is transported mechanically as particulate oxide and is found in the matrix of the fly ash (these are Klein's (1975) Class I metals). If the metal oxide and chloride stabilities are similar, the metal is transported by both mechanical processes and by volatilisation/condensation. Examples of these metals include Fe and Mn. If the chloride is more stable than the oxide, the metal is transported in the chloride form by volatilisation; it condenses on the surfaces of fly ash particles and is highly soluble (these are Klein's Class II metals). Not surprisingly, concentrations of soluble, recondensed chlorides are highest where the particle size of fly ash is smallest, providing the biggest surface area for condensation (Davison *et al.*, 1974). The smallest particles have the longest atmospheric residence times and can be transported long distances from the source, where they can provide soluble inputs of metals to soils. The smallest particles are also those which are readily inhaled, with soluble metal condensates being liberated into the respiratory system (e.g. Natusch and Wallace, 1974). The pattern of metal fallout summarised from the above theory and a range of empirical studies indicates that the oxides and sulphides of metals can be deposited as dust close to incinerator stacks (Pacyna, 1987), metal halides are transported further and condense on particle surfaces, to become soluble in soil (e.g. Fernandez *et al.*, 1992), while Hg remains volatile. The fate of deposited metals and metal complexes in soils and waters depends very much on ambient conditions of soil and water pH, organic matter status, redox and temperature. Some of their fates and transformations are discussed in Chapter 3. In soils, free metal ions and complex metal species can become strongly adsorbed onto clays, Fe and Mn oxides and soil organic matter, while in waters and in the soil solution, metals tend to form organic chelates with humic and fulvic acids (Table 1.11).

1.4 CONCLUSIONS

A vast amount of information on quantities of toxic metals in the environment is now available for different parts of the world. Kabata-Pendias and Pendias and Pendias (1992) review published world data on toxic metals in a very wide range of soils and plants. In a review of the historical perspective of heavy metal pollution from atmospheric input, Levitt (1988) reports evidence from north-western Europe which indicates metal pollution from 2000 years BP, with steep increases in the past 200 years. Metal pollution has a shorter history in North America, of around 100 years.

In many cases it is difficult to make comparisons of metal pollution between sites or to make general conclusions about the fate of toxic metals in different soils or ecosystems since studies have their own assumptions, methods vary greatly and results are often acquired in different ways. However, available data on total metals and mean concentrations in soils, sediments, waters, plants, soil organisms and other environmental materials, allow some estimation of the accumulation effects of air pollution, sewage sludge applications and fertiliser additions. Subsequent chapters of this volume discuss these issues in more detail.

REFERENCES

Aaby, B. and Jacobsen, J. (1979) Changes in biotic conditions and metal deposition in the last millennium as reflected in ombrotrophic peat in Draved Mose, Denmark. *Danm. Geol. Unders. Arbog*, **1978**, 5–43.

Allaway, W. H. (1968) Agronomic controls over the environmental cycling of trace elements. *Advances in Agronomy*, **20**, 235–274.

Alloway, B. J. (1990*a*) The origin of heavy metals in soils. In: Alloway, B.J. (Ed.) *Heavy Metals in Soils*, pp. 29–39. Blackie, London; Wiley, New York.

Alloway, B. J. (1990*b*) Cadmium. In: Alloway, B. J. (Ed.) *Heavy Metals in Soils*, pp. 100–124. Blackie, London; Wiley, New York.

Baker, D. E. (1990) Copper. In: Alloway, B. J. (Ed.) *Heavy Metals in Soils*, pp. 151–176. Blackie, London; Wiley, New York.

Berrow, M. L. and Reaves, G. A. (1986) Total chromium and nickel contents of Scottish soils. *Geoderma*, **37**, 15–27.

Bertine, K. K. and Goldberg, E. D. (1971) Fossil fuel combustion and the major sedimentary cycle. *Science*, **173**, 233–235.

Biddappa, C. C., Chino, M. and Kumazawa, K. (1982) Migration of heavy metals in two Japanese soils. *Plant and Soils*, **66**, 299–316.

Bohn, H. L., McNeal, B. L. and O'Connor, G. A. (1985) *Soil Chemistry*. Wiley Interscience, Wiley, New York.

Bolter, E., Hemphill, D., Wixson, B., Butherus, D. and Chen, R. (1972) Geochemical and vegetation studies of trace substances from lead smelting. In: Hemphill, D. D. (Ed.) *Trace Substances in Environmental Health*, Vol. 6, pp. 79–86. University of Missouri, Columbia.

Bowen, H. J. M. (1979) *Environmental Chemistry of the Elements*. Academic Press, London.

Brabec, E., Cudlin, P., Rauch, O. and Skoda, M. (1983) Lead budget in a smelter-adjacent grass stand. In: *Heavy Metals in the Environment*, Vol. 2, pp. 1150–1153. Heidelberg International Conference, CEP Consultants, Edinburgh.

Brooks, R. R. (1987) *Serpentine and its Vegetation*. Dioscorides Press, Portland.

Brooks, R. R., Reeves, R. D., Baker, A. J. M., Rizzo, J. A. and Ferreira, H. D. (1990) The Brazilian serpentine plant expedition (BRASPEX), 1988. *National Geographic Research*, **6**(2), 205–219.

Buchauer, M. J. (1973) Contamination of soil and vegetation near a zinc smelter by zinc, cadmium, copper and lead. *Environmental Science and Technology*, **7**, 131–135.

Burkitt, A., Lester, P. and Nickless, G. (1972) Distribution of heavy metals in the vicinity of an industrial complex. *Nature*, **238**, 327–328.

Campbell, P. G. C., Stokes, P. M. and Galloway, J. N. (1983) The effect of atmospheric deposition on the geochemical cycling and biological availability of metals. In: *Heavy Metals in the Environment*. Vol. 2, pp. 760–763. Heidelberg International Conference. CEP Consultants, Edinburgh.

Colbourn, P. and Thornton, I. (1978) Lead pollution in agricultural soils. *Journal of Soil Science*, **29**, 513–526.

Davies, B. E. (1990) Lead. In: Alloway, B. J. (Ed.) *Heavy Metals in Soils*, pp. 177–196. Blackie, London; Wiley, New York.

Davison, R. L., Natusch, D. F. S., Wallace, J. R. and Evans, C. A. (1974) Trace elements in fly ash. Dependence of concentration on particle size. *Environmental Science and Technology*, **13**, 1107–1113.

Fergusson, J. E. (1990) *The Heavy Elements: Chemistry, Environmental Impact and Health Effects*. Pergamon Press, London.

Fernandez, M. A., Martinez, L., Segarra, M., Garcia, J. C. and Espiell, F. (1992) Behaviour of heavy metals in the combustion gases of urban waste incinerators. *Environmental Science and Technology*, **26**, 1040–1047.

Harrison, R. M. (1987) Physico-chemical speciation and chemical transformations of toxic metals in the environment. In: Coughtrey, P. J., Martin, M. H. and Unsworth, M. H. (Eds) *Pollutant Transport and Fate in Ecosystems*, pp. 239–247. Blackwell Scientific Publishers, Oxford.

Hemphill, D. D., Wixson, B. G., Gale, N. L. and Clevenger, T. E. (1983) Dispersal of heavy metals into the environment as a result of mining activities. In: *Heavy Metals in the Environment*. Vol. 2, pp. 915–924. Heidelberg International Conference. CEP Consultants, Edinburgh.

Hutchinson, T. C. (1979) Copper contamination of ecosystems by smelter activities. In: Nriagu, J. O. (Ed.) *Copper in the Environment. Part I: Ecological Cycling*, pp. 451–502. Wiley, New York.

Hutchinson, T. C. and Whitby, L. M. (1974) Heavy metal pollution in the Sudbury mining and smelting region of Canada. I. Soil and vegetation contamination by nickel, copper and other metals. *Environmental Conservation*, **1**, 123–132.

John, M. K., van Laerhoven, C. J. and Cross, C. H. (1975) Cadmium, lead and zinc accumulation in soils near a smelter complex. *Environmental Letters*, **10**, 25–35.

Kabata-Pendias, A. and Pendias, H. (1984) *Trace Elements in Soils and Plants*. CRC Press, Boca Raton, Florida.

Kabata-Pendias, A. and Pendias, H. (1992) *Trace Elements in Soils and Plants*. 2nd Edition, CRC Press, Boca Raton, Florida.

Klein, D. H., Andren, A. W., Carter, J. A., Emery, J. F., Feldman, C., Fukerson, W., Lyon, W. S., Ogle, J.C., Talmi, Y., van Hook, R.I. and Bolton, N. (1975) Pathways of thirty-seven trace elements through coal-fired power plant. *Environmental Science and Technology*, **9**, 973–979.

Koh, T. S. and Judson, G. J. (1986) Trace elements in sheep grazing near a lead–zinc smelting complex at Port Pirie, South Australia. *Bulletin of Environmental Contamination and Toxicology*, **37**, 87–95.

Levinson, A. A. (1974) *Introduction to Exploration Geochemistry*. Applied Publishing, Calgary.

Levitt, E. A. (1988) Geochemical monitoring of atmospheric heavy metal pollution: theory and applications. *Advances in Ecological Research*, **18**, 65–177.

Levitt, E. A., Lee, J. A. and Tallis, J. H. (1979) Lead, zinc and copper analyses of British blanket peats. *Journal of Ecology*, **67**, 865–891.

Lindberg, S. E. and Harriss, R. C. (1989) The role of atmospheric deposition in an eastern U.S. deciduous forest. *Water, Air and Soil Pollution*, **16**, 13–31.

Lindquist, O., Jernelov, A., Jonhansson, K. and Rodhe, H. (1983) *Mercury Pollution of Swedish Lakes. Global and Local Sources*. Swedish Environmental Protection Board, Solna, Sweden.

Lindsay, W. L. (1979) *Chemical Equilibria in Soils*. Wiley Interscience, Wiley, New York.

Little, P. and Martin, M. H. (1972) A survey of lead, zinc and cadmium in soil and natural vegetation around a smelting complex. *Environmental Pollution* **3**, 241–254.

Martin, M. H. and Bullock, R. (1994) The impact and fate of heavy metals in an oak woodland ecosystem. In: Ross, S.M. (Ed.) *Toxic Metals in Soil–Plant Systems*, pp. 327–365. Wiley, Chichester.

Martin, M. H. and Coughtrey, P. J. (1982) *Biological Monitoring of Heavy Metal Pollution: Land and Air*. Applied Science Publishers, London.

Martin, M. H., Coughtrey, P. J. and Ward, P. (1979) Historical aspects of heavy metal pollution in the Gordano Valley. *Proceedings of the Bristol Natural History Society*, **37**, 91–97.

McGrath, S. P. and Smith, S. (1990) Chromium and Nickel. In: Alloway, B. J. (Ed.) *Heavy Metals in Soils*, pp. 125–150. Blackie, London, Wiley, New York.

Merry, R. H. and Tiller, K. G. (1978) The contamination of pasture by a lead smelter in a semi-arid environment. *Australian Journal of Experimental Agriculture and Animal Husbandry*, **18**, 89–96.

Mitchell, R. L. (1964) Trace elements in soil. In: Bear, F. E. (Ed.) *Chemistry of the Soil*, pp. 320–368. Reinhold Publishing Corporation, New York; Chapman and Hall, London.

Natusch, D. F. S. and Wallace, J. R. (1974) Toxic trace elements: preferential concentration in respirable particles. *Science*, **183**, 202–204.

Nikiforova, E. M. and Smirnova, R. S. (1975) Metal technophility and lead technogenic migration. In: *Proceedings of the International Conference on Metals in the Environment, Toronto*, pp. 94–96. Toronto, 1975.

Norton, S. A. (1986) A review of the chemical record in lake sediments of energy-related air pollution and its effects on lakes. *Water, Air and Soil Pollution*, **30**, 331–345.

Nriagu, J. O. (1980) (Ed.) *Cadmium in the Environment. Part 1: Ecological Cycling*. John Wiley, New York.

Nriagu, J. O. and Pacyna, J. M. (1988) Quantitative assessment of worldwide contamination of air, water and soils by trace metals. *Nature*, **333**, 134–139.

Nriagu, J. O., Pacyna, J. M., Milford, J. B. and Davidson, C. I. (1988) In: Nriagu, J. O. and Nieboer, E. (Eds) *Chromium in the Natural and Human Environment*. pp. 125–172. Wiley, New York.

Pacyna, J. M. (1986) Atmospheric trace elements from natural and anthropogenic sources. In: Nriagu, J. O. and Davidson, C. I. (Eds) *Toxic Metals in the Atmosphere*, pp. 33–52. Wiley, New York.

Pacyna, J. M. (1987) Atmospheric emissions of arsenic, cadmium, lead and mercury from high temperature processes in power generation and industry. In: Hutchinson, T. C. and Meema, K. M. (Eds). *Lead, Mercury, Cadmium and Arsenic in the Environment*, pp. 69–87. SCOPE, Vol. 31. Wiley, New York.

Pacyna, J. M., Semb, A. and Hanssen, J. E. (1984) Emission and long-range transport of trace elements in Europe. *Tellus (Series B)*, **36B**(3), 163–178.

Post, J. E. and Buseck, P. R. (1985) Quantitative energy-dispersive analysis of lead halide particles from the Phoenix urban aerosol. *Environmental Science and Technology*, **19**, 682–685.

Ragaini, R. C., Ralston, R. and Roberts, N. (1977) Environmental trace metal contamination in Kellog, Idaho, near a lead smelting complex. *Environmental Science and Technology* **11**, 773–781.

Rutherford, G. K. and Bray, C. R. (1979) Extent and distribution of soil heavy metal contamination near a nickel smelter at Coniston, Ontario. *Journal of Environmental Quality*, **8**, 219–222.

Steinnes, E. (1987) Impact of long-range atmospheric transport of heavy metals to the terrestrial environment in Norway. In: Hutchinson, T. C. and Meema, K. M. (Eds) *Lead, Mercury, Cadmium and Arsenic in the Environment*, pp. 107–117. SCOPE, Vol. 31. Wiley, New York.

Swaine, D. J. (1955) The trace element content of soils. *Commonwealth Bureau Soil Science Technical Communication No. 48*, CAB, Farnham Royal, Bucks, England.

Ter Haar, G. L. and Bayard, M. A. (1971) Composition of airborne lead particles. *Nature*, **232**, 553–554.

Tiller, K. G. (1989) Heavy metals in soils and their environmental significance. *Advances in Soil Science*, **9**, 113–142.

Tiller, K. G. and de Vries, M. P. C. (1977) Contamination of soils and vegetables near the lead–zinc smelter, Port Pirie, by cadmium, lead and zinc. *Search* **8**, 78–79.

US Environmental Protection Agency (1979) *Criteria for Classification of Solid Waste Disposal Facilities and Practices. Federal Register 44*, 53 438–53 468.

Wheeler, G. L. and Rolfe, G. L. (1979) The relationship between daily traffic volume and the distribution of lead in roadside soil and vegetation. *Environmental Pollution*, **18**, 265–274.

2 The Meaning of Metal Toxicity in Soil–Plant Systems

SHEILA M. ROSS
University of Bristol, UK
KATHERINE J. KAYE
Oxford University, UK

ABSTRACT

It is important to define what is meant by metal toxicity and to assess exactly when and how metals can become toxic to plants and soil organisms. Acute and chronic toxicity are defined and methods for describing the effects of toxicity experiments, in terms of thresholds and lethal concentrations, are outlined.

Mechanisms of metal uptake by plants are introduced. Plants can be classified as metal excluders, index (or indicator) plants, or accumulators, depending on whether they restrict metal uptake or actively concentrate metals in their tissues. Metal ion uptake is divided into two types of process: *passive*, apoplastic uptake, and *active*, symplastic uptake. Plants show a range of different mechanisms for protecting themselves against metal uptake. These include the production of phytochelatins, subcellular metal compartmentalisation and organic ligand exudation.

Toxic levels of metals in soils can detrimentally affect the number, diversity and activity of soil organisms, with knock-on effects on soil organic matter decomposition and nitrogen mineralisation processes. Soil organisms can also accumulate metals in their tissues to concentrations up to 50 times higher than in surrounding soil.

There is no uniform procedure for setting acceptable contamination standards, or thresholds, for heavy metals in soil. Many of the available contamination classifications are based on the intended future use of the land. Three main types of soil ameliorative techniques are available: thermal, washing and biological treatment. As many as 35 000 metal contaminated sites in Germany and 50 000–100 000 sites in the UK are in need of soil restoration due to metal pollution.

2.1 INTRODUCTION

Baker's (1983) description of plants as "... *miners* of [nutrients from] the earth's crust ..." is a highly appropriate image for us to use since it reminds us that plants are the major selective accumulators of all types of inorganic

Toxic Metals in Soil–Plant Systems. Edited by S. M. Ross
© 1994 John Wiley & Sons Ltd

nutrients, including metals, on which other life forms depend. Several trace metals are essential micronutrients for plant and animal growth. Tyler (1981) suggests that Cu, Zn, Fe and Mn general requirements are around $1-100\ \mu g\,g^{-1}$ dry biomass. Higher levels are often toxic. The margin between adequacy and toxicity can be narrow, often as little as a factor of 2 (Bowen, 1966). Having said this, studies on both micronutrient requirements and toxicity effects of trace metals on both soil organisms and native plants in field conditions, are extremely limited. More commonly, the effects of metals on key test organisms and plants have been examined in sterile and much simplified laboratory conditions, often in ill-conceived and poorly interpreted glasshouse pot experiments. How applicable the results of such experiments are to true, complex field conditions, has been the subject of much debate. Much research has been directed at understanding the effects of metals on food plant production and, until recently, rather less interest has been focused on issues of trace metal cycling in natural ecosystems.

A simplified illustration of the storage compartments and fluxes involved in the cycling of metals in a soil–plant system is given in Figure 2.1. Metal

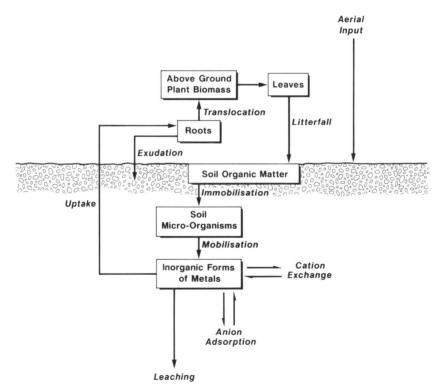

Figure 2.1. Storage compartments and fluxes involved in the cycling of metals in soil–plant systems

compartmentalisation in different ecosystems has received some attention, but, unlike more commonly studied macronutrients such as N, P and K, remarkably little is known about many metal transformation and trans-location *processes*, particularly (i) chemical transformations in both the soil matrix and within plant cells, (ii) bioavailability and uptake of different metal chemical species, (iii) mobility of different organic and inorganic metal chemical species in both soils and plants, and (iv) adsorption and exchange processes of different metal chemical species in soil. The following discussions of metal toxicity in connection with soil–plant systems will first define what is understood by toxicity; secondly, the phytotoxic effects of metals on higher plants will be introduced; and thirdly, toxic effects of metals on soil biological processes will be discussed. Finally, the problems associated with the setting of toxicity thresholds, or soil contamination limits, will be introduced.

2.2   TOXICITY OF METALS

In toxicology studies, two forms of approach dominate: (i) dose-response rela-tionships and (ii) epidemiological studies of toxicity patterns in populations through case-control or cohort studies. Five toxicity characteristics must be considered in any toxicological study:

(1) quantity (of toxin)
(2) route of exposure (ingestion, root uptake, etc.)
(3) distribution of dose/exposure time
(4) type and severity of injury
(5) time needed to produce that injury

Environmental scientists encounter numerous problems analysing these characteristics of any metal toxicity story: how do we determine how much of what went where, when, with what result showing up how much later? In environmental and ecological toxicology, we must define the organisational level of the toxin action: individual organisms, species level, sub-ecosystem populations, or ecosystems. Care must be taken to use the terms "environ-mental toxicology" and "ecotoxicology" in their existing strict definitions:

— *Environmental toxicology*: the toxicology of the built environment, specifically within-building spaces, such as homes or workplaces.
— *Ecotoxicology*: the study of toxins on population dynamics.

To understand the effects of toxic metals on soil–plant systems, much greater definition and precision of a number of aspects of the metals is required. Some of these key details are listed in Table 2.1. These categorisations show what kinds of data might be applicable to the environmental study of a toxic metal, but leave questions of how the data might be collected, analysed and related to a wider context. A very large number of factors, both environmental and

Table 2.1. Required information on the nature and action of toxic metals in soil–plant systems

---

Characteristics of the toxic metal
— metal chemical forms
— bioavailability
— residence time(s) in the organism(s) under study
— distribution of "dose" over time
— route (method) of exposure
Mechanisms and effects of toxic metal action
— lethal or sub-lethal?
— are they reversible?
— interactions with other metals?
— cumulative effects?
— acute effects at lethal levels but none at sub-lethal?
— chronic effects?
— dependent on developmental stage of the entity affected?
Targets of toxic metal action
— reproductive functions
— respiratory functions
— photosynthetic processes
— genetic material
— increases/decreases in populations

---

biological, influence the toxicity of metals in ecosystems. Bryan (1976) attempts to list these for aquatic ecosystems in Table 2.2. For the toxicity of metals in the soil, Table 2.2 must have an additional section, dealing with influencing soil factors, including proportion of clay content and clay mineralogy, soil organic matter content, cation exchange capacity and redox potential.

In soil–plant systems, one of our interests is in defining metal toxicity and phytotoxicity under "natural" field conditions. *Toxicity* is usually defined as a poisonous effect on a living organism. While toxicity testing techniques have primarily come from public health assessments and fisheries, and probably have limited relevance for plants growing in field soils, the concepts of toxicity are similar. Two categories of toxic effects are usually defined (e.g. Alderdice, 1967):

(a) *Acute toxicity*: a large dose of poison for short duration, which is usually lethal.

(b) *Chronic toxicity*: a low dose of poison over a long period of time, which can be lethal or sublethal.

The quantitative effects of toxicity studies can be described in a number of ways, for example, those listed in Table 2.3. Pollutant and toxicity testing has almost exclusively focused on humans and on higher animals. While work in agriculture and forestry has gone some way to linking soil metal concentrations

Table 2.2. Factors influencing the toxicity of heavy metals in solution

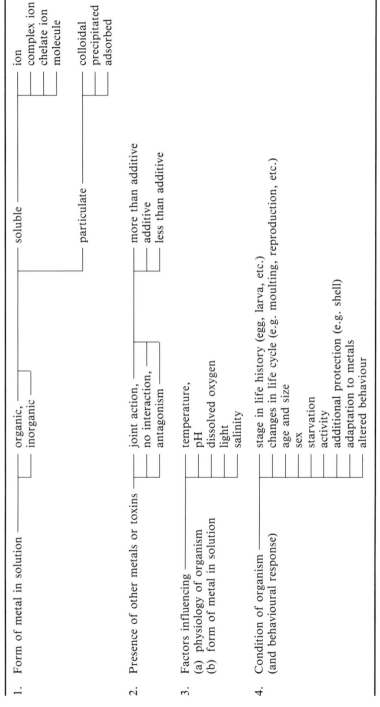

Reproduced by permission of Academic Press (London) Ltd from Bryan (1976).

Table 2.3. Methods of describing the effects of toxicity studies

---

— *Lethal concentration* (LC): death is the criterion for toxicity. The results are expressed with a number which describes the percentage of organisms killed at a particular concentration (LC_{50}, LC_{75}, etc.). The time of exposure must also be stated, e.g. 24-hour LC_{50} is the concentration of a toxic material which kills 50% of the test organism in 24 hours.

— *Effective concentration* (EC): used to describe toxic effects other than death. In microbial and animal studies this could include developmental abnormalities, respiratory stress and behavioural changes; in plants the effects could include growth retardation, reproductive abnormalities, or altered foliar development and discoloration, such as chlorosis or necrosis. Results are expressed in a similar manner to lethal concentration, e.g. 48-hour EC_{50}.

— *Phytoxicity threshold* (PT): plant leaf tissue concentration corresponding to a defined growth reduction. PT_{50} defines a 50% growth reduction.

— *Incipient lethal level* is the concentration at which acute toxicity ceases. (Usually taken as the concentration at which 50% of the population of test organisms can live for an indefinite time).

— *Safe concentration* is the maximum concentration of a toxic substance that has no observable effect on a species after long-term exposure over one or two generations.

— *Maximum allowable toxicant concentration* is the concentration of a toxic waste which may be present in a receiving water without causing harm to its productivity and its uses.

Sources: American Public Health Association (1976) and Chang *et al.* (1992).

to plant metal uptake and growth abnormalities, more success appears to have been achieved in the past in deficiency studies of essential trace metals than in toxicity studies. A major difficulty for plant ecologists is the relative paucity of data relating levels of potentially toxic metals in soils or waters to the health of *native* plants. Conservationists are interested in protecting communities at risk from metal pollution, either from the air or in water, while environmental managers interested in the ecological restoration of derelict sites and mine wastes require information on the toxicities of metals on native species to allow them to plant successful revegetation programmes.

A range of ecotoxicological issues are currently at the forefront of debates. Three main topics requiring further research are: (i) biomagnification, i.e. the transferal and accumulation of metals up the foodchain; (ii) developing more useful, relevant and statistically acceptable criteria for selection of metal toxicity thresholds in soil; and (iii) empirical study of the effects of multiple (simultaneous) metal toxicities which may cause shifts in the toxicity thresholds of plants and organisms for individual metals. In terms of bio-magnification, there is evidence to suggest that not all metals are equally mobile in terrestrial food chains. Laskowski (1991), for example, provides a statistical analysis of available data to illustrate that Cu and Cd are differ-entially mobile in food chains and that all metals are not biomagnified through the food chain. These processes, however, are likely to be specific for different species and different ecosystems (van Straalen and Ernst, 1991). The second topic, the selection of relevant toxicity thresholds for soil, has attracted much attention recently. One of the main debates surrounds the need to test and develop toxicity thresholds for individual species, populations or groups of species, in experimental conditions and then apply them to complex field soil conditions. Hopkin (1990, 1993; Chapter 8, this volume) discusses the scientific and ethical background for the adoption of particular thresholds. On the third topic, that of multiple toxicity effects, extremely little is known in practice for different organisms or plants. Wallace and Berry (1983) provide a theoretical basis for understanding the interaction of metal toxicities, including synergistic, additive and antagonistic effects.

Perhaps one of the most serious research gaps to date has been the lack of identification of the actual chemical *forms* of metals, particularly in soils, which can be taken up by different soil organisms and by plants. Soil processes affecting which chemical forms of metals occur under different soil conditions are reviewed in Chapter 3. The next important steps in the soil–plant toxicity jigsaw puzzle must be: (i) to properly identify which chemical forms of metals are phytoavailable; and (ii) to develop soil analytical techniques (perhaps microbial and biochemical as well as chemical) which properly target only those chemical forms of metals "known" to be phytoavailable. Only once these linkages have been made will *specifications* of "toxic" levels for plants, or *phytotoxic thresholds* of metals in soils be possible.

There are two possible approaches to understanding how plants respond to the presence of toxic metals in the soil. First, it is necessary to identify damage or injury to the plant and its component parts. Secondly, it is possible to look for plant defence mechanisms which may impart tolerance. These two aspects of metal toxicity in plant studies are discussed in the two following sections.

2.2.1 DEFINING METAL TOXICITY AND PHYTOTOXICITY

At the *cellular level*, Ochiai (1987) identifies several possible mechanisms by which toxic metals may effect damage:

(i) blocking functional groups of biologically important molecules, such as enzymes, polynucleotides, or transport systems for nutrient ions;
(ii) displacing and/or substituting essential metal ions from biomolecules and functional cellular units;
(iii) denaturing and inactivating enzymes; and
(iv) disrupting cell and organelle membrane integrity.

In higher plants, root cells are likely to experience these damaging effects first. How much of an effect is necessary, or what combinations of effects are necessary to produce visible or measurable symptoms of toxicity, or to cause death, will be different for each plant species, each combination of soil and environmental conditions and each ecotype. In lower plants and unicellular organisms it is a little simpler to assess the degree of effect required to be lethal. Plants experiencing various degrees of phytotoxicity exhibit a range of different symptoms during the course of their growth. There may be changes in susceptibility to toxicity as well as changes in the site of action or impact of the metal. Toxicity symptoms are a result of specific *injurious effects* that metals can have on plant *physiological processes*, including: inhibition of photosynthesis and respiration (e.g. Carlson *et al.*, 1975); alteration of plant–water relations, causing water stress and wilting (e.g. Poschenrieder *et al.*, 1989); increased permeability of root cell plasma membrane, rendering roots less ion selective (e.g. Loneragan *et al.*, 1987) through changes in cytoplasm pH and plasma membrane electrical potential (Cumming and Tomsett, 1992); and adverse effects on the activities of metabolic enzymes.

Whether the disorganisation or disruption of vital physiological processes results in detectable symptoms or not will depend on many things, including the species of plant, how stressed the plant is by other environmental factors such as drought and how sensitive is the detection technique. Chang *et al.* (1992) suggest that four symptom criteria should be fulfilled if metal toxicity is to be confirmed in a plant:

  (i) plants show *sustained* injuries;
 (ii) a potentially phytotoxic metal has accumulated in the plant tissues;
(iii) observed abnormalities are not due to other disorders of plant growth; and
(iv) the biochemical mechanisms that cause the metal to be harmful to plants are observed during the course of growth.

There is a real shortage of necessary published data on points (iii) and (iv) above which unequivocally demonstrate metal phytotoxicity. Since soil-derived chemicals accumulate in plant tissues, foliar chemical analysis has been widely used as a good indication of soil fertility. Foliar analysis has also been used as an indication of the levels of trace metals in soils, particularly for perennial crops such as forestry. We might expect to be able to use this technique to determine threshold plant tissue concentrations of particular metals which result in phytotoxicity (Beckett and Davies, 1977). Five main problems arise. First, phytotoxicity thresholds differ with plant species. Secondly, soil properties influence the rates at which metals transfer to plants. Thirdly, roots may sequester metals and prevent or reduce translocation to the leaves. Fourthly, no chemical or toxicant interactions are taken into account. Fifthly, changes in foliar chemistry may be influenced by other environmental factors such as water availability, pH, redox or salinity.

2.2.2 MECHANISMS OF METAL UPTAKE

We might expect that plants growing on metal-contaminated sites need to develop some degree of tolerance to metal toxicity in order to survive. Since all plants contain at least some metal in their tissues, they clearly are incapable of completely *excluding* potentially toxic elements, but simply of *restricting* their uptake and/or translocation within the plant. It is possible to classify plants into three main groups, according to their metal uptake characteristics:

(i) *Excluders*: plants with restricted uptake of toxic metals or restricted translocation into the shoot over a wide range of soil metal concentrations.

(ii) *Index plants*: plants in which uptake and translocation reflect soil metal concentrations.

(iii) *Accumulators*: plants which actively concentrate metals in their tissues.

Brooks *et al.* (1977) identify an extreme form of accumulators, described as hyperaccumulators, in which the tissue metal concentration can exceed 1000 μg metal g^{-1}. There are many plants which sequester metals in their roots, preventing translocation to the shoots. *Agrostis stolonifera* growing on serpentine soils in Shetland is a good example (Shewry and Peterson, 1976). Concentrations of chromium and nickel were approximately twice and five times greater respectively in the roots than in the shoots (Figure 2.2). The

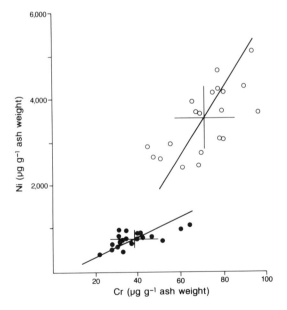

Figure 2.2. The relationship between chromium and nickel concentrations in roots ($\circ$) and shoots ($\bullet$) of plants of *Agrostis stolonifera* collected from the north-west slope of the Keen of Hamar, Unst, Shetland. Regression lines are drawn and the arithmetic means are marked with crosses, the limits of which show the standard deviations. Reproduced by permission of Blackwell Scientific Publications from Shewry and Peterson (1976)

actual processes contributing to restricted translocation of metal in higher plants are still largely unknown. In order to understand some of the contributary factors, mechanisms by which metals are taken up into plant cells are outlined below, followed by a brief account of some of the metal exclusion or restriction processes thought to be important in conveying metal tolerance.

In soil–plant systems, heterotrophic soil bacteria, fungi, soil micro and macrofauna, must obtain their nutrition by breaking down complex organic molecules into absorbable forms, through some form of digestion or enzymatic hydrolysis. Metals in leaf litter and soil organic matter may be released in this way. Plants and cyanobacteria, which are autotrophs, survive in an entirely inorganic environment because they manufacture their own organic compounds from water-soluble inorganic raw materials. Heterotrophic digestion makes simple forms of nutrients and metals available for autotrophic uptake. Toxic metal ions are thought to enter cells by means of the same uptake processes that move essential micronutrient metal ions, such as Cu and Zn. All materials must ultimately pass through some form of biological barrier before they can be metabolised, although single-celled organisms may first engulf materials through pinocytosis. In plants there are two stages in ion uptake and it is generally agreed that these also operate for metal ions: (i) passive uptake via the *apoplast*, and (ii) active uptake via the *symplast*. In higher plants, solute cations move passively into the root cortical cell walls, which form a hydrated free space continuum between the external bathing solution and the cortical cell membranes (apoplast). Since primary cell walls consist of a network of cellulose, hemicellulose (including pectins) and glucoprotein, negative charges, generated on carboxylic groups ($-R-COO^-$) for example, act as cation exchangers and anion repellers. Plants differ in their cation exchange capacities, with that of dicotyledonous species generally much higher than that of monocotyledonous species. Di- and polyvalent cations are preferentially attracted to these cation exchange sites. Specific components of proteins, particularly nitrogen, sulphur and oxygen, can also interact chemically with some metals, binding them to cell walls. The affinities of different metals for inorganic and organic ligands is determined by their thermodynamic characteristics, as discussed by Pearson (1968a,b) and more recently in a biological toxicity context by Nieboer and Richardson (1980). These chemical principles are outlined in Section 3.2.1, Chapter 3. Class A metals (e.g. K, Ca, Mg) preferentially bind with oxygen-rich ligands (e.g. carboxylic groups), class B metals (e.g. Hg, Pt, Au) preferentially bind with sulphur- and nitrogen-rich ligands (e.g. amino acids), and borderline metals (e.g. Cd, Pb, Cu, Zn) show intermediate preferences, with the heavier metals tending towards class B characteristics.

Botanists have thought that plant cell walls play an important part in metal tolerance since Turner (1967, 1969) found that cell walls of Cu/Zn-tolerant *Agrostis tenuis* plants had a higher metal binding capacity than those of non-tolerant plants. Turner and his co-workers realised that metal binding in the cell wall could protect plants from toxic effects of metals by restricting metal transport within the plant and preventing metals from reaching sensitive sites in the roots. Several substances in cell walls, including pectates (e.g. Baker, 1978), amino acids (e.g. Wyn Jones *et al.*, 1971) and carboxylic acids have been suggested as compounds for metal binding. Evidence of chemical binding of metal ions to components of the plant cell wall has also been shown by van Cutsem and Gillet (1982) for example, who reported that copper was bound to nitrogen-containing groups of either glycoproteins or proteins of ectoenzymes in the cell wall of the stonewort, *Nitella*.

Figure 2.3 illustrates soil and plant processes contributing to bioavailability and uptake of metals by plants. The apoplast (cell wall continuum) consists of two zones: the *water free space*, which is freely available to ions equilibrating from the external solution, and the *Donnan free space*, where cation exchange and anion repulsion take place. These two zones make up the *apparent free space*. The inner limit of this free space, and hence of the passive, apoplastic radial transport of ions across the root to the stele, is the endodermis. For metal ions to be metabolised, they must pass through the plasma membrane and this can only occur by active transport processes. Baker (1983) thus visualises the plasma membrane of root cortical cells (symplast) to be in contact with a very large surface area of cell wall free space from which active and selective ion uptake can occur. Passive uptake, involving ion diffusion driven by the ionic concentration gradient, can occur in both live and dead cells. The same form of Fick's law of diffusion can be used for diffusion into the apoplast as is used for metal ion diffusion through the bulk soil solution (see Section 3.3.7). Theoretically it is possible to calculate the rate of entry of metal ions into cells. In practice, there are major problems with this simplistic approach, not least of which are:

(i) the complexity and dynamism of the soil solution chemistry,
(ii) the difficulty of measuring accurately concentrations of simple metal ions and metal ligands both within intact cells and in bathing solutions; and
(iii) the lack of empirically derived diffusion coefficients even for simple metal ions, far less those for more complex metal ligands.

Active transport of ions across the plasma membrane, often against the solute concentration gradient, must be facilitated by some form of carrier mechanism. The mechanisms appear to depend on ion type, size, valency and the geometry of the carrier site. The chemical species of metals present in the soil solution and available for uptake are determined by surrounding soil conditions, including pH, redox, soil organic matter and concentration of other ions in solution (see Chapter 3). pH in particular affects the ionisation state of metals in solution. Ionisation states at known soil solution pH can be predicted using the Henderson–Hasselbach equation (see Section 3.3.7). Most authors now agree that active uptake processes involve ion-specific carriers and the consumption of energy, probably as energy-rich phosphates in the

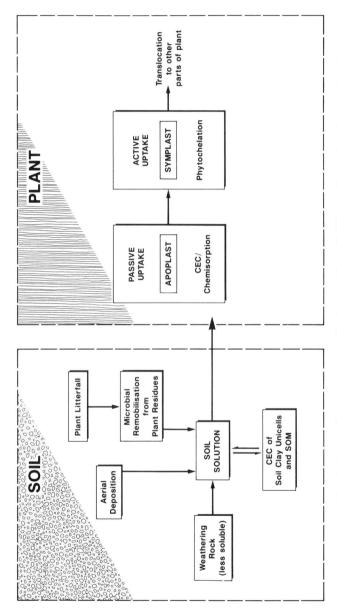

Figure 2.3. Soil and plant cell processes contributing to bioavailability and uptake of metals by plants

form of ATP. For potassium uptake in plants, Leonard (1983) suggests that ATPase in the plasma membrane of plant cells acts as a pump to exchange hydrogen ions from the cell interior with potassium ions on the outside of the cell. A similar process has been suggested for fungi and yeasts (Borst-Pauwels, 1981; White and Gadd, 1987). In higher plant cells, further transport across membranes, into the vacuole (across the tonoplast membrane) or into organelles such as chloroplasts or mitochondria, are thought to require similar, energy-consumptive active carrier processes. Although no such processes specific for toxic metals have yet been positively identified, evidence is accumulating to indicate that a range of biological and biophysical characteristics of the plasma membrane, including surface charge character- istics, fatty acid composition and sterol content, affect metal uptake. Damage, in the form of chemical changes to the membrane as a response to the presence of toxic metals, has been alluded to by some authors, but remains unproven. Cummings and Tomsett (1992) give a review of the evidence for active transport of metals across the plasma membrane in a range of organisms and higher plants.

The kinetics of monovalent cation uptake has been shown in several studies to follow a relationship similar to the Michaelis–Menten equation which is used in analysing the kinetics of enzyme reactions:

$$v = \frac{CV_{max}}{K_s + C}$$

where:

v = ion uptake velocity

C = ion concentration in external solution

V_{max} = maximum rate of uptake when all available carrier sites are loaded (i.e. maximum transport rate)

K_s = Michaelis constant (which equals the substrate ion concentration that gives half the maximum transport rate)

Reviewing kinetic data from a number of studies of metal uptake in yeasts and fungi, Gadd and White (1989) note that both V_{max} and K_s values range widely, even in different studies of uptake of the same metal by the same organism. Other authors have also observed that plots of $1/v$ versus $1/C$ for metal uptake frequently yield a "curve" comprising two or more straight line portions, from which different K_s and V_{max} can be obtained. This has been interpreted as evidence of the presence of different carriers within the membrane (see discussion by Baker, 1983). Despite the use of these theoretical analogies for simple cations, studies of the processes and kinetics of toxic metal uptake into root cells of higher plants is in its infancy and invoking similar carrier processes and kinetics is speculative. In addition, toxic metals can produce disruptive effects on the structure and function of the plasma membrane and this can alter the kinetics of metal uptake.

Metal uptake by yeasts, fungi and bryophytes has been shown to follow similar patterns of passive (non-metabolic) and active (carrier-mediated) uptake to that of higher plant cells. Passive uptake processes can take place even when cells are dead. ATP energised processes also appear to operate at carrier sites of active uptake in yeasts and fungi (Gadd and White, 1989). Fungi show several mechanisms for retaining metals in particular parts of the cell, or for sequestering metals outside the cell. First, metals may be retained in the cell wall by biosorption onto wall constituent molecules such as melanins, glucans, mannans, chitin and chitosan (Gadd and White, 1989), which are active even when the fungal cells are dead. These binding mechanisms make dead fungal biomass a useful option for cleaning up waste industrial effluents. Within the cell, fungi and yeasts also produce metallo-thioneins, which are metal-binding proteins, in response to elevated concentrations of some metals, such as copper (e.g. Butt and Ecker, 1987). These binding compounds act to regulate the concentration of metal in the cell and prevent further metal transport into the organism. Some fungi also secrete organic compounds into the surrounding soil to remove metals from solution by chelation. High affinity iron-binding molecules, called *siderophores*, have been identified in soil, produced by some rhizosphere bacteria (Raymond *et al.*, 1984), soil fungi and some mycorrhizal fungi (e.g. Morselt *et al.*, 1986). Such compounds may also have some function in protecting both the fungus and the plant from metal toxicity. Gildon and Tinker (1983) found that vesicular–arbuscular mycorrhizae (VAM) infected plants showed reduced uptake of Zn and Cu compared to non-VAM plants if metal concentrations were high in the soil. Similar effects of ectomycorrhizae have been observed on *Calluna vulgaris* growth in contaminated soils (Bradley *et al.*, 1981) and on *Quercus robur* growth in Cd, Ni and Pb polluted soils (Dixon, 1988).

There is now evidence to suggest that some higher plants also have an armoury of metal protection mechanisms. Proposed metal tolerance mechanisms include: (a) metal sequestration by specially produced organic compounds; (b) sequestration in certain cell compartments; (c) metal ion efflux; and (d) organic ligand exudation. These will be discussed briefly.

(a) Phytochelatins

Apart from the metal-binding properties of some cell wall constituents such as amino acids and proteins, some plants appear to be able to produce metal-binding polypeptides to sequester metals and render them unavailable for transport within the plant. Stiffens (1990) calls these organic compounds *phytochelatins*. Grill *et al.* (1987) have shown that a range of metals can induce phytochelatin formation, including: Cd, Pb, Sb, Ag, Ni, Hg, Cu, Sn and Au. Cd is a strong inducer while Zn is a weak inducer. It must be pointed out that Grill's experiments appear to be confined to observing phytochelatin production in cell cultures, not in live, intact plants. Salt *et al.* (1989) and Scheller *et al.* (1987), however, identified phytochelatin production in response to high metal concentrations in *Mimulus* and tomato respectively. Research into the mode of action of plant polypeptide compounds is still in its infancy and the extent and utility of the phenomenon in the plant kingdom is relatively unknown.

(b) Subcellular compartmentalisation

A second defence mechanism in response to toxic metals is compartmentalisation of the metal in certain parts of the cell, particularly the vacuole. Brookes *et al.* (1981) found that Zn became concentrated in the vacuole of zinc tolerant *Deschampsia caespitosa* compared to non-tolerant plants when exposed to elevated Zn concentrations. Mathys (1977) suggested that malate is produced by tolerant plants to move Zn from the cytoplasm into the vacuole where it is complexed by organic ligands including citrate, oxalate, mustard oils and anthocyanins. There is some evidence in other plants that ligand sequestration of Ni (Brooks *et al.*, 1981) and Cd (Vogeli-Lange and Wagner, 1990) also occur in the cell vacuole.

(c) Active metal efflux

Active pumps for excreting metals have been identified in prokaryotes (e.g. Nies and Silver, 1989), but while similar mechanisms have not been proven in higher plants, efflux pumps have been suggested as a process for controlling metal accumulation, particularly for aluminium (e.g. Zhang and Taylor, 1989).

(d) Organic ligand exudation

Organic molecules exuded by root cells, as well as mucilages at the root tip, can bind, or chelate, with metals in the rhizosphere, and render them unavailable or less available for root uptake. Exudation is thought to be due to changes in cell electrical potential as a result of metal toxicity (e.g. Otto *et al.*, 1980). Since metal chelation outside the root, depending on the organic ligands involved, can reduce metal uptake, exudation can confer metal tolerance in these plants. Merckx *et al.* (1986), studying ^{14}C labelled, water-soluble exudates released from wheat and maize roots, showed that ionic forms of Co, Zn and Mn in the soil were converted into organic complexes in the rhizosphere. Mycorrhizas have also been shown to produce organic substances which complex metals. Denny and Wilkins (1987) found that Zn tolerance in birch was partly helped by the production of polysaccharides by the associated ectomycorrhiza, as well as metal binding to hyphal cells, which reduced Zn uptake by the plant. A similar response was found in pine by Cumming and Weinstein (1990). It must be pointed out, however, that organic chelates have

Table 2.4. Generalised pattern of metal toxicities published for a range of different organisms

	Organism	Toxicity sequence	Source
I Animals			
	Protozoa (*Paramecium*)	Hg, Pb > Ag > Cu, Cd > Ni, Co > Mn > Zn	Shaw (1954)
	Annelida (Polychaete worm)	Hg > Cu > Zn > Pb > Cd	Reish *et al.* (1976)
	Vertebrata (Stickleback)	Ag > Hg > Cu > Pb > Cd > Zn > Ni > Cr	Jones (1939)
II Bacteria			
	N-mineralising bacteria	Ag > Hg > Cu > Cd > Pb > Cr > Mn > Zn = Ni > Sn	Liang and Tabatabai (1977)
III Plants			
	Algae (*Chlorella vulgaris*)	Hg > Cu > Cd > Fe > Cr > Zn > Ni > Co > Mn	Sakaguchi *et al.* (1977)
	Fungi	Ag > Hg > Cu > Cd > Cr > Ni > Pb > Co > Zn	Lukens (1971)
	Higher plants (Barley)	Hg > Pb > Cu > Cd > Cr > Ni > Zn	Oberlander and Roth (1978)

Source: Modified from Nieboer and Richardson (1980).

also been used to make soil metals more soluble for plant uptake in cases of micronutrient deficiency. Brown (1969), for example, used EDTA and other similar synthetic chelates to alleviate iron chlorosis in plants. EDTA made soil iron more soluble and hence more accessible for plant uptake.

Despite serious gaps in knowledge, there are some generalisations and conclusions which can be made about metal toxicity in different organisms. First, it can be seen from the example toxicity patterns presented in Table 2.4 for empirical studies, that there are similarities between toxic metal sequences even for quite different organisms, with class B type metal ions more toxic than borderline metal ions, which are more toxic than class A type metal ions (Neiboer and Richardson, 1980). Secondly, certain organic molecules, both within cell walls and secreted in response to high metal concentrations, are able to bind and chelate metals, effectively immobilising them and preventing further metal transport. Thirdly, the pH of both the soil and the bathing soil solution have a very important effect in controlling both passive and active uptake mechanisms. Analogous to the dissociation which occurs on hydroxyl groups on the surfaces of clay micelles in the soil, increased soil pH causes carboxylic and phenolic groups on cell walls to dissociate, elevating cell wall cation exchange capacity. Gadd and Griffiths (1980), for example, found that the amount of copper binding to fungal cell walls was increased at higher soil pH. There is some evidence that energy-dependent uptake is also increased at higher soil pH, with maximum active transport at pH 6.0–7.0 (e.g. Gadd and White, 1985). How analogous metal tolerance mechanisms and metal transport and uptake in fungi, yeasts, bryophytes and microalgae are to processes in higher plants is still unclear.

2.3   EFFECTS OF TOXIC METALS ON SOIL BIOLOGY

The influence of high concentrations of trace metals on soil biological processes has been reviewed by Tyler (1981). An extensive summary tabulation of critical metal concentrations reported to affect soil microbial numbers and microbially mediated soil processes, including nitrogen mineralisation, litter decomposition and enzyme activities, has been published by Baath (1989). A large amount of data suggest that the relative toxicities of trace metals to soil organisms (on a $\mu g\,g^{-1}$ soil basis) decreases in the order Hg > Cd > Cu > Zn > Pb (see also Table 2.4). Chang and Broadbent (1981) compiled *threshold concentrations* for soil population respiration (representing C_{10}, or the metal concentration required to produce a 10% reduction in CO_2 evolution), and *toxicity indices* (where T_{10} and T_{50} represent the slope of the graph of CO_2 production versus soil extractable metal). Their data, using two different extractants—nitric acid and DTPA (diethylenetriaminepentaacetic acid)—is given in Table 2.5. Results are expressed in polarities to allow comparisons between metals whose atomic weights are quite different. From the suite of

Table 2.5. Threshold concentrations (C_{10}), toxicity indices (T_{10} T_{50}) and metal loading rates of soils that would produce threshold soil concentrations

	Metal loading rate to produce C_{10}		C_{10}(nm g^{-1})		Toxicity index	
	ppm	nm g^{-1}	DTPA	HNO$_3$	T_{10}	T_{50}
Cd	8.7	77.4	22.1	48.0	0.456	0.235
Cr	8.6	165	14.5	73.4	6.46	1.35
Cu	11.8	186	65.6	339	0.607	0.049
Pb	26.8	129	13.6	98.6	0.916	—
Zn	11.7	179	96.2	266	1.04	0.014

Reproduced by permission of Williams & Wilkins from Chang and Broadbent (1981).

metals studied, it can be seen that soil respiration, a general indicator of soil microbial activity, is most sensitive to Cd. Duxbury (1985) classified metals into three categories according to their potential toxicity for soil organisms and soil microbial processes: *extremely toxic* for metals such as Hg, *intermediate toxicity* for metals such as Cd, and *relatively low toxicity* for metals such as Cu, Ni and Zn.

Some examples of the effects of toxic metals on the abundance of soil organisms and on key soil microbial processes such as respiration, organic matter decomposition and nitrogen transformations, are given in the following sections.

2.3.1 SOIL ORGANISM ABUNDANCE

Both abundance and diversity of soil organisms appear to be affected by high concentrations of heavy metals. However, the effects of metals on the soil population are not simple and not all organisms are equally affected. Studying soil organisms in the vicinity of a lead smelter, Bisessar (1982) found that increasing copper concentrations resulted only in decreasing soil bacterial numbers. Increasing concentrations of cadmium and lead resulted in decreased numbers of bacteria, fungi, actinomycetes, nematodes and earthworms. Increased metal concentrations in the soil of Pb/Zn mine waste have been reported by Williams *et al.* (1977) to reduce numbers of arthropods (especially mites) and fungi, to cause no differences in numbers of bacteria or actinomycetes and to increase numbers of springtails. It is likely that these results reflect ecological interactions in the soil organism food web. The patterns are similar to those reported by Edwards (1969) for the effects of pesticides on the soil organism population. The increase in number of springtails may be due to more of the mites that prey on springtails being killed by metal toxicity, resulting in reduced predation pressure.

2.3.2   ORGANIC MATTER DECOMPOSITION AND RESPIRATION

There is ample evidence to indicate that organic matter decomposition rates are slowed down by high levels of trace metals in soils. Superficial examples of this include the presence of deeper litter accumulation horizons in contaminated woodlands (e.g. Coughtrey et al., 1979; Freedman and Hutchinson, 1980). The effects of Cd and Zn on decomposition rates was shown by Coughtrey et al. (1979) to be more important than that of Pb, Cu or pH. Later stages in the decomposition of leaf litter once fragmentation had occurred were found to be more sensitive to metal contamination than the early stages. Reduced rates of litter decomposition have also been linked directly to high levels of Pb (e.g. Mikkelsen, 1974;), Zn (e.g. Berg et al., 1991), Cd (e.g. Spalding, 1979) and to metals in combination (e.g. Ruhling and Tyler, 1973).

Soil respiration is often used as a general indication of microbial activity. Soil respiration studies indicate that metal additions in low amounts can have a variety of effects, depending on prevailing soil conditions. Brookes and McGrath (1984) found no significant decrease in microbial respiration when metal-contaminated sewage sludge was applied to agricultural soil. This may have been because the sludge also provided a valuable carbon energy source. Metals have also been shown to stimulate microbial respiration and CO_2 evolution, either immediately after application (e.g. Ausmus et al., 1978) or after a short lag (Nordgren et al., 1988). High levels of metals reduce CO_2 evolution, as in the case of Pb and Zn (Jackson and Watson, 1977), Cu (Mathur et al., 1979), Cd or Ni (Lighthart et al., 1983). In the metal-contaminated soil around the Cu/Ni smelter at Sudbury, Ontario, Freedman and Hutchinson (1980) found that Cu had a greater influence on CO_2 evolution than had equal amounts of Ni. This same pattern was reported by Nordgren et al. (1983).

Several studies have shown severely reduced soil microbial biomass in relation to toxic levels of metals in soils. In general, the effects are more severe in acidic soils. Dumontet and Mathur (1989), working on heavily Cu-contaminated acidic soils in northern Quebec, found microbial biomass reductions of 44% and 36% in organic and mineral soils respectively, compared to uncontaminated soils. Baath (1989) also reported reduced microbial biomass in acidic forest humus containing 750 $\mu g\,g^{-1}$ Cu and 1000 $\mu g\,g^{-1}$ Zn. Accumulation of metals in sludge-amended soils has also been reported to affect microbial biomass amounts. Brookes and McGrath (1984) found a 55% microbial biomass reduction in agricultural soil which had received 20 years of sewage sludge application, averaging 40–90 $\mu g\,g^{-1}$ Cu and 5–10 $\mu g\,g^{-1}$ Ni.

2.3.3   BACTERIALLY MEDIATED SOIL NITROGEN PROCESSES

Reduction in nitrogen mineralisation and nitrification rates with increased trace metal loadings have been reported in several studies. Liang and

Tabatabai (1977) found very different effects of heavy metals on nitrogen mineralisation in different soils. Mercury caused a much greater inhibition of N mineralisation in acid soil (73%) than in alkaline soil (32–35%) while copper showed the opposite effect: higher inhibition in alkaline soil (82%) and lower in acid soil (20%). In many studies the method of assessing inhibited nitrogen mineralisation has generally been the use of laboratory incubation studies to which trace metals have been added. Under these conditions, increased NO_3-N accumulation at low metals concentrations and decreased NO_3-N accumulation at high metals concentration has been reported (e.g. Rother et al., 1982). By simply measuring the end result of trace metal treatment, namely nitrate production, it is impossible to identify which of a series of possible chemical or biological processes may be affected by metal addition. Contributing processes to nitrate accumulation include: (i) mineralisation and the production of ammonium nitrogen; (ii) exchange, release or fixation of NH_4^+ on clay micelles; (iii) the successive oxidations of Nitrosomonas (NO_2 production) and Nitrobacter (NO_3 production); (iv) denitrification (NO_3 and NO_2 reduction) and volatile losses of N_2O. Since Chang and Broadbent (1982) report reduced denitrification rates with increased trace metals in soil, low NO_3-N accumulation is unlikely to be due to inhibited bacterial reduction. This result would suggest that nitrification processes may be inhibited by high metal concentrations. McKenney and Vriesacker's (1985) denitrification studies indicate that nitrite reduction was more sensitive to Cd additions than was nitrate reduction.

In studies of the effects of metals on symbiotic nitrogen fixation, Rother et al. (1982) found little effect of Cd, Pb or Zn on nitrogenase activity, as measured by acetylene reduction. Subsequently, other authors have suggested that metal effects on symbiotic nitrogen-fixing associations are more related to the higher plant response than to either the bacterial fixation or the symbiosis itself. Porter (1983), for example, suggested that metal toxicity caused leaf chlorosis in alfalfa, resulting in impaired chlorophyll synthesis and reduced photosynthesis. The lack of photosynthates was thought to be responsible for inhibited dinitrogen fixation.

2.3.4 METAL ACCUMULATION BY SOIL ORGANISMS

It has been known for some time that soil organisms are capable of binding and accumulating heavy metals in their cells. Soil microbial food chains are thus capable of significant metal *biomagnification*. Martin and Coughtrey (1982) compiled a review of metal concentrations in the tissues of earthworms living in metal-contaminated soils. Ranges of metal concentrations in various soil fauna are given in Table 2.6. It can be seen that remarkably high concentrations of metals can be accumulated in organisms' tissues. The relationship between soil concentrations and equivalent organism concentrations is not always reported. Martin and Coughtrey (1982) plotted available data for both

Table 2.6. Ranges of tissue metal concentrations in various soil organisms

Organism	Metal concentration ($\mu g\,g^{-1}$)			
	Cd	Pb	Zn	Cu
Earthworm	$0.18 \pm 0.02 - 35$	$0.31 \pm 0.09 - 100 \pm 5$	$68 - 914$	$0.2 \pm 0.03 - 13.1 \pm 5.8$
Woodlice	$1.12 - 232 \pm 10$	$1.58 \pm 0.5 - 1190 \pm 77$	$54.3 \pm 1.7 - 1930$	$70 \pm 9 - 538 \pm 261$
Snail	$5 - 76.1$	$10.7 - 365 \pm 65$	$85.6 - 714 \pm 90$	$30.4 \pm 15 - 86.7 \pm 25$
Slug	$8.3 - 139$	$8.5 - 315$	$516 - 1868$	—

Source: Data summarised from Martin and Coughtrey (1982).

Table 2.7. Mean concentration factors for heavy metals in various soil organisms

Organism	Metal concentration factor			
	Cd	Pb	Zn	Cu
Woodlice	0.28 –48.9	0.01 –19	0.05 –7.49	0.27 –35.8
Terrestrial snails	1.37 –4.77	0.04 –0.52	0.14 –1.28	—
Slugs	0.75 –8.18	0.001 –0.42	0.14 –4.83	—

Source: Data summarised from a large number of soil types in Martin and Coughtrey (1982).

soil and earthworm metal concentrations in a range of studies. They found good relationships between soil Pb and Cd and concentration of these metals in earthworms. They concluded that earthworms could be used as monitors of Pb and Cd contamination in soil. The relationship between soil and earthworm Zn concentrations was found to be much more variable. Studying soils to which sequential additions of sewage sludge had been added, Helmke *et al.* (1979) compared earthworm tissue metal concentrations, metals in earthworm casts and soil metal concentrations, to differentiate between those metals actually bioaccumulated in organism tissue and those metals present in the gut. These authors concluded that earthworms could be useful monitors of metal bioavailability in soils as long as both tissue and cast metal concentrations were measured.

The metal-concentrating abilities of some soil fauna have been reviewed by Martin and Coughtrey (1982). From Table 2.7 it can be seen that in their studies, woodlice were capable of the highest concentration, accumulating Cd to concentrations almost 50 times higher than in the surrounding soil, and Cu to almost 36 times higher. These figures indicate the potential problems which exist in soil food webs in which accumulated metals are potentially passed on from animal to animal.

2.4 SETTING ENVIRONMENTAL CONTAMINATION LIMITS

A clearer realisation of the polluting effects on soils and waters of aerially transported emissions, industrial effluents and other anthropogenically derived metals has alerted politicians and governmental bodies to the urgent need for regulations on acceptable levels of metal contamination in the environment. There are serious difficulties in setting acceptable levels, since they must be based on metal exposure and risk assessments. Component parts of the assessment are estimating the magnitude, frequency, duration and route of exposure. Metal toxicity assessments must thus examine metal sources, pathways of exposure, concentrations, speciation, the affected populations, mode of action of the metal and any toxin interactions (e.g. metal interactions).

Two main methods of assessing metal exposure and risk have been dose-response testing and simulation modelling. Dose-response toxicity testing suffers from the problems of non-representativeness of using individual, test organisms in simplified and unrealistic environments, often with unrealistic doses of metal toxin. There are also problems in assessing differences in acute versus chronic exposure and risk. Nevertheless, the dose-response approach has been widely used to set acceptable metal contamination levels in potable waters. Mathematical models to simulate the likely response of a system to toxic metal input can be useful in cases where suitable good metal quantification is lacking, as can be the case where analyses are difficult or costly, or because concern over the exposure arose retrospectively (Jones, 1991). Transport and food chain modelling are primarily designed to predict exposure and fate of metals and hence provide evidence for decision making. Urgently required, however, are accurate field data for model calibration and validation.

2.4.1 SETTING TOXICITY LIMITS BASED ON TOXICITY TESTING

There is no *uniform* procedure throughout the European Union for setting contamination thresholds for heavy metals in soils. Perhaps one of the main problems associated with designation of toxicity limits is the origin and form of the metal in soil. Different metal origins (e.g. mining compared to sewage sludge application) and different metal species will not be equally mobile or bioavailable. A second major difficulty is the intended future use of the site. Different metal concentrations probably mean different degrees of toxicity for different uses and users. This is partly due to different routes of metal exposure and partly due to different "toxicity thresholds" for the exposed plants, animals and people. For sites with identical soil metal concentrations, the toxicity potential for land uses such as domestic vegetable gardening or a childrens' playground would likely be very much higher than if the land use was building construction or further industrial development. These different land-use requirements are recognised in metal contamination guidelines laid down for developments in the United Kingdom by both the BSI (British Standards Institute, 1988) and the ICRCL (Inter-Departmental Committee on the Redevelopment of Contaminated Land, 1987) for *sensitive* and *less sensitive* land developments. A more detailed scheme for land uses under the jurisdiction of the Greater London Council was developed by Kelly (1979). Table 2.8 attempts to summarise currently used soil metal contamination classifications for the United Kingdom. In both cases cited in Table 2.8, it is unclear on exactly what the soil metal contamination categories are based. The ICRCL categories are described as "trigger thresholds", based on

Table 2.8. Soil toxic metal contamination classifications for use in land development projects in the United Kingdom (all values $\mu g\,g^{-1}$ soil)

Metal (totals)	ICRCL trigger value[a]		GLC–Kelly values[b]			
	Residential use	Hard cover use	Slight class 1	Contaminated class 2	Heavy contamination class 3	Unusually heavy class B
Sb	—	—	30–50	50–100	100–500	> 500
Cd	3	15	1–3	3–10	10–50	> 50
Cr	600	1000	100–200	200–500	500–2500	> 2500
Pb	500	2000	500–1000	1000–2000	2000–10 000	> 10 000
Hg	1	20	1–3	3–10	10–50	> 50
Cu	130	130	100–200	200–500	500–2500	> 2500
Ni	70	70	20–50	50–200	200–1000	> 1000
Zn	300	300	250–500	500–1000	1000–5000	> 5000

Source: [a]ICRCL (1987); [b]Kelly (1979).

theoretically perceived hazard, rather than on (a) dose-response tests for soil organisms or plants, or (b) observed *in situ* toxicity indications such as the absence of plant growth, severely reduced plant species diversity, or elevated metal concentrations in plant biomass. Since quoted ICRCL "threshold triggers" are total metal concentrations measured in soil with a pH of 6.5, there are also serious problems associated with comparing metal concentrations measured by different techniques and in soils with different pH conditions.

In several countries, the first step to solving the problem has been to compile a register of contaminated land, for former industrial land uses, listing all known sites and their associated contamination characteristics. Once registered, a suite of land and soil assessment and rehabilitation techniques are available. One drawback of such a contaminated land register is that the designation would remain with the land, despite subsequent ameliorative and restoration measures. For this reason, the proposed compilation of a contaminated land register in the UK was cancelled (early in 1993), to the dismay of environmentalists.

Rehabilitation of contaminated land must involve: (i) containment of the source of toxin, and (ii) site and soil ameliorative techniques, prior to (iii) revegetation. Franzius (1987) points out that since containment measures alone may provide only a temporary solution to a toxicity problem, more lasting procedures are required for true rehabilitation. Rehabilitation measures can be divided into on-site contamination removal (e.g. digging out affected soil) or off-site soil cleaning and replacement. In either case, suitable groundwater drainage schemes must be installed to ensure predictable soil moisture and redox conditions. There are three main soil treatment operations:

— *Thermal* treatment: heating soil, or incineration. Generally used to get rid of organic contamination; not suitable for metal contamination.
— *Extractive*, or *washing* treatment: chemical or physical extraction of pollutants. Can be used for organic or inorganic contamination. Main problems are (i) expense, and (ii) that contaminated solutions or sediment fines must subsequently be disposed of.
— *Biological* treatment: microbial treatment of organic media or of leachates. Not really suitable for soil, and bioaccumulated contaminants must be disposed of.

We can thus see that few currently available techniques are suitable for large-scale soil treatment. A detailed review of the metal leaching and bioaccumulation abilities of microorganisms, particularly bacteria, for the reclamation of metals from contaminated media, is given by Hutchins *et al.* (1986).

There is clearly an urgent need for standardised metal toxicity assessments in contaminated soils and for the development and long-term monitoring of *in situ* contaminated soil treatments and for land and habitat restorations. Regulatory agencies, such as the National Rivers Authority (NRA) in the UK, are increasingly requiring developers to carry out leachate tests on soils to determine the mobility of pollutants and to undertake post-treatment validation assessments to ensure compliance with agreed clean-up standards. In addition, the NRA is compiling a register of landfill sites which are common sources of metal-contaminated leachate to groundwaters. Internationally, the scale of the contaminated soil restoration problem is huge; illustrated by an estimate of 35 000 industrial or waste contaminated sites in Germany (Franzius, 1987) and 50 000–100 000 such sites in the UK (Haines, 1991).

2.4.2 MODELLING TOXICITY IN SOIL-PLANT SYSTEMS

Three main reasons for modelling the mobilisation and fate of metals in soil–plant systems can be identified (e.g. Bonazountas, 1987):

(1) assessment of environmental quality,
(2) assessment of human exposure, and
(3) decision-making, including the testing of control strategies for environmental and human protection.

Most of these modelling approaches are designed to study major exposure pathways that arise from hazardous waste disposal sites, but their application could be much wider. Research effort has concentrated attention on modelling contaminants in groundwater and leachates, surface waters, residual wastes and air, but least of all in contaminated soils. A review of chemical contaminant modelling in these systems is given by Bonazountas (1987). The complexities of combined soil–plant systems have received least attention of all. In soil–plant systems (depicted in Figure 2.1), we identified five main pathways of toxic metal transport and transformation:

— *absorption*, via root uptake and possibly via stomata and foliar absorption
— *redistribution*, via xylem and phloem transport
— *metabolism*, in biosynthesis
— *excretion*, in litterfall and root decay
— *recycling*, via decomposition and "mobilisation" processes

Attempts to measure and quantify these processes in natural ecosystems are in their infancy. Many of the studies which profess to do so have simply analysed the *spatial scale* of the problem (i.e. the compartments of the system), not the *temporal scale* (i.e. the fluxes between compartments). Longer term studies of the fate of metals in ecosystems (e.g. Martin and Coughtrey, 1987; Levine *et al.*, 1989), agriculture (e.g. McGrath, 1987) and particularly the distribution of metals in soils (e.g. Harmsen, 1992) are beginning to add detail to models of metal cycling. Metals tend to accumulate in topsoils through aerial inputs and by vegetation uptake, assimilation and subsequent litter fall. More studies quantifying the accumulation of metals in topsoils are required to provide evidence for understanding how soil biological processes respond over time.

One of the central issues of this book is that of *prediction*: predicting the harmful effects of metals, of toxic levels of different forms of those metals, and of the mode and site of action of those metals in a given soil–plant system. One approach to the problems of difficult and highly complex field conditions, is to model, or mathematically simulate, the system to be studied. Thermodynamic modelling of metal speciation in soils is reviewed by Ross in Chapter 3. Other ecosystem hydrology and nutrient models could show potential for toxic metal modelling if adequate field data on the compartments and fluxes were available for the type of ecosystem to be simulated. More widely applied are models employed by environmental regulatory agencies. These approaches are based on geochemical calculations and biomagnification tests (i.e. the accumulation of metals by test organisms). The point of applying the model is to set an Environmental Quality Standard (EQS): a concentration of an element which, if exceeded, will result in an increasing risk of significant harm to living resources. The source of the element, if of human origin, is regulated in theory by (i) the notion of the EQS for a particular pollutant, (ii) the EU directives protecting particular species, and (iii) the Uniform Emission Standards set by the EU for particular industries. In the UK, Her Majesty's Inspectorate of Pollution (HMIP) regulate metal emissions to the atmosphere and in water. For water quality, regulatory modelling appears to be advancing, but modelling metal contamination competes for developmental funds with modelling organic and synthesised chemical waste. The regulatory task is considerable. The idea of combining ecosystem and geochemical modelling within a regulatory framework is still far from reality.

In recent years, planners and environmentalists have attempted to control levels of metal pollution by setting legal acceptable limits on metal contamination of soils. This type of approach to pollution prevention can only work where: (i) metal contamination has an identifiable and controllable origin, such as in the case of sewage sludge applications; (ii) enough is known about soil conditions such as pH, Eh and soil organic matter content to realistically assess the retention and mobility of applied trace metals; and (iii) simple biological systems operate, such as agricultural monocultures, in which exposure routes are simplified and more easily quantified. When metal contamination comes from several sources or from an uncontrollable source, where soil conditions are unusual or unknown, or where complex ecosystems operate (e.g. deciduous woodlands, wetlands, etc.), the fate and manipulation of trace metal processes, for habitat restoration or for pollution prevention, is not simple.

2.5 CONCLUSIONS

A variety of definitions of metal toxicity have been suggested for soil–plant systems. Which definitions are appropriate may depend on which part of the system is being studied:

(1) Classical *dose-response toxicity approaches* may be suitable for assessing the impact of specific metals or metal species on individual soil organisms. Similar approaches may be useful for assessing metal toxicity thresholds for higher plants, but only in the absence of soil, which is likely to complicate the availability of the metal.

(2) Examining *symptoms of injury in plants* may provide data which could be used in ranking metals and different metal species according to their degree of toxicity. This approach, however, may not provide the sensitivity required for subsequent toxicity questions, such as when do crop plants or herbage contain metal concentrations which would be toxic for herbivores, including man. This is because plants may accumulate toxic levels of metals without exhibiting symptoms of injury.

(3) *Epidemiological and cohort studies* might be required to examine effects of toxic metals in population ecology. These data should be used to answer questions about the fitness of affected species, comprising survivorship (how long can individuals survive?) and fecundity (how many seeds can individuals produce?).

The most problematic metal toxicity issues to manage and solve are soil–plant systems in which metals are accumulating, such as sites subjected to aerial fallout, or fields with regular sewage sludge applications. In such cases assessments must be made of the rate of accumulation, compared to perceived toxicity to the organisms and plants of the habitat. In the USA and in the European Union, legislation has been formulated to limit sewage sludge applications to land (CEC, 1986; USEPA, 1989). These limits are based on metal contents of sludges, in combination with "no adverse effect" and phytotoxicity threshold (PT) data (Chang *et al.*, 1992) from studies of sludge application on agricultural land. Chang *et al.* suggest calculating a metal toxicity "safety factor" (SF) associated with sludge application to land. A regulatory authority may select a desired PT level, based on analytical studies of growth retardation in relation to foliar metal content. Figure 2.4(a) illustrates the relationship for zinc in bush bean. If a PT_{50} is selected, then the foliar Zn concentration must be reduced to 375 mg Zn kg^{-1}. To do this, Chang *et al.* suggest that the Zn loading from sludge must be reduced from 3500 to 1750 kg ha^{-1}. Their estimation is based on data accumulated from a limited number of field experiments on bush bean and modified using data for corn. Using data accumulated from different published sources, Chang *et al.* (1992) showed that the relationship between sludge metal loading rate and foliar metal concentrations in crop plants was extremely variable (Figure 2.4(b)). This relationship is clearly the weakest link in the exposure and risk assessment procedure for metals in sewage sludges. Metal toxicity SF values vary for different plants, different plant uses (e.g. crops or amenity), different

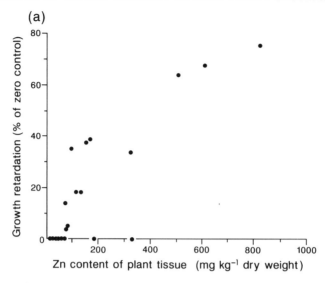

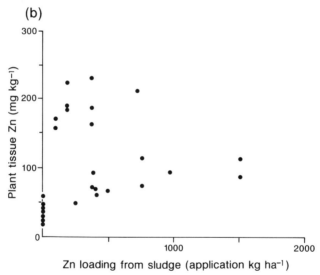

Figure 2.4. (a) Cause and effect relationship between plant tissue Zn and growth retardation in bush bean, and (b) Zn loading from sludge application and its effect on Zn content of bush bean grown in sludge-amended soils. Reproduced by permission of the American Society of Agronomy from Chang *et al.* (1992)

soils and different metals. Chang *et al.*, for example, found that for zinc, PT_{50} occurred at around 2200, 475 and 375 mg Zn kg^{-1} in the foliage of corn, lettuce and bush bean respectively. Hence safety factors must be directly tailored to the soil–plant system being assessed. Clearly this is rather difficult if crop rotation is being practised.

A growing international awareness of inherited problems of metal contamination in soils from mine wastes or industrial derelict sites has led to the compilation in many countries of registers of contaminated land. The criteria used to assess whether levels of metals in soils are present in toxic amounts are not universally defined and vary for different countries and land-use purposes. A major problem in setting metal contamination criteria for soils is that it is not necessarily relevant to set the same toxicity criteria for: (i) different soil types, (ii) different metals and different metal species, (iii) different vegetation types, and (iv) different intended land-use purposes. Further field comparative data are required to allow comparisons of metal toxicities in different, complex soil–plant systems.

REFERENCES

Alderdice, D. F. (1967) The detection and measurement of water pollution—biological assays. Canadian Department of Fisheries: Canadian Fisheries Report No. 9, pp. 33–39.

American Public Health Association (1976) *Standard Methods for the Examination of Water and Wastewater*, 14th edn. American Public Health Association, Americal Waterworks Association, Water Pollution Control Federation. Washington.

Ausmus, B. S., Dodson, G. J. and Jackson, D. R. (1978) Behaviour of heavy metals in forest microcosms. 3. Effects on litter and soil carbon metabolism. *Water, Air and Soil Pollution*, **10**, 19–26.

Baath, E. (1989) Effects of heavy metals in soil on microbial processes and populations (A review). *Water, Air and Soil Pollution*, **47**, 335–379.

Baker, A. J. M. (1978) The uptake of zinc and calcium from solution culture by zinc-tolerant and non-tolerant *Silene maritima* With. in relation to calcium supply. *New Phytologist*, **81**, 321–330.

Baker, D. A. (1983) Uptake of cations and their transport within the plant. In: Robb, D. A. and Pierpoint, W. S. (Eds) *Metals and Micronutrients: Uptake and Utilization by Plants. Annual Proceedings of the Phytochemical Society of Europe*, No. 21, pp. 3–19. Academic Press, London.

Beckett, P. H. T. and Davies, R. D. (1977) Upper critical levels of toxic elements in plants. *New Phytologist*, **79**, 95–106.

Berg, B., Ekbohm, G., Soderstrom, B. and Staaf, H. (1991) Reduction of decomposition rates of Scots Pine needle litter due to heavy-metal pollution. *Water, Air and Soil Pollution*, **59**, 165–177.

Bisessar, S. (1982) Effect of heavy metals on microorganisms in soils near a secondary lead smelter. *Water, Air and Soil Pollution*, **17**, 305–308.

Bonazountas, M. (1987) Chemical fate modelling in soil systems: a state-of-the-art review. In: Barth, H. and L'Hermite, P. (Eds) *Scientific Basis for Soil Protection in the European Community*, pp. 487–566. Elsevier Applied Scientific Publishers, London.

Borst-Pauwels, G. W. F. H. (1981) Ion transport in yeast. *Biochimica et Biophysica Acta*, **650**, 88–127.

Bowen, H. J. M. (1966) *Trace Elements in Biochemistry*. Academic Press, New York.

Bradley, R., Burt, A. J. and Read, D. J. (1981) Mycorrhizal infection and resistance to toxicity in *Calluna vulgaris*. *Nature*, **292**, 335–337.

British Standards Institute (1988) *Draft for Development: Code of practice for the identification of potentially contaminated land and its investigation.* British Standards Institute DD175:1988.

Brookes, A., Collins, J. C. and Thurman, D. A. (1981) The mechanisms of zinc tolerance in grasses. *Journal of Plant Nutrition*, **3**, 695–705.

Brookes, P. C. and McGrath, S. P. (1984) Effects of metal toxicity on the size of the soil microbial biomass. *Journal of Soil Science*, **35**, 341–346.

Brooks, R. R., Lee, J., Reeves, R. D. and Jaffre, T. (1977) Detection of nickeliferous rocks by analysis of herbarium specimens of indicator plants. *Journal of Geochemical Exploration*, **7**, 49–57.

Brooks, R. R., Shaw, S. and Asensi Marfil, A. (1981) The chemical form and physical function of nickel in some Iberian *Alyssum specum*. *Physiologia plantarum*, **51**, 167–170.

Brown, J. C. (1969) Agricultural use of synthetic metal chelates. *Proceedings of Soil Society of America*, **33**, 59–61.

Bryan, G. W. (1976) Heavy metal contamination in the sea. In: Johnston, R. (Ed.) *Marine Pollution*, pp. 185–302. Academic Press, London.

Butt, T. R. and Ecker, D. J. (1987) Yeast metallothionein and applications in biotechnology. *Microbiological Reviews*, **51**, 351–364.

Carlson, R. W., Bazzaz, F. A. and Rolfe, G. L. (1975) The effect of heavy metals on plants. II Net photosynthesis and transpiration of whole corn and sunflower plant treated with Pb, Cd, Ni and Tl. *Environmental Research*, **10**, 113–121.

CEC (Commission of the European Communities) (1986) Council Directive on the protection of the environment, and in particular of the soil, when sewage sludge is used in agriculture. *Official Journal of the European Communities*, No. L181/6 (86/278/EEC).

Chang, A. C., Granato, T. C. and Page, A. L. (1992) A methodology for establishing phytotoxicity criteria for chromium, copper, nickel and zinc in agricultural land application of municipal sewage sludges. *Journal of Environmental Quality*, **21**, 521–536.

Chang, F.-H. and Broadbent, F. E. (1981) Influence of trace metals on carbon dioxide evolution from a Yolo soil. *Soil Science*, **132**, 416–421.

Chang, F.-H. and Broadbent, F. E. (1982) Influence of trace metals on some soil nitrogen transformations. *Journal of Environmental Quality*, **11**, 1–4.

Coughtrey, P. J., Jones, C. H., Martin, M. H. and Shales, S. W. (1979) Litter accumulation in woodlands contaminated by lead, zinc, cadmium and copper. *Oikos*, **39**, 51–60.

Cumming, J. R. and Tomsett, A. B. (1992) Metal tolerance in plants: signal transduction and acclimation mechanisms. In: Adriano, D. C. (Ed.) *Biogeochemistry of Trace Metals*. Lewis Publishers, Boca Raton, Florida.

Cumming, J. R. and Weinstein, L. H. (1990) Aluminium–mycorrhizal interactions in the physiology of pitch pine seedlings. *Plant and Soil*, **125**, 7–18.

Cutsem, P. van and Gillet, C. (1982) Activity coefficient and selectivity values of Cu^{2+}, Zn^{2+} and Ca^{2+} ions adsorbed in the *Nitella flexilis* L. cell wall during triangular ion exchanges. *Journal of Experimental Botany*, **33**, 847–853.

Denny, H. J. and Wilkins, D. A. (1987) Zinc tolerance in *Betula* spp. IV. The mechanism of ectomycorrhizal amelioration of zinc toxicity. *New Phytologist*, **106**, 545–553.

Dixon, R. K. (1988) Response of ectomycorrhizal *Quercus rubra* to soil cadmium, nickel and lead. *Soil Biology and Biochemistry*, **20**, 555–559.

Dumontet, S. and Mathur, S. P. (1989) Evaluation of respiration based methods for measuring microbial biomass in metal contaminated acidic mineral and organic soils. *Soil Biology and Biochemistry*, **21**, 431–436.

Duxbury, T. (1985) Ecological aspects of heavy metal responses in microorganisms. *Advances in Microbial Ecology*, **8**, 185–235.

Edwards, C. A. (1969) Soil pollutants and soil animals. *Scientific American*, **220**(4), 88–99.

Franzius, V. (1987) Impacts on soils related to industrial activities: Part I—Effects of abandoned waste disposal sites on the soil: Possible remedial measures. In: Barth, H. and L'Hermite, P. (Eds) *Scientific Basis for Soil Protection in the European Community*, pp. 247–257. Elsevier Applied Science, London.

Freedman, B. and Hutchinson, T. C. (1980) Effect of smelter pollutants on forest leaf litter decomposition near a nickel–copper smelter at Sudbury, Ontario. *Canadian Journal of Botany*, **58**, 1722–1736.

Gadd, G. M. and Griffiths, A. J. (1980) Influence of pH on toxicity and uptake of copper in *Aureobasidium pullulans*. *Transactions of the British Mycological Society*, **75**, 91–96.

Gadd, G. M. and White, C. (1985) Copper uptake by *Penicillium ochro-chloron*: influence of pH on toxicity and demonstration of energy-dependent copper influx using protoplasts. *Journal of General Microbiology*, **131**, 1875–1879.

Gadd, G. M. and White, C. (1989) Heavy metal and radionuclide accumulation and toxicity in fungi and yeasts. In: Poole, R. M. and Gadd, G. M. (Eds) *Metal–Microbe Interactions*, pp. 19–38. IRL Press, SGM 26.

Gildon, A. and Tinker, P. B. (1983) Interactions of vesicular arbuscular mycorrhizal infection and heavy metals in plants. I—The effect of heavy metals on the development of vesicular arbuscular mycorrhizas. *New Phytologist*, **95**, 247–261.

Grill, E., Winnacker, E.-L. and Zenk, M. H. (1987) Phytochelatins, a class of heavy-metal-binding peptides from plants, are functionally analogous to metallothioneins. *Proceedings of the National Academy of Science, USA*, **84**, 439–443.

Haines, R. (1991) Scale and extent of contaminated land in the UK—an opportunity for the construction industry. In: *Contaminated Land: A practical examination of the technical and legal issues*. Proceedings of a Conference arranged by IBC, London.

Harmsen, K. (1992) Long-term behaviour of heavy metals in agricultural soils: a simple analytical model. In: Adriano, D. C. (Ed.) *Biogeochemistry of Trace Metals*, pp. 217–247. Lewis Publishers, Boca Raton, Florida.

Helmke, P. A., Robarge, W. P., Karotev, R. L. and Schomberg, P. J. (1979) Effects of soil-applied sewage sludges on concentrations of elements in earthworms. *Journal of Environmental Quality*, **8**, 322–327.

Hopkin, S. P. (1990) Critical concentrations, pathways of detoxification and cellular ecotoxicology of metals in terrestrial arthropods. *Functional Ecology*, **4**, 321–327.

Hopkin, S. P. (1993) Ecological implications of "95% protection levels" for metals in soil. *Oikos*, **66**, 137–141.

Hutchins, S. R., Davidson, M. S., Brierly, J. A. and Brierly, C. L. (1986) Micro-organisms in reclamation of metals. *Annual Review of Microbiology*, **40**, 311–336.

Inter-Departmental Committee on the Redevelopment of Contaminated Land (1987) *Guidance on the Assessment and Redevelopment of Contaminated Land*. ICRCL 59/83, 2nd edn. Code EPTSSP187-5RH3, CDEP/EPTS, Romney House, 43 Marsham Street, London SW1P 3PY.

Jackson, D. R. and Watson, A. P. (1977) Disruption of nutrient pools and transport of heavy metals in a forested watershed near a lead smelter. *Journal of Environmental Quality*, **6**, 331–338.

Jones, J. R. E. (1939) The relation between electrolytic solution pressure of the metals and their toxicity to the stickleback (*Gasterosteus aculeatus* L.). *Journal of Experimental Biology*, **16**, 425–437.

Jones, K. C. (1991) Transport and food chain modelling and its role in assessing human exposure to organic chemicals. *Journal of Environmental Quality*, **20**, 317–329.

Kelly, G. (1979) Greater London Council: Guidelines for contaminated soils. In: *Site Investigation and Material Problems. Proceedings of Conference on Reclamation of Contaminated Land*, Eastbourne, Oct. 1979. Society of the Chemical Industry, London.

Laskowski, R. (1991) Are the top carnivores endangered by heavy metal biomagnification? *Oikos*, **60**, 387–390.

Leonard, R. T. (1983) Potassium transport and the plasma membrane-ATPase in plants. In: Robb, D. A. and Pierpoint, W. S. (Eds) *Metals and Micronutrients: Uptake and Utilization by Plants*. Annual Proceedings of the Phytochemical Society of Europe, No. 21, pp. 71–86. Academic Press, London.

Levine, M. B., Hall, A. T., Barrett, G. W. and Taylor, D. H. (1989) Heavy metal concentrations during ten years of sludge treatment to an open-field community. *Journal of Environmental Quality*, **18**, 411–418.

Liang, C. N. and Tabatabai, M. A. (1977) Effects of trace elements on nitrogen mineralisation in soils. *Environmental Pollution*, **12**, 141–147.

Lighthart, B., Baham, J. and Volk, V. V. (1983) Microbial respiration and chemical speciation in metal-amended soils. *Journal of Environmental Quality*, **12**, 543–548.

Loneragan, J. F., Kirk, G. J. and Webb, J. (1987) Translocation and function of zinc in roots. *Journal of Plant Nutrition*, **10**, 1247–1254.

Lukens, R. J. (1971) *Chemistry of Fungicidal Action*. Springer Verlag, New York.

Martin, M. H. and Coughtrey, P. J. (1982) *Biological Monitoring of Heavy Metal Pollution. Land and Air*. Applied Science Publishers, London.

Martin, M. H. and Coughtrey, P. J. (1987) Cycling and fate of heavy metals in a contaminated woodland ecosystem. In: Coughtrey, P. J., Martin, M. H. and Unsworth, M. (Eds) *Pollutant Transport and Fate in Ecosystems*, pp. 319–336. Blackwell Scientific, Oxford.

Mathur, S. P., Hamilton, N. A. and Levesque, M. P. (1979) The mitigating effect of residual fertilizer copper on the decomposition of an organic soil *in situ*. *Soil Science Society of America, Journal*, **43**, 200–203.

Mathys, W. (1977) The role of malate, oxalate and mustard oil glucosides in the evolution of zinc-resistance in herbage plants. *Plysiologia plantarum*, **40**, 130–136.

McGrath, S. P. (1987) Long-term studies of metal transfers following application of sewage sludge. In: Coughtrey, P. J., Martin, M. H. and Unsworth, M. H. (Eds) *Pollutant Transport and Fate in Ecosystems*, pp. 301–317. Blackwell Scientific, Oxford.

McKenney, D. J. and Vriesacker, J. R. (1985) Effect of cadmium contamination on denitrification processes in Brookston Clay and Fox Sandy Loam. *Environmental Pollution*, **38**, 221–233.

Merckx, R., van Ginkel, J. H., Sinnaeve, J. and Cremers, A. (1986) Plant-induced changes in the rhizosphere of maize and wheat. II. Complexation of cobalt, zinc and manganese in the rhizosphere of maize and wheat. *Plant and Soil*, **96**, 95–107.

Mikkelsen, J. P. (1974) Indvirkning af bly pa jordbundens mikrobiologiske aktivitet. *Tidsskrift fur Planteavl.*, **78**, 509–516.

Morselt, A. F. W., Smits, W. T. M. and Limonard, T. (1986) Histochemical demonstration of heavy metal tolerance in ectomycorrhizal fungi. *Plant and Soil*, **96**, 417–420.

Nieboer, E. and Richardson, D. H. S. (1980) The replacement of the nondescript term "heavy metals" by a biologically and chemically significant classification of metal ions. *Environmental Pollution (Series B)*, **1**, 2–26.

Nies, D. H. and Silver, S. (1989) Plasmid-determined inducible efflux is responsible for resistance to cadmium, zinc and cobalt in *Alcaligenes eutrophus*. *Journal of Bacteriology*, **171**, 896–900.

Nordgren, A., Baath, E. and Soderstrom, B. (1983) Microfungi and microbial activity along a heavy metal gradient. *Applied Environmental Microbiology*, **45**, 1829–1837.

Nordgren, A., Baath, E. and Soderstrom, B. (1988) Evaluation of soil respiration characteristics to assess heavy metal effects on soil microorganisms using glutamic acid as a substrate. *Soil Biology and Biochemistry*, **20**, 949–954.

Oberlander, H. E. and Roth, K. (1978) Die wirking der schwermetalle chrom, nickel, kupfer, zink, cadmium, quecksilber und blei auf die aufnahme und verlagerung von kalium und phosphat bei jungen gerstepflanzen. *Zeitschrift fur Pflanzenernahrung, Dungen und Bondenkunde*, **141**, 107–116.

Ochiai, E. I. (1987) *General Principles of Biochemistry of the Elements*. Plenum Press, New York.

Otto, R., Sonnenberg, A. S. M., Veldkamp, H. and Konings, W. N. (1980) Generation of an electrochemical proton gradient in *Streptococcus cremoris* by lactate efflux. *Proceedings of the National Academy of Science, USA.*, **77**, 5502–5506.

Pearson, R. (1968a) Hard and soft acids and bases. HSAB, Part I Fundamental principles. *Journal of Chemical Education*, **45**, 581–587.

Pearson, R. (1968b) Hard and soft acids and bases. HSAB, Part II Underlying theories. *Journal of Chemical Education*, **45**, 643–648.

Porter, J. R. (1983) Variation in the relationship between nitrogen fixation, leghaemoglobin, nodule numbers and plant biomass in alfalfa (*Medicago sativa*) caused by treatment with arsenate, heavy metals and fluoride. *Physiologia plantarum*, **68**, 143–148.

Poschenrieder, C., Gunse, B. and Barcelo, J. (1989) Influence of cadmium on water relation, stomatal resistance and abscissic acid content in expanding bean leaves. *Plant Physiology*, **90**, 1365–1371.

Raymond, K. N., Muller, G. and Mayzanke, B. F. (1984) Complexation of iron by siderophores. A review of their solution and structural chemistry and biological function. *Topics in Current Chemistry*, **123**, 49–102.

Reish, D. J., Martin, J. M., Piltz, F. M. and Word, J. Q. (1976) The effect of heavy metals on laboratory populations of two polychaetes with comparisons to the water quality conditions and standards in Southern California marine waters. *Water Research*, **10**, 299–302.

Rother, J. A., Millbank, J. W. and Thornton, J. (1982) Effects of heavy metal additions on ammonification and nitrification in soils contaminated with cadmium, lead and zinc. *Plant and Soil*, **69**, 239–258.

Ruhling, A and Tyler, G. (1973) Heavy metal pollution and decomposition of spruce needle litter. *Oikos*, **24**, 402–416.

Sakaguchi, T., Horikoshi, T. and Nakajima, A. (1977) Uptake of copper ions by *Chlorella regularis*. *Journal of Agricultural Chemistry Society of Japan*, **51**, 497–505.

Salt, D. E., Thurman, D. A., Tomsett, A. B. and Sewell, A. K. (1989) Copper phytochelatins of *Mimulus guttatus*. *Proceedings of the Royal Society, Series B*, **236**, 79–89.

Scheller, H. V., Huang, B., Hatch, E. and Goldsbrough, P. B. (1987) Phytochelatin synthesis and glutathione levels in response to heavy metals in tomato cells. *Plant Physiology*, **85**, 1031–1414.

Shaw, W. H. R. (1954) Toxicity of cations towards living systems. *Science*, **120**, 361–363.

Shewry, P. R. and Peterson, P. J. (1976) Distribution of chromium and nickel in plants and soil from serpentine and other sites. *Journal of Ecology*, **64**, 195–212.

Spalding, B. P. (1979) The effect of biocidal treatments on respiration and enzymatic activities of Douglas fir needle litter. *Soil Biology and Biochemistry*, **10**, 537–543.

Steffens, J. C. (1990) The heavy-metal binding peptides of plants. *Annual Review of Plant Physiology and Plant Molecular Biology*, **41**, 553–575.

Turner, R. G. (1967) *Experimental studies on heavy metal tolerance.* PhD Thesis, University of Wales, Cardiff.

Turner, R. G. (1969) Heavy metal tolerance in plants. In: Rorison, I. H., Bradshaw, A. D. and Chadwick, M. J. (Eds) *Ecological Aspects of the Mineral Nutrition of Plants*. Proc. 9th Symposium of the British Ecological Society, pp. 399–410. Blackwell Scientific, Oxford.

Tyler, G. (1981) Heavy metals in soil biology and biochemistry. In: Paul, E. A. and Ladd, J. N. (Eds) *Soil Biochemistry*, **5**, 371–414.

US Environmental Protection Agency (1989) Standards for the disposal of sewage sludge. *EPA Proposed Rule 40, CFR Part 503. Federal Resister 54*, 5746–5902.

van Straalen, N. M. and Ernst, W. H. O. (1991) Metal biomagnification may endanger species in critical pathways. *Oikos*, **62**, 255–256.

Vogeli-Lange, R. and Wagner, G. J. (1990) Subcellular localization of a transport function for CD-binding peptides. *Plant Physiology*, **92**, 1086–1093.

Wallace, A. and Berry, W. L. (1983) Shift in threshold toxicity levels in plants when more than one trace metal contaminates simultaneously. *The Science of the Total Environment*, **28**, 257–268.

White, C. and Gadd, G. M. (1987) Inhibition of H^+ influx and K^+ uptake and induction of K^+ efflux in yeast by heavy metals. *Toxicity Assessment*, **2**, 437–447.

Williams, S. T., McNeilly, T. and Wellington, E. M. H. (1977) The decomposition of vegetation growing on metal mine waste. *Soil Biology and Biochemistry*, **9**, 271–275.

Wyn Jones, R. G., Sutcliffe, G. M. and Marshall, C. (1971) Physiological and biochemical basis for heavy metal tolerance in clones of *Agrostis tenuis*. In: Samish, R. M. (Ed.) *Recent Advances in Plant Nutrition*, Vol. 2, 575–581. Gordon and Breach, New York.

Zhang, G. and Taylor, G. J. (1989) Kinetics of aluminium uptake by excised roots of aluminium-tolerant and aluminium-sensitive cultivars of *Triticum aestivum* L. *Plant Physiology*, **91**, 1094–1099.

3 Retention, Transformation and Mobility of Toxic Metals in Soils

SHEILA M. ROSS

University of Bristol, UK

ABSTRACT

Three types of approaches can be used to characterise metal speciation, complexation and fractionation in soils: (i) geochemical theory, (ii) a good understanding of the soil properties and conditions which influence metal transformations, and (iii) laboratory analyses, including the use of sequential soil extractions. Geochemical approaches to calculating metal speciation in the soil solution involve two different thermodynamic methods, using equilibrium constants, or Gibbs free energies. Several examples of equilibrium constant calculations are illustrated, indicating how such techniques are the basis of geochemical modelling of metal ion speciation, using models such as GEOCHEM and SOILCHEM. The main process associated with the retention and mobility of metals in soils are (1) weathering, (2) dissolution and solubility, (3) precipitation, (4) uptake by plants, (5) immobilisation by soil organisms, (6) exchange on soil cation exchange sites, (7) specific adsorption and chemisorption, (8) chelation, and (9) leaching. The principal factors influencing solubility of metals in soil are pH, soluble organic matter and redox. For most trace metals, apart from Cd and Zn, non-specific cation exchange is a less important retention mechanism than is specific sorption by Fe and Mn oxides and by soil organic matter. Soil metal retention processes are generally much more important than metal leaching processes. Topsoils subjected to sewage sludge additions or aerial metal pollution inputs tend to accumulate metals. The risk of metal leaching to groundwaters is generally small. One of the principal techniques used widely to quantify different metal fractions in soil is sequential extraction. A range of sequential extraction schemes are outlined and the success of predicting plant metal availability from individual extractions is assessed.

3.1 INTRODUCTION

The fate of toxic metals in soils is not only dependent on environmental and edaphic conditions such as pH, waterlogging and soil organic matter content, but also on the initial chemical form of the metal and the types of plants and animals in the system. The origin and initial form of toxic metals in soils is given in Chapter 1 and phytoavailability of toxic metals is discussed in

Chapter 4. This chapter is concerned with the soil compartment: the solid, exchangeable and aqueous phases of toxic metals and their solubility, chemical transformation, complexation, adsorption and mobility characteristics. Three main approaches are available to us for the characterisation of metal speciation, metal complexation and different metal fractions in contaminated soils:

(1) *geochemical theory* with computer-based modelling;
(2) clear understanding of the *soil processes and conditions* which control the reactions and transformations of metal species and hence influence metal mobility; and
(3) *laboratory analyses* of different metal fractions in contaminated soils, using a sequence of chemical extractants.

Individually, each of the three approaches provides an indication of the processes associated with toxic metals in soil and their relative retentions and mobilities. Although in recent years Sposito and his collaborators have attempted comparisons of laboratory approaches with those from computer-based simulations using chemical theory (some examples are given by Sposito, 1983), there are remarkably few examples, if any, of ecosystem and soil studies that have utilised a combination of these three approaches in an attempt to provide more detailed interpretations of metal processes in soils and a clearer picture of their short and long term fate. Numerous studies, particularly agricultural and ecological applications, have used chemical extractants to quantify different soil metal fractions. A smaller number of studies have attempted to characterise metal phytoavailability by correlating soil extractable metal fractions with plant metal uptake. These laboratory and glasshouse experimental approaches will be discussed later in Section 3.4. The intention in the next section is to review the three approaches outlined above and to evaluate their scope and limitations in predicting the fate of metals in different ecosystems.

3.2 GEOCHEMICAL THEORY OF METAL SPECIATION IN SOILS

Before considering the influence of environmental, particularly edaphic, controls on the fate of metals in soil–plant systems, it is useful to review briefly the basic physico-chemical principles that control metal speciation in environmental systems.

Using basic chemical theory and information from the periodic table, a simple framework can be constructed which allows preliminary prediction of how different types of metals might react in environmental systems. Calculations of this type have been used quite widely to discuss trace metal speciation in fresh and salt waters (e.g. Sibley and Morgan, 1975; Stumm and

Morgan, 1981; Andreae, 1986; Turner, 1987; Fergusson, 1991). Similar work in soil and soil solutions has taken some time to appear. Morgan (1987) and Sposito (1986a) have gone a long way in outlining these basic geochemical principles for metal speciation in environmental systems, often based on chemical theory that has been available for over 50 years. This is the basis of computer-based geochemical modelling approaches presented by Sposito and Mattigod (1980) and developed by Sposito and Coves (1988) specifically for soils. Chemical fractionation techniques, using chemical extractant solutions designed to remove particular and identifiable soil metal fractions, have provided empirical data for testing the output of geochemical speciation computations. Sposito and colleagues have found good relationships between such studies and their computer simulations. These scenarios are beginning to help in predicting trace metal speciation, particularly in soils to which sewage sludges have been applied.

Although much of the current effort in geochemical theory is directed at improving existing models for calculating species equilibria in soil solutions, the same chemical "rules" used in such complex calculations can aid the soil scientist directly in predicting which metal species to expect in different environmental and soil systems. Some basic geochemical principles are outlined below.

3.2.1 CHEMICAL PRINCIPLES OF METAL SPECIATION

3.2.1.1 "Rules" governing the formation and stability of metal complexes

The speciation of metals in the environment may be predicted from a knowledge of the three main types of electronic structures they form:

(i) *Coordination reactions*, involving the sharing of electrons, e.g.

$$A + B = AB$$

(ii) *Electron transfer reactions*, such as redox reactions, in which electrons are transferred between oxidant–reductant pairs:

$$Ox_1 + Red_1 = Ox_2 + Red_2$$

(iii) *Free radical reactions*

Chemists classify metals according to their atomic configurations into electron-pair donors (Lewis bases) and electron-pair acceptors (Lewis acids). Pearson (1968a,b) subsequently defined "hard" and "soft" donors and acceptors on the basis of their formation of stable compounds. Metal cations of groups Ia and IIa in the Periodic Table (including Na^+, K^+, Ca^{2+}, Mg^{2+}, but also Mn^{3+}, Al^{3+}, Fe^{3+}, etc.) are considered as "hard" acceptors (classified as class A metals by Nieboer and Richardson, 1980), while metal

Table 3.1. Classification of Pearson's "hard" and "soft" acceptors and donors, and Nieboer and Richardson's Class A and B metals

Pearson (1968a,b)

Hard acceptor
Na^+, K^+, Mg^{2+}, Ca^{2+}, Mn^{2+}, Al^{3+}, Cr^{3+}, Co^{3+}, Fe^{3+}, As^{3+}

Intermediate acceptor
Fe^{2+}, Co^{2+}, Ni^{2+}, Cu^{2+}, Zn^{2+}, Pb^{2+}

Soft acceptor
Cu^+, Ag^+, Au^+, Hg_2^{2+}, Cd^{2+}, Pt^{2+}, Hg^{2+}

Hard donor
H_2O, OH^-, F^-, Cl^-, PO_4^{3-}, SO_4^{2-}, CO_3^{2-}, O^{2-}, NO_3^-

Intermediate donor
Br^-, NO_2^-, SO_3^{2-}

Soft donor
SH^-, S^{2-}, CN^-, SCN^-, CO

Nieboer and Richardson (1980)

Class A metals
K^+, Na^+, Ca^{2+}, Mg^{2+}, Al^{3+}

Borderline metals
Pb^{2+}, Sn^{2+}, Cd^{2+}, Cu^{2+}, Fe^{2+}, Co^{2+}, Ni^{2+}, Cr^{2+}, Zn^{2+}, Mn^{2+}, Fe^{3+}

Class B metals
Au^+, Ag^+, Cu^+, Hg^{2+}, Pt^{2+}

cations such as Cu^+, Ag^+, Cd^{2+}, Pt^{4+}, Au^+, and Hg^{2+} are considered "soft" acceptors (class B metals of Nieboer and Richardson, 1980) (Table 3.1). Many of the trace metals are borderline, neither soft nor hard. Nieboer and Richardson (1980) place Cd into the borderline category along with most of the the other trace metals. Hard acceptors are characterised by low polarizability, low electronegativity and large positive charge density (high oxidation state and small radius), while hard donors have low polarizability, high electronegativity and high negative charge density. The opposite characteristics are true of soft acceptors and donors. Among the borderline metals, "soft" or class B characteristics increase in the order: $Mn^{2+} > Zn^{2+} > Ni^{2+} > Fe^{2+} \geqslant Cd^{2+} > Cu^{2+} > Pb^{2+}$ (Nieboer and Richardson, 1980). Soft metals form strong complexes with heavier electron donors from the third, fourth and fifth rows of the Periodic Table: P, S, Cl, Br and I, while hard metals form strong complexes with electron donors from the second row of the Periodic Table: N, O and F. For hard metal complexes, orders of affinity preference and hence stability are:

$$CO_3^{2-} > NO_3^-$$
$$PO_4^{3-} \gg SO_4^{2-} \gg ClO_4^-$$

while for soft metal complexes, orders of affinity preference and hence stability are:

$$I > Br > Cl \gg F$$
$$Se > S \gg O$$
$$As > P \gg N$$
$$S > N > O$$

This explains the greater insolubilities of "soft" metal sulphides, for example, compared to hydroxides, carbonates or phosphates and also explains why some metals are found in the earth's crust predominately as sulphide ores (Pb, Zn, Hg) while others are found mainly as oxides and carbonates (e.g. Al, Ca, Mg). A fuller discussion of these principles is given by Morgan (1987).

3.2.1.2 Ionic potential and solubility

The ionic potential of elements (the ratio of ionic charge (Z) to ionic radius (r)) is a useful indication of the relative solubilities of ions, with low values indicating higher solubilities. In Figure 3.1, the alkali cations (Na, K) have Z/r ratios lower than 30 and high ionic potentials. They are very soluble, easily weathered to form hydrated cations and are easily leached from soils. Ions of transition metals (Mn, Fe, Al) and some of the potentially toxic metals, with Z/r ratios greater than 30 and intermediate ionic potentials, are not very soluble and when weathered tend to precipitate as oxyhydroxides (Bohn et al., 1985).

68

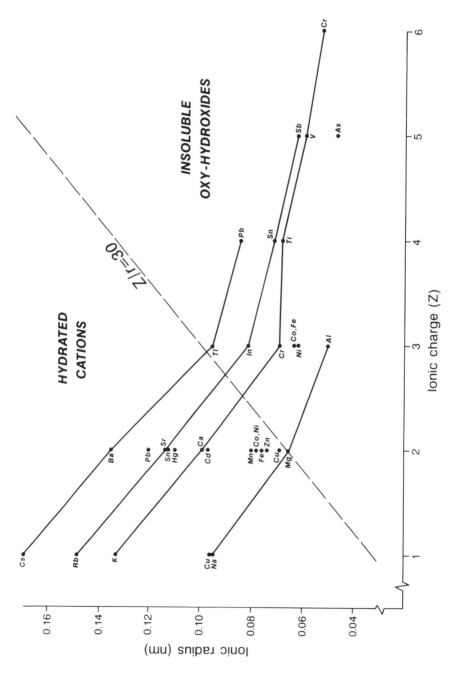

Figure 3.1. Ionic potentials of metals found in soils and soil solutions

Two different thermodynamic methods can be used to calculate ion speciation: using *equilibrium constants*, or using *Gibbs free energies*. Both are subject to the conditions of equilibrium and mass balance. Chemical equilibrium defines the most stable condition for a given suite of reactants and products and can be calculated from equilibrium constants and free energies. Mass balance dictates that the sum of the concentrations of ions and complexes must be equal to the total concentration. In theory, the same thermodynamic calculations can be applied to both inorganic and organic metal complexes. In practice, insufficient is known about both structure and binding mechanisms of organic ligands and we do not have reliable thermodynamic data for realistic soil organic species.

Calculation of ionic speciation using equilibrium constants

The solubility equilibrium equation of a two-component metal solid can be given as:

$$M_a L_{b(s)} = a M_{(aq)}^{m+} + b L_{(aq)}^{l-} \qquad (3.1)$$

where:

M = trace metal
L = ligand
a, b = stoichiometric coefficients
m, l = valencies of metal and ligand respectively

The equilibrium constant for this reaction (K_r) is:

$$K_r = \frac{(M^{m+})^a (L^{l-})^b}{(M_a L_{b(s)}}\qquad (3.2)$$

This expression can be rewritten in log form for the metal:

$$\log M^{m+} = \frac{1}{a} \left[\log(M_a L_{b(s)}) - b \log(L^{l-}) + \log K_r \right] \qquad (3.3)$$

and shows the two factors which regulate the activity of the metal cation in the soil solution—namely the activities of $M_a L_b$ and L^{l-}. The contribution of L^{l-} is a critical factor since L can participate in a variety of precipitation, adsorption and complexation processes in soil. Which of a range of possible metallic solid complexes is most likely to form in given soil conditions, and hence which solid phases are most likely to control the activity of the metal cation in solution, can be predicted using activity ratio graphs. These plot the log [(solid phase)/(free metal cation)] against any important soil characteristic, such as pH or P_{CO_2} where brackets ( ) refer to thermodynamic activity. An example of the procedure for lead phosphate solid phases is explained by Sposito (1983).

Dissolution reactions for three lead phosphates at standard temperature and pressure are given below, with the calculations for the dissolution constant and the activity ratio given in steps 2 and 3:

— *Step 1*: **Dissolution reactions** for three lead phosphates (at 298.15 $^\circ$K) are given by: (log K° values are calculated from Lindsay, 1979)

$$
\begin{array}{lrr}
 & \log K^\circ & \\
Pb_3(PO_4)_2 + 4H^+ \approx 3Pb^{2+} + 2H_2PO_4^- & -1.80 & (3.4a) \\
Pb_5(PO_4)_3Cl + 6H^+ \approx 5Pb^{2+} + 3H_2PO_4 + Cl^- & -5.06 & (3.4b) \\
Pb\, Al_3(PO_4)_2(OH)_5.H_2O + 9H^+ \approx & & \\
\quad Pb^{2+} + 3Al^{3+} + 2H_2PO_4^- + 6H_2O & 9.74 & (3.4c)
\end{array}
$$

— *Step 2*: The **dissolution constant** for each reaction is calculated (the example below is for equation (3.4a))

$$
K^\circ = \frac{[Pb^{2+}]^3 . [H_2PO_4^-]^2}{[Pb_3(PO_4)_2] . [H^+]^4} \tag{3.5}
$$

— *Step 3*: The **activity ratio** is calculated for each part of equation (3.4) (the example below is for equation (3.4a))

$$
\log K^\circ = 3 \log Pb^{2+} + 2 \log H_2PO_4^- - \log Pb_3(PO_4)_2 + 4pH
$$

$$
\log \left[\frac{Pb_3(PO_4)_2}{3Pb^{2+}} \right] = 1.8 + \frac{2}{3} \log H_2PO_4^- + \frac{4}{3} pH \tag{3.6}
$$

To plot an activity ratio diagram, one of the soil solution parameters must be chosen as the independent variable. The others must be held constant by inserting in the equation values which represent a typical soil solution concentration for the desired species. Levels of $H_2PO_4^-$ in soil are controlled by the presence of iron phosphates (strengite) at low pH and by calcium phosphates at high pH. Thus for $H_2PO_4^-$, concentrations in the soil solution over pH 3–7 range from $10^{-5.5}$ to $10^{-6.5}$ Molar. Using a concentration of 10^{-6} M, a value of -6 is inserted in place of log H_2PO_4. Thus, equation (3.6) becomes:

$$
\log [(Pb(PO_4)_{0.67})/(Pb^{2+})] = -2.2 + 4/3pH \tag{3.7a}
$$

A constant of 1 is commonly used for the activity of water, and Sposito (1983) uses a concentration of $10^{-2.9}$ M for Cl^-, and the dissolution reaction of kaolinite ($Al_2Si_5O_5(OH)_4$) to calculate Al^{3+} activity. Thus, he calculates

$$
\log(Al^{3+}) = 7.48 - 3pH
$$

Using these insertions, equations (3.4b) and (3.4c) similarly become:

$$
\log [(Pb(PO_4)_{0.6}Cl_{0.2})/(Pb^{2+})] = 0.88 + \frac{6}{5} pH \tag{3.7b}
$$

$$
\log [(Pb\, Al_3(PO_4)_2(OH)_5.H_2O)/Pb^{2+})] = 0.70 \tag{3.7c}
$$

Equations (3.7a), (3.7b) and (3.7c) are plotted in Figure 3.2. For a chosen value of the soil characteristic under examination, for example, soil pH, and assuming that solid phases are in the Standard State, the solid in a soil mixture which produces the largest $\log[(\text{solid})/(\text{free})]$ value is the most stable and hence the only one that will form at equilibrium. The activity ratio is largest when the activity of the trace metal cation is the smallest. For the conditions stated in the calculations of equations (3.7a) and (3.7b), Figure 3.2 shows that chloropyromorphite ($Pb_5(PO_4)_3Cl$) is the most stable solid across the pH range of 3–7.

Calculation of ion speciation using Gibbs free energies

A second method for assessing the stability of compounds and the dissolution sequence of mixtures of compounds is to calculate the change in Gibbs free energy for selected processes. For an equation such as that in equation (3.1) above, the free energy for the reaction is calculated as:

$$\Delta G_r^\circ = \Delta G[\text{products}] - \Delta G[\text{reactants}] \tag{3.8}$$

If the result of this calculation is negative, then the reactants are unstable and the reaction will occur spontaneously. If the ΔG_r° is positive, the reaction is less likely to occur without other additions to the system. The relationship between K° (reaction activity constant) and ΔG_r° (change in free energy for the reaction) is given by:

$$\Delta G_r^\circ = -RT \ln K^\circ \tag{3.9}$$

where R = universal gas constant and T = absolute temperature ($^\circ$K).
At 25 $^\circ$C, equation (3.9) reduces to:

$$\Delta G_r^\circ = -1.364 \log K^\circ$$

or, more usefully,

$$\log K^\circ = \frac{-\Delta G_r^\circ}{1.364} \tag{3.10}$$

This equation is useful for calculating equilibrium constants for chemical weathering reactions in which thermodynamic data are difficult to measure by conventional methods. Lindsay (1979) uses this method to calculate the $\log K^\circ$ values of an extremely large number of likely and unlikely trace metal reactions in soil. Lindsay (1979) also develops qualitative diagrams which provide graphic summaries of mineral sequences which might be expected to occur if equilibrium were attained.

These types of thermodynamic calculations are useful tools for predicting the general sequence of groups of reactions, but are considered unreliable since they suffer from several important problems: (a) ΔG calculations are based on chemical species in the Standard State and these conditions are rarely achieved

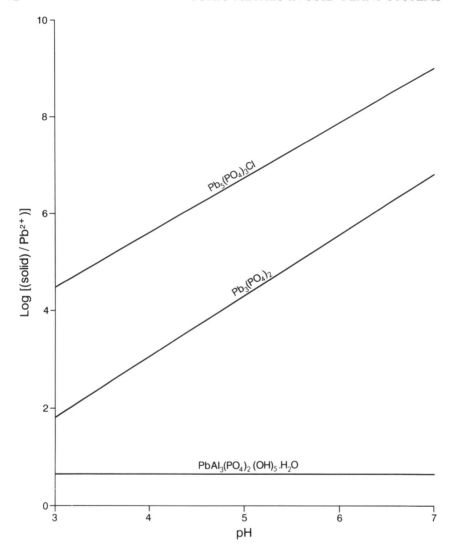

Figure 3.2. Activity ratio diagram for lead phosphate solid phases in soil, under the conditions: $(H_2PO_4) = 10^{-6}$, $(Cl^-) = 10^{-3}$, $(H_2O) = 1$, and control of (Al^{3+}) by kaolinite with $(Si(OH)_4^0) = 10^{-4.5}$ Reproduced by permission of Academic Press Ltd from Sposito (1983)

in field conditions; (b) there are difficulties in writing equations which realistically represent complicated reactions and interactions occurring in field soils; (c) the continually changing chemical and physical conditions of field soils and the fact that many mineral transformation processes proceed extremely slowly mean that a true equilibrium is rarely achieved; (d) for many trace metals in soils, reactions involve organic complexes and often very little is known about their formation, structure or interactions with inorganic metal ligands and free ions; and (e) soil organics are continually being produced and decomposed and hence are in a constant state of flux. Despite these drawbacks, thermodynamic calculations provide at least some estimate of element speciation and concentrations in complex systems, and also in cases where concentrations of metal ions and ligands are below detection levels for conventional analytical methods such as atomic absorption spectrophotometry (AAS). Such calculations are also the basis of geochemical modelling strategies outlined below. There are currently more reliable and available equilibrium constant data than free energy values and while Gibbs free energy approaches may be adequate for simple systems, equilibrium constant approaches are preferred for large complex systems such as the soil (Nordstrom et al., 1979).

3.2.2 MODELLING METAL SPECIATION IN SOILS

Computer models have been developed over the last decade to calculate the equilibrium partitioning of soluble metal ions, metal ligands and organo-metallic complexes in aqueous solutions and in soils. All are based on standard chemical thermodynamic data. Several publications have tabulated thermo-dynamic data in standard states: dissociation constants, solubility products, free energies and enthalpies for different solute forms of trace metals (e.g. Feitknecht and Schindler, 1963; Sillen and Martell, 1964; Hogfeldt, 1979; Robie et al., 1979) but problems in using geochemical modelling as a tool for simulating trace metal speciation are sometimes due to inconsistencies in the quoted values of these data. Other problems arise through the need to characterise realistically the amounts and speciation of the major elements in solution to allow correction for the ionic strength activity of the solution. For variable valence elements (especially many trace metals), a knowledge of redox potential is also required in order to determine their valency states.

Geochemistry models for aqueous systems have been available for some time, including WATEQ (Truesdell and Jones, 1974), SOLMNEQ (Kharaka and Barnes, 1973) and REDEQL2 (McDuff and Morel, 1973). These have had the capacity to calculate the concentrations and particular forms of major and some trace elements in solution and to estimate the effects of changing solution conditions, such as pH, Eh, ionic strength, CO_2 pressure, or the concentrations of particular ions, on the solubility and speciation of a chosen chemical element in solution. Development of models capable of handling the geochemistry of soil solutions in contact with solid soil particles requires simulation of processes such as solubility equilibria, including precipitation and dissolution reactions, specific adsorption on "oxide-like" surfaces and cation exchange. First GEOCHEM (Sposito and Mattigod, 1980) and subsequently SOILCHEM (Sposito and Coves, 1988) are models written specifically to calculate the speciation of chemical elements among the aqueous solution, solid and adsorbed forms in soil. GEOCHEM and SOILCHEM are later developments of the REDEQL model (Morel and Morgan, 1972) which is based on the equilibrium constant approach. GEOCHEM is an iterative program which uses stability constants corrected for ionic strength to calculate the metal and ligand species present. Its main advantages over previous models is its inclusion of a large number of metal–ligand complexes. GEOCHEM uses a large database of 2000 aqueous species and 889 organic species. In addition, routines have been added to the model to handle ion exchange, adsorption/resorption processes and clay mineral solubility. Since geochemical simulations are based on thermodynamic calculations, they suffer from the same limitations and benefits as expressed in Section 3.2.1 above. Simulating concentrations of metal ions and ligands which are below instrument detection levels is advantageous while the problems of non-equilibrium are limiting. Although soils do not reflect equilibrium conditions, it is possible to consider "steady state" for some of these soil components that are known to react with one another on a short-term and nearest-neighbour basis. This steady state can be characterised by knowing (i) the soil solids that affect mineral solubility on a short-term basis, and (ii) the type and quantity of organic ligands that are present in *macro*amounts. For instance, the addition of citric and oxalic acids to soils should, at least temporarily, transform some water-insoluble Al and Fe into water-soluble Al and Fe.

The main purpose of geochemical models is the prediction of species composition and concentrations in defined soil conditions. Several applications of the GEOCHEM model have indicated its relatively accurate prediction of metal speciation in various soil contamination conditions. Mattigod and Sposito (1979), for example, modelled the effects on soil chemistry of geothermal brine spillage. In another application, Sposito and Bingham (1981) showed an overall relationship between total soil Cd, and Cd uptake by sweetcorn. They then computed Cd speciation in saturated soil extracts and found a good correlation between plant Cd uptake and concentrations of $CdCl^+$ in the soil solution. No relationship was found between Cd in the plant and Cd^{2+} in the soil solution.

Lighthart *et al.* (1983) used GEOCHEM to predict the types and concentrations of Cd and Cu species in soil and the levels that inhibit soil microbial respiration. The model predicted four soil Cd fractions (Cd^{2+}, adsorbed, organic and other) and five soil Cu fractions (Cu^{2+}, hydroxides, carbonates, organic, adsorbed). At a soil total Cd concentration of 0.5 mmol kg^{-1}, the Cd^{2+} concentration was calculated to be 10^{-5} M, and just within the range considered to be inhibitory to microbial respiration (Figure 3.3(a)). Increasing free Cd^{2+} in solution correlated well with decreasing respiration. The predominant Cd phase in soil was thought to be $CdCO_3$, which accounted for 40% of the Cd at the 0.5 mmol kg^{-1} total soil Cd. The speciation of soil Cu is controlled by organic complexation and a strong adsorption affinity. Cu did not have an inhibitory effect until total soil Cu concentrations of 5–50 mmol kg^{-1} were reached. These levels were computed to have solution Cu^{2+} concentrations of 0.01–0.1 μM respectively (Figure 3.3(b)).

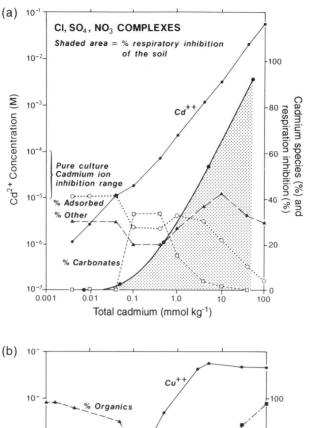

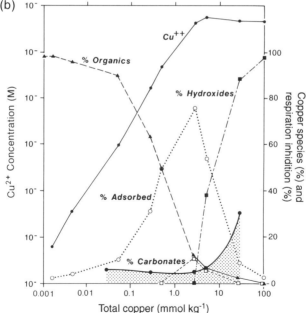

Figure 3.3. GEOCHEM computer simulation of (a) Cd and (b) Cu speciation using measured solution constituents of soil amended with (a) Cd and (b) Cu at pH 6.2. Shaded area shows per cent respiratory inhibition of the soil. Reproduced by permission of the American Society of Agronomy from Lighthart *et al*. (1983)

Several papers report the modelling of trace metal species in soils to which sewage sludge was added. Mahler *et al.* (1980) used GEOCHEM to calculate the distribution of Cd species in saturation extracts of soils amended with various levels of Cd. Free ionic Cd^{2+} accounted for 64–73% and 64–68% of total Cd in acid (pH 5.1) and calcareous (pH 7.6) soils respectively. Cd was mainly associated with sulphate, chloride and fulvate ligands. Behel *et al.* (1983) carried out a similar study of the effect of sewage sludge application on metal speciation in soil. For rates of sludge application ranging from 50 to 800 t ha^{-1}, they calculated that ionic strength of the soil solutions was directly related to the amount of sludge applied. Values ranged from 7.8×10^{-4} M in untreated soil, to 3.0×10^{-2} M in 800 t ha^{-1} treatments. Greater than 80% of the Cd, Zn, Ni and Mn was present in free cationic form in acid soil conditions of pH 5–6 (Table 3.2). Inorganic complexes typically comprised <5% of the soluble metals. This pattern of results, with high proportions of free Cd^{2+} and Zn^{2+} cations in solution after sludge amendment was also reported by Mullins and Sommers (1986). Sludge application also increased DTPA extractable (see Section 3.4.2) Cd and Zn.

Field validations of GEOCHEM are less numerous. Hirsch and Banin (1990) used GEOCHEM to test Cd speciation in the soil solution of arid forest soils using specific ion electrodes (ISE). They concluded that most of the Cd added to calcareous soils is very rapidly adsorbed and soil solution concentrations of Cd remain low. At pH values of 7.5–8, they found only 40–45% of total soluble Cd in free Cd^{2+} form, with bicarbonate and carbonate complexes important at higher pH. As much as 35–40% total soluble Cd was present in the $CdHCO_3^+$ form. Other inorganic Cd complexes in the

Table 3.2. Forms of toxic metals present in soil to which sewage sludge was added (as predicted using the simulation model, GEOCHEM)

Metal		Form of metal present in soil
Zn	87–97%	present as Zn^{2+}
	5–8%	present as sulphates
	<0–1%	simple inorganic ligands (PO_4, CO_3, Cl, OH)
	<2%	fulvate ligands
Cd	91–92%	present as Cd^{2+} (untreated soil)
	79–89%	present as Cd^{2+} (sludge–treated)
	2.7–13%	sulphate complexes
	2.9–7.5%	chloro complexes
	<3%	fulvate ligands
Mn	95–98%	as Mn^{2+}
	2–8%	sulphates
	9.5–17.9%	fulvates
Ca	80%	fulvates

Source: Behel *et al.* (1983).

soil solution were much less important, decreasing in the order: $CdSO_4^0 > CdCl^+ > CdOH^+$.

Despite modifications to GEOCHEM, including the alteration of pH from a concentration term to an activity term (Parker *et al.*, 1987) to improve its general applicability and potential accuracy, there are many criticisms of this kind of approach. Five main *limitations* to soil geochemical modelling have been recognised and summarised by Campbell and Tessier (1987):

1. *shortage of reliable thermodynamic data* for many of the species suspected to be present in natural systems;
2. *inconsistencies and inadequacies of the equations used* to correct ion activity coefficients for changes in ionic strength;
3. *problems in defining the redox* status of natural systems;
4. *lack of kinetic data* for many chemical and biological processes, especially those involving changes in oxidation states; and
5. *difficulties in characterising the organic ligands* present in natural waters and soil solutions (primarily due to the complexity of these chemical systems).

Improvements in geochemical modelling can undoubtedly be achieved when improved thermochemical data are available for both inorganic and organic species and when soil organic compounds can be more accurately characterised.

Soil scientists working at the landscape pollution scale have opted for simpler models of toxic metal speciation, linked to more detailed models of soil erosion and solute transport. Some of these approaches are discussed in Section 3.3.5.

3.3 THE EFFECT OF VARYING SOIL CONDITIONS ON METAL PROCESSES IN SOILS

McBride (1989) suggested that the most valuable way to assess the likely mobility of metals in soils was through a clear understanding of the soil properties and conditions which affect the long and short term fate of metals in soils. The main problem associated with this approach is the potentially massive number of influential factors and their interaction in soil. Soil properties that are likely to influence metal reactions, transformations and mobility in soils include: particle size distribution and particle surface area, bulk density, temperature, aeration and redox status, pH, ion exchange capacity, quantity and quality of organic matter, type and amount of Fe, Mn and Al oxides, and the type and amount of clay minerals. Important databases of soil information exist in many countries and there is relatively good understanding of soil properties and their interrelationships. An ever increasing body of data on trace metals in soils is accumulating, but more detailed information on factors controlling metal speciation and soil retention processes is necessary before best use can be made of existing soil information systems.

In this section the soil processes associated with metals in soils will be reviewed, outlining in each case the influence of changes in environmental conditions and soil properties.

Below are listed some of the main processes associated with toxic metals in soils:

(i) *weathering* of *in situ* parent material
(ii) *dissolution and solubility* of minerals and complexes, accompanied by *precipitation* and *co-precipitation* of inorganic insoluble species, such as carbonates and sulphides
(iii) *uptake* by plant roots and *immobilisation* by soil organisms (some current ideas about uptake are discussed in Chapters 2 and 4)
(iv) *exchange* onto cation exchange sites of clays or soil organic matter
(v) specific *chemisorption* and *adsorption/desorption* on oxides and hydroxides of iron, aluminium and manganese
(vi) *chelation* and complexation by different fractions of soil organic matter
(vii) *leaching* of mobile ions and soluble organo-metallic chelates

Whether any one of the above processes dominates over any other in controlling soil solution metal speciation and concentration, depends entirely on the toxic metal in question, its speciation, and a whole range of soil properties and conditions, including soil pH, redox, organic matter amount and composition, clay content and Fe, Mn and Al oxide content. It is worth clarifying the use of terminology here. The term "metal fractionation" will be used for the soil fractions in which the metal is located (e.g. easily exchangeable, chemisorbed, or organically bound), while the term "metal speciation" will be used for the different chemical forms or species in which the metal can exist, such as different oxidation states or the hydroxide or sulphide ligands. In this section the aim will be to review each of the possible processes associated with heavy metals in soil, in light of the major influencing environmental and soil factors.

3.3.1 WEATHERING OF PARENT MATERIALS

Generally the trace metal input to soil from *in situ* weathering of parent rock is low and only likely to produce potentially toxic metal concentrations locally in areas of oxide-rich deposits, ores and other lithologies high in trace metals, such as ultramafic rocks, including serpentine. In soils developed on serpentine, for example, extremely high concentrations of nickel (up to $500–1000\ \mu g\ Ni\ g^{-1}$ total Ni) have been recorded (see Chapter 12, this volume). Such metal-enriched soils are often characterised by very typical plants, including "hyperaccumulator" species, which can have foliar Ni concentrations as high as 1000 $\mu g\,g^{-1}$ (see Chapter 12, this volume). The mineral chemical weathering processes—dissolution, hydration, hydrolysis, oxidation, reduction and carbonation—are described by Ollier (1984). The chemical weathering processes—solution, reduction and oxidation—are particularly important processes for trace metals in soils and will be discussed in more detail in the following sections. Background mineralogy and chemical theory relevant to chemical weathering and metal mobility are reviewed by Ollier (1984) and Paton (1978).

3.3.2 DISSOLUTION/PRECIPITATION, SOLUBILITY AND FREE IONS IN SOLUTION

Precipitation/dissolution mechanisms and adsorption/desorption mechanisms (Section 3.3.3) are the main physico-chemical processes that control concentrations of metal species in the soil solution. Once in solution, trace metal ions, whether simple or complex, exhibit typical exchange behaviour on silicate clay minerals, with the strength of metal bonding dependant on ion charge and hydration characteristics. In the soil solution, trace metal cations such as Cd^{2+}, Pb^{2+} and Cu^{2+} compete with more abundant soil cations such as Ca^{2+} and Na^+ for cation exchange sites. For this reason, strong partitioning of trace metals onto cation exchange sites is not normally seen. Instead, many trace metals are specifically adsorbed, or chemisorbed, onto amorphous oxides of Al, Fe and Mn, and also onto soil organic matter (see Section 3.3.3). The rates and direction of both precipitation/dissolution processes and adsorption/desorption processes are strongly influenced by acidity and redox potential. Theoretically precipitation/dissolution processes should occur at a given pH, and, unlike adsorption and ion exchange processes, precipitation/dissolution is less dependent on the amount of reactant or the different mineral surfaces present in soil. Apart from Fe and Mn, trace metal solubility in soils is not controlled by the solubility product of a pure solid phase (Brummer et al., 1983). This is partly because adsorption of metal cations from the very low soil solution concentrations is able to maintain solution solubility at a level too low for precipitation to occur. At high metal loadings, and in alkaline and calcareous soils, precipitation processes may begin to control metal ion concentrations in the soil solution. In acid mineral soils and in organic soils, precipitation of metals as hydroxides or carbonates is highly unlikely, even with high metal loadings. Brummer et al. (1983) point out the very close interrelations which exist between ion exchange/adsorption and precipitation/dissolution processes in the generally low metal concentration environment of the soil.

An extremely important breakthrough in understanding mineral and metal solubility in soil was made in 1979 with the publication of the seminal work on soil chemical equilibria by Lindsay (1979), in which he calculated the solubility relationships for a very large number of minerals and metal complexes in soils. He presents graphs of the solubilities of Al, Fe, Mn, Zn, Cu, Cd, Pb and Hg minerals, which help in predicting which minerals are likely to control the solubility of these metals in pure solutions of differing pH, redox and ionic composition. Some examples of these calculations are given later in this section. Their applicability to complex soil solutions is questionable, but an understanding of solubility theory is a vital precursor to estimating how metal cation concentrations in the soil solution may vary in relation to other pedological factors.

The kinetics of precipitation–dissolution reactions in soils have been little studied, particularly for metal-containing minerals. Much less is known about controls on rates of solubility processes than for adsorption processes and the lack of kinetic data for solubility processes is a real limitation for modelling metal speciation.

3.3.2.1 Influence of acidity on metal solublity in soil

One of the most important factors controlling metal solubility in soils is acidity. Soils in humid temperate climates naturally become more acidic through time by leaching, unless lithological replenishment or anthropogenic inputs of mineral cations occurs. Acid precipitation exacerbates this problem. Other soil processes, particularly organic matter decomposition and root ion uptake, can also contribute, even if only locally, to soil acidification. A summary of the relative mobilities of trace metals in relation to pH and redox (Eh) is given in Table 3.3. According to Plant and Raiswell (1983), many metals are relatively more mobile under acid, oxidising conditions and are retained very strongly under alkaline and reducing conditions.

Three types of evidence have been used to indicate how metal solubility increases with increased soil acidity. The *first* of these is direct measurement of metal ion activities in soils of different pH. Field and pot experiments extracting soil solutes from a range of soil types maintained at different pH conditions have shown that Zn, Cd, Cu and to a lesser extent Pb, are much more soluble at pH 4–5 than in the pH range 5–7 (Brummer and Herms, 1983). Solution metal concentrations increased in the order Cd > Zn ≫ Cu > Pb with decreasing pH. The *second* type of evidence is provided by correlations between metal uptake in plants and pH of rhizosphere soil. Sarkar and Wyn Jones (1982) found that with increasing soil rhizosphere acidity, french beans took up increasing amounts of Zn, Fe and Mn. Acidity caused by rhizosphere metabolic products, such as H_2CO_3, was also suggested by Xian and Shokohifard (1989) as a possible reason for increased metal uptake of metal carbonates which become increasingly soluble as soil pH declines. Different plants have differing abilities to acidify their rhizospheres. Youssef and Chino (1991) found that soybean has a greater ability than barley to solubilise Zn, Mn and Fe in the rhizosphere. The *third* type of evidence is derived from chemical thermodynamic calculations of solid-solution metal equilibria in soils. These types of calculations are discussed in more detail below.

Table 3.3. Relative mobilities of trace metals as a function of soil Eh and pH

Relative mobility	Soil condition			
	Oxidising	Acid	Neutral–alkaline	Reducing
Very high	—	—	—	—
High	Zn	Zn	—	—
Medium	Cu, Co, Ni, Hg, Ag, Au	Cu, Co, Ni, Hg, Ag, Au		
	Cd	Cd	Cd	—
Low	Pb	Pb	Pb	—
Very low to immobile	Fe, Mn, Al, Sn, Pt, Cr, Zr	Al, Sn, Pt, Cr	Al, Sn, Cr	
	—	—	Zn, Cu, Co, Ni, Hg, Ag, Au	Zn, Co, Cu, Ni, Hg, Ag, Au, Cd, Pb

Source: Reproduced from Plant and Raiswell (1983) by permission of Academic Press Ltd.

Geochemical theory (see Section 3.2.1) can be used to write equilibration reactions for the solubility of soil minerals in the form of equations. Illustrations of these equations in relation to controlling factors such as pH or pH + pe can allow estimation of which mineral solids might be important in controlling metal cation concentrations in the soil solution. The solubility product approach allows identification of the least soluble mineral able to precipitate from the soil solution. Theoretically, this least soluble mineral under the given soil conditions, is likely to control metal ion concentrations in the soil solution. The numerous examples given by Lindsay (1979) confirm that pH plays an important part in many of the reactions. The solubility product approach to estimating which solid phases control metal ion concentrations suffers from a few problems (McBride, 1989):

(i) Soil metals are usually present in trace quantities which make it impossible to identify discrete mineral phases.

(ii) It is not always possible to measure the free metal ion concentration in solution (separately from metal–ligand complexes) and hence it is not possible to accurately estimate the ion activity product.

(iii) It is doubtful whether equilibrium is ever attained in complex field soils.

(iv) Pure minerals may not control solution metal ion concentrations because mixed oxides are less soluble and may maintain low metal solubilities.

(v) Co-precipitation of metals as impurities in Al, Fe and Mn oxides, sometimes with organic matter, complicates the simple solubility product model.

Table 3.4. Activities of several metal cations in well-oxidised soils

Metal	Reaction	$\log K^{\circ}$	Transformed equation
Zn	Soil-Zn $+ 2H^+ \approx Zn^{2+}$	5.8	$\log Zn^{2+} = 5.8 - 2\,pH$
Cu	Soil-Cu $+ 2H^+ \approx Cu^{2+}$	2.8	$\log Cu = 2.8 - 2\,pH$
Fe	Soil-Fe $+ 3H^+ \approx Fe^{3+}$	2.7	$\log Fe^{3+} = 2.7 = 3\,pH$
Cd	Soil-Cd $\approx Cd^{2+}$	-7.0	$\log Cd^{2+} = -7.0$
Pb	Soil-Pb $\approx Pb^{2+}$	-8.50	$\log Pb^{2+} = -8.5$
$< pH\ 7$:			
Ca	Soil-Ca $\approx Ca^{2+}$	-2.50	$\log Ca^{2+} = -2.5$
Mg	Soil-Mg $\approx Mg^{2+}$	-3.0	$\log Mg^{2+} = -3.0$
$> pH\ 7$:			
Ca	$CaCO_3 + 2H_2^+ \approx Ca^{2+} + CO_2 + H_2O$	9.72	$\log Ca^{2+} = 9.72 - \log CO_2^{a,b} - 2\,pH$
Mg	$MgCaCO_3 + 2H_2^+ \approx Mg^{2+} + CO_2$ $+ H_2O + CaCO_3$	8.70	$\log Mg^{2+} = 8.70 - \log CO_2^{a,b} - 2\,pH$

Source: Reproduced by permission from Lindsay (1979). Copyright © John Wiley & Sons Inc.
[a] $\log CO_2$ at 0.003 atmos $= -2.52$; [b] $\log CO_2$ at 0.0003 atmos $= -3.52$.

These difficulties have limited the use of the solubility product approach for predicting and explaining metal ion mobilities in soils. However, to overcome the difficulties in identifying trace quantities of minerals, Lindsay (1979) suggests that solubility measurements can be made directly on soils, and the results expressed empirically as soil-Pb, soil-Zn and soil-Cd. The activities of several cations in well oxidised soils are given in Table 3.4 and illustrated, in relation to soil pH, in Figure 3.4. At pH values above 7, activities of Ca^{2+} and Mg^{2+} are controlled by the solubilities of calcium carbonate ($CaCO_3$) and dolomite ($MgCa(CO_3)_2$) respectively. The solubilities of these minerals are controlled by CO_2 concentration, as illustrated in Figure 3.4.

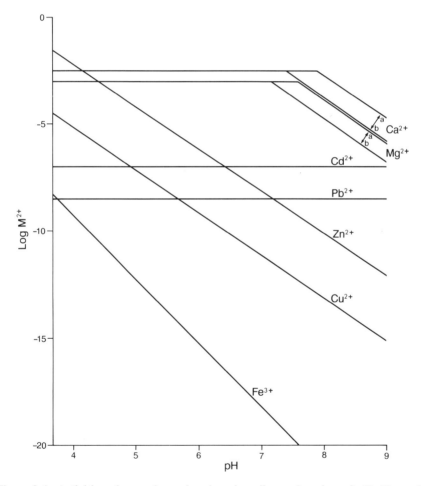

Figure 3.4. Activities of several metal cations in soil as a function of pH. From the equations of Lindsay (1981). Key to Ca^{2+} and Mg^{2+} solubilities at pH > 7: [a] = log CO_2 at 0.003 atmos. = -2.52 [b] = log CO_2 at 0.0003 atoms. = -3.52

Figures 3.5, 3.6 and 3.7 are graphic illustrations of the relationship between the metal cations Zn^{2+}, Pb^{2+}, Cd^{2+} and soil solution pH for a number of selected soil minerals. The log $K°$ values used to construct these graphs are given in Table 3.5 (taken from Lindsay, 1979). Other conditions used in graph construction are given in the keys to each graph. Illustrations of this type can be used to suggest which soil minerals are likely to control soil solution metal ion activity over a given range of soil pH conditions and under the influence of certain controlling factors. For the formation of metal phosphates, for example, the factors which control soil solution H_2PO_4 concentration will also

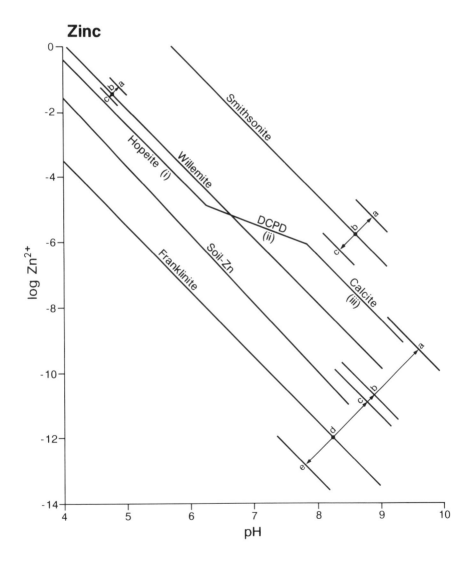

control the formation of metal phosphates. At low pH, soil H_2PO_4 concentrations are controlled by iron phosphates, while at high pH, H_2PO_4 concentrations are controlled by calcium phosphates. The controls of soil iron and calcium phosphates on the stability of Zn, Pb and Cd phosphates and hence on metal ion activity in the soil solution, are illustrated in Figures 3.5, 3.6 and 3.7.

The Zn hydroxides, oxides and carbonates are very soluble and will dissolve if added to the soil. In Figure 3.5, smithsonite (Zn carbonate) is the most soluble mineral illustrated. Willemite (Zn silicate) is of intermediate stability and franklinite (a Zn-ferric oxide) is very stable in soil and is probably the most important mineral to control Zn^{2+} activity in soil (Sadiq, 1991). The solubility of Zn^{2+} in equilibrium with franklinite is controlled by Fe^{3+} activity, which in turn is affected by the presence in soil of different iron minerals. In Figure 3.5 the solubility of franklinite is illustrated for five different forms of soil iron. The solubilities of all Zn species illustrated in Figure 3.5 decrease 100-fold for every pH unit increase (Lindsay, 1979).

In soil, Cd hydroxides, oxides, sulphates and silicates are all extremely soluble. In calculations of Cd stability in soil, Lindsay (1979) shows that $CdCO_3$ (octavite) is the main Cd mineral to control Cd^{2+} activity in the soil solution (Figure 3.7). At elevated CO_2 concentrations, octavite decreases Cd^{2+} solubility 100-fold for every pH unit increase above pH 7.5. Cd phosphates may also influence Cd^{2+} activity in calcareous soils. In the case of both lead and manganese, both pH and redox are important controls on ion solubility. In the range of pH experienced in soils, the most insoluble of the soil lead phosphate minerals, $Pb_5(PO_4)_3Cl$ (chloropyromorphite) has the capability to control Pb^{2+} concentrations in solution (Figure 3.6). Most of the other lead

Smithsonite	$ZnCO_2$	$a = CO_2 = 0.03$ atmos. $b = CO_2 = 0.003$ atmos. $c = CO_2 = 0.0003$ atmos.
Willemite	Zn_2SiO_4	a = Quartz-Si b = Soil-Si c = Amorphous-Si
Hopeite	$Zn_2(PO_4)_2 \cdot 4H_2O$	(i) = Strengite/soil-Fe (ii) = Soil-Ca (iii) = Calcite-Ca
Franklinite	$ZnFe_2O_4$	a = Goethite-Fe b = Lepidocrocite-Fe c = Maghemite-Fe d = Soil-Fe e = Amorphous-Fe ($Fe(OH)_3$)

Figure 3.5. The solubilities of several zinc minerals compared to soil-Zn. Reproduced by permission from Lindsay (1979). Copyright © John Wiley & Sons, Inc.

inorganic ligands, such as the silicates, sulphates, carbonates and hydroxides, are all too soluble to form for any length of time in soils (Lindsay, 1979). A summary of the likely minerals controlling the activities of Zn^{2+}, Cd^{2+} and Pb^{2+} in soils is given in Table 3.6. It is possible to conclude that different minerals may be responsible for controlling metal ion activities in acid soils compared to calcareous soils. This is not always directly related to pH, but

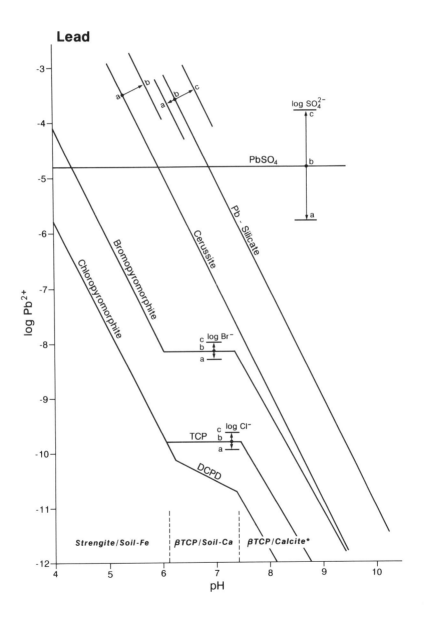

frequently to factors controlling phosphate concentrations, CO_2 concentrations or redox conditions. The principal chemical species of trace metals in acid and alkaline soil solutions under oxidising conditions are summarised in Table 3.7. From this great variety of species in cationic, anionic and uncharged conditions, we begin to understand why predicting metal speciation in soil solutions is so problematic.

In addition to ion activity graphs constructed from equations of solubility reactions for inorganic metal ligands, it is possible to draw similar illustrations of the reactions of metals with organic ligands. There are, however, several problems with these types of calculations. First, the structure of soil organic matter and organic fractions is continually changing through decomposition processes; secondly, our knowledge of the structure of soil organics is far from complete and, although we have some information about metal binding sites of organic molecule surfaces, in many cases we can only speculate about the actual mechanisms of metal binding. To obtain some idea of metal ion activities in equilibrium with chelates, and to predict likely effects of changing soil pH conditions, Lindsay (1979) developed stability–pH graphs for model organic chelates such as EDTA (ethylenediaminetetraacetic acid) and DTPA (diethylenetriaminepentaacetic acid). It could be argued that such predictions are meaningless since they are highly simplified compared to reactions occurring in the soil and since they deal with model organics that probably never occur in soil. However, their value lies in providing an idea of how soil organics influence other metal equilibrium reactions in the soil solution. When a chelating agent is added to the soil, a proportion of metal cations will react with it to form metal chelates. Soil solution metal ion concentrations are

Cerussite	$PbCO_3$	$a = CO_2 = 0.003$ atmos.
		$b = CO_2 = 0.0003$ atmos.
Pb-Silicate	$PbSi_3$	$a =$ Amorphous Si
		$b =$ Soil-Si
		$c =$ Quartz
Chloropyromorphite		$a = \log Cl^- = -2$
$Pb_5(PO_4)_3Cl$		$b = \log Cl^- = -3$
		$c = \log Cl^- = -4$
Bromopyromorphite		$a = \log Br^- = -2$
$Pb_5(PO_4)_3Br$		$b = \log Br^- = -3$
		$c = \log Br^- = -4$
Pb-Sulphate	$PbSO_4$	$a = \log SO_4^{2-} = -2$
		$b = \log SO_4^{-2} = -3$
		$c = \log SO_4^{-2} = -4$

< pH 6:	phosphate activity fixed by strengite and soil-Fe.
pH 6–pH 7.3:	phosphate activity fixed by tricalcium phosphate (TCP) and soil-Ca.
> pH 7.3:	phosphate activity fixed by TCP, calcite and $[CO_2]$ at 0.003 atmos.

Figure 3.6. The solubilities of several lead minerals at given concentrations of SO_4^{2-}, Cl^-, Br^- and CO_2, indicating the various solid phases which control solution phosphate concentration. Compiled from Lindsay (1979)

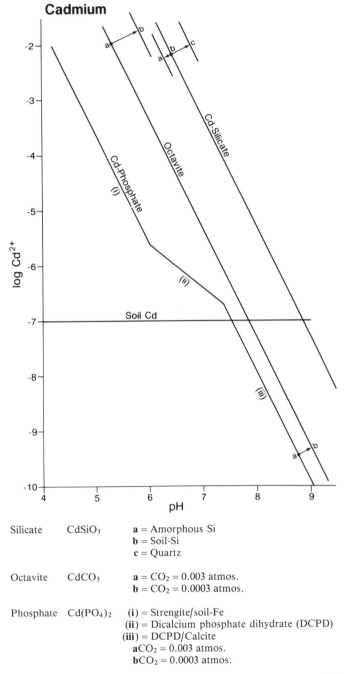

Cadmium

Silicate	CdSiO₃	a = Amorphous Si
		b = Soil-Si
		c = Quartz

Silicate $CdSiO_3$ a = Amorphous Si
                      b = Soil-Si
                      c = Quartz

Octavite $CdCO_3$ a = CO_2 = 0.003 atmos.
                      b = CO_2 = 0.0003 atmos.

Phosphate $Cd(PO_4)_2$ (i) = Strengite/soil-Fe
                         (ii) = Dicalcium phosphate dihydrate (DCPD)
                        (iii) = DCPD/Calcite
                              aCO_2 = 0.003 atmos.
                              bCO_2 = 0.0003 atmos.

Figure 3.7. The solubility of several cadmium minerals compared to soil-Cd 10^{-7} M. Compiled from Lindsay (1979)

Table 3.5. Values of log K° used in the construction of Figures 3.5, 3.6 and 3.7

			log K°
Zn			
Willemite	$Zn_2SiO_4 + 4H^+$	$2\,Zn^{2+} + H_4SiO_4$	13.15
Hopeite	$Zn_3(PO_4)_2.4H_2O + 4H^+$	$3\,Zn^{2+} + 2H_2PO_4 + 4H_2O$	3.80
Franklinite	$ZnFe_2O_4 + 8H^+$	$Zn^{2+} + 2Fe^{3+} + 4H_2O$	9.85
Smithsonite	$ZnCO_3 + 2H^+$	$Zn^{2+} + CO_2 + H_2O$	7.91
Cd			
Cd-phosphate	$Cd_3(PO_4) + 4H^+$	$3Cd^{2+} + 2H_2PO_4$	1.00
Cd-slicate	$CdSiO_3 + 2H^+ + H_2O$	$Cd^{2+} + H_4SiO_4$	7.63
Octavite	$CdCO_3 + 2H^+$	$Cd^{2+} + CO_2 + H_2O$	6.16
Pb			
Pb-silicate	$PbSiO_3 + 2H^+ + H_2O$	$Pb^{2+} + H_2SiO_4$	5.94
Cerussite	$PbCO_3 + 2H^+$	$Pb^{2+} + CO_2 + H_2O$	4.65
Bromopyromorphite	$Pb_5(PO_4)_3Br + 6H^+$	$5Pb^{2+} + 3H_2PO_4 + Br^-$	-19.49
Chloropyromorphite	$Pb5(PO_4)_3Cl + 6H^+$	$5Pb^{2+} + 3H_2PO_4 + Cl^-$	-25.05
PbSO$_4$	$Pb^{2+} + SO_4^{2-}$	$PbSO_4^\circ$	2.62

Source: Reproduced by permission from Lindsay (1979). Copyright © John Wiley & Sons, Inc.

momentarily depleted and this causes exchangeable cations to dissociate and solid phases to dissolve to replenish the ions which are chelated. It is this complex plethora of metal equilibration reactions which the SOILCHEM and GEOCHEM models, with their 889 organic species, are designed to calculate, *en route* to predicting metal ion activities in the soil solution.

An ability to predict the effects of pH change on soil metal mobilities has been vitally important in recent years with high levels of acid precipitation recorded in many parts of the world. In a review of the effects of acid deposition on geochemical cycling of trace elements, Campbell *et al.* (1983) conclude that Al, Mn, Zn and possibly Cd and Ni can be mobilised in acidified soils, while Pb and Cu would not be expected to be mobilised. The influence of acid rain on metal mobility and transport are discussed in Section 3.3.5.

3.3.2.2   Influence of redox potential on metal solubility in soil

Several potentially toxic metals occur in soils in more than one oxidation state, most importantly: Fe, Mn, Cr, Cu, As, Ag, Hg and Pb. Redox equilibria in soil are governed by the pe (the negative logarithm of the aqueous free electron activity) in the soil solution. Table 3.8 lists the critical pe values for redox transformations of selected metals in the normal ranges of soil pe (-6.8 to 13.5) and pH (4 to 9).

When soil redox conditions change, the ratio of oxidised to reduced species in solution will also change. In an example where there are two minerals in

Table 3.6. Metal compounds and their solubility in soil

Metal	Likely soil compounds	Factors controlling solubility
Zn	$ZnCl$, $ZnOH$, $ZnNO_3$	All *too soluble* to be present in soil.
	$Zn(OH)_2.ZnSO_4$	*Too soluble* to be present in soil even at high $[SO_4]$.
	$ZnCO_3$ (smithsonite), Zn_2SiO_4 (willemite)	Solubilities controlled by $[CO_2]$ and $[H_4SiO_4]$ respectively. Solubilities of all above minerals decrease 100-fold for each unit increase in pH.
	$ZnFe_2O_4$ (franklinite)	Solubility controlled by $[Fe^{3+}]$ in solution, which in turn is controlled by iron minerals such as goethite and maghemite.
	$Zn_3(PO_4)_2.4H_2O$ (hopeite)	Solubility controlled by phosphate minerals such as strengite ($FePO_4.2H_2O$) and brushite ($CaHPO_4.2H_2O$) which control the soil $[H_2PO_4]$ at low and high pH respectively.
Cd	CdO, $CdOH_2$, $CdSO_4.2CdOH_2$, $CdSiO_3$ $CdCO_3$ (octavite)	All are *too soluble* to be present in soil. At $[CO_2] > 0.003$ atmos., octavite lowers Cd^{2+} activity 100-fold for every unit increase in pH > 7.84.
	$Cd_3(PO_4)_2$	Solubility controlled by the soil minerals which control the concentration of soil $[H_2PO_4]$: — at pH < 6, strengite ($FePO_4.2H_2O$) — at pH > 8, tricalcium phosphate (TCP), calcite and CO_2 — at intermediate pH, TCP and soil-Ca. $Ca_3(PO_4)_2$ is too soluble in acid soils to account for $[Cd^{2+}]$. In calcareous soils, octavite ($CdCO_3$) is more stable than $Ca_3(PO_4)_2$ in equilibrium with any of the mineral phosphates.
Pb	$PbSO_4$, $PbCO_3$, PbO, $PbCO_3.2PbO$	All are generally *too soluble* to be important in soils. $PbSO_4$ (anglesite) limits Pb^{2+} at $10^{-4.79}$ M, and its solubility increases one log unit for every log unit increase in soil $[SO_4]$. $PbSO_4$ is stable below pH 6, while $PbCO_3$ (cerussite) is stable above pH 6.
	Pb_2SiO_4, $PbSiO_3$, PbO, PbO_2, Pb_3O_4 Pb phosphates	Minerals are *too soluble* to be expected in soils. Solubilities all increase with decreasing redox. All solubilities are controlled by the minerals in soils which control soil $[H_2PO_4]$. $Pb_5(PO_4)_3Cl$ (chloropyromorphite) is the most insoluble of the lead phosphates and could be important in controlling soil $[Pb^{2+}]$ throughout the pH range of soils.
	Pb halides	PbI^+, $PbBr^+$, PCl^- and PbF^- may all contribute to total soluble lead in soil, where they are locally important.

Source: Summarised from Lindsay (1979).   [ ] represent concentrations in solution.

Table 3.7. Principal chemical species of trace metals in acid and alkaline soil solutions (oxidising conditions)

Metal	Principal species	
	Alkaline soils	Acid soils
Cr	$CrOH_2^+$, CrO_4^{2-}	CrO_4^{2-}, $Cr(OH)_4^-$
Mn	Mn^{2+}, $MnSO_4^\circ$, Org.[a]	Mn^{2+}, $MnSO_4^\circ$, $MnCO_3^\circ$, $MnHCO_3^+$, $MnB(OH)_4^+$
Fe	Fe^{2+}, $FeSO_4$, $FeH_2PO_4^+$	$FeCO_3^\circ$, Fe^{3+}, $FeHCO_3^+$, $FeSO_4^\circ$
Ni	Ni^{2+}, $NiSO_4^\circ$, $NiHCO_3^+$, Org.[a]	$NiCO_3^\circ$, $NiHCO_3^+$, Ni^{2+}, $NiB(OH)_4^+$
Cu	Org.[a], Cu^{2+}	$CuCO_3^\circ$, Org.[a], $CuB(OH)_4^+$, $Cu(B(OH)_4)_2^\circ$
Zn	Zn^{2+}, $ZnSO_4^\circ$	$ZnHCO_3^+$, $ZnCO_3^\circ$, Zn^{2+}, $ZnSO_4^\circ$, $ZnB(OH)_4^+$
Cd	Cd^{2+}, $CdSO_4^\circ$, $CdCl^+$	Cd^{2+}, $CdCl^+$, $CdSO_4^\circ$, $CdHCO_3^+$
Pb	Pb^{2+}, Org.[a], $PbSO_4^\circ$, $PbHCO_3^+$	$PbCO_3^\circ$, $Pb(CO_3)_2^{2-}$, $PbOH^+$

Sources: Sposito (1983) and Alloway (1990).
[a] Organic complexes, e.g. humic, fulvic acid complexes.

Table 3.8. Critical values of pe for changes in the redox speciation of trace metals in soils at 25.15 °C

Oxidation states	Reduction half-reaction	pe crit (Eh(v))	
		pH 5	pH 8
Cr^{VI}/Cr^{III}	$Cr_2O_7^{2-} + 8H^+ + 6e^- = 2Cr(OH)_{3(s)} + H_2O_{(1)}$	13.0 (0.1770)	8.0 (0.472)
Mn^{III}/Mn^{II}	$MnOOH_{(s)} + 3H^+ + e^- = Mn^{2+} + 2H_2O_{(1)}$	15.9 (0.9381)	6.9 (0.4071)
Mn^{IV}/Mn^{II}	$MnO_{2(s)} + 4H^+ + 2e^- = Mn^{2+} + 2H_2O_{(1)}$	14.7 (0.8673)	8.7 (0.5133)
Fe^{III}/Fe^{II}	$Fe(OH)_{3(s)} + 3H^+ + e^- = Fe^{2+} + 3H_2O_{(1)}$	8.9 (0.5251)	-0.1 (-0.0059)
Cu^{II}/Cu^I	$CuO_{(s)} + 2H^+ + e^- = Cu^+ + H_2O_{(1)}$	8.3 (0.4859)	2.3 (0.1357)
Hg^{II}/Hg^I	$2 HgO_{(s)} + 4H^+ + 2e^- = Hg_2^{2+} + 2H_2O_{(1)}$	11.8 (0.6962)	5.8 (0.3422)

Source: Reproduced by permission of Oxford University Press from Sposito (1981).

soil, containing the same element, but in different oxidation states, reducing the soil (i.e. adding electrons) will cause the oxidised mineral to dissolve and the reduced mineral to precipitate. If the soil becomes oxidised, the opposite will occur. Metals are generally less soluble in their higher oxidation states. Iron and manganese are the two most important soil elements to occur in different oxidation states in soil. The ability of Mn oxides, and to a lesser extent Fe oxides, to directly oxidise other metals or to catalyse metal oxidation by O_2 may play an important part in lowering trace metal solubility (McBride, 1989). Co and Pb may be oxidised in this way at Mn oxide surfaces. McBride (1989) also describes a number of redox reactions involving the different oxidation states of iron, such as the reduction of ferric oxides to ferrous iron in the presence of organic molecules. Whether this type of reduction occurs with trace metals in soil has not yet been established.

In most redox mineral transformations, both redox and pH play important roles in trace metal solubility. This means that both electrons and protons take part in the reactions. This can be illustrated in the case of manganese. The solubility of Mn^{2+} in soil is governed by β-MnO_2 (pyrolusite) in soil conditions with pe + pH > 16.62, by γ-$MnOOH$ (manganite) at slightly lower redox, and by $MnCO_3$ (rhodochrosite) at still lower redox, depending on CO_2 concentration (Lindsay and Sadiq, 1983). The effect of redox is potentially much greater than that of acidity in determining the activity of soluble Mn in soil. However, at any given redox state, decreasing the pH will result in increased levels of soluble Mn. The dissolution/solution of galena (PbS) in soil is also controlled by (pe + pH). Figure 3.8 illustrates the redox (pe + pH) at which various metal sulphides precipitate. In Figure 3.8, galena (PbS), for example, cannot form at pe + pH > 4.59 in acid soils, and below this redox value, Pb^{2+} solubility is controlled by galena and whatever other SO_4^{2-} or other sulphur species persists. Experimental evidence indicates that Fe^{2+} activity in the soil solution is controlled by siderite ($FeCO_3$) below pe + pH = 8.5 and above this value, by ferrosic hydroxide ($Fe_3(OH)_8$) (Schwab and Lindsay, 1983). At low pH, these authors showed that lepidocrocite (γ-$FeOOH$) or goethite (α-$FeOOH$) was the controlling solid phase of solution Fe^{2+}.

Soil organic molecules play an important role in redox reactions. Strong complexation with functional groups in organic molecules may stabilise certain oxidation states of metals, inhibiting redox reaction which would otherwise occur rapidly (McBride, 1989). Many metal redox transformations are abiotic processes. Some reactions can only be activated in the presence of biological activity, because of the requirement for oxidisable organic substrates, but they are not in themselves biological processes.

Two main types of chemical transformations may be responsible for lowering the mobility and perhaps the availability of trace metals under waterlogging and reducing soil conditions: (i) formation of insoluble sulphides, and (ii) formation of discrete carbonate, metal oxides or hydroxides of low solubility. There is much evidence of increased solubility and plant uptake of Mn and Fe in waterlogged soils. This effect is reported for a wide variety of plants and soil situations, including *Erica cinerea* and *Erica tetralix* in peat bog soils in England (Jones and Etherington, 1970), rice in paddy fields (e.g. Weerarantna, 1969; Benckiser *et al.*, 1984) and crop plants in flooded soils (Iu *et al.*, 1982). Increased Co uptake by french beans and sweet corn in waterlogged soils has also been reported (Iu *et al.*, 1982). A series of trace metal deficiencies and reduced metal uptake have been observed in plants growing

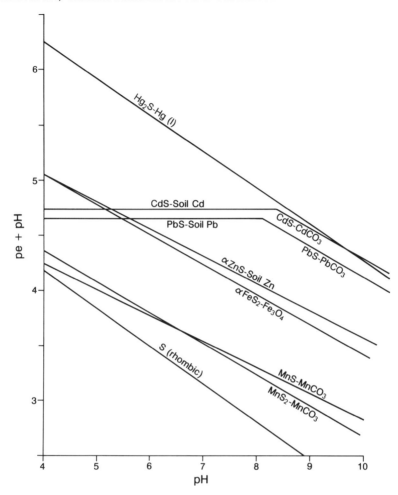

Figure 3.8. The redox and pH (pH + pe) at which various metal sulphides precipitate in soils when SO_4^{2-} is 10^{-3} M. Reproduced by permission from Lindsay (1979). Copyright © John Wiley & Sons, Inc.

in waterlogged soils. Lowered solubility and uptake of Cd (e.g. Bingham *et al.*, 1976; Reddy and Patrick, 1977), Hg (Kothny, 1973) and Zn (e.g. Forno *et al.*, 1975) have been observed. The biggest problem relates to Zn deficiency in rice grown in flooded and reduced paddy fields (e.g. Gangwar and Mann, 1972; Forno *et al.*, 1975; Mikkelsen and Brandon, 1975; IRRI, 1977; Sajwan and Lindsay, 1986). Several theories have been proposed to explain the cause of Zn deficiency in flooded soils, including the formation of insoluble zinc sulphides (e.g. Kittrick, 1976), the influence of added phosphorus, reducing Zn uptake through mineral complex formation (e.g. Seatz and Jurinak, 1957; Stanton and Berger, 1970), or the influence of reduced Fe and Mn complexing with Zn to form franklinite-like minerals (Sajwan and Lindsay, 1986).

3.3.3 ION EXCHANGE, ADSORPTION AND CHEMISORPTION

To adequately describe the mobility of metal ions in the soil solution we must first understand the partitioning of metal ions between the solid and liquid phases. It is now known that a whole series of time-related exchange and adsorption processes operate at particle surfaces. Mechanisms removing ions from solution include ion exchange at charged surfaces such as silicate clays and organic molecules, and non-exchangeable adsorption or chemisorption at mineral surfaces such as iron and manganese oxides, hydroxides and amorphous aluminosilicates. The permanent charge sites of silicate clay minerals and the pH-dependent charge sites of soil organic matter retain metal cations by non-specific electrostatic attraction. In the absence of high pH, which would favour hydrolysis, divalent and trivalent transition and metal cations will exchange onto layer silicates. Cu^{2+}, Co^{2+}, Ni^{2+}, Mn^{2+}, VO^{2+} and Cr^{2+} can be attracted to smectite clay exchange sites by electrostatic forces only and, as with the alkaline earth cations, the strength of metal attraction will depend on the charge and degree of hydration of the metal cation.

Since metal ions such as Pb^{2+}, Cu^{2+} or Cd^{2+} must compete for cation exchange sites with the much more abundant Ca^{2+} and Mg^{2+} ions, strong electrostatic attraction of metal ions onto exchange sites is rarely expected. Instead, more specific sorption mechanisms are needed to explain the retention, low solubility and difficulty of extraction of metal ions reported in studies of metal mobility in soils. Bummer et al. (1983) compared the cation exchange capacity and Zn adsorption capacity of several soil materials in $CaCO_3$ buffered systems (Table 3.9). These data indicate that the mechanisms and binding sites for cation exchange of alkaline and alkaline earth cations are quite different from those of heavy metal adsorption.

Table 3.9. Cation exchange capacity (CEC) and zinc adsorption capacity of different substances in $CaCO_3$ buffered systems

Soil material	CEC pH 7.6 (μmol g^{-1})	Zn adsorption capacity (μmol g^{-1})
$CaCO_3$	—	0.44
Bentonite	450	44
Amorphous Fe-oxide	160	1190
Amorphous Al-oxide	50	1310
βMnO_2	230	1540
Humic acid	1700	842

Source: Reproduced by permission of Kluwer Academic Publishers from Brummer et al. (1983).

Since most studies of heavy metal retention in mineral soil indicate non-exchangeable complexing, these more specific retention processes have been widely described as adsorption, chemisorption or, more generally, sorption processes. McBride (1989) lists three main types of evidence for the formation of specific metal–mineral surface bonding:

(i) the high degree of specificity shown by oxides for particular metals (Kinniburgh et al., 1976);
(ii) the release of as many as two H^+ ions for each M^{2+} ion adsorbed (Forbes et al., 1976); and
(iii) changes in the surface charge properties of the oxide as a result of adsorption (Stumm and Morgan, 1981).

Spectroscopic techniques, particularly ESR (Electron Spin Resonance) and UV-visible procedures, have also confirmed specific adsorption of metals. There is evidence for the exchange and adsorption of a wide range of forms of metals, including free aquo ions, such as Cu^{2+}, Cd^{2+} and Zn^{2+}, but also ion pairs and complex species, such as $CuCl^+$, $CuOH^+$, $ZnOH^+$ and $ZnCl^+$, hydrolytic species, such as $Cu(OH)_2$, or even other metal complexes such as $NiNO_3$ (Mattigod et al., 1979, 1981; McBride, 1989). There is also evidence for a range of adsorbing surfaces, including silica, amorphous aluminosilicates and Al hydroxides as well as perhaps the most important adsorbing surfaces, the iron and manganese oxides. Amorphous oxides of Si and Al and layer silicates appear to be secondary in importance to Fe and Mn oxides as metal adsorbing surfaces. Jenne (1977) suggested that the primary role of layer silicates is as a substrate on which amorphous Fe and Mn oxides precipitate. Robert and Terce (1989) have also confirmed that gels or coatings of Fe and Al compounds on clay surfaces have important metal cation adsorption properties. The complexity of metal adsorption processes in soil is due to the very large number of possible metal species, adsorbing surfaces and bonding mechanisms. A review of current research on metal–mineral adsorption mechanisms is given by McBride (1989) and a discussion of the problems of over-simplifying adsorption processes is given by Barrow (1989).

3.3.3.1 The order of metal affinities during adsorption

A generally accepted view is that specific adsorption of heavy metals by soil minerals is related to metal ion hydrolysis. Increasing ability of metals to form hydroxy complexes increases their likelihood of specific adsorption. Brummer (1986) lists the order of increasing heavy metal hydrolysis as Cd < Ni < Co < Zn ≪ Cu < Pb < Hg and proposes that this is the same order for increasing specific adsorption. The close relationship between metal ion hydrolysis and specific adsorption is especially well established for goethite and haematite, although these two minerals do not behave in an identical manner. Kinniburgh et al. (1976) reported orders of metal affinities for amorphous Fe and Al hydroxides which are a little different from the order of hydrolysis (see Figure 3.9), as did McKenzie (1980) for Mn oxides and the Fe minerals goethite and haematite. These graphs show that in all cases, metal adsorption (and desorption) is controlled by pH. The results of Kinniburgh *et al.* (1976) showed that a number of divalent transition and heavy metal cations can be adsorbed by Fe and Al hydrous gels at considerably lower pH than are the alkaline earth cations. These findings have important implications for metal retention in acid soils and in soils affected by acid precipitation. The results of several studies designed to elucidate the metal affinity sequences for different soil components are illustrated in Table 3.8(a). These sequences indicate that soil

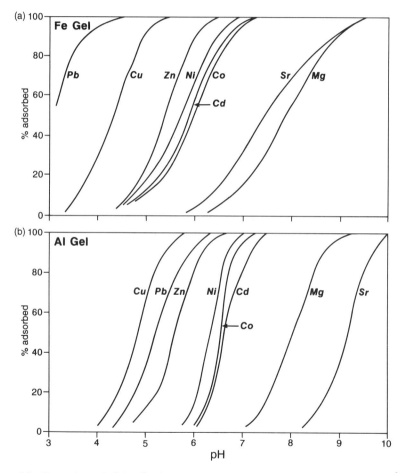

Figure 3.9. Retention of eight divalent metal cations (each present at 0.125×10^{-3} M) in relation to pH in a 1 M NaOH system: (a) fresh Fe gel (0.093 MFe), and (b) fresh Al gel (0.093 M Al). Reproduced from Kinniburgh *et al.* (1976) by permission of the Soil Science Society of America

oxides and organic fractions preferentially adsorb Pb, Cu and Zn compared to Cd, Ni and Co. Predicted metal affinity sequences, based on chemical theory, are given by McBride (1989) in Table 3.10(a). All of these sequences are quite different from experimental data in Table 3.10(b), perhaps confirming the specificity of many trace metal sorption mechanisms, rather than electrostatic ion exchange for many divalent trace metals.

3.3.3.2 The nature of metal adsorption

Oxide-metal bonding is not entirely electrostatic, since it is not possible to predict exactly the bonding order based on the ionic potential (Z/r), where Z is the ion charge and r is the ionic radius. Electrostatic factors alone would place Mg high and Pb low in the affinity sequence (McBride, 1989). The covalent contribution (electronegativity or "softness" parameters) to bonding is not the main factor either, since this would, for example, give a higher

Table 3.10(a). Predicted metal affinity series for divalent metal cations, based on metal properties

Property	Predicted order of affinity
Z^2/r	Ni > Mg > Cu > Co > Zn > Cd > Sr > Pb
Electronegativity (Pauling)	Cu > Ni > Co > Pb > Cd > Zn > Mg > Sr
"Softness"	Pb > Cd > Co > Cu > Ni > Zn > Sr > Mg
Irving-Williams series	Cu > Ni > Zn > Co > Mg > Sr

Source: Reproduced by permission of Springer-Verlag from McBride (1989).

Table 3.10(b). Empirical heavy metal affinity series for soil components

Material	Affinity sequence	Source
Amorphous Al oxides	Cu > Pb > Zn > Ni > Co > Cd	Kinniburgh et al. (1976)
Amorphous Al oxides	Cu > Pb > Zn > Cd	Leckie et al. (1980)
Amorphous Fe oxides	Pb > Cu > Zn > Ni > Cd > Co	Kinniburgh et al. (1976)
Amorphous Fe oxides	Pb > Cu > Zn > Cd	Leckie et al. (1980)
Goethite (FeOOH)	Cu > Pb > Zn > Cd	Forbes et al. (1974)
Goethite (FeOOH)	Cu > Pb > Zn > Co > Ni > Mn	McKenzie (1980)
Haematite	Pb > Cu > Zn > Co > Ni > Mn	McKenzie (1980)
Mn oxide (birnessite)	Pb > Cu > Mn = Co > Zn > Ni	McKenzie (1980)
Mn oxides	Cu > Zn	Murray (1975)
Mn-SiO_2	Pb > Cu > Zn > Cr > Cd	Leckie et al. (1980)
Fulvic acid (pH 5)	Cu > Pb > Zn	Schnitzer and Skinner (1967)
Humic acid (pH 4–7)	Zn > Cu > Pb	Verloo and Cottenie (1972)
Humic acid (pH 4–6)	Cu > Pb ≫ Cd > Zn	Stevenson (1977)

affinity for Cd^{2+} than for Zn^{2+}. McBride concludes that transition metals classified as "hard" (see Table 3.1) are bonded more strongly than "soft" transition metals. However, "soft" non-transition metals (e.g. Pb^{2+}) are preferred over "harder" non-transition metals (e.g. Cd^{2+}, Mg^{2+}). Zn^{2+} displays intermediate behaviour.

The kinetics of sorption processes are not well understood but a little more evidence is available than for solubility reactions. Christensen (1984) found that Cd sorption from low Cd concentration conditions in soils, was extremely rapid, with 95% of the sorption occurring within the first 10 min of a sorption isotherm experiment. Other studies have indicated a two-step sorption process, with an initially rapid first step, representing adsorption onto highly accessible sites on the adsorbing surface, followed by a second, slower type of sorption, characteristic of modified surfaces, co-precipitation associated with Fe and Mn oxides, and solid state diffusion processes (Mattigod *et al.*, 1981). These second stage sorption processes may occur over a period of days. Much further research is required to properly understand the sorption kinetics of different potentially toxic metals on different surfaces under different soil conditions.

In addition to problems in understanding the rates of sorption reactions, other concurrent processes, which are closely associated with metal adsorption, complicate the sorption story:

(i) precipitation at the adsorbing surface
(ii) formation of ternary complexes
(iii) metal diffusion into the mineral surface

Precipitation

Both adsorption and precipitation of metals may occur at mineral surfaces. These processes are sometimes difficult to distinguish. Sposito (1986*b*) has highlighted this problem for Mn and Fe oxides. Mn oxides, for example, promote the oxidation of Mn and Fe. The resulting close association of Mn with Fe and other heavy metals may be due to co-precipitation rather than adsorption at the oxide surface. Gleam and McBride (1986), working with rather unusual titanium oxides, showed that Cu and Mn adsorption and precipitation were pH-related. At pH < 6, adsorption predominated while at pH > 6, precipitation predominated. A similar pattern was reported for Pb, with adsorption occurring below pH 6 and the precipitation of lead carbonates occurring at pH > 6 (Harter, 1979).

Ternary complexes

Metal sorption in soil can be complicated by the formation of stable surface-metal–ligand (ternary) complexes. Sposito (1983) identifies four categories of ligand effects:

(a) the ligand has a high affinity for the metal, forms a stable complex and the complex has either
   (i) a high affinity, or
   (ii) a low affinity for the adsorbing surface
(b) the ligand has a high affinity for the adsorbing surface and is adsorbed, then the adsorbed ligand has either
   (i) a high affinity, or
   (ii) a low affinity for the metal.

Categories (a(i)) and (b(i)) above would result in enhanced adsorption of the metal while categories (a(ii)) and (b(ii)) would result in decreased metal adsorption. Both organic and inorganic ligands can enhance metal adsorption in this way. An organic ligand example of (a(i)) above is provided by glutamic acid which increases Cu^{2+} adsorption on amorphous iron oxide (Davis and Leckie, 1978). Not all organic ligands promote metal adsorption. Farrah and Pickering (1977) found that EDTA prevented the adsorption of Cd on clay minerals over a wide pH range. Haas and Horowitz (1986) found a similar depression of Cd adsorption on kaolinite in the presence of EDTA, but Cd sorption was slightly promoted in the presence of alginic acid and humic acid (see Figure 3.11(e)). Chairidchai and Ritchie (1990) have subsequently shown that Zn sorption in soil is depressed in the presence of several organic ligands found in the rhizosphere. This effect may increase availability of Zn to plants. Phosphate and sulphate complexion with metals at the surface of soil minerals is thought to be an important adsorption process for Cd on Fe and Mn oxides (Benjamin and Leckie, 1982), Zn on Fe and Al oxides (e.g. Bolland et al., 1977; Shuman, 1986) and Cu on allophane (Clark and McBride, 1985). The effect of different levels of soil extractable P on Zn sorption is illustrated in Figure 3.11(f). Since sulphate and phosphate tend to be present in soils in higher concentrations than trace metals, they have the potential, through ternary complex formation and precipitation, of controlling the availability of trace metals to plants.

Metal diffusion at the mineral surface

Metal sorption reactions may not be restricted to the mineral surface, but metal diffusion into the mineral structure can occur. Gerth and Brummer (1983) reported the diffusion of Zn, Ni and Cd into goethite, with the rates of these three metals decreased in the order Ni < Zn < Cd, paralleling their ionic radii of 0.35, 0.37 and 0.49 Å respectively. These authors suggest that metal adsorption is determined by three different steps: surface adsorption, diffusion into the mineral and fixation at positions within the mineral.

Similar metal diffusion processes have been observed for manganese oxides (McKenzie, 1980) and for illite and smectite clays (Gerth, 1985).

3.3.3.3 Metal desorption and lability

The critical issue of whether adsorbed metals can subsequently become desorbed, and hence available for plant uptake, has been studied using ion exchange experiments. Experimental evidence and chemical theory suggest that metal sorption on Fe and Al oxides is an inner sphere complexion and that these metals are unlikely to desorb or be exchanged by cations with no specific

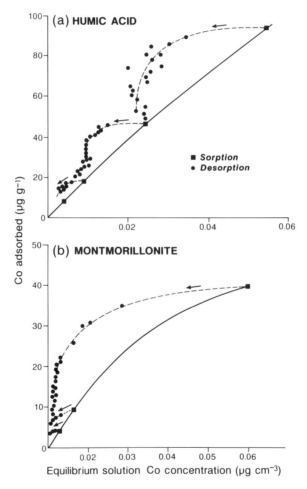

Figure 3.10. Desorption of Co from soil components: sorption (■); desorption (•). Reproduced by permission of Blackwell Scientific Publications from McLaren *et al.* (1986)

affinity for the oxide; only by other metal cations which have an affinity (McBride, 1989), or by H^+. Herms and Brummer (1984), for example, found big increases in the concentration of Zn and Cd in the equilibrium solutions of soils samples with decreasing pH. Clark and McBride (1984) also report that Co^{2+} and Cu^{2+} adsorbed on allophane could not be exchanged by Ca^{2+}, but that Pb^{2+} displaced most of the Cu^{2+}. These and other results indicate that metal sorption on soil minerals is not wholly reversible and sorbed metals are not generally labile. McBride (1989) lists six measures of lability for assessing metal availability:

(i) exchangeability by cations that do not specifically adsorb
(ii) exchangeability by specifically adsorbing cations
(iii) pH reversibility of adsorption
(iv) isotopic exchangeability
(v) desorbability by chelating agents
(vi) dissolution by strong acids

Patterns of metal desorption have generally differed from patterns of metal sorption. McLaren et al. (1986), for example, found that substantial amounts of Co could be desorbed from montmorillonite clay, humic acid and soil oxide, but that the shape of the desorption isotherm differed markedly from that of the adsorption isotherm in each case (Figure 3.10). Barrow (1989) warned that differences in sorption and desorption should alert us to the fact that sorption mechanisms are complicated, often slow and changing in rate over time, not achieving equilibration, and hence extremely difficult to model.

3.3.3.4 Modelling metal adsorption

Ion adsorption processes have generally been described using adsorption isotherms. Selected examples of metal adsorption isotherms are given in Figure 3.11, illustrating the effects of six different soil conditions: (a) pH, (b) the order of metal selectivity, (c) the presence of competing metal cations, (d) metal ionic strength, (e) presence of organic ligands, and (f) the presence of an inorganic ligand (phosphate). There are two types of mathematical descriptions of isotherms: equilibrium models and kinetic models. The linear Langmuir (e.g. Figure 3.11(d)) and Freundlich (e.g. Figure 3.11(f)) models are the most widely used equilibrium models. Kinetic models include first-, second- and nth-order reversible and irreversible reactions. All of these approaches are single-reaction or single-site models. Since most metal adsorption mechanisms are complex, often involving more than one process at the adsorbing surface and often different time-dependent steps, attempts to describe patterns of adsorption using simple, single-site adsorption isotherms and first-order kinetics have generally been rejected. Amacher et al. (1986) found that none of the simple, single-reaction kinetic models adequately described the kinetics of Cr, Cd and Hg retention for a range of metal

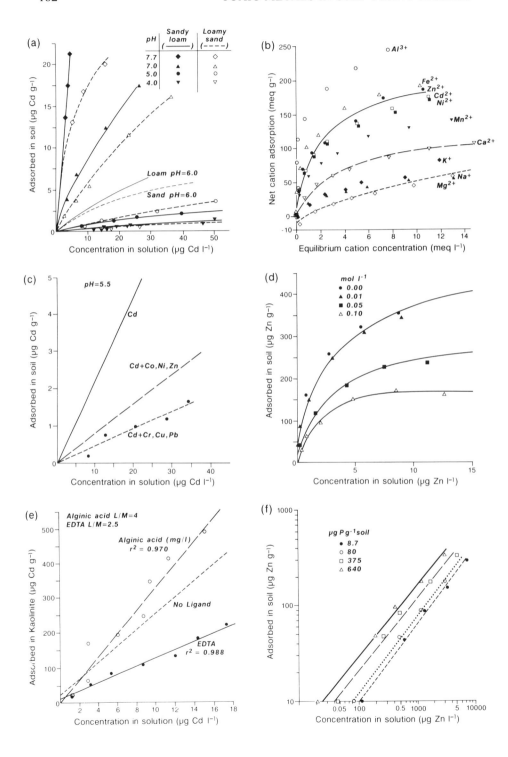

concentrations and soil types. A number of multi-reaction and multi-site models have been developed to more accurately describe the retention, and particularly the slow release, of solutes. A two-site Langmuir equilibrium model, which partitions adsorption into two reaction types or soil fractions, was one of the first multi-site models to be developed (Sposito, 1982). Subsequently, more complex, non-linear, multi-site adsorption models have been developed (e.g. Amacher et al., 1988). A difficulty with all modelling studies is that a successful verification of a model prediction does not provide sufficient evidence to prove that the processes modelled actually occur in the soil. Using the two-site Langmuir model, Sposito (1982) showed that even if the model and data fit perfectly, this does not prove that there are two types of soil reaction sites.

3.3.3.5 Factors influencing metal sorption in soil

A large number of factors have been identified to have an effect on metal sorption. Apart from the types and amounts of soil colloids (clay minerals, soil oxides and organic matter), the principal controls on metal sorption are: pH, bathing solution ionic concentration, metal cation concentration, the presence of competing metal cations and the presence of organic and inorganic ligands. The examples in Figure 3.11(a)–(f) illustrate the effects of several of these factors.

In general, very clear reductions in metal sorption occur as pH is reduced (Figure 3.11(a)). The loamy sand, with a slightly more coarse texture than the sandy loam, also shows relatively lower sorption at all pH values. The order of metal sorption on *Sphagnum* peat in Figure 3.11(b) follows the pattern of lower sorption for small monovalent cations and higher sorption for di- and trivalent metal cations. The competing ions Co, Ni and Zn have a bigger impeding effect on Cd sorption (Figure 3.11(c)) than do Cr, Cu and Pb. While

---

Figure 3.11. Metal adsorption isotherms for a range of soil conditions and metal cations. (a) **Effect of pH**: Cd adsorption isotherms for sandy loam and sand at five different pH values (determined in 10^{-3} M $CaCl_2$. Reproduced by permission of Kluwer Academic Publishers from Christensen (1984). (b) **Cation selectivity**: Metal adsorption isotherms for 10 cations, binding to *Sphagum* peat. Three curves represent best-fit Langmuir equations for the Zn^{2+}, Ca^{2+} and Na^+ data (units in mille-equivalents to allow comparison of metals). Reproduced by permission of Kluwer Academic Publishers from Wieder (1990). (c) **Competing ions**: Cd adsorption isotherms for a sandy loam in the presence of mixtures of Co, Ni, Zn and Cr, Cu and Pb. Reproduced by permission of Kluwer Academic Publishers from Christensen (1987). (d) **Ionic strength**: Zn adsorption isotherms of a clay soil, using four ionic strengths of Zn. Reproduced by permission of the Soil Science Society of America from Shuman (1986). (e) **Presence of organic ligands**: Cd adsorption isotherm for Na-saturated kaolinite with fixed ratios of alginic acid : Cd and EDTA : Cd. L/M = ligand/metal ratio. Reproduced by permission of Kluwer Academic Publishers from Haas and Horowitz (1986). (f) **Presence of phosphate**: Zn adsorption isotherms (with Freundlich equations) for a clay soil with four levels of soil-extractable phosphate (note log × log axes). Reproduced by permission of the Soil Science Society of America from Saeed and Fox (1979)

Christensen (1987) does not given an explanation of this result, it may be due to the fact that Cd sorption in this soil is mainly by cation exchange and that Co, Ni and Zn are better competitors for electrostatic, non-specific sites than are Cu and Pb which predominantly adsorb specifically on Fe/Mn oxides and organic matter. It is important to note that adsorption equilibrations are carried out in a bathing electrolyte solution, representative of ionic concentrations found in the soil solution. In Christensen's (1984) experiments, changing the salt bathing solution concentration from 10^{-3} M to 10^{-2} M $CaCl_2$ resulted in a five to six times decrease in Cd sorption capacity since Ca competed for available adsorption sites. Shuman's (1979) data in Figure 3.11(d) indicate that the influence of higher metal ionic strengths in the soil solution is to depress metal sorption. The effects of organic (EDTA and alginic acid) and inorganic (P) ligands (Figures 3.11(e) and (f)) were discussed earlier (p. 99), when introducing the concept of ternary complex formation during adsorption.

3.3.3.6 Quantifying metal adsorption in soil fractions and in whole soils

For whole soils, there are several approaches to quantifying the contribution of different sorbed phases to trace metal cation sorption:

  (i) sequential extractions using a range of chemical extractants of differing strengths (e.g. Tessier *et al.*, 1979) (see Section 3.4);
 (ii) analysis of statistical relationships between sorption and soil properties (e.g. Christensen, 1989; King, 1988);
(iii) observed similarities in adsorption on single sorbents and in whole soils (e.g. Tiller *et al.*, 1963);
 (iv) comparisons of sorption on natural and treated samples (e.g. Lion *et al.*, 1982).

Each of these techniques has limitations. Some of the problems associated with sequential extractions are discussed in Section 3.4. Statistical approaches suffer from two main problems: (a) the assumption that soil characteristics are independent, when in fact they are generally highly correlated; and (b) the assumption that the independent variable (i.e. the measure of soil properties) is without error, which is impossible (Zachara *et al.*, 1992). These difficulties have limited the number of studies that attempt to aportion soil metal sorption to different soil fractions of whole soils. Instead, more attention has focused on determining the adsorption characteristics of individual, isolated fractions.

There is a vast literature on the metal-adsorbing characteristics of individual soil mineral fractions, particularly the Fe, Mn and Al oxides, the layer silicates and different Fe minerals, such as goethite and haematite. Experimental results, however, frequently provide conflicting evidence. There is quite good agreement, however, between studies of relative metal affinity sequences for soil materials (Table 3.10(b)). These indicate that, in general, Pb and Cu are

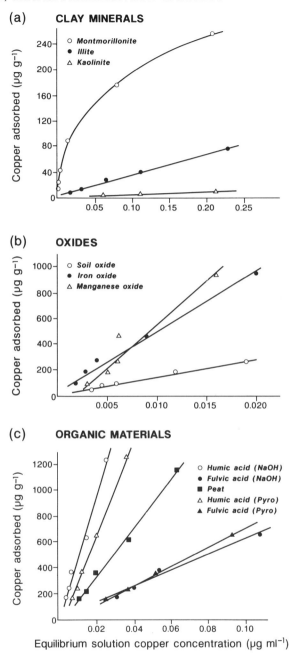

Figure 3.12. Cu adsorption isotherms for different soil materials: (a) clay minerals, (b) soil oxides, (c) organic matter. Reproduced by permission of Blackwell Scientific Publications from McLaren *et al.* (1981)

more strongly retained by many soil colloids than are Cd, Zn and Ni. Cd and Zn have greater cation exchange abilities on layer silicates, but the relative strength of their adsorption on soil oxides and organic fractions (Table 3.10(b)) is less. More than the other divalent metal ions, Cu^{2+} is strongly adsorbed by organic matter and soil oxides. McLaren *et al.* (1981), measuring Cu adsorption on a range of soil clay minerals, oxides and organic matter, found that Cu sorption on Fe and Mn oxides and humic acids could be as much as five to six times higher than on soil clay minerals, while sorption on fulvic acids was only around two times higher (Figure 3.12).

Table 3.11 is an attempt to summarise the salient characteristics of soil-metal adsorption for Cd, Cu, Zn and Pb. Comparing the soil sorption characteristics of these four metals, there are some clear similarities and some clear differences in behaviour. First, the effect of soil pH on metal adsorption is strong for Cd, Zn and Pb, but less so for Cu. The *general* retention pattern for most trace metals in relation to soil pH is similar to that described by Yong *et al.* (1990) for lead:

(i) cation exchange processes at low pH (pH 2–4), with differences due to valence and ionic size,

(ii) formation of soluble hydroxy species at intermediate pH (pH 4–6) which are adsorbed on clay surfaces, and

(iii) when pH exceeds that required for hydroxide formation (pH > 5), precipitation processes dominate.

Ca^{2+} has been shown to be important in inhibiting divalent metal cation sorption for Cd^{2+}, Zn^{2+} and Pb^{2+}. Theoretically, this could reduce metal sorption in calcareous soils, but in practice, the higher pH of such soils elevates metal retention overall and Cd (and possibly other metals) has been shown to be specifically sorbed by calcite (Alloway *et al.*, 1988). The sorption of metals by organic matter is a particularly important process for Cu and for Pb. Cu, Pb and Zn form stable soluble organic chelates in water, and in the soil solution and in sludge-treated soils, Neal and Sposito (1986) found that soluble organic chelates of Cd reduced metal sorption. When soluble sludge organics were removed from soil by washing, Cd adsorption on soil mineral materials was increased.

Summarising the above information would lead us to suggest that Cd and Zn might be more mobile in soils than Cu or Pb since these two ions tend to be sorbed by cation exchange, and competing ions in the soil solution (including H^+, through pH effects) are likely to successfully replace these ions on exchange complexes. Cd and Zn migration and mobility through soil may thus be enhanced by soil acidification. Soils high in Fe, Mn and Al oxides and in organic matter, such as brown forest soils or podsols, could potentially adsorb all divalent metals, but preferentially retain Cu, Pb and, to a lesser extent, Zn. In soils to which organic sludges have been applied, Cu, Pb and Zn would be the most likely organo-metallic soluble chelates in leachate

Table 3.11. Summary of soil adsorption of trace metals

Metal	Major adsorption characteristics
Cd	**Factors influencing:** increased pH = increased Cd sorption[a,b,c] increased CEC = increased Cd sorption[d] (associated with layer silicates)[e] increased OM = increased Cd sorption[d] increased CaCO₃ = increased Cd sorption[f] **Competing cations:** Ca^{2+}, Co^{2+}, Cr^{2+}, Ni^{2+}, Zn^{2+}, Pb^{2+}—can inhibit Cd sorption[g] **Organic Cd complexes:** Cd-humic acid complexes are less stable than Pb or Cu[h]
Cu	**Factors influencing:** pH changes—less effect on Cu sorption at low concentrations than other metal ions[i,j] organic matter and Fe/Mn oxides are the most important controls on Cu sorption[i] clay minerals and CEC are less important for Cu sorption[i] **Competing cations:** Ca^{2+} much less effective at releasing Cu^{2+} into solution than for Cd^{2+}[j] **Organic Cu complexes:** humic and fulvic acids bind Cu^{2+} strongly[k] soluble Cu—chelates are important Cu species in the soil solution[i]
Zn	**Factors influencing:** increased pH = increased Zn sorption[c,l] increased CEC = increased Zn sorption[m] increased clay and soil organic matter = increased Zn sorption[l] **Competing cations:** Ca^{2+} inhibits Zn^{2+} sorption[m] Phosphate enhances Zn sorption on variable charge colloids (Fe/Mn oxides)[n,o] **Organic Zn complexes:** Soluble Zn—fulvates are important Zn species in the soil solution[p]
Pb	**Factors influencing:** increased pH = increased Pb retention, but probably mainly by precipitation as lead carbonate at high pH[q] or as adsorption of the hydrolysed species: $PbOH^+$ at intermediate pH[r] Fe, Mn and Al oxides—all have strong affinity for Pb[s,t] increased CEC = increased Pb sorption at intermediate pHs[r] **Competing cations:** Ca^{2+} inhibits Pb sorption at intermediated pHs[r] **Organic Pb complexes:** Stronger association of Pb with organic matter at high pHs[u] Humic and fulvic—Pb complexes more stable at high pH[v] Soluble Pb-chelates are important Pb species in the soil solution[u]

Sources: [a]Christensen (1984), [b]Eriksson (1989), [c]Harter (1983), [d]Levi-Minzi et al. (1976), [e]Zachara et al. (1992), [f]Alloway et al. (1988), [g]Christensen (1987), [h]Stevenson (1977), [i]McLaren et al. (1981), [j]Cavallaro and McBride (1978), [k]Senesi et al. (1989), [l]Shuman (1985), [m]Kiekens (1990), [n]Bolland et al. (1977), [o]Gerritse and van Driel (1984), [p]Geering and Hodgson (1969), [q]Griffin and Shrimp (1976), [r]Harter (1979), [s]Kinniburgh et al. (1976), [t]McKenzie (1980), [u]Gregson and Alloway (1984), [v]Schnitzer and Skinner (1967).

waters. The *desorption* processes associated with Cd, Pb, Zn and Cu, and their controlling factors, have been less studied than sorption processes. Information on metal *sorption* was sought to allow estimations of acceptable metal loadings to soils. Indications that metal desorption patterns are different from sorption patterns confirms that implying desorption patterns and quantities from studies of metal adsorption would provide an inadequate estimation. Experimentation designed to produce information specifically on metal *desorption* is now required for predicting soil *outputs* and subsequent likely pollution inputs to adjacent watercourses.

3.3.4 ORGANIC COMPLEXATION AND CHELATION

Organic complexation of metals in soils and waters is thought to be one of the most important factors governing solubility and bioavailability of metals in soil–plant systems. It is important to differentiate between naturally occurring organic compounds in soil and those compounds derived from man's activities. Senesi (1992) divides into three main classes the organic compounds in soil which could form metal complexes:

  (i) naturally occurring soil organic molecules of known structure and chemical properties, including aliphatic acids, polysaccharides, amino acids, polyphenols;

 (ii) anthropogenically derived organic chemicals from agriculture, industrial and urban activities;

(iii) humic and fulvic acid substances which accumulate in soil but whose structures are unknown in detail.

By far the largest source of organic matter in soils is the decomposition of plant residues. Soil organic matter (SOM) is an incredibly complex mixture of different organic compounds whose detailed composition is never completely known, partly because of the difficulty in determining the exact molecular configurations of lignin and its aromatic decay products, and partly because organic matter decomposition continually changes its composition. In general, the composition of SOM is dominated by large molecular weight humin and humic acid (HA) compounds and lower molecular weight fulvic acids (FA). Schnitzer and Khan (1972) define humic acids as molecules having molecular weights of around 10 000–2000, and fulvic acids as having molecular weights around 2000–500. Organic compounds that are present in the smallest amounts are those which are broken down easily—the carbohydrates, such as cellulose, hemicelluloses and polysaccharides. Nitrogen-containing compounds, including proteins and amino acids, are also fairly easily decomposed. Plant-protecting compounds such as lignins, tannins and waxes, are broken down only with difficulty and thus remain in soil for longer periods. Apart from studies of simple organic ligands in plant root exudates, studies of organo-metallic complexation in soil have focused on the more persistent HA and FA fractions since these compounds potentially affect the longer term fate of metals in soil–plant systems.

The first studies of metal–organic complexation in soils aimed to elucidate mechanisms of metal, particularly iron and aluminium, migration during podsolisation. The earliest studies implicated the role of polyphenols from tree foliar drip in chelating and mobilising soil Fe and Al (e.g. Bloomfield, 1957; Coulson et al., 1960; Malcolm and McCracken, 1968). A second school of thought, headed by Schitzer and his colleagues in the USA, emphasised the importance of humic and fulvic acids in Fe and Al mobility and translocation (e.g. Schnitzer and Desjardins, 1969). Subsequent investigators of toxic metals and their mobility in soils have analogised from these early studies. After a waning of interest from pedologists studying podsolisation, recently attention has swung back to the metal-chelating properties of fresh litter organics and of smaller molecular weight organic molecules. One of the reasons for this increased interest is the influences of acid rain and leaf litter acidification on metal mobility in soil. McColl and Pohlman (1986), for example, found that organic acids from *Pinus ponderosa* litter leached Al and other metals faster than mineral acids at comparable pH and concentration. They identified oxalic, malic, citric, protocatechuic and salicylic acids as chelating organic acids. In terms of binding and transporting capacity of potentially toxic metals, it is now generally agreed that fulvic acids are the most important organic fraction in soils. Humic acids do have important metal-binding capacity, but their molecular size and configuration means that they are generally less mobile in the soil pore-space and less likely to be leached down the soil profile. Humic acids are thus regarded more as the metal-immobilising fraction of soil organic matter.

It would be wrong to underestimate the importance of organo-metallic complexation in maintaining in solution metal ions which under the more normal pH ranges of soils (around neutrality and in slightly neutral conditions) would normally be converted into insoluble compounds. Enhanced solubility of potentially toxic metals through organic chelation can cause environmental problems in, for example, leachates from disposal sites, or from sewage sludge applications. Soluble organic chelates provide the mobile, transport phase for metal migration, particularly contributing to the contamination of surface and groundwaters. Where organo-metallic complexes are resistant to bio-degradation, their persistence in the environment makes them extremely important in the transport of metals. Francis et al. (1992) confirmed this suggestion even for very small molecular weight citrate–metal complexes which they found to be resistant to bacterial decay.

As well as increasing metal mobility through enhanced solubility, metal chelates may also affect the bioavailability of toxic metals to plants. Piccolo (1989), for example, found that the addition of humic acid to both organic and mineral soils acted to immobilise soluble and exchangeable forms of potentially toxic metals. The effect was more pronounced in mineral soils which had lower organic matter contents at the outset. To quantify amounts of bioavailable metals after HA treatment, Piccolo (1989) determined extractable metals in DTPA (diethylenetriaminepentaacetic acid) extracts. The effectiveness of added HA in decreasing metal extractability followed the order: Pb > Cu > Cd > Ni > Zn. This is more or less the same order as the stability constants of humic substances: Cu > Pb ≫ Cd (Takamatsu and Yoshida, 1978). Since added organic substances influence both metal mobility and bioavailability, and the effects may be different, depending on the original amounts and fractions of soil organic matter, the additions of sewage sludges and other organic slurries and leachates to soils should be planned with great care and with suitable attention to *in situ* soil conditions.

Soluble metal chelates can also reduce the concentrations of toxic metal ions in solution. Much attention has been focused on organo-Al complexes in soil since these are likely to limit transport of toxic Al^{3+} to groundwaters and adjacent watercourses. Bloom *et al.* (1979) has shown that the chelation of Al^{3+} by humic acids in acid soils controls the soil solution concentration of toxic Al^{3+}. Similar results have been reported by Hue *et al.* (1986) who showed that Al^{3+} chelation with short chain carboxylic acids, such as citric, oxalic and tartaric acids, is an important Al detoxifying process in acid subsoils where deep roots would otherwise suffer Al^{3+} toxicity. The complexing of Al^{3+} with HA in acid topsoils has even been shown to reduce the need for liming (Hoyt, 1977).

As with inorganic ligands the classification of metals according to Pearson (1968 *a,b*) and Nieboer and Richardson (1980; see Table 3.1) indicates their organic ligand complexing tendencies. Senesi (1992) has tabulated the preferred electrostatic interactions of class A, class B and intermediate metals (Table 3.12). From this table it can be predicted that most potentially toxic metals, classified as borderline metals by Nieboer and Richardson, will have a stronger tendency to form organic complexes with N-containing amide and amino ligands: Mn^{2+}, Zn^{2+} and Ni^{2+} show a tendency to bind also with sulphur ligands. Borderline metals can also bind with carboxylic, alcoholic and phenolic-OH ligands but their ability to compete for these binding sites with class A metals depends on the relative concentrations of metals in the bathing solution.

The selectivity and strength of metal binding by humic acids depends on the amount of metals available for binding and the amount of metal bound. Davies *et al.* (1969) found that the strength of binding of Cu^{2+} to HA from a sedge peat increased as the Cu content decreased. This trend was confirmed by Zunino *et al.* (1979) who found that the strength of Zn^{2+} binding to the COOH and phenolic-OH functional groups of synthetic FA- and HA-like polymers was increased at lower metal concentration.

Although the exact structure of humic and fulvic acids is unknown, the bonding associated with their large number of functional groups can be studied and modelled. The amounts of metals associated with HA and FA

Table 3.12. Major organic complexing functional groups in soil, classified according to their preference for class A, class B and intermediate metals

Ligands preferred by class A metals	Ligands preferred by intermediate metals	Ligands preferred by class B metals	
Carboxylic	Amino groups	Sulphydril,	Sulphide
$-C\overset{\displaystyle O}{\underset{\displaystyle O^-}{\diagup\!\!\!\!\diagdown}}$	$NH_2\!=\!NH\!\equiv\!N$	$-SH$	$-S^-$
Alcoholic	Amide	Disulphide,	Thioether
$-OH^-$	$-NH-C\overset{\diagup\!\!\!\!/\,O}{\diagdown}$	$-S-S-$	$-S-$
Phenolic			
⟨O⟩$-OH$			
Carbonyl			
$\overset{\diagdown}{\underset{\diagup}{}}C=$			
Phosphate			
$-O-\overset{\displaystyle O}{\underset{\displaystyle OH}{\overset{\|}{P}}}-O^-$ and $-O-\overset{\displaystyle O}{\underset{\displaystyle O^-}{\overset{\|}{P}}}-OH$			
Sulphate			
$-O-SO_3^-$			

fractions in soils have also been quantified. The major functional sites for metal bonding are oxygen-containing ligands, including carboxyl, phenol, alcohol and carbonyl groups. The configuration of functional groups, such as carboxyl groups, on the molecule also appears to be important in metal binding. Two adjacent carboxyl groups on an *aliphatic* molecular structure, and phthalate and salicylate configurations on an *aromatic* molecular structure are thought to be three of the most important. Typical metal complexing configurations are illustrated in Figure 3.13. Exactly how complexation occurs will depend on the electronic status of the ligand site, the associated aliphatic chains, aromatic rings and geometry of the functional site. The environment around the complexing site will also affect the process, particularly pH, ionic strength of the bathing solution and the metal species

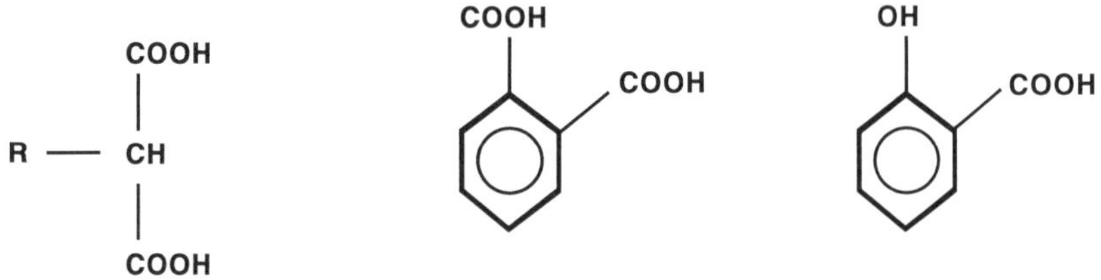

COOH
|
R —— CH
|
COOH

(two adjacent carboxyls
on an aliphatic chain)

COOH
|
⬡ , COOH

Phthalate-type
(two adjacent carboxyls
on an aromatic ring)

OH
|
⬡ , COOH

Salicylate-type
(both carboxyl and
hydroxyl groups on an
aromatic ring)

Figure 3.13. Functional group configurations important in metal bonding by fulvic acid. Reproduced by permission of Routledge and Kegan Paul from Ross (1989)

taking part in the process. Intra- and intermolecular bonding can also confer a "gel"-like condition on metal complexes (e.g. Van Dijk, 1971), making them less mobile in the soil.

Two main branches of investigations into metal–organic matter complexation developed from these early studies: (i) mathematical modelling, based on simple assumptions about the main types of binding sites; and (ii) development of spectrophotometric techniques for studying the structure and geometry of organic matter fractions and their associated functional binding sites.

3.3.4.1 Modelling metal binding on organic matter

The modelling of metal binding to soil organic matter has concentrated on the humic and fulvic acid fractions. The general approach in modelling has been to predict binding and mobility of organo-metallic complexes in environmental systems rather than to gain further understanding about the detailed chemistry of the organic matter or the binding mechanisms. Two main types of models have been used for this purpose: (i) those identifying a small number of discrete binding sites (multi-site models), and (ii) those which use a continuum of binding sites (electrostatic, affinity spectrum or continuous stability function models) (Campbell and Tessier, 1987). Since the detailed structure of FA and HA components of SOM is unknown, the principal modelling approach adopted for simulating and predicting metal complexing on these fractions, and the most successful, has been to assume that metal binding takes place only at major proton-dissociating functional groups: COOH and phenolic-OH, for example. Murray and Linder (1983) used a random molecular simulation model to predict the concentration of metal-binding sites per unit mass of fulvic acid. Developing their model to simulate the binding of Mg^{2+}, Ca^{2+}, Mn^{2+}, Zn^{2+}, Cu^{2+} and Fe^{3+}, Murray and Linder (1984) showed that phthalate binding sites were important for all metals except Fe^{3+}, and that the salicylate type sites were the most important binding sites for Fe^{3+}. The affinity of metal ions for fulvic acid as a whole was found to decrease in the order:

$$Fe^{3+} > Cu^{2+} > Zn^{2+} > Mn^{2+} > Ca^{2+} > Mg^{2+}$$

This sequence is very similiar to that generated by Tipping and Hurley (1992) who modelled metal binding on humic substances. Their affinity sequence of metals for undefined "humic substances" followed the order:

$$Cu^{2+} > Pb^{2+} > Zn^{2+} = Ni^{2+} > Co^{2+} > Cd^{2+} > Mn^{2+} > Ca^{2+} > Mg^{2+}$$

These patterns of metal affinity agree very well with that reported in an empirical study of soil fulvic acid with an ionic strength of 0.1 and pH 3 (Schnitzer and Harmsen, 1970):

$$Fe^{3+} > Al^{3+} > Cu^{2+} > Ni^{2+} > Co^{2+} > Pb^{2+} = Ca^{2+} > Mn^{2+} > Mg^{2+}$$

and agrees well with the theory that class B and borderline metals should show stronger binding affinities with soil organic ligands than should class A metals.

3.3.4.2 Spectrophotometric techniques for studying metal binding on organic matter

A large number of spectroscopy techniques have been valuable in elucidating how metal complexation occurs in HA and FA fractions. Most important of these spectroscopic techniques are infrared (IR), electron spin resonance (ESR) and nuclear magnetic resonance (NMR). While the techniques will not be discussed here, some of the major results will be used to illustrate current understanding of complexation processes. It is valuable to review the mechanisms of metal binding to soil organic fractions since these provide information on the activities of functional groups, and on the exchangeability and potential mobility of metals. A comprehensive review of spectrophotometric and other techniques for assessing metal bonding in HA and FA from different environmental origins is given by Senesi (1992).

IR spectroscopy has shown how important COOH groups on HA are in metal complexing. IR results confirmed the earlier work of Schnitzer and Skinner (1965) who chemically blocked the acid carboxyl (COOH) and the phenolic hydroxyl (OH) groups on soil fulvic acid and showed that both ligand types were important sites of metal binding during chelation. Piccolo and Stevenson (1982) found that the quantity of metal ion available for complexing determined the type of binding which occurred on the dissociated COO^- of soil FA and HA fractions. At low levels of Cu^{2+}, covalent bonds are preferentially formed, while at higher metal concentrations, bonding becomes increasingly ionic. Perhaps the biggest insights into these mechanisms of metal and HA ligand binding has come from studies using ESR spectrophotometry.

These techniques have indicated how two "levels" of metal binding are possible:

(i) *inner sphere complexes* in which metals form bonds of covalent character with HA ligands, and both are completely or partially dehydrated; and
(ii) *outer sphere complexes* in which metals are electrostatically attracted to HA functional ligands, and the metal ion remains hydrated.

Very simple examples of these two types of complexation are illustrated in Figure 3.14. Inner sphere complexes are properly called *chelates* since functional groups donate electron pairs to the metal ion. In outer sphere complexation, the ligand does not interact directly with the electrons of the metal ion. Authors have implied from these two types of complexation that metals bound to outer sphere complexes may be more readily available for exchange, and for plant and microbial uptake.

Lakatos *et al.* (1977) has added more detail to understanding of inner and outer sphere binding mechanisms using ESR spectrophotometry. These authors classified HA ligands into three categories, depending on the predominant type of complexation that occurred in experiments with Mn^{2+}, VO^{2+} and Cu^{2+}. Weak complexing groups, including sulphonic, phosphoric, carboxylic and aliphatic groups, formed weak outer sphere coordinations. Strong complexing ligands, including iminodiacetate acetylacetonate, peptide and porphyrin groups, formed inner sphere complexes. Lakatos *et al.* (1977) suggested that carboxyl groups are a third category, which are borderline in their completing abilities, since they tended to form inner sphere complexes

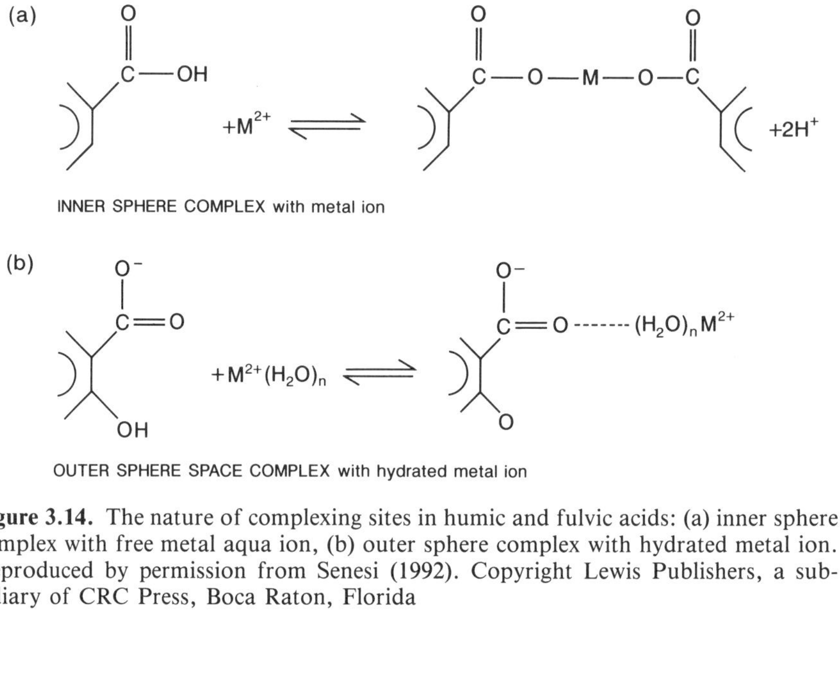

Figure 3.14. The nature of complexing sites in humic and fulvic acids: (a) inner sphere complex with free metal aqua ion, (b) outer sphere complex with hydrated metal ion. Reproduced by permission from Senesi (1992). Copyright Lewis Publishers, a subsidiary of CRC Press, Boca Raton, Florida

with Cu^{2+} and VO^{2+}, but form outer sphere complexes with Mn^{2+} and Fe^{2+}. Since this early work, a very large number of studies have examined the way in which metals bond with the functional groups of humic and fulvic acids. A comprehensive review is given by Senesi (1992) and a selection of example results is given in Table 3.13. In the main, O-containing ligands appear to be the most important groups for metal binding, particularly carboxylic and phenolic-OH. Fe^{2+}, Fe^{3+}, Cu^{2+} and VO^{2+} all tend to form inner sphere complexes. There are conflicting reports of mechanisms for Mn^{2+} complexation, with some authors reporting inner sphere complexes (e.g. Lakatos et al., 1977), and others reporting outer sphere complexes (e.g. Cheshire et al., 1977). McBride (1982) suggests that differences reported for Mn^{2+} binding are probably due to differences in pH and temperature, which control exchangeability between outer and inner complexation. McBride (1982) found no evidence for inner sphere complexing of Mn^{2+} by fulvic acid ligands below pH 8, which, realistically, is the upper pH likely in most soils. This result confirms that under most soil conditions, Mn^{2+} is more likely to have purely electrostatic association in outer sphere complexes, but that stronger bonding of Mn^{2+} can be induced by raising the soil pH.

A general conclusion from studies of organo-metal complexation is that most divalent metal ions, with the exception of Cu^{2+}, Pb^{2+}, Fe^{2+} and VO^{2+}, are bound in outer sphere complexes, with the more tightly bound inner sphere metal ions likely to be in a hydrated condition (Bloom, 1981). Spectrophotometric work on the FA and HA complexation of other potentially toxic metals in soil is in its infancy. Existing work on the above metals suggests that if most potentially toxic metal ions complex with soil and sludge organic fractions in outer sphere complexes, they may be more exchangeable and hence bioavailable than if inner sphere chelates were formed.

An increasingly important source of toxic metals, and of organo-metallic complexes in agricultural soils is in sewage sludge additions. Sewage sludges not only contain elevated levels of potentially toxic metals (see Table 1.6) but also organic constituents whose structure and composition are different from native soil organic matter. The characterisation of sludge organics is a vital precursor to answering important questions about metal binding and metal release, mobility and potential bioavailability during organic matter decomposition. Using a ^{14}C labelling technique, Terry et al. (1979) showed that a large fraction (up to 60%) of the sludge consisted of humic and fulvic acids. Studies of metal complexes in sludges and in sludge-amended soils have focused attention on organic fractions and binding sites which are not generally present in soil in great quantities. In particular, sludge organic matter is high in protein materials, sulphur-containing compounds and phenolic compounds. Boyd et al. (1979) used infrared spectrophotometry to show that amide-N and amide-O sites were important for metal binding in sewage sludges. Senesi et al. (1989) found that aridisols in California, to which composted sewage sludge had been added, had higher HA fractions, with

Table 3.13. Types of metal–organic ligand complexes formed in soil organic fractions

Metal	Organic fraction	Organic ligand type		Inner or outer sphere complex	Source
Fe^{3+}	soil HA, FA	carboxylic, polyphenolic	[O]	inner	Senesi et al. (1977)
Fe^{3+}	soil FA	carboxylic phenol-OH	[O]	inner	Schnitzer and Ghosh (1982)
Cu^{2+}	soil HA	carboxylate, phenolate, carbonyl	[O]	inner	Senesi et al. (1986)
Cu^{2+}	soil, sludge, HA and FA	porphyrin	[N]	inner	Senesi et al. (1985)
Fe^{2+}, Cu^{2+}	leaf litter	—	[O]	inner	Senesi and Sposito (1989)
VO^{2+}	soil HA	carboxylate, phthalate, salicylate, catechol	[O]	inner	Templeton and Chasteen (1980)
Mn^{2+}	peat HA	carboxylate, phenol-OH, carbonyl	[O]	inner	Lakatos et al. (1977)
Mn^{2+}	soil HA, FA	carboxylate, phenolate	[O]	outer	Cheshire et al. (1977)
Mn^{2+}	leaf litter	—	[O]	outer	Senesi and Sposito (1989)

higher S and N contents, lower C:N ratios, higher phenolic-OH content and higher intensities of aside infrared adsorption bands. Senesi *et al.* (1989) also found that metals derived from the sludge were selectively adsorbed onto soil HA, with Cu, Ni, Zn, Fe and Cr strongly adsorbed in preference to Mn, V, Ti and Mo. In the same study, Senesi *et al.* found that the effects of sludge application on soil organic matter, and especially the HA fraction, were not long lasting and had more or less disappeared after 14 months. On the other hand, metals added in sludges became strongly adsorbed onto soil HA and both total and NaOH-extractable metal content of sludge amended soils increased with increasing sludge addition. These results indicate the cumulative effects that repeated applications of sewage sludges can have on soil metal retention, with possible future toxicity problems as adsorption thresholds are exceeded. This is especially true for strongly adsorbed metals such as Cu and Ni (Senesi *et al.*, 1989). There is also evidence that sludge application to soil increases the water-soluble concentrations of metals, organic ligands and metal–ligand complexes (e.g. Baham and Sposito, 1983). Care must be taken to prevent groundwater and surface water contamination by these compounds, by limiting both quantity and frequency of sludge addition to coarse textured soils, and by curtailing applications in wet conditions.

3.3.4.3 Mobility of HA and FA fractions and their metal chelates

Three characteristics of fulvic and humic acids determine their mobility in soil: (i) viscosity in the soil solution, (ii) size of molecule, and (iii) shape of molecule. The relative viscosity can be defined as the ratio of the solution viscosity to that of pure water. Viscosity is influenced by changes to the molecular configuration of the polymer, and hence is influenced by metal binding. Adding an electrolyte to a humic acid solution causes an increase in viscosity. This is due to ionisation occurring on the polymer and resulting in molecular expansion due to repulsion by similarly charged functional groups along the polymer chain (Hayes and Swift, 1978). The shape and size of FA and HA complexes affects their mobility in the soil pore-space because frictional forces act on the macromolecular surfaces. Molecular chain branching, which is thought to occur in humic acids, results in a higher molecular density and a molecule that is more compact than that of a single linear chain of similar molecular weight. Thus, many-branched polymer complexes are likely to have smaller frictional surfaces and may be more mobile (Cameron *et al.*, 1972).

If the complexation by metals affects the size and shape of HA molecules, HA mobility in soil may be altered. Richie and Posner (1982) used Al, Fe and La to study the effects of metal binding and pH on the transport of HA in soil. Binding with Fe, Al and La significantly increased the molecular weight of both low and high molecular weight organic fractions, while increasing pH significantly decreased molecular weight of both size fractions. Richie and Posner, however, suggest that it is not only the type of metal cation, the degree of dissociation or the molecular weight of the organic molecule which determines solubility and mobility of humic complexes. Humic acids tend to form insoluble complexes with the metal species M^{3+} regardless of the molecular weight of the HA. As the cation becomes increasingly hydrolysed, the complex that forms is more soluble. The shape of HA complexes was studied by Ghosh and Schnitzer (1981) who found that the flexibility of soil fulvic acid molecule was affected by Cu^{2+} and Fe^{3+} binding.

3.3.5 MOBILITY, LEACHING AND TRANSPORT

The role of soil processes in mobilising and transporting toxic metals is of prime interest in environmental science, given their central position in the hydrological cycle between atmospheric inputs and surface water outputs. The possible regulatory role of soils in preventing the conveyance of metals to groundwaters and potable waters has attracted significant attention. Controls on the solubility, exchange and adsorption processes which determine solid-solution partitioning of metals in soils were discussed earlier. Once soluble metal species are present in the soil solution, their mobility and transport is determined by the same processes as other soil solutes, namely diffusion and mass flow. These processes will be discussed briefly in terms of modelling metal transport in soil.

Four aspects of the leaching of metals in soils will be addressed in this section: (a) the downprofile elution of metals, particularly associated with waste sludge applications and the influence of acid rain and resultant soil acidification, (b) the possibility of metal contamination of groundwaters, and (c) approaches to modelling metal mobility and transport in soils.

3.3.5.1 Downprofile leaching of metals in soils

Many studies have been made of the potential leaching of trace metals in different soils under different ambient conditions. Studies divide into four main types:

   (i) *column leachate studies* on repacked and usually sieved and prepared soils (e.g. Korte *et al.*, 1976; Boyle and Fuller, 1987; Welch and Lund, 1987),
  (ii) *intact soil cores* removed from the field (e.g. Sommers *et al.*, 1979; Miller *et al.*, 1983; Sheppard *et al.*, 1987),
 (iii) *field in situ studies*, often using lysimeters (e.g. Hinesly and Jones, 1977; Parker *et al.*, 1978; Berggren, 1992), and
  (iv) *downprofile soil sampling*, followed by chemical extraction of metals (e.g. Williams *et al.*, 1980; Scokart *et al.*, 1983; Dowdy *et al.*, 1991).

A modified version of approach (iv) is provided by Dreiss (1986) who used *in situ* ceramic cup suction samplers to obtain downprofile soil water samples which she measured for their trace metal contents. Soil leaching studies have been used to examine two main types of potential metal pollution problems: the transport of metals and their speciation (a) with sewage sludge or other waste and leachate applications, or (b) with acid rain leachates.

Metal leaching in soil after sludge applications

A wide range of different agricultural, urban and industrial wastes are added to soil both as a method of disposal and, since they contain organic matter, as a method of soil conditioning for soil structure improvement. Many of these wastes, including sewage sludge, composted refuse, or leachates from solid waste, contain potentially toxic metals. Around 60% of sewage sludge produced in the UK is applied to the land, amounting annually to around 750 000 t of dry matter, or 18 000 000 t wet matter (McGrath, 1987). As legislation changes to restrict dumping at sea, more land application of sewage sludge is likely to occur.

Sludge and wastewater application to soils and croplands, with the potential for metal recycling, and the possibilities of metal migration through soil to groundwaters has heightened public concern. In the European Union, three methods of control of metal additions to soils are: limits on sludge metal concentrations, limits on soil metal concentrations and 10-year metal loading rates (Commission of the European Communities, 1982). Current limits for concentration of metals in sewage sludge applied to soil are: 40 mg Cd kg^{-1}, 400 mg Ni kg^{-1}, 1000 mg Pb kg^{-1}, 3000 mg Zn kg^{-1} and 1500 mg Cu kg^{-1} (Commission of the European Communities, 1982). Environmental agencies have sought methods of predicting the likely soil leaching of sludge metals and bioavailability, and have generally based their guidelines for sludge applications on the cation exchange capacity (CEC) of the soil (e.g. Commission of European Communities, 1982; Giordano, 1986). From the review in Section 3.3.3 on metal sorption in soil, and from statistical studies of soil properties and soil metals (e.g. King, 1988), it is clear that other soil factors that might be more appropriate predictors of metal retention and mobility include soil pH, Fe and Mn oxides and soil organic matter content (e.g. Korte *et al.*, 1976; King, 1988). Korte *et al.* (1976) found that soil texture, surface area, per cent free iron oxides and pH were the best predictors of Cd, Be, Zn and Ni migration through the seven main soil orders of the USA. Inclusion of CEC in Korte *et al.*'s statistical analysis did not improve their predictions. The results of these studies introduce doubt as to whether current environmental guidelines for sludge applications are adequate.

Both field trials and column leaching studies have indicated that metals in sludges tend to accumulate in topsoils (e.g. Parker *et al.*, 1978; Chang *et al.*, 1984*a*; Schirado *et al.*, 1986; Dowdy *et al.*, 1991). Metals applied in sludge rarely migrate to any depth in soil and virtually never reach groundwaters via the process of soil leaching. McGrath (1987), summarising the results of 15 field studies on arable soils treated with sewage sludge, found that in nine studies no metal leaching occurred below 20–30 cm, in 11 studies the metal uptake by plants remained constant over the study period of 3–13 years, and in four studies which reported extractable soil metals, no changes occurred over the study period. These results indicate that for a large range of soil types and sludge application rates in the short to medium term, metals are not actively mobilised and transported below the ploughed layer. McGrath's main concern in reviewing these works was that most metal leaching studies are rather short term, yielding information which may not be very meaningful in the long term. In a more recent field study, reporting the results of 14 years of heavy sludge application (Dowdy et al., 1991), not only were very small amounts of the applied Cd and Zn moved out of topsoils over the period (only 4.1% Cd and 4.8% Zn), but more than 50% of both metals was found to be so strongly adsorbed that it could not be extracted with 4 M nitric acid. Dowdy et al. (1991) suggested that sludge metals may even have diffused into soil minerals and clay lattices over the application period. Soil and sludge characteristics are important factors in controlling the retention and mobility of sludge metals. In particular, metal mobility might be enhanced if sludges add soluble organic matter or reduce the soil pH. Neal and Sposito (1986), for example, showed that the soluble organic matter in sewage sludge could reduce soil Cd adsorption and hence increase the mobility of Cd in sludge treated soil. However, using ^{109}Cd labelled leachates, Cline and O'Connor (1984) found that soils freshly amended with sludge showed only slightly reduced abilities to adsorb further additions of Cd. Welch and Lund (1987) examined the influence of soil moisture conditions on metal mobility after sewage sludge additions. In a 13 month study they found that less Ni leaching occurred under saturated conditions (3.3% of applied Ni) than under unsaturated conditions (10.7%). For other potentially toxic metals the effect of high soil moisture contents on mobility would undoubtedly be different, particularly for Mn, whose mobility is strongly influenced by soil redox condition.

The results of many studies using different experimental techniques indicate that metals in sewage sludge and other urban wastes do not leach significantly in soil, but instead accumulate in the topsoil. This highlights a second hydrological route which is more likely to transport sludge metals to adjacent watercourses: overland flow and soil erosion, particularly if these processes occur soon after sludge applications. Such processes transport soil and sludge, and both soluble and particulate forms of metals to surface waters. McGrath and Lane (1989) have suggested that lateral movement of metal-contaminated soil materials as a result of tillage during a long-term sludge treatment experiment may account for significant removal of metals from the site of application. Soil erosion processes have operated extensively in the past, by transporting metal-contaminated sediments into rivers. This can be seen from the high concentrations of metals in alluvial river terraces adjacent to areas of past metal mining activities (e.g. Macklin and Smith, 1990).

Metal leaching in soil as a result of acid rain and soil acidification

Evidence of desorption and increased solubility of metals with decreasing pH has caused speculation that acid rain and resultant soil acidification would increase metal mobilities in soil and cause water pollution problems and enhanced uptake of metals by plants. Much of this speculation has centred on geographical regions subject to aerial pollution inputs, including the north-east USA and Canada, Scandinavia and north-west and central Europe. The greatest body of evidence for metal leaching is for aluminium, particularly in already acidic forest soils (e.g. Matzner *et al.*, 1986; Bergkvist *et al.*, 1989). Mineral soils are rich in aluminosilicate clays, which provide a source of mobilisable Al at pH values below 5. The fine roots of plants and mycorrhizae are adversely affected by high levels of inorganic forms of Al, including Al^{3+}, $AlOH^{2+}$, $Al(OH)_2^+$, $Al(OH)_3$, and $AlSO_4^+$ (Foy, 1984; Andersson, 1988). Al toxicities are exacerbated in soils which are low in Ca. In soils with organic matter and organic acids, Al complexation occurs and chelates can be less mobile and less bioavailable than free ions (Bartlett and Riego, 1972). While some authors suggest that the cause of forest declines in areas suffering acid rain inputs is Al toxicity (e.g. Ulrich *et al.*, 1988), others suggest that it is due to metal deficiencies (Ca, Mg, K and Zn) caused by excess leaching as a result of soil acidification (e.g. Zottl and Huttl, 1986). A seasonality of Al leaching was reported by Mulder *et al.* (1987), working in deciduous woodland soils in the Netherlands, who found highest soil solution Al concentrations in summer, coinciding with highest SiO_2 and NO_3-N leaching rates. Coincidence of high Al and SiO_2 concentrations indicated congruent dissolution of the vermiculite and smectite clay minerals. The coincidence of high Al and NO_3-N concentrations has been attributed to increased mobilisation of Al due to nitrification of NH_4-N and organic N (van Breemen and Jordens, 1983), since these are acidifying processes, generating hydrogen ions:

$$NH_4^+ + 2O_2 \geqslant NO_3^- + H_2O + 2H^+$$

While highest concentrations of soil solution Al were recorded in summer, greatest Al transport through soils was observed in winter when soil water fluxes were highest (Mulder *et al.*, 1987).

For metals other than Al, significant increases in leaching of Mg, Ca, Mn, Zn and Cd have been observed with decreases in soil pH (e.g. Campbell *et al.*, 1983; Bergkvist, 1986; LaZerte, 1986). Campbell *et al.* (1983) ranked 18 metals according to their geochemical mobility in environments subjected to acid rain. They reported the following order of mobility with increasing acidity:

Al, Mn, Zn (high) > Cd, Co, Cu, Ni > Pb, V (low)

with insufficient evidence to rank Ag, As, Be, Hg, Mo, Se, Sn, Te and Tl. They divided metals mobilised under acid conditions into two groups: (i) those showing significantly increased *solubility* with pH decrease (Al, Mn), and

(ii) those metals whose speciation is sensitive to pH changes, particularly in the presence of organic ligands (Al, Be, Cu). Bergkvist *et al.* (1989) identified a similar classification of metals in terms of their mobility in soil, with increased leaching of Ca, Cd, Mg, Mn, Na, Ni, Zn and Al being associated with increasing acidity, and the leaching of Cr, Cu, Fe, Pb and Al being associated with solubility of organic matter. In summary, acidification of soils may be expected to mobilise and enhance leaching of Al, Mn, Zn and perhaps Cd and Ni, but not Cu and Pb, particularly if these metals are strongly associated with soil organic matter.

The above general patterns are by no means the only responses to acid rain. Miller *et al.* (1983), for example, working on sandy forest soils in east Chicago, found that even soil leachates simulating 10 years of acid rain did not significantly mobilise soil trace metals that had accumulated in the upper 5 cm from aerial pollution. Berggren (1992) studied the differing responses of Al and Cd to acidification in two different Swedish forest soil types: podsols and cambisols. In podsols, large quantities of organic matter released from the surface mor horizon, formed organic chelates of Al and Cd and enhanced the leaching of these metals in the E horizon. Organic matter was not a major control on metal mobility in cambisols. Berggren (1992) concluded that pH was the most important factor determining the leaching of Al in both soils, but that soluble organic matter was more important for Cd. Dumontet *et al.* (1990) studied the distribution of metals downprofile in acid peat near a smelter in Quebec. As with all other aerial pollution studies, the highest accumulations of metals were in the immediate surface soil horizon. Even at pH values as low as 2.9–4.4, little or no migration of Cu, Zn, Ni, Cd or Pb was observed, even with extremely high metal concentrations in the peat (5525 μg Cu g^{-1}; 884 μg Zn g^{-1}; 1617 μg Pb g^{-1}).

3.3.5.2 Possibilities of groundwater contamination by metals leached through soils

Studies of sludge-amended soils indicate that sludge metals are unlikely to leach through soils and into groundwaters even in the long term. Even where liquid wastes, such as landfill leachate (e.g. Fuller *et al.*, 1976), industrial metal leachate (e.g. Dreiss, 1986), or simulated waste leachates (e.g. Biddappa *et al.*, 1982) have been applied to test soils in very high quantities with high metal loadings, metals have not been significantly leached. It is possible to identify soil and sludge conditions which could be most at risk for metal leaching. Such soils might be coarse textured and lacking in organic matter. On the basis of existing field and experimental evidence, more attention should be paid to preventing overland routes of surface water contamination via water runoff and both soil and sludge erosion.

3.3.5.3 Modelling metal mobility and transport in soils

A number of modelling approaches, of varying degrees of complexity, have been used to predict mobility and transport of metals in soils. All approaches are based on modifications of existing soil solute transport models. Introductions to soil solute transport and its modelling are given by Wild (1981), Smettem (1986) and Ross (1989) and an excellent review of types of soil solute transport models is given by Addiscott and Wagenet (1985). Solute transport models are based on a combination of soil water transport equations and chemical transport equations, based on the solute processes of mass flow (or convection) and diffusion. Mass flow describes the movement of dissolved species in the soil water, where the rate of flow is driven by the hydraulic gradient. Rates of solute diffusion, on the other hand, are unique for each species and are determined by solute characteristics and solute concentration gradients. Difficulties in modelling realistically solute transport processes lie in failure to account properly for the immense complexity of field soils. In addition, modifications are required to use solute transport models for predicting metal mobility, particularly the inclusion of suitable parameters for adsorption, ion exchange, precipitation and chelation processes. As we have seen earlier, the exact operation of some of these processes is not understood. We have also seen that very little is known about the kinetics of these processes. For these reasons, simplified "black-box" or empirical approaches must be used to represent these processes when testing metal transport models.

Despite the above difficulties, models of differing scale and complexity have been developed for predicting metal transport in soil and in soil systems. At the small scale of predicting ion migration in soil cores, Sheppard and Sheppard (1987) and Amoozegar-Fard et al. (1984) have achieved some success in simulating the movement of the metal ions Cd, Ni and Zn and the nucleotides U, Cr, Fe and Tc respectively. van Riemsdijk and van der Zee (1989) modelled the enhanced mobility of metals in the presence of organic ligands, including EDTA. To account for strong metal retention in sludge-treated forest soils, Sidle et al. (1977) used the one-dimensional solute transport model of van Genuchten and Wierenga (1974), incorporating Freundlich isotherms to characterise the adsorption of Cu, Zn and Cd. A recent model which attempts to account for many of the aforementioned problems, is based on long-term studies of the fate of Cd in agricultural soils of the Netherlands (Harmsen, 1992). This model pays particular attention to specific and non-specific adsorption of metals. Harmsen (1992) uses the model to predict the effects on soil Cd accumulation of increased annual Cd inputs in aerial deposition, inorganic fertilisers and organic manures and sludges. His results are long-term predictions, which suggest that taking land out of agricultural production could result in increased Cd mobility. Larger scale modelling approaches have been used to simulate metal transport in whole landscape systems. These include the model of Sanden (1991), simulating metal transport from old mine tailings into the adjacent watercourse of a whole catchment, and that of Wagenet and Grenney (1983) simulating the aerial transport, deposition and soil/water transport of metals from power generating plants.

Modelling approaches can be designed either for prediction, or to help in understanding how component processes of a system function. In the case of metal transport modelling, the first objective—prediction—has been the primary aim of modelling projects to date. In the case of sludge applications to soils, modelling the retention or mobility of metals provides a management tool for making decisions about sludge disposal. More recent experimental studies, such as the use of breakthrough curve (BTC) experiments and model simulations for studying the adsorption characteristics of particular metal ions under controlled soil conditions (e.g. Selim *et al.*, 1992) are exploiting the process-elucidation aspect of modelling.

3.4 EXPERIMENTAL METHODS FOR DETERMINING METAL SPECIATION AND METAL FRACTIONS IN SOILS

Previous sections in this review have indicated that metals are present in soils in an extremely large range of forms. A vast amount of effort has been expended in attempting to quantify metals held in different soil fractions, particularly those fractions thought to be mobile and bioavailable, since these fractions can potentially leach to pollute groundwaters or can pass through food chains from plant uptake. Several types of analysis have been used in determining the *physical* and the *chemical* characteristics of metals and their complexes in soils. Since the type of metal complex present in waters and in the soil solution is thought to be related to particle size, a range of techniques, including ultrafiltration, dialysis, gel permeation chromatography, liquid chromatography and ion-exchange chromatography have been used to determine these different fractions in solution. The most popular and simplest physical separation technique used on soil extracts is to filter solutions through micropore membranes of pore size 0.45 μm, thus rather crudely differentiating between "soluble" and "particulate" metals. The electronic structure of metal complexes, particularly organo-metallic complexes, has been examined using many different spectroscopy techniques, including UV-visible, infrared, electron spin resonance (ESR) and nuclear magnetic resonance (NMR). A review of these procedures and discussion of their value in understanding soil organo-metal interactions is given by Senesi (1992). Various chemical properties of metals and metal complexes have been examined, including adsorption isotherms and ion exchange properties. Anodic stripping voltametry (ASV) has been used to determine electrochemical reactivity of various metal species, but this technique does not differentiate between free metal aqua ions and soluble inorganic and organic ligands (Sposito, 1983). Various chromatographic separation techniques have also been applied to the fractionation of soil metals (a useful review of these techniques is given by Cottenie *et al.*, 1979).

It is possible to summarise the principal forms of mobile and mobilisable toxic metals in soil (Berrow and Burridge, 1980) as:

(i) in the *soil solution*: ionic, molecular, chelated and colloidal forms;
(ii) at the *exchange interface*: readily exchangeable ions in inorganic or organic fractions;
(iii) in *adsorption complexes*: more firmly bound ions;
(iv) incorporated in *precipitated* sesquioxides and *insoluble* salts; and
(v) fixed in *crystal lattices* of secondary clay minerals.

A scheme to illustrate the potential availability of these fractions is illustrated in Figure 3.15. In order to quantify these fractions and to assess the mobility and bioavailability of different fractions, soil scientists, ecologists and botanists have attempted to extract these fractions using a range of different chemical extractants. These relatively simple approaches have generally been preferred, rather than the direct and frequently highly technical determinations such as ESR and ASV. In addition, correlations between extractable fractions and uptake by test plants have been used to estimate bioavailability of soil metals. Despite the apparent elegance of such schemes, there are several major difficulties with sequential extraction procedures which will be discussed in Section 3.4.1.

In addition to conceptual problems with extraction procedures, all studies of metals in waters and soil solutions suffer from two major problems which must be appreciated and accounted for when designing and reporting research:

(i) finding suitable techniques for detecting extremely low soluble concentrations; and
(ii) differentiating between free metal ions in solution and soluble organo-metallic chelates.

In too many studies, only passing reference is made to *ad hoc* concentration techniques used to bring solutions up to detection limits of available instrumentation, and frequently no reference is made at all to the fact that chelated metals or even ion complexes may be measured along with free metal cations in solution.

3.4.1 CHEMICAL EXTRACTANT TECHNIQUES FOR ASSESSING METAL FRACTIONS IN SOILS

A vast body of literature exists which describes experiments designed to characterise the different fractions of metals in soil. One purpose of such work is to assess the different bonding or retention strengths of toxic metals present

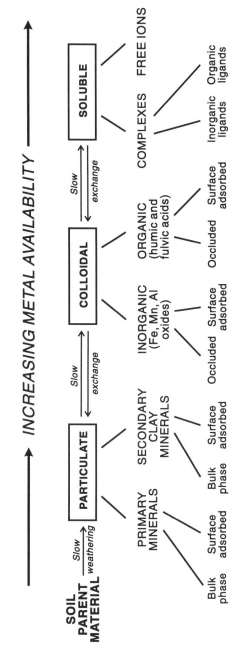

Figure 3.15. Forms of potentially toxic metals in soils

in soil, and hence to imply their potential mobility and bioavailability. The most widely used approach is to choose a chemical extractant or a series of extractants which are thought to remove particular chemical phases of metals in the soil, especially extractable metal fractions which correlate well with amounts of metals taken up by plants grown in the test soil. These studies aim to predict the bioavailability of toxic metals in soils, particularly polluted sites or soils treated with sewage or industrial sludges. It is important to determine the likely plant uptake of metals under these circumstances to prevent contamination up the food chain into crops and perhaps into animals or humans. Of equal importance would be an ability to predict the likely mobility of metals in soil systems, since this determines their transport in hydrological systems and their transfer into streams, surface water bodies and groundwaters. Once present in watercourses, metals can be taken up by freshwater plants and animals and passed up the freshwater food chain, or can contaminate animal or human drinking water supplies. Earlier studies have suggested a general correlation between mobile and bioavailable metal fractions but this relationship requires confirmation for different metals in soil.

Excellent reviews of extraction techniques used for determination of soil metal fractions are given by Lake *et al.* (1984) and Beckett (1989). Single chemical extractions are generally used to determine "available" amounts of soil metals and usually aim to extract the water-soluble, easily exchangeable and some of the organically bound metals. Other metal fractions may become available over time through chemical weathering or organic matter decomposition, but metals occluded by stable secondary minerals may not become available in the short or medium term. Moderately dilute and weak acids as well as synthetic chelating agents, such as EDTA, have been used to determine "available" metals in soil (see Section 3.4.2). More commonly used to quantify the different fractions of metal retention in soils are sequences of different chemical extractions, usually starting with the weakest, least aggressive and ending with the strongest and most aggressive. Sequential extractants are generally used to characterise at least five or six fractions of toxic metals in soils. The earliest, weakest extractants in the system are most specific, the later, stronger extractants the least specific, but because they come late in the sequence, there may remain only one or two groups of compounds that they can dissolve (Beckett, 1989). A very large number of sequential extraction schemes have been used for soils, generally attempting to identify metals held in any of nine (or more) fractions:

   (i) soluble
  (ii) extractable (non-selectively sorbed)
 (iii) adsorbed
  (iv) organically bound
   (v) Mn oxide occluded

Table 3.14. Four principal sequential extraction schemes for assessing metal fractionation in soils. Numbers refer to order of each stage in the extraction scheme

Fraction	Tessier et al. (1979)	Miller et al. (1986)	Sposito et al. (1982)	McLaren and Crawford (1973)
Soluble		**1** H_2O		
Exchangeable	**1** 1 M $MgCl_2$	**2** 0.5 M $Ca(NO_3)_2$	**1** 0.5 M KNO_3	**1** 0.05 M $CaCl_2$
Acid Soluble		**4** HOAc 0.1 M $Ca(NO_3)_2$		
Adsorbed		**3** $Pb(NO_3)_2$	**2** H_2O	**2** 2.5% HOAc
Organic	**4** 0.02 M HNO_3 30% H_2O_2 3.2 M NH_4OAc	**5** 0.1 M $K_4P_2O_7$	**3** 0.5 M NaOH	**3** 1 M $K_4P_2O_7$
Fe Mn-Oxide	**3** 0.04 M $NH_2OH.HCl$ in 25% HOAc			
Mn Oxide		**6** 0.01 M $NH_2OH.HCl$ + 0.1 M HNO_3		
Fe Oxide		**7** Ammonium Oxalate Acid in UV light		**4** Ammonium Oxalate Acid
Carbonate	**2** 1 M NaOAc		**4** 0.05 M Na_2–EDTA	
Residual	**5** $2 \times 70\%$ HNO_3 40% $HF/72\%$ $HClO_4$	**8** HNO_3 + HF	**5** 4 M HNO_3	**5** Conc. HF

(vi) amorphous Fe oxide occluded
(vii) crystalline Fe oxide occluded
(viii) carbonate
(ix) residual (total)

The chemical extractant chosen for quantifying each metal fraction is ordered into a sequence, but this requires consideration of selectivity in order to avoid, or at least minimise, solubilisation of multiple fractions in one extract (Miller et al., 1986). Possible approaches for the evaluation of extractant selectivity are given by Campbell and Tessier (1987):

1. *use of pure solids*:
   — alone
   — in model soils
   — spiked into natural soils
2. *analysis of extracts and/or residual soils for various correlated properties*:
   — organic C and/or inorganic C
   — total S and/or acid volatile sulphide
   — Al, Si
3. *successive extractions with the same extractant*

Disappointingly few evaluations of this type have been published. The most commonly used schemes are based on that designed by Tessier et al. (1979), which was actually designed for extracting heavy metals from contaminated river sediments. This technique together with three of the most commonly used sequential extraction schemes are given in Table 3.14. Despite the apparent simplicity of these schemes, a vast range of alternative sequential chemical extractants have been published for different soils and toxic metal problems, a selection of which are summarised in Table 3.15. It is very clear from this list that there have been some differences of opinion concerning appropriate use of extractants. However, across all the cited studies, there is general agreement about the *kinds* of extractants needed for each fraction. For example, the soluble and easily exchangeable fractions are commonly extracted using dilute salt solutions of replacing cations, such as $MgCl_2$, $CaCl_2$, NH_4NO_3 or NH_4-acetate. Organically bound fractions are released using oxidising agents such as pyrophosphate or hydrogen peroxide or, in the case of Kuo et al. (1983) and Shuman and Hargrove (1985), sodium perchloride. Reducing agents such as ammonium oxalate has been widely used to release metals bound in Fe and Mn oxides, while strong, concentrated or boiling nitric acid, with or without hydrofluoric acid or perchloric acid, have been used to assess *residual* or occluded metals in soils. While some authors' objective has been to attempt characterisation of as many soil fractions as possible without omitting any metal forms (e.g. Tessier et al., 1979; Salomons and Forstner, 1984; Xian and Shokohifard, 1989), others have specialised in the detailed extraction and characterisation of a "single" fraction, such as the

Table 3.15. Sequential extraction techniques employed to fractionate trace metals in soils

Metal extracted	Chemical form extracted								Source
	Soluble	Extractable (exchangeable) (non-specific)	Adsorbed (specific)	Organically bound	Fe and Mn oxides	Carbonate (precipitated)	Occluded	Residual	
Cu		$CaCl_2$	CH_3COOH	$K_4P_2O_7$			Cu-oxalate	HF	McClaren & Crawford (1973)
Cd	Deionized H_2O	$CaCl_2$		$K_4P_2O_7$			$HONH_2HCl$	Conc HNO_3	Alloway et al. (1979)
Cd Cu Ni Zn		$MgCl_2$		HNO_3/H_2O	Hydroxylamine hydrochloride + CH_3COOH	$NaCOOH$		digest HNO_3	Tessier et al. (1979)
Cd Cu Ni Zn	Saturated extract	KNO_3		$NaOH$		Na—EDTA		HNO_3	Emmerich et al. (1982)
Cd Ni Zn		$CH_3CO_2NH_4$		$(CH_3CO_2)_2Cu$			HNO_3		Soon & Bates (1982)
Cd Cu Ni Zn		$CH_3CO_2NH_4$		$(CH_3CO_2)_2Cu$	$(NH_4)_2C_2O_4$		CBD[a]		Hickey & Kittrick (1984)
Cd Cu Ni Zn		$CH_3CO_2NH_4$		$(CH_3CO_2)_2Cu$	$(NH_4)_2C_2O_4$		CBD[a]		Mandal & Mandal (1987)
Cd Cu Pb Zn		NH_4NO_3		NH_4—EDTA			boiling HNO_3		Davies et al. (1987)
Cd Zn Pb		$MgCl_2$		HNO_3/H_2O_2	NH_2OH/HCl/CH_3COOH	$NaCOOH$		$HNO_3 + HClO_4$	Xian & Shokohifard (1989)
Cu Zn Cd Fe Mn		$MgCl_2$		Na OCl	oxalate			$HNO_3 + HClO_4$	Kuo et al. (1983)
Mn Cu Fe Zn		$Mg(NO_3)_2$		Na OCl	3 fractions[b]			$HF + HNO_3 + HCl$	Shuman & Hargrove (1985)
Cu Mn Fe	H_2O	$Ca(NO_3)_2$	$Pb(NO_3)_2$	$K_4P_2O_7$	3 fractions[b]			$HF + HNO_3$	Miller et al. (1985)
Ni Cu Cd Zn Pb		$H_2O + KNO_3$		$NaOH$		EDTA	HNO_3		Sposito et al. (1982)
Zn	$H_2O + KNO_3$		Na OH		DTPA	EDTA	HNO_3		LeClaire et al. (1984)
Cu		NH_4Cl		$Na_4P_2O_7$		EDTA	HNO_3		Berggren (1992)
Cu Mn Zn		$MgCl_2$		H_2O_2 digest	$NH_4O_x + HO_x$	EDTA + $NaCOOH$	HNO_3		Shuman (1979)

[a] CBD = citrate-bicarbonate-dithionite.
[b] Three fractions of metal oxides extracted separately: (i) Mn-oxides: $NH_2OH.HCl$; (ii) amorphous Fe-oxides: $(NH_4)_2C_2O_4 + H_2C_2O_4$ (oxalate reagent); (iii) crystalline Fe-oxides: oxalate reagent + UV irradiation.

toxic metals associated with iron and manganese oxides (Shuman and Hargrove, 1985), the metal-organic fractions associated with sewage sludge application (e.g. Keefer *et al.*, 1984), or the so-called "bioavailable" fraction (Neilsen *et al.*, 1987). Warden and Reisenauer (1991) designed a sequential extraction technique specifically for soil Mn, using a $Ca-DTPA-B_4O_7$ reagent to extract weakly adsorbed Mn, and avoiding the use of extractants that would reduce oxide-Mn. Extractants specifically designed for quantifying plant-available forms of metals in soils and their relative success are discussed in Section 3.4.2.

The examples in Table 3.15 and numerous other experiences have allowed assessment of a range of limitations and problems associated with sequential extraction techniques. There are at least six types of difficulties associated with soil extractions generally. These are: (i) whether the soil has been oven dried or not, (ii) the time of soil–extractant contact, (iii) the ratio of soil to extracting solution, (iv) the volume of the extracting vessel, (v) the method and degree of agitation of extraction and (vi) extraction temperature. The major problem specific to soil metals is that extraction procedures only crudely differentiate between the various forms of toxic metals in soil; there are areas of overlap and uncertainty in the way that sequential extractants remove chemical forms, i.e. extractants are not as specific as often stated. A further important difficulty is that the use of such a large number of different extracting techniques means that it is extremely difficult to compare results from different studies. Apart from this, there are at least three serious problems that must be considered when using sequential extraction procedures (Salomons and Forstner, 1984):

1. Labile metal phases could be transformed during sample preparation.
2. Readsorption or precipitation processes could occur during extraction.
3. Duration of the extraction procedure and the soil/solution ratios play important roles in the amount of metals extracted.

Salomons and Forstner (1984) list another seven chemical problems specific to certain extracting solutions. Soil organic matter and Fe and Mn oxides are mainly responsible for soil metal retention. However, these soil components are commonly laid down in gels or surface coatings, and in complex mixtures, on particle or pore surfaces. To identify metal retention by different Fe, Mn or organic fractions requires a careful ordering of chemical extractants, and this order may be different for different soils (Miller *et al.*, 1986). Different researchers have used different orders for the extractants of Fe and Mn oxides, carbonates and organic matter (see, for example, Table 3.14). The extractant order could be especially important for characterising metals in waterlogged soils or soils treated with waste sludges where remobilisation and deposition of reduced Fe and Mn, and translocation and decomposition of organic matter, may have occurred. Despite a seemingly endless list of difficulties, sequential extractions remain the most popular technique for characterising different soil metal fractions.

3.4.1.1 Sequential extractions for quantifying metal fractions in soils

Metal fractionations using sequential extraction techniques have primarily been used to identify the fate of metals applied in waste sludges and aerially deposited metals from smelters or air pollution. An early fractionation study which identified Cu, Zn and Mn present in four soil fractions—exchangeable ($MgCl_2$), organic (H_2O_2), Fe oxides (NH_4-oxalate in oxalic acid) and residue (1 M HNO_3)—also separated the metals contained in sand, silt and clay fractions (Shuman, 1979). In this study, Shuman examined eight different soil types, chosen to represent the major groups of the south-eastern United States. His results confirmed the theory outlined in Section 3.3, namely that in fine-textured soils, the largest proportions of Zn and Cu were present in the clay fraction, but in coarse-textured soils, the largest metal proportions were in the organic matter fraction.

In sewage-sludge-treated soil, we might expect a predominance of the same chemical forms of metals that were present in the original sludge. Stover *et al.* (1976) used a five-fraction sequential extraction technique to separate the metals in sewage sludge into exchangeable (1 M KNO_3), adsorbed (0.5 M KF), organic (0.1 M $Na_4P_2O_7$), carbonate (0.1 M EDTA) and sulphide (1 M HNO_3). Carbonates made up 61% of Pb, 49% of Cd, and 32% of Ni of the sludge metals. 35% of the Cu was in sulphide form and 50% of the Zn was bound organically. Adsorbed and exchangeable fractions of Cd, Cu, Pb and Zn were < 17%. Despite the problems of selectivity of extractants for different metal fractions, the above results suggest that sludge metals are mainly present in carbonate and sulphide forms, with substantial amounts of organically bound metals. In sandy loam soils treated with sewage sludges, Sposito *et al.* (1982) found that Zn, Cd and Pb were mainly present in carbonate form, Cu was mainly present in organic form and Ni was mainly present in sulphide form. Irrespective of sludge application rate, exchangeable amounts of metals were very low, averaging between 1.1 and 3.7% (Sposito *et al.*, 1982). These patterns of metal fractionation in sludge-treated soil have been confirmed elsewhere. Dudka and Chlopecka (1990), for example, found that residual forms of Cd, Cu, Ni and Zn were the most important metal fractions in untreated sandy loams. In sludge-amended soils, Zn was mainly present as carbonates and Fe oxides, Ni was mainly bound in the residual fraction, Cu was mainly present as Fe oxides and in organic forms, while Cd was mainly present in residual and carbonate forms. Only Zn was present in any quantity in exchangeable form (Table 3.16). Both of the above studies suggest that treating soils with sewage sludges shifts the solid phases of soil metals away from being strongly held, residual forms that are completely immobile, to forms that are potentially more mobile, labile and available to soil organisms and plants.

Fractionation studies on soils polluted with metals from smelting or other forms of aerial input indicate only small differences in soil metal distribution

Table 3.16. Fractions of soil metals present in soils near a Cu smelter and in sewage-sludge-treated soils, determined using a sequential extraction procedure (all values in $\mu g\,g^{-1}$, with % values in brackets, calculated on the basis of total soil metal contents)

Fraction	Cd Control[a]	Cd Sludge[a]	Cd Sludge[b]	Cd Smelter[b]	Cu Control[a]	Cu Sludge[a]	Cu Sludge[b]	Cu Smelter[b]	Ni Control[a]	Ni Sludge[a]	Ni Sludge[b]	Ni Smelter[b]	Zn Control[a]	Zn Sludge[a]	Zn Sludge[b]	Zn Smelter[b]
Exchangeable $(CaCl_2)$[a] $(MgCl_2)$[b]	0.15 (13)	0.12 (15)	14.0 (30)	6.6 (35)	0.11 (1)	0.16 (0.1)	1.0 (0.4)	9.6 (0.5)	0.6 (5)	1.4 (7)	0.7 (0.5)	0.2 (0.4)	2.1 (3)	112.8 (26)	21.7 (3)	47 (7)
Carbonates (NaOAc)	0.08 (7)	0.37 (25)	14.7 (32)	2.9 (15)	0.2 (1)	2.7 (2)	69 (26)	247 (12)	1.8 (14)	3.0 (16)	12.9 (10)	0.9 (1.4)	11.2 (16)	178.9 (41)	226 (36)	93 (14)
Fe oxides $(NH_2OH.HCl)$	0.2 (17)	0.2 (14)	9.2 (20)	5.7 (30)	0.16 (1)	77.2 (56)	75 (28)	895 (43)	1.8 (14)	3.4 (18)	41.6 (34)	8.3 (13)	18.0 (26)	145.0 (33)	277 (44)	301 (44)
Organics $(HNO_3 + H_2O_2)$	0.15 (13)	0.15 (10)	0.7 (2)	0.4 (2)	1.8 (10)	46.6 (34)	44 (17)	681 (33)	1.5 (12)	1.7 (9)	5.4 (4)	2.2 (3)	29.0 (42)	61.6 (14)	32.7 (5)	41 (6)
Residual $(HF + HCl_4)$	0.5 (43)	0.5 (34)	4.5 (10)	3.9 (20)	16.5 (34)	21.7 (16)	55 (20)	228 (11)	10.0 (80)	14.0 (74)	62.3 (50)	51.3 (80)	25.2 (36)	28.5 (6)	103 (16)	185 (27)
Σ fractions	1.08 (94)	1.34 (91)	43.1 (94)	19.4 (102)	18.8 (104)	148 (108)	244 (91)	2060 (97)	15.7 (126)	23.5 (124)	123 (99)	63 (98)	85.5 (124)	526.8 (120)	660 (104)	666 (98)
Total	1.15	1.47	46.1	18.9	18.1	137.7	269	2070	12.5	19.0	124	64	69.2	439.0	635	682

[a] Total metal contents were determined following digestion in a mixture of 10 ml conc. HF and 3 ml conc. $HClO_4$. [b] Total metal contents were determined by bomb-digestion using 4 ml conc. HNO_3, 1 ml (60%) $HClO_4$ and 6 ml (48%) HF.
Sources: [a] Dudka and Chlopecka (1990); [b] Hickey and Kittrick (1984).

compared to sludge-treated soils. A comparison of five soil metal fractions, extracted by identical sequential techniques, is given in Table 3.16 for sludge-treated and smelter-affected soils. Each of the two cited studies report results for sewage-sludge-treated soil, but at different metal loadings. Dudka and Chlopecka's data do indicate a big increase in exchangeable Zn after sludge application and a smaller increase in exchangeable Ni, but these may reflect characteristics of the sludge used. It is not surprising to see differences in metal fractions between the two sludge studies with different sewage sludge materials, containing different initial metal concentrations and at different land application rates. Comparing sludge-treated soil and soils near a Cu smelter, there are three interesting points. First, sludge-treated soils contain higher carbonate fractions of all four metals. This general finding of higher carbonate metal fractions in sludge-treated soils is also confirmed elsewhere (e.g. Chang et al., 1984b; Sposito et al., 1982). Secondly, Cu in both types of contaminated soils is present mainly in Fe oxide and organic fractions. The third point is that in both types of soil, Ni is present mainly in a residual form. There is little consistency in either over- or underestimation of "total" soil metal, based on the sum of individual fractions, as compared to strong acid digest, but the majority of fraction sums are within 10% of the totals.

The effect of pH on distribution of soil metal fractions has important implications for metal retention and metal mobility in contaminated acid soils. Xian and Shokohifard (1989) showed that a reduction in soil pH from 7.0 to 4.55 caused increases in exchangeable fractions of Cd, Zn and Pb, and small decreases in the carbonate fractions of these metals.

3.4.2 ASSESSING "PLANT AVAILABLE" AMOUNTS OF METALS IN SOIL

For almost 50 years, extractants have been tested specifically for the estimation of "plant-available" soil metals. Developments of soil metal availability tests were mainly, although not exclusively, in response to the need to monitor metal uptake and accumulation associated with increased aerial pollution, sewage sludge loading to agricultural soil, and metals in herbicide residues. Table 3.17 lists some of the main extractants which authors have assessed as "successful" in their abilities to extract soil metals in quantities similar to those taken up and measured in bioassay experiments with test plants.

Whether toxic metals are phytoavailable or not depends on the form of metal and on the species of plant. Many authors have reported good correlations between soil-extractable metals (using various of the chemicals listed in Table 3.17) and metal uptake by different test plants. Three main types of chemicals used for extracting "available" soil metals are: weak replacement ion salts (mainly $MgCl_2$ and $CaCl_2$), weak acids (especially acetic acid), and weak chelating agents (especially EDTA and DTPA). Good examples of successful predictions of plant metal uptake using replacement cations include

Table 3.17. Chemical extractants used to measure "plant available" soil metals in comparison with metals taken up by test plants

Extractant	Metal(s)	Bioassay	Source
2 M $MgCl_2$	Zn	Japanese millet	Stewart and Berger (1965)
		Wisconsin agricultural soils	
0.25 M $MgCl_2$	Zn	Navy beans, orchard soils	Neilsen et al. (1987)
0.005 M $CaCl_2$	Zn	Clover. wheat	Tiller et al. (1972)
0.005 M $CaCl_2$	Cd	Lettuce, cabbage, sewage sludge soils	Jackson and Alloway (1991)
1 M NH_4NO_3	Cd	Radish	Symeonides and McRae (1977)
0.42 M CH_3COOH			Berrow and Webber (1972)
1 M CH_3COOH	Zn, Ni, Cd, Cu	Swiss chard (Beta vulgaris)	Haq et al. (1980)
		Contaminated Ontario Soils	
5% CH_3COOH	Zn, Pb, Cd, Cu	Radish	Davies et al. (1987)
		Soil in metal mining areas of the UK	
$(NH_4)_2$-EDTA	Zn, Pb, Cd, Cu	Radish	Davies et al. (1987)
		Soil in metal mining areas of the UK	
0.05 M HCl + 0.025 H_2SO_4	Zn	Zea mays, sorghum	Wear and Evans (1968)
0.005 M DTPA + 0.1 M Triethanolamine + 0.01 M $CaCl_2$, pH 7.3	Fe, Zn, Mn, Cu	Sorghum	Lindsay and Norvell (1978)
		77 major soils of Colorado state	
0.005 M DTPA + 0.1 M Triethanolamine + 0.01 M $CaCl_2$, pH 7.3	Zn, Pb, Cd, Cu		Silviera and Sommers (1977)
0.005 M DTPA	Cu	Snapbeans	Minnich et al. (1987)
		Sewage sludge soils	
0.005 M DTPA	Zn, Ni, Cd, Cu	Zea mays	Rappaport et al. (1988)
0.005 M DTPA + 1 M NH_4HCO_3, pH 7.6	Zn	Barley, sorghum sewage sludge soils	LeClaire et al. (1984)

the use of MgCl$_2$-extractable Zn in relation to Zn uptake in orchards (Neilsen *et al.*, 1987), CaCl$_2$ to predict Cd uptake in agricultural crops (Jackson and Alloway, 1991), or NH$_4$NO$_3$ to predict the Cd uptake of radish (Symeonides and McRae, 1977). The extremely good correlation between 1 M NH$_4$NO$_3$-extractable Cd in 25 different soil types from Kent, England, and uptake of Cd by radish grown on these soils, is illustrated in Figure 3.16. Of the weak chelating agents, DTPA (diethylenetriaminepentaacetic acid) has been used successfully as an extractant for characterising bioavailability of both native soil metals (e.g. Gough *et al.*, 1980; Lee *et al.*, 1983) and metals added to soil in sewage sludges (e.g. Minnich *et al.*, 1987). Metals removed by these extractants are thought to represent soluble and easily exchangeable metals. Some authors have also reported negative results from DTPA tests, with insignificant relationships between extracted soil metals and test plant metal concentrations (e.g. Haq and Miller, 1972; Rappaport *et al.*, 1988),

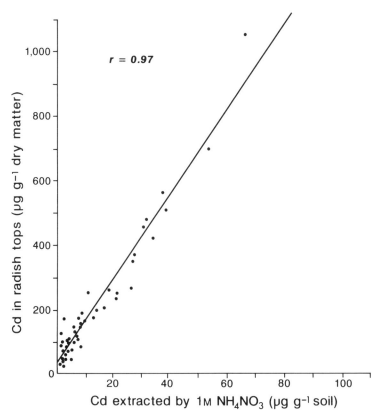

Figure 3.16. Relationship between Cd extracted by 1 M NH$_4$NO$_3$ from 25 Kent soils and uptake by radish plants. Reproduced by permission of the American Society of Agronomy from Symeonides and McRae (1977)

while Latterell *et al.* (1978) found that metal concentration in DTPA extracts was a good predictor of foliar Zn in snapbeans, but not of Cu or Cd. Rappaport *et al.* (1988) found good correlations between DTPA-extractable soil metals and yield of *Zea mays* test crops on sewage-sludge-treated soils, but not with crop metal concentrations. Despite these results, this extraction has been widely adopted as a "standard" metal availability technique (Baker and Amacher, 1982). The DTPA test was originally developed to identify near-neutral to calcareous soils with insufficient available Zn, Fe, Mn or Cu for maximum yields of crops (Lindsay and Norvell, 1978). O'Connor (1988) has subsequently identified a whole series of "misuses" of the DTPA test which probably account for failures of the test to predict metal availability. With so many reports of insignificant correlations between DTPA-extractable metals and metal concentrations in plants, it is worrying that so many researchers use DTPA-extractable quantities of metals to imply plant availability, often in acid soils with high metal loadings and without testing their thesis on plants.

Davies *et al.* (1987) found that the best correlations between plant uptake and amounts of toxic metals in contaminated soil were found with strong extractants such as EDTA or with "total" soil metal determinations using strong acids such as hot or boiling HNO_3. Similar conclusions were drawn by Soon and Bates (1982) who also obtained good correlations between concentrations of Cd and Zn in *Zea mays* plants and soil Cd and Zn held more strongly and determined in 1 M HNO_3 extracts. In several studies, some success has been achieved in predicting plant metal uptake using linear regressions based on the metal content of single chemical extractions (e.g. Wear and Evans, 1968; Soon and Bates, 1982; LeClaire *et al.*, 1984; Minnich *et al.*, 1987), or equations based on the metal content of a single chemical extractant in combination with individual soil properties, such as pH, organic matter and per cent clay content (e.g. Haq *et al.*, 1980; Mahler *et al.*, 1980; Soon and Bates, 1982; Lee *et al.*, 1983). Other studies which include all the sequential fractions as independent variables in the prediction of plant metal content (e.g. Iyengar *et al.*, 1981), while apparently producing good results, are questionable, since multicollinearity of the variables has been ignored.

Comparisons of the plant availability of metals applied to soils as inorganic salts and in sewage sludges showed that for Cd, Zn and Cu, consistently higher DTPA-extractable fractions were found when the metals were added in inorganic salts (Korcak and Fanning, 1985), implying greater phytoavailability. Minnich *et al.* (1987) found almost the opposite result, with snapbeans growing on sludge treated soils taking up greater quantities of Cu than those growing on metal salt treated soil. Tadesse *et al.* (1991) showed that while *yields* of wheat grown on sludge treated soils were higher than for wheat grown on metal salt treated soils, uptake of metals was not significantly different. Tadesse *et al.* (1991) did find that more Zn was taken up from sludge-treated sandy soils than from sludge-treated silty clay loams. This may be due to greater exchange of Zn onto cation exchange sites in the clayey soil.

Part of metal availability differences between metal salt and sludge treated soils could be due to differences in soil pH. Xian and Shokohifard (1989) found that Cd, Zn and Pb concentrations in shoots, roots and stems of kidney bean plants all increased steeply when the soil pH was reduced from 7.0 to 4.5. In the study conducted by Korcak and Fanning (1985), for example, metal salt treatments were accompanied by liming, while sludge treatments were not. However, there were no significant differences in measured pH between sludge and salt treated soils at the end of the experiment.

It is extremely difficult to summarise the findings and to estimate the value of the very large number of studies that have reported experiments on metal extractants, sequential extractions of metals and plant availability of metals, since so many use different test solutions, different bioassay test plants and, of course, different soils, sludges or metal salts, to say nothing about different conditions of the extraction techniques. Comparisons between studies are made virtually impossible. The more widespread use of DTPA reagent (see Baker and Amacher, 1982) as an extractant for "availability" of Ni, Cu, Zn and Cd has meant that at least this technique is now more standardised. However, many researchers use this extractant to quantify plant-variable metals without confirming their results on test plants. A further problem relates to the use of ideal test plants, such as radish and lettuce, in evaluations of soil metal availability. Only a few studies report metal availability results, determined from chemical extractions, for native species under real field conditions (e.g. Gough *et al.* (1980) for rangeland species, and Lee *et al.* (1983) for wetland macrophytes). In the future, research which combines evidence on metal availability from different sources, such as using sequential extractions in parallel with other separation techniques, including dialysis or chromatography, may provide useful corroboration of the exact role of extractants (Cottenie *et al.*, 1979). Similarly, the use of X-ray diffraction to identify solubilised metal fractions during chemical extraction may provide useful evidence of the selectivity of extractants. Studies such as these may even point towards the development of more specific extracting reagents.

3.5 CONCLUSIONS

The complexity of soil metal reactions and transformations is the reason why it is so difficult to predict soil metal bioavailability, mobility and retention. In this chapter the main soil conditions which influence transformation and adsorption processes have been discussed. It is clear that the most important soil characteristics influencing soil-metal processes are pH, organic matter quality and quantity, Fe and Mn oxides, and the per cent clay content. In the case of Mn and Fe, soil redox is also extremely important. Since some trace metals are sorbed more by non-specific ion exchange processes (e.g. Cd and Zn) while others are more specifically sorbed (e.g. Cu and Pb), it is not possible to make sweeping generalisations about "heavy metal adsorption and retention in soil". More care must be taken to identify the soil characteristics and processes associated with individual metals, their free aqua ions and also other metal species present in soil solution or colloidal fractions.

The question of whether metal "mobility" and "bioavailability" are synonymous, remains generally unresolved. There are two main difficulties in resolving this issue. First, it is necessary to determine whether chelated metals are phytoavailable, and if so, in what forms and under what soil and plant conditions. Related to this problem is the difficulty, introduced in Section 3.4, of finding analytical techniques capable of differentiating between free metal ions and soluble organo-metallic chelates. Secondly, the more chemical theory-based work on DTPA-extractable metals (e.g. Norvell and Lindsay, 1969; Norvell and Lindsay, 1972; Sadiq, 1983) has been separated from studies which compare DTPA-extractable metals with metals taken up by plants. Combined studies of metal retention and transport (perhaps using mathematical modelling) with studies of soil metal phytoavailability and chemical simulations of metal speciation using programs such as GEOCHEM, would be extremely helpful in linking together different parts of the metal retention–mobility–phytoavailability story. Some steps towards such integrated approaches are illustrated by the work of Emmerich et al. (1982), Sadiq (1983), Dudley et al. (1987) and Harmsen (1992), while other advances are being made in understanding the influence of different soil properties on metal sorption, transformation, mobility and bioavailability (e.g. Xian and Shokohifard, 1989; Brummer and Herms, 1983; Korcak and Fanning, 1985; Davies et al., 1987; King, 1988) and in modelling metal mobility and uptake (e.g. van Riemsdijk and van der Zee, 1991; Youssef and Chino, 1991; Selim et al., 1992).

REFERENCES

Addiscott, T. M. and Wagenet, R. J. (1985) Concepts of solute leaching in soils: a review of modelling approaches. *Journal of Soil Science*, **36**, 411–424.

Alloway, B. J. (1990) The origin of heavy metals in soils. In: Alloway, B. J. (Ed.) *Heavy Metals in Soils*, pp. 29–39. Blackie, London; Wiley, New York.

Alloway, B. J., Gregson M., Gregson, S. K., Tanner, R. and Tills, A. (1979) In: *Management and Control of Heavy Metals in the Environment*, pp. 545–548. International Conference, London. CEP Consultants Ltd., Edinburgh.

Alloway, B. J., Thornton, I., Smart, G. A., Sherlock, J. and Quinn, M. J. (1988) Metal availability. *Science of the Total Environment*, **75**, 41–69.

Amacher, M. C., Kotuby-Amacher, J., Selim, H. M. and Iskandar, I. K. (1986) Retention and release of metals by soils—evaluation of several models. *Geoderma*, **38**, 131–154.

Amacher, M. C., Selim, H. M. and Iskandar, I. K. (1988) Kinetics of chromium (VI) and cadmium retention in soils; a nonlinear multireaction model. *Soil Science Society of America*, **52**, 398–408.

Amoozegar-Fard, A., Fuller, W. H. and Warrick, A. W. (1984) An approach to predicting the movement of selected polluting metals in soils. *Journal of Environmental Quality*, **13**, 290–297.

Andersson, M. (1988) Toxicity and tolerance of aluminuim in vascular plants. *Water, Air and Soil Pollution*, **39**, 439–462.

Andreae, M. O. (1986) Chemical species in seawater and marine particulates. In: Bernhard, M., Brinckman, F. E. and Sadler, P. T. (Eds) *Speciation in Environmental Processes*, Dahlem Konferenzen, pp. 301–335. Springer-Verlag, Berlin.

Baham, J. and Sposito, G. (1983) Chemistry of water-soluble organic ligands extracted from sewage sludge. *Journal of Environmental Quality*, **12**, 96–100.

Baker, D. E. and Amacher, M. C. (1982) In: Page, A. L. (Ed.) *Methods of Soil Analysis*, Vol. 2. *Chemical and Microbiological Properties*, 2nd edn. Agronomy No. 9, pp. 323–336. Americal Society of Agronomy and Soil Science Society of America, Madison, Wisconsin.

Barrow, N. J. (1989) Suitability of sorption-desorption models to simulate partitioning and movement of ions in soils. In: Bar-Yosef, B., Barrow, N. J. and Goldshmid, J. (Eds) *Inorganic Contaminants in the Vadose Zone*, pp. 18–32. Springer-Verlag, Berlin.

Bartlett R. J. and Riego, D. C. (1972) Effect of chelation on the toxicity of aluminium. *Plant and Soil*, **37**, 419–423.

Beckett, P. H. R. T. (1989) The use of extractants in studies on trace metals in soils, sewage sludges and sludge-treated soils. *Advances in Soil Science*, **9**, 143–176.

Behel, D., Nelson, J. R. and Sommers, L. E. (1983) Assessment of heavy metal equilibria in sewage sludge-treated soil. *Journal of Environmental Ouality*, **12**, 181–186.

Benckiser, G., Ottow, J. C. G., Watanabe, I. and Santiago, S. (1984) The mechanism of excessive iron-uptake (iron toxicity) of wetland rice. *Journal of Plant Nutrition*, **7**, 177–185.

Benjamin, M. M. and Leckie, J. O. (1982) Effects of complexation by Cl, SO_4 and S_2O_3 on adsorption behaviour of Cd on oxide surfaces. *Environmental Science and Technology*, **16**, 162–170.

Berggren, D. (1992) Speciation and mobilization of aluminium and cadmium in podsols and cambisols of S. Sweden. *Water, Air and Soil Pollution*, **62**, 125–156.

Bergkvist, B. (1986) Leaching of metals from a spruce forest as influenced by experimental acidification. *Water, Air and Soil Pollution*, **31**, 901–916.

Bergkvist, B. (1987) Soil solution chemistry and metal budgets of spruce forest ecosystems in South Sweden. *Water, Air and Soil Pollution*, **33**, 131–154.

Bergkvist, B., Folkeson, L. and Berggren, D. (1989) Fluxes of Cu, Zn, Pb, Cd, Cr and Ni in temperate forest ecosystems. A literature review. *Water, Air and Soil Pollution*, **47**, 217–286.

Berrow, M. J. and Webber, J. (1972) Trace elements in sewage sludges. *Journal of the Science of Food and Agriculture*, **23**, 93–100.

Berrow, M. L. and Burridge, J. C. (1980) Trace element levels in soil: effects of sewage sludge. In: *Inorganic Pollution and Agriculture*, pp. 159–190. MAFF Reference Book No. 326. HMSO, London.

Biddappa, C. C., Chino, M. and Kumazawa, K. (1982) Migration of heavy metals in two Japanese soils. *Plant and Soil*, **66**, 299–316.

Bingham, F. T., Page, A. L., Mahler, R. J. and Ganje, T. J. (1976) Cadmium availability to rice in sludge-amended soil under "Flood" and "Nonflood" cultures. *Soil Science Society of America*, **40**, 715–719.

Bleam, W. F. and McBride, M. B. (1986) The chemistry of adsorbed Cu(II) and Mn(II) in aqueous titanium dioxide suspensions. *Journal of Colloid Interface Science*, **110**, 335–346.

Bloom, P. R. (1981) Metal–organic matter interactions in soil. In: Stelly, M. (Ed.) *Chemistry in the Soil Environment*. American Society of Agronomy Special Publication, pp. 129–149.

Bloom, P. R., McBride, M. B. and Weaver, R. M. (1979) Aluminium organic matter in acid soils: buffering and solution aluminium activity. *Soil Society of America Journal*, **43**, 488–493.

Bloomfield, C. (1957) The possible significance of polyphenols in soil formation. *Journal of the Science of Food and Agriculture*, **8**, 389–392.

Bohn, H., McNeal, B. and O'Connor, G. (1985) *Soil Chemistry*, 2nd edn. Wiley-Interscience, New York.

Bolland, M. D. A., Posner, A. M. and Quirk, J. P. (1977) Zinc adsorption by goethite in the absence and presence of phosphate. *Australian Journal of Soil Research*, **15**, 279–286.

Boyd, S. A., Sommers, L. E. and Nelson, D. W. (1979) Infrared spectra of sewage sludge fractions: evidence for an amide metal binding site. *Soil Science Society of America*, **43**, 893–899.

Boyle, M. and Fuller, W. H. (1987) Effect of municipal solid waste leachate composition on zinc migration through soils. *Journal of Environmental Quality*, **16**, 357–360.

Brummer, G. W. (1986) Heavy metal species, mobility and availability in soils. In: Bernhard, M., Brinckman, F. E. and Sadler, P. J. (Eds) *The Importance of Chemical Speciation in Environmental Processes*, pp. 169–192. Dahlem Konferenzen 1986. Springer-Verlag, Berlin, Heidelberg.

Brummer, G. W. and Herms, U. (1983) Influence of soil reaction and organic matter on solubility of heavy metals in soils. In: Ulrich, B. and Pankrath, J. (Eds) *Effects of Accumulation of Air Pollutants in Forest Ecosystems*, pp. 233–243. D. Reidel Publishing Company, Dordrecht, Germany.

Brummer, G. W., Tiller, H. G., Herms, U. and Clayton, P. M. (1983) Adsorption–desorption and/or precipitation-dissolution processes of zinc in soils. *Geoderma*, **31**, 337–354.

Cameron, R. S., Thornton, B. K., Swift, R. S. and Posner, A. M. (1972) Molecular weight and shape of humic acid from sedimentation and diffusion measurements on fractionated extracts. *Journal of Soil Science*, **23**, 394–408.

Campbell, P. G. C. and Tessier, A. (1987) Current status of metal speciation studies. In: Patterson, J. W. and Passino, R. (Eds) *Metals Speciation, Separation and Recovery*, pp. 201–224. Lewis Publishers, Michigan.

Campbell, P. G. C., Stokes, P. M. and Galloway, J. N. (1983) Effects of atmospheric deposition on the geochemical cycling and biological availability of metals. In: *Heavy Metals in the Environment*, Proceedings of an International Conference, Heidelberg, Vol. 2, pp. 760–763. CEP Consultants, Edinburgh.

Cavallaro, N. and McBride, M. B. (1978) Copper and cadmium adsorption characteristics of selected acid and calcareous soils. *Soil Science Society of America*, **42**, 550–556.

Chairidchai, P. and Ritchie, G. S. P. (1990) Zinc adsorption by laterite soil in the presence of organic ligands. *Soil Science Society of America*, **54**, 1242–1248.

Chang, A. C., Warneke, J. A., Page, A. L. and Lund, L. J. (1984*a*) Accumulation of heavy metals in sewage-sludge treated soils. *Journal of Environmental Quality*, **13**, 87–91.

Chang, A. C., Page, A. L., Warneke, J. A. and Grgurevic, E. (1984*b*) Sequential extraction of soil heavy metals following a sludge application. *Journal of Environmental Quality*, **13**, 33–38.

Cheshire, M. V., Berrow, M. L., Goodman, B. A. and Mundie, C. M. (1977) Metal distribution and nature of some Cu, Mn and V complexes in humic and fulvic acid fractions of soil organic matter. *Geochimica Cosmochimica Acta*, **41**, 1131–1138.

Christensen, T. H. (1984) Cadmium sorption at low concentrations. I—Effect of time, cadmium load, pH and calcium. *Water, Air and Soil Pollution*, **21**, 105–114.

Christensen, T. H. (1987) Cadmium sorption at low concentrations. V—Evidence of competition by other heavy metals. *Water, Air and Soil Pollution*, **34**, 293–303.

Christensen, T. H. (1989) Cadmium soil sorption at low concentrations: VIII—Correlations with soil parameters. *Water, Air and Soil Pollution*, **44**, 71–82.

Clark, C. J. and McBride, M. B. (1984) Chemisorption of Cu(II) and Co(II) on allophane and imogolite. *Clays and Clay Minerals*, **32**, 300–310.

Clark, C. J. and McBride, M. B. (1985) Adsorption of Cu(II) by allophane as affected by phosphate. *Soil Science*, **139**, 412–421.

Cline, G. R. and O'Connor, G. A. (1984) Cadmium sorption and mobility in sludge-amended soils. *Soil Science*, **138**, 248–254.

Commission of the European Communities (1982) Proposal for a Directive on the use of sewage sludge in agriculture. *COM(82) 527*, final, Brussels.

Cottenie, A., Camerlynck, R., Verloo, M. and Dhaese, A. (1979) Fractionation and determination of trace elements in plants, soils and sediments. *Pure and Applied Chemistry*, **52**, 45–53.

Coulson, C. B., Davies, R. I. and Lewis, D. A. (1960) Polyphenols in plant, humus and soil. II Reduction and transport by polyphenols of iron in model soil columns. *Journal of Soil Science*, **11**, 30–44.

Davies, B. E., Lear, J. M. and Lewis, N. J. (1987) Plant availability of heavy metals in soils. In: Coughtrey, P. J., Martin, M. H. and Unsworth, M. H. (Eds) *Pollutant Transport and Fate in Ecosystems*, pp. 267–275. British Ecological Society Special Publication No. 6. Blackwell Scientific Publishers, Oxford.

Davies, R. I., Cheshire, M. V. and Graham-Bryce, I. J. (1969) Retention of low-level of copper by humic acids. *Journal of Soil Science*, **20**, 65–71.

Davis, J. A. and Leckie, J. O. (1978) Effect of adsorbed, complexing ligands on trace metal uptake by hydrous oxides. *Environmental Science and Technology*, **12**, 1309–1315.

Dowdy, R. H., Lattrell, J. J., Hinesly, T. D., Grossman, R. B. and Sullivan, D. L. (1991) Trace metal movement in an Aeric Ochraqualf following 14 years of annual sludge applications. *Journal of Environmental Quality*, **20**, 119–123.

Dreiss, S. J. (1986) Chromium migration through sludge-treated soils. *Ground Water*, **24**, 312–321.

Dudka, S. and Chlopecka, A. (1990) Effect of solid-phase speciation on metal mobility and phytoavailability in sludge-amended soil. *Water, Air and Soil Pollution*, **51**, 153–160.

Dudley, L. M., McNeal, B. L., Baham, J. E., Coray, C. S. and Cheng, H. M. (1987) Characterisation of soluble organic compounds and complexation of copper, nickel and zinc in extracts of sludge-amended soils. *Journal of Environmental Quality*, **16**, 341–348.

Dumontet, S., Levesque, M. and Mathur, S. P. (1990) Limited downward migration of pollutant metals (Cu, Zn, Ni and Pb) in acid virgin peat soils near a smelter. *Water, Air and Soil Pollution*, **49**, 329–342.

Emmerich, W. E., Lund, L. J., Page, A. L. and Chang, A. C. (1982) Solid phase forms of heavy metals in sewage sludge-treated soils. *Journal of Environmental Quality*, **11**, 174–178.

Eriksson, J. E. (1989) The influence of pH, soil type and time on adsorption and uptake by plants of Cd added to soil. *Water, Air and Soil Pollution*, **48**, 317–335.

Farrah, H. and Pickering, W. F. (1977) The sorption of lead and cadmium species by clay minerals. *Australian Journal of Chemistry*, **29**, 1177–1184.

Feitknecht, W. and Schindler, P. (1963) *Solubility Constants of Metal Oxides, Metal Hydroxides and Metal Hydroxide Salts in Aqueous Solution.* Butterworth, London.

Fergusson, J. E. (1990) *The Heavy Elements. Chemistry, Environmental Impact and Health Effects.* Pergamon Press, Oxford.

Forbes, E. A., Posner, A. M. and Quirk, J. P. (1974) The specific adsorption of inorganic Hg(II) species and Co(III) complex ions on goethite. *Journal of Colloid Interface Science,* **49**, 403–409.

Forbes, E. A., Posner, A. M. and Quirk, J. P. (1976) The specific adsorption of divalent Cd, Co, Cu, Pb and Zn on goethite. *Journal of Soil Science,* **27**, 154–166.

Forno, D. A., Yoshida, S. and Asher, C. J. (1975) Zinc deficiency in rice. II—Soil factors associated with the deficiency. *Plant and Soil,* **42**, 537–550.

Foy, C. D. (1984) Physiological effects of hydrogen, aluminium and manganese toxicities on acid soils. In: Adams, F. (Ed.) *Soil Acidity and Liming,* 2nd edn. Agronomy 12, American Society of Agronomy, Madison, Wisconsin.

Francis, A. J., Dodge, C. J. and Gillow, J. B. (1992) Biodegradation of metal citrate complexes and implications for toxic-metal mobility. *Nature,* **356**, 140–142.

Fuller, W. H., Korte, N. E., Niebla, E. E. and Alesii, B. A. (1976) Contribution of the soil to the migration of certain common and trace elements. *Soil Science,* **122**, 223–235.

Gangwar, M. S. and Mann, J. S. (1972) Zinc nutrition of rice in relation to iron and manganese uptake under different water regimes. *Indian Journal of Agricultural Science,* **42**, 1032–1035.

Geering, H. R. and Hodgson, J. H. (1969) Micronutrient cation complexes in soil solution. III Characterisation of soil solution ligands and their complexes with Zn^{2+} and Cu^{2+}. *Soil Science Society of America,* **33**, 54–59.

Gerritse, R. G. and Van Driel, E. (1984) The relationship between adsorption of trace metals, organic matter and pH in temperate soils. *Journal of Environmental Quality,* **13**, 197–204.

Gerth, J. (1985) Untersuchungen zur adsorption von nickel, zink und cadmium durch bodentonfractionen unterschiedlichen stoffbestandes und verschiedene bodenkomponenten. Dissertation, University of Kiel.

Gerth, J. and Brummer, G. (1983) Adsorption und festlegung von nickel, zink und cadmium durch goethit (-FeOOH). *Fresenius Zeitschrift fur Anal. Chem.,* **316**, 616–620.

Ghosh, K. and Schnitzer, M. (1981) Fluorescence excitation spectra and viscosity behaviour of a fulvic acid and its copper and iron complexes. *Soil Science Society of America,* **45**, 25–29.

Giordano, P. M. (1986) Current guidelines and regulations for the agricultural use of sewage sludge in the southern region—an overview. In: King, L. D. (Ed.) *Agricultural Use of Municipal and Industrial Sludges in the Southern United States,* pp. 41–44. Southern Cooperative Service Bulletin 314. North Carolina State University, Raleigh.

Gough, L. P., McNeal, J. M. and Severson, R. C. (1980) Predicting native plant copper, iron, manganese and zinc levels using DTPA and EDTA soil extractants, Northern Great Plains. *Soil Science Society of America,* **44**, 1030–1036.

Gregson, S. K. and Alloway, B. J. (1984) Gel permeation chromatography studies on the speciation of lead in solutions of heavily polluted soils. *Journal of Soil Science,* **35**, 55–61.

Griffin, R. A. and Shrimp, N. F. (1976) Effect of pH on exchange-adsorption or precipitation of lead from landfill leachates by clay minerals. *Environmental Science and Technology,* **10**, 1256–1261.

Haas, C. H. and Horowitz, N. D. (1986) Adsorption of cadmium to kaolinite in the presence of organic matter. *Water, Air and Soil Pollution,* **27**, 131–140.

Haq, A. U. and Miller, M. H. (1972) Prediction of available soil Zn, Cu and Mn using chemical extractants. *Agronomy Journal*, **64**, 779–782.

Haq, A. U., Bates, T. E. and Soon, Y. K. (1980) Comparison of extractants for plant-available zinc, cadmium, nickel and copper in contaminated soils. *Soil Science Society of America*, **44**, 772–777.

Harmsen, K. (1992) Long-term behaviour of heavy metals in agricultural soils: a simple analytical model. In: Adriano, D. C. (Ed.) *Biogeochemistry of Trace Metals*, pp. 217–247. Lewis Publishers, Boca Raton.

Harter, R. D. (1979) Adsorption of copper and lead by A_p and B_2 horizons of several northeastern United States soils. *Soil Science Society of America*, **43**, 679–683.

Harter, R. D. (1983) Effect of soil pH on adsorption of lead, copper, zinc and nickel. *Soil Science Society of America*, **47**, 47–51.

Hayes, M. H. B. and Swift, R. S. (1978) *The Chemistry of Soil Constituents*, pp. 179–320. Wiley, Chichester, UK.

Herms, U. and Brummer, G. (1984) Einflussgrossen der Schwermetall-Loslichkeit und Bindung in Boden. *Zeitschrift fur Pflanzenernahr Bodenkund*, **147**, 400–424.

Hickey, M. G. and Kittrick, J. A. (1984) Chemical partitioning of cadmium, copper, nickel and zinc in soils and sediments containing high levels of heavy metals. *Journal of Environmental Quality*, **13**, 372–376.

Hinesly, T. D. and Jones, R. L. (1977) Heavy metal content in runoff and drainage waters from sludge-treated field lysimeter plots. In: *Proceedings of the National Conference on Disposal of Residues on Land*, pp. 27–44. USEPA, St Louis.

Hirsch, D. and Banin, A. (1990) Cadmium speciation in soil solutions. *Journal of Environmental Quality*, **19**, 366–372.

Hogfeldt, E. (1979) *Stability Constants of Metal-Ion Complexes. Part A: Organic Ligands*. Pergamon Press, Oxford.

Hoyt, P. B. (1977) Effects of organic matter content on exchangeable aluminium and pH dependent acidity of very acid soils. *Canadian Journal of Soil Science*, **57**, 221–222.

Hue, N. V., Craddock, G. R. and Adams, F. (1986) Effect of organic acids on aluminium toxicity in subsoils. *Soil Science Society of America*, **50**, 28–34.

IRRI (1977) *International Rice Research Institute. Annual Report. 1976*. The International Rice Research Institute, Los Banos, Phillipines.

Iu, K. L., Pulford, I. D. and Duncan, H. J. (1982) Influence of soil waterlogging on subsequent plant growth and trace metal content. *Plant and Soil*, **66**, 423–427.

Iyengar, S. S., Martens, D. C. and Miller, W. P. (1981) Distribution and plant availability of soil zinc fractions. *Soil Science Society of America*, **45**, 735–739.

Jackson, A. P. and Alloway, B. J. (1991) The bioavailability of cadmium to lettuce and cabbage in soils previously treated with sewage sludges. *Plant and Soil*, **132**, 179–186.

Jenne, E. A. (1977) Trace element sorption by sediments and soils. Sites and processes. In: Chappell, W. and Peterson, K. (Eds) *Symposium on Molybdenum in the Environment*, Vol. 2, pp. 425–553. Marcel Dekker, New York.

Jones, R. and Etherington, J. R. (1970) Comparative studies of plant growth and distribution in relation to waterlogging. III—The response of *Erica cinerea* L. and *Erica tetralix* L. and its apparent relationship to iron and manganese uptake in waterlogged soil. *Journal of Ecology*, **58**, 487–496.

Keefer, R. F., Codling, E. E. and Singh, R. N. (1984) Fractionation of metal-organic components extracted from a sludge-amended soil. *Soil Science Society of America*, **48**, 1054–1059.

Kharaka, Y. K. and Barnes, I. (1973) *Solution–Mineral Equilibrium Computations*. Contribution, NTIS Report PB, 215-899, US Geological Survey.

Kiekens, (1990) Zinc. In: Alloway, B. J. (Ed.) *Heavy Metals in Soils*, pp. 197–279. Blackie, London, Wiley, New York.

King, L. D. (1988) Retention of metals by several soils of the southeastern United States. *Journal of Environmental Quality*, 17, 239–246.

Kinniburgh, D. G., Jackson, M. L. and Syers, J. K. (1976) Adsorption of alkaline earth, transition and heavy metal cations by hydrous gels of iron and aluminium. *Soil Science Society of America*, 40, 796–799.

Kittrick, J. A. (1976) Control of Zn^{2+} in the soil solution by sphalerite. *Soil Science Society of America, Proceedings*, 40, 314–317.

Korcak, R. F. and Fanning, D. S. (1985) Availability of applied heavy metals as a function of type of soil material and metal source. *Soil Science*, 140, 23–34.

Korte, N. E., Skopp, J., Fuller, W. H., Niebla, E. E. and Alesh, B. A. (1976) Trace element movement in soils: influence of soil physical and chemical properties. *Soil Science*, 122, 350–359.

Kothny, E. L. (1973) The three-phase equilibrium of mercury in nature. In: Kothny, E. L. (Ed.) *Trace Elements in the Environment*. Advances in Chemistry Series, 123, pp. 48–80. Americal Chemical Society, Washington DC.

Kuo, S., Heilman, P. E. and Baker, A. S. (1983) Distribution and forms of copper, zinc, cadmium, iron and manganese in soils near a copper smelter. *Soil Science*, 135, 101–109.

Lakatos, B., Tibai, T. and Meisel, J. (1977) EPR spectra of humic acids and their metal complexes. *Geoderma*, 19, 319–338.

Lake, D. L., Kirk, P. W. W. and Lester, J. N. (1984) Fractionation, characterization and speciation of heavy metals in sewage sludge and sludge-amended soils: a review. *Journal of Environmental Quality*, 13, 175–183.

Latterell, J. J., Dowdy, R. H. and Larson, W. E. (1978) Correlation of extractable metals and metal uptake of snap beans grown on soil amended with sewage sludge. *Journal of Environmental Quality*, 7, 435–440.

LaZerte, B. (1986) Metals and acidification: an overview. *Water, Air and Soil Pollution*, 31, 569–576.

LeClaire, J. P., Chang, A. C., Levesque, C. S. and Sposito, G. (1984) Trace metal chemistry in arid-zone field soils amended with sewage sludge: IV—Correlations between zinc uptake and extracted soil zinc fractions. *Soil Science Society of America*, 48, 509–513.

Lee, C. R., Folsom, B. L. and Bates, D. J. (1983) Prediction of plant uptake of toxic metals using a modified DTPA soil extraction. *The Science of the Total Environment*, 28, 191–202.

Levi-Minzi, R., Soldatini, G. F. and Riffaldi, R. (1976) Cadmium sorption by soils. *Journal of Soil Science*, 27, 10–15.

Lighthart, B., Baham, J. and Volk, V. V. (1983) Microbial respiration and chemical speciation in metal-amended soils. *Journal of Environmental Quality*, 12, 543–548.

Lindsay, W. L. (1979) *Chemical Equilibria in Soils*. Wiley, New York.

Lindsay, W. L. (1981) Solid phase-solution equilibria in soils. In: Dowdy, R. H., Ryan, J. A., Volk, V. V. and Baker, D. E. (Eds) *Chemistry in the Soil Environment*, pp. 183–202. American Society of Agronomy and American Soil Science Society Special Publication No. 40. American Society of Agronomy, Wisconsin, USA.

Lindsay, W. L. and Norvell, W. A. (1978) Development of a DTPA soil test for zinc, iron, manganese and copper. *Soil Science Society of America*, 42, 421–428.

Lindsay, W. L. and Sadiq, M. (1983) Use of pe and pH to predict and interpret metal solubility relationships in soils. *The Science of the Total Environment*, 28, 169–178.

Lion, L. W., Altmann, R. S. and Leckie, J. O. (1982) Trace-metal adsorption characteristics of estuarine particulate matter: Evaluations of contributions of Fe/Mn oxides and organic surface coatings. *Environmental Science and Technology*, **16**, 660–666.

Macklin, M. G. and Smith, R. S. (1990) Historic riparian vegetation development and alluvial metallophyte plant communities in the Tyne Basin, North East England. In: Thornes, J. B. (Ed.) *Vegetation and Erosion*, pp. 239–256. Wiley, Chichester.

Mahler, R. J., Bingham, F. T., Sposito, G. and Page, A. L. (1980) Cadmium-enriched sewage sludge application to acid and calcareous soils: relation between treatment, cadmium in saturation extracts and cadmium uptake. *Journal of Environmental Quality*, **9**, 359–364.

Malcolm, R. L. and McCracken, R. J. (1968) Canopy drip: a source of mobile soil organic matter for mobilisation of iron and aluminium. *Soil Science Society of America Proceedings*, **32**, 834–838.

Mandal, L. N. and Mandal, B. (1987) Fractionation of applied zinc in rice soils at two moisture regimes and levels of organic matter. *Soil Science*, **144**, 266–273.

Mattigod, S. V. and Sposito, G. (1979) Chemical modelling of trace metal equilibria in contaminated soil solutions using the computer program GEOCHEM. In: Jenne, E. A. (Ed.) *Chemical Modelling in Aqueous Systems. Speciation, Sorption, Solubility and Kinetics*, pp. 837–850. American Chemical Society Symposium Series No. 93. Washington DC.

Mattigod, S. V., Gibaldi, A. L. and Page, A. L. (1979) Effect of ionic strength and ion pair formation on the adsorption of nickel by kaolinite. *Clays and Clay Minerals*, **27**, 411–416.

Mattigod, S. V., Sposito, G. and Page, A. L. (1981) Factors affecting the solubilities of trace metals in soils. In: Dowdy, R. H., Ryan, J. A., Volk, V. V. and Baker, D. E. (Eds) *Chemistry in the Soil Environment*, pp. 203–221. American Society of Agronomy and American Soil Science Society Special Publication No. 40. American Society of Agronomy, Wisconsin, USA.

Matzner, E., Murach, D. and Fortmann, H. (1986) Soil acidity and its relationship to root growth in declining forest stands in Germany. *Water, Air and Soil Pollution*, **31**, 273–282.

McBride, M. B. (1982) Electron spin resonance investigation of Mn^{2+} complexation in natural and synthetic organics. *Soil Science Society of America*, **46**, 1137–1143.

McBride, M. B. (1989) Reactions controlling heavy metal solubility in soils. *Advances in Soil Science*, **10**, 1–56.

McColl, J. G. and Pohlman, A. A. (1986) Soluble organic acids and their chelating influence on Al and other metal dissolution from forest soils. *Water, Air and Soil Pollution*, **31**, 917–927.

McDuff, R. E. and Morel, F. M. M. (1973) Description and use of the chemical equilibrium program REDEQL2. Technical Report EQ-73-02, W. M. Keck Laboratory, California Institute of Technology, Pasedena, California.

McGrath, S. P. (1987) Long-term studies of metal transfers following application of sewage sludge. In: Coughtrey, P. J., Martin, M. H. and Unsworth, M. H. (Eds) *Pollution Transport and Fate in Ecosystems*. British Ecological Society, Special Publication No. 6, pp. 301–317. Blackwell Scientific Publishers, Oxford.

McGrath, S. P. and Lane, P. W. (1989) An explanation for the apparent losses of metals in a long-term field experiment with sewage sludge. *Environmental Pollution*, **60**, 235–256.

McKenzie, R. M. (1980) The adsorption of lead and other heavy metals on oxides of manganese and iron. *Australian Journal of Soil Research*, **18**, 61–73.

McLaren, R. G. and Crawford, D. V. (1973) Studies on soil copper. I—The fractionation of copper in soils. *Journal of Soil Science*, **24**, 172–181.

McLaren, R. G., Swift, R. S. and Williams, J. G. (1981) The adsorption of copper by soil materials at low equilibrium solution concentrations. *Journal of Soil Science*, **32**, 247–256.

McLaren, R. G., Lawson, D. M. and Swift, R. S. (1986) Sorption and desorption of cobalt by soils and soil components. *Journal of Soil Science*, **37**, 413–426.

Mikkelsen, D. S. and Brandon, D. M. (1975) Zinc deficiency in California rice. *California Agriculture*, **29**, 8–9.

Miller, W. P., McFee, W. W. and Kelly, J. M. (1983) Mobility and retention of heavy metals in sandy soils. *Journal of Environmental Quality*, **12**, 579–584.

Miller, W. P., Martens, D. C. and Zelazny, L. W. (1985) Effect of sequence in extraction of trace metals from soils. *Soil Science Society of America*, **50**, 598–601.

Minnich, M. M., McBride, M. B. and Chaney, R. L. (1987) Copper activity in soil solution: II Relation to copper accumulation in young snapbeans. *Soil Science Society of America*, **51**, 573–578.

Morel, F. and Morgan, J. J. (1972) A numerical method for computing equilibria in aqueous systems. *Environmental Science and Technology*, **6**, 58–67.

Morgan, J. J. (1987) General affinity concepts, equilibria and kinetics in aqueous metals chemistry. In: Patterson, J. W. and Passino, R. (Eds) *Metal Speciation, Separation and Recovery*, pp. 27–61. Lewis Publishers, Michigan, USA.

Murder, J., van Grinsven, J. J. M. and van Breemen, N. (1987) Impacts of atmospheric deposition on woodland soils in the Netherlands: III—Aluminium chemistry. *Soil Science Society of America*, **51**, 1640–1646.

Mullins, G. L. and Sommers, L. E. (1986) Characterization of cadmium and zinc in four soils treated with sewage sludge. *Journal of Environmental Quality*, **15**, 382–387.

Murray, J. W. (1975) The interaction of metal ions at the manganese dioxide solution interface. *Geochimica Cosmochimica Acta*, **39**, 505–519.

Murray, K and Linder, P. W. (1983) Fulvic acids: structure and metal binding. I—A random molecular model. *Journal of Soil Science*, **34**, 511–523.

Murray, K. and Linder. P. W. (1984) Fulvic acids: structure and metal binding. II—Predominant metal binding sites. *Journal of Soil Science*, **35**, 217–222.

Neal, R. H. and Sposito, G. (1986) Effects of soluble organic matter and sewage sludge amendments on cadmium sorption by soils at low cadmium concentrations. *Soil Science*, **142**, 164–172.

Neilsen, D., Hoyt, P. B. and MacKenzie, A. F. (1987) Measurement of plant-available zinc in British Columbia orchard soils. *Communications in Soil Science and Plant Analysis*, **18**(2), 161–186.

Nieboer, E. and Richardson, D. H. S. (1980) The replacement of the nondescript term "heavy metals" by a biologically and chemically significant classification of metal ions. *Environmental Pollution (Series B)*, **1**, 2–26.

Nordstrom, D. K. *et al.* (1979) A comparison of computerized chemical models for equilibrium calculations in aqueous systems. In: Jenne, E. A. (Ed.) *Chemical Modelling in Aqueous Systems. Speciation, Sorption, Solubility and Kinetics*, pp. 857–892. American Chemical Society Symposium Series, No. 93.

Norvell, W. A. and Lindsay, W. L. (1969) Reactions of EDTA complexes of Fe, Zn, Mn and Cu with soils. *Soil Science Society of America, Proceedings*, **33**, 86–91.

Norvell, W. A. and Lindsay, W. L. (1972) Reactions of DTPA chelates of iron, zinc, copper and manganese in soils. *Soil Science Society of America, Proceedings*, **36**, 778–783.

O'Connor, G. A. (1988) Use and misuse of the DTPA soil test. *Journal of Environmental Quality*, **17**, 715–718.

Ollier C. (1984) *Weathering*. Geomorphology Texts No. 2, 2nd edn. Longman, London.

Parker, D. R., Zelazny, L. W. and Kinraide, T. B. (1987) Improvements to the program GEOCHEM. *Soil Science Society of America*, **51**, 488–491.

Parker, G. R., McFee, W. W. and Kelly, J. M. (1978) Metal distribution in forested ecosystems in urban and rural northwestern Indiana. *Journal of Environmental Quality*, **7**, 337–342.

Paton, T. R. (1978) *The Formation of Soil Material*. George Allen and Unwin, London.

Pearson, R. (1968a) Hard and soft acids and bases. HSAB, Part I—Fundamental principles. *Journal of Chemical Education*, **45**, 581–587.

Pearson, R. (1968b) Hard and soft acids and bases. HSAB, Part II—Underlying theories. *Journal of Chemical Education*, **45**, 643–648.

Piccolo, A. (1989) Reactivity of added humic substances towards plant available heavy metals. *The Science of the Total Environment*, **81/82**, 607–614.

Piccolo, A. and Stevenson, F. J. (1982) Infrared spectra of Cu^{2+}, Pb^{2+} and Ca^{2+} complexes of soil humic substances. *Geoderma*, **27**, 195–208.

Plant, J. A. and Raiswell, R. (1983) Principles of environmental geochemistry. In: Thornton, I. (Ed.) *Applied Environmental Geochemistry*, pp. 1–39. Academic Press, London.

Rappaport, B. D., Martens, D. C., Reneau, R. B. and Simpson, T. W. (1988) Metal availability in sludge-amended soils with elevated metal levels. *Journal of Environmental Quality*, **17**, 42–47.

Reddy, C. N. and Patrick, W. H. Jr. (1977) Effect of redox potential and pH on the uptake of Cd and Pb by rice plants. *Journal of Environmental Quality*, **6**, 259–262.

Richie, G. S. P. and Posner, A. M. (1982) The effect of pH and metal binding on the transport properties of humic acids. *Journal of Soil Science* , **33**, 233–247.

Robert, M. and Terce, M. (1989) Effects of gels and coatings on clay mineral chemical properties. In: Bar-Yosef, B., Barrow, N. J. and Goldshmid, J. (Eds) *Inorganic Contaminants in the Vadose Zone*, pp. 57–71. Springer-Verlag, Berlin.

Robin, R. A., Hemingway, B. S. and Risher, J. R. (1979) Thermodynamic properties of minerals and related substances at 298.15 $^{\circ}$C and 1 bar (10^5 Pascals) pressure and at higher temperature. *US Geological Survey, Bulletin No. 1452*.

Ross, S. M. (1989) *Soil Processes: A Systematic Approach*. Routledge, London.

Sadiq, M. (1983) Complexing of lead by DTPA in calcareous soils. *Water, Air and Soil Pollution*, **20**, 247–255.

Sadiq, M. (1991) Solubility and speciation of zinc in calcareous soils. *Water, Air and Soil Pollution*, **57–58**, 411–421.

Saeed, M. and Fox, R. L. (1979) Influence of phosphate fertilization on zinc sorption by tropical soils. *Soil Science Society of America*, **43**, 683–686.

Sajwan, K. S. and Lindsay, W. L. (1986) Effects of redox on zinc deficiency in paddy rice. *Soil Science Society of America*, **50**, 1264–1269.

Salomons, W. and Forstner, U. (1984) *Metals in the Hydrocycle*. Springer-Verlag, New York.

Sanden, P. (1991) Estimation and simulation of metal mass transport in an old mining area. *Water, Air and Soil Pollution*, **57–58**, 387–397.

Sarkar, A. N. and Wyn Jones, R. G. (1982) Effect of rhizosphere pH on the availability and uptake of Fe, Mn and Zn. *Plant and Soil*, **66**, 361–372.

Schirado, T., Vergara, I., Schalscha, E. B. and Pratt, P. F. (1986) Evidence for movement of heavy metals in a soil irrigated with untreated wastewater. *Journal of Environmental Quality*, **15**, 9–12.

Schnitzer, M. and Skinner, S. I. M. (1967) Organo-metallic interactions in soils. 7. Stability constants of Pb^{2+}, Ni^{2+}, Mn^{2+}, Co^{2+}, Cu^{2+}, and Mg^{2+} fulvic acid complexes. *Soil Science*, **103**, 247–252.

Schnitzer, M. and Desjardins, J. G. (1969) Chemical characteristics of a natural soil leachate from a humic podzol. *Canadian Journal of Soil Science*, **49**, 151–158.

Schnitzer, M. and Ghosh, K. (1982) Characteristics of water-soluble fulvic acid-copper and fulvic acid-iron complexes. *Soil Science*, **134**, 354–363.

Schnitzer, M. and Harmsen, E. H. (1970) Organo-metallic interactions in soils: 8. An evaluation of methods for the determination of stability constants of metal-fulvic acid complexes. *Soil Science*, **109**, 333–340.

Schnitzer, M. and Khan, S. U. (1972) *Humic Substances in the Environment*. Marcel Dekker, New York.

Schnitzer, M. and Skinner, S. I. M. (1965) Organo-metallic interactions in soils. 4. Carboxylic and hydroxyl groups in organic matter and metal retention. *Soil Science*, **99**, 278–284.

Schwab, A. P. and Lindsay, W. L. (1983) Effect of redox on the solubility and availability of iron. *Soil Science Society of America*, **47**, 201–205.

Scokart, P. O., Meeus-Verdinne, K. and de Borger, R. (1983) Mobility of heavy metals in polluted soils near zinc smelters. *Water, Air and Soil Pollution*, **20**, 451–463.

Seatz, L. F. and Jurinak, J. J. (1957) *The 1957 Yearbook of Agriculture*. USDA, Washington DC.

Selim, H. M., Buchter, B., Hinz, C. and Ma, L. (1992) Modelling the transport and retention of cadmium in soils: multireaction and multicomponent approaches. *Soil Science Society of America*, **56**, 1004–1015.

Senesi, N. (1992) Metal–humic substance complexes in the environment. Molecular and mechanistic aspects by multiple spectroscopic approach. In: Adriano, D. C. (Ed.) *Biogeochemistry of Trace Metals*, pp. 429–496. Lewis Publishers, Boca Raton, Florida.

Senesi, N. and Sposito, G. (1989) Characterization and stability of transition metal complexes of chestnut (*Castanea sativa* L.) leaf litter. *Journal of Soil Science*, **40**, 461–472.

Senesi, N., Griffith, S. M., Schnitzer, M. and Townsend, M. G. (1977) Binding of Fe^{3+} by humic materials. *Geochimica Cosmochimica Acta*, **41**, 969–976.

Senesi, N., Bocian, D. F. and Sposito, G. (1985) Electron spin resonance investigation of copper(II) complexation by fulvic-acid extracted from sewage sludge. *Soil Science Society of America*, **49**, 119–126.

Senesi, N., Sposito, G. and Martin, J. P. (1986) Copper(II) and iron(III) complexation by soil humic acids: an IR and ESR study. *Science of the Total Environment*, **55**, 351–362.

Senesi N., Sposito, G., Holtzclaw, K. M. and Bradford, G. R. (1989) Chemical properties of metal-humic acid fractions of a sewage sludge-amended aridisol. *Journal of Environmental Quality*, **18**, 186–194.

Sheppard, M. I. and Sheppard, S. C. (1987) A soil solute transport model evaluated on two experimental systems. *Ecological Modeling*, **37**, 191–206.

Sheppard, M. I., Thibault, D. H. and Mitchell, J. H. (1987) Element leaching and capillary rise in sandy soil cores: experimental results. *Journal of Environmental Quality*, **16**, 273–284.

Shuman, L. M. (1979) Zinc, manganese and copper in soil fractions. *Soil Science*, **127**, 10–17.

Shuman, L. M. (1985) Fractionation method for soil microelements. *Soil Science*, **140**, 11–22.

Shuman, L. M. (1986) Effect of ionic strength and anions on zinc adsorption by two soils. *Soil Science Society of America*, **50**, 1438–1442.

Shuman, L. M. (1990) The effect of soil properties on zinc adsorption by soils. *Soil Science Society of America*, **39**, 454–458.

Shuman, L. M. and Hargrove, W. L. (1985) Effect of tillage on the distribution of manganese, copper, iron and zinc in soil fractions. *Soil Science Society of America*, **49**, 1117–1121.

Sibley, T. H. and Morgan, J. J. (1975) Equilibrium speciation of trace metals in fresh water sea water mixtures. In: *International Conference on Heavy Metals in the Environment, Toronto*, pp. 319–338.

Sidle, R. C., Kardos, L. T. and van Genuchten, M. Th. (1977) Heavy metals transport model in a sludge-treated soil. *Journal of Environmental Quality*, **6**, 438–443.

Sillen, L. G. and Martell, A. E. (1964) *Stability constants of metals–iron complexes. I—Inorganic ligands: II—Organic ligands.* The Chemical Society, London, Special Publication No. 17.

Silviera, D. J. and Sommers, L. E. (1977) Extractability of Cu, Zn, Cd and Pb in soils incubated with sewage sludge. *Journal of Environmental Quality*, **6**, 47–52.

Smettem, F. R. J. (1986) Solute movement in soils. In: Trudgill, S. J. (Ed.) *Solute Processes*, pp. 141–165. Wiley, Chichester.

Somers, L. E., Nelson, D. W. and Silviera, D. J. (1979) Transformations of carbon, nitrogen and metals in soils treated with waste materials. *Journal of Environmental Quality*, **8**, 287–294.

Soon Y. K. and Bates, T. E. (1982) Chemical pools of cadmium, nickel and zinc in polluted soils and some preliminary indications of their availability to plants. *Journal of Soil Science*, **33**, 477–488.

Sposito, G. (1981) The Thermodynamics of Soil Solutions. Oxford University Press, Oxford.

Sposito, G. (1982) On the use of the Langmuir equation in the interpretation of "adsorption" phenomena. II The "two-surface" Langmuir equation. *Soil Science Society of America*, **46**, 1147–1152.

Sposito, G. (1983) The chemical form of trace metals in soils. In: Thornton, I. (Ed.) *Applied Environmental Geochemistry*, pp. 123–170. Academic Press Geology Series, London.

Sposito, G. (1986a) Distribution of potentially hazardous trace metals. In: Sigel, H. (Ed.) *Metal Ions in Biological Systems*, Vol. 20. Marcel Dekker, New York.

Sposito, G. (1986b) Distinguishing adsorption from surface precipitation. In: Davis J. A. and Hayes, K. F. (Eds) *Geochemical Processes at Mineral Surfaces*, pp. 217–228. Americal Chemical Society Symposium Series No. 323. Washington DC.

Sposito, G. and Bingham, F. T. (1981) Computer modelling of trace metal speciation in soil solutions: correlation, with trace metal uptake by higher plants. *Journal of Plant Nutrition*, **3**, 81–108.

Sposito, G. and Coves, J. (1988) *SOILCHEM: A Computer Program for the Calculation of Chemical Speciation in Soils.* The Kearney Foundation of Soil Science, University of California, Riverside and Berkeley, USA.

Sposito, G. and Mattigod, S. V. (1980) *A Computer Program for the Calculation of Chemical Equilibria in Soil Solutions and Other Natural Water Systems.* Kearney Foundation of Soil Science, University of California, Riverside, USA.

Sposito, G., Lund, L. J. and Chang, A. C. (1982) Trace metal chemistry in arid-zone field soils amended with sewage sludge. I—Fractionation of Ni, Cu, Zn, Cd and Pb in solid phases. *Soil Science Society of America*, **46**, 260–264.

Stanton, D. A. and Berger, R. D. (1970) Studies of zinc in selected range Free State soils. 5. Mechanisms for the reaction of zinc with iron and aluminium oxides. *Agrochemophysica*, **2**, 65–76.

Stevenson, F. J. (1977) Nature of divalent transition metal complexes of humic acids as revealed by a modified potentiometric titration method. *Soil Science*, **123**, 10–17.

Stevenson, F. J. (1986) Stability of Cu^{2+}, Pb^{2+} and Cd^{2+} complexes with humic acids. *Soil Science Society of America*, **40**, 665–672.

Stewart, J. A. and Berger, K. C. (1965) Estimation of soil zinc using magnesium chloride as extractant. *Soil Science*, **100**, 244–250.

Stover, R. C., Sommers, L. E. and Silviera, D. J. (1976) Evaluation of metals in wastewater sludge. *Journal of the Water Pollution Control Federation*, **48**, 2165–2175.

Stumm, W. and Morgan, J. J. (1981) *Aquatic Chemistry*. Wiley, New York.

Symeonides, C. and McRae, S. G. (1977) The assessment of plant-available cadmium in soils. *Journal of Environmental Quality*, **6**, 120–123.

Tadesse, W., Shuford, J. W., Taylor, R. W., Adriano, D. C. and Sajwan, K. S. (1991) Comparative availability to wheat of metals from sewage sludge and inorganic salts. *Water, Air and Soil Pollution*, **55**, 397–408.

Takamatsu, T. and Yoshida, T. (1978) Determination of stability constants of metal–humic acid complexes by potentiometric titration and ion-selective electrodes. *Soil Science*, **125**, 377–386.

Templeton, G. D. and Chasteen, N. D. (1980) Vanadium-fulvic acid chemistry: conformational and binding studies by electron spin probe techniques. *Geochimica Cosmochimica Acta*, **44**, 741–752.

Terry, R. E., Nelson, D. W. and Sommers L. E. (1979) Carbon cycling during sewage sludge decomposition in soils. *Soil Science Society of America*, **43**, 494–499.

Tessier, A., Campbell, P. G. C. and Bisson, M. (1979) Sequential extraction procedure for the speciation of particulate trace metals. *Analytical Chemistry*, **51**(7), 844–851.

Tiller, K. G., Hodgson, J. F. and Pesch, M. (1963) Specific adsorption of cobalt on soil clays. *Soil Science Society of America*, **15**, 392–399.

Tiller, K. G., Honeysett, J. L. and de Vries, M. P. C. (1972) Soil zinc and its uptake by plants. II Soil chemistry in relation to prediction of availability. *Australian Journal of Soil Research*, **10**, 165–182.

Tipping, E. and Hurley, M. A. (1992) A unifying model of cation binding by humic substances. *Geochimica et Cosmochimica Acta*, **56**, 3627–3641.

Truesdell, A. H. and Jones, B. F. (1974) A computer program for calculating chemical equilibria of natural waters. *Journal of the Research of the US Geological Survey*, **2**(2), 233–248.

Turner, D. R. (1987) Speciation and cycling of arsenic, cadmium, lead and mercury in natural waters. In: Hutchinson, T. C. and Meena, K. M. (Eds) *Lead, Mercury, Cadmium and Arsenic in the Environment*. SCOPE 31, pp. 175–186. Wiley, Chichester.

Ulrich, B., Mayer, R. T. and Khanna, K. (1988) Chemical changes due to acid precipitation in a loess-derived soil in Central Europe. *Soil Science*, **130**, 193–199.

van Breemen, N. and Jordens, E. R. (1983) Effects of atmospheric ammonium sulphate on calcareous and non-calcareous soils of woodlands in the Netherlands. II—Nitrogen-transformations. *Soil Science Society of America*, **51**, 1634–1640.

Van Dijk, H. (1971) Colloidal chemical properties of humic matter. In: McLaren, A. D. and Skujins, J. (Eds) *Soil Biochemistry, Vol. 2*, pp. 16–35. Marcel Dekker, New York.

van Genuchten, M. T. and Wierenga, P. J. (1974) Simulation of one-dimensional solute transfer in porous media. New Mexico Agricultural Experimental Station Bulletin 638.

van Riemsdijk, W. H. and van der Zee, S. E. A. T. M. (1989) Multicomponent transport modelling of enhanced metal leaching using synthetic ligands. *Geoderma*, **44**, 143–158.

Verloo, M. and Cottenie, A. (1972) Stability and behaviour of complexes of Cu, Zn, Fe, Mn and Pb with humic substances of soils. *Pedologie*, **22**, 174–184.

Wagenet, R. J. and Grenney, W. J. (1983) Modelling the terrestrial fate of heavy metals. In: Jorgensen, S. E. and Mitsch, W. J. (Eds) *Applications of Ecological Modelling in Environmental Management, Part B*, 7–34.

Warden, B. T. and Reisenauer, H. M. (1991) Fractionation of soil manganese forms important to plant availability. *Soils Science Society of America*, **55**, 345–349.

Wear, J. I. and Evans, C. E. (1968) Relationship of zinc uptake by corn to soil zinc measured by three extractants. *Soil Science Society of America*, **32**, 543–546.

Weeraratna, C. S. (1969) Adsorption of manganese by rice under flooded and unflooded conditions. *Plant and Soil*, **30**, 121–125.

Welch, J. E. and Lund, L. J. (1987) Soil properties, irrigation water quality and soil moisture level influences on the movement of nickel in sewage sludge-treated soils. *Journal of Environmental Quality*, **16**, 403–410.

Wieder, R. K. (1990) Metal cation binding to *Sphagnum* peat and sawdust: relation to wetland treatment of metal-polluted waters. *Water, Air and Soil Pollution*, **53**, 391–400.

Wild, A. (1981) Mass flow. In: Greenland, D. J. and Hayes, M. H. B. (Eds) *The Chemistry of Soil Processes*, pp. 37–80. Wiley, Chichester.

Williams, D. E., Vlamis, J., Pukite, A. H. and Corey, J. E. (1980) Trace element accumulation, movement and distribution in the soil profile from massive applications of sewage sludge. *Soil Science*, **129**, 119–132.

Xian, X. and Shokohifard, G. (1989) Effect of pH on chemical forms and plant availability of cadmium, zinc, and lead in polluted soils. *Water, Air and Soil Pollution*, **45**, 265–273.

Yong, R. N., Warkentin, B. P., Phadungchewit, Y. and Galvez, R. (1990) Buffer capacity and lead retention in some clay soils. *Water, Air and Soil Pollution*, **53**, 53–67.

Youssef, R. A. and Chino, M. (1991) Movement of metals from soil to plant roots. *Water, Air and Soil Pollution*, **57–58**, 249–258.

Zachara, J. M., Smith, S. C., Resch, C. T. and Cowan, C. E. (1992) Cadmium sorption to soil separates containing layer silicates and iron and aluminium oxides. *Soil Science Society of America*, **56**, 1074–1084.

Zottl, H. W. and Huttl, R. F. (1986) Nutrient supply and forest decline in Southwest Germany. *Water, Air and Soil Pollution*, **31**, 449–462.

Zunino, H., Aguilera, M., Caiozzi, M., Peirano, P., Borie, F. and Martin, J. P. (1979) Metal-binding organic macromolecules in soil: 3. Competition of Mg(II) and Zn(II) for binding sites in humic and fulvic-type model polymers. *Soil Science*, **128**, 257–266.

4 The Responses of Plants to Heavy Metals

ANDREW P. TURNER

Wye College, University of London, Ashford, Kent, UK

ABSTRACT

The evolution of tolerance in populations of short-lived plant species through the natural selection of a few tolerant individuals, present in normal populations on uncontaminated soils, is well understood. However, it is still far from clear as to why so few species are able to evolve tolerance and survive in soils containing normally phytotoxic concentrations of metals. There is a lack of knowledge on the limits of metal tolerance, with little information on the actual metal concentration above which no further adaptation is possible. Very few reports exist of metal tolerance in populations of long-lived trees and a characteristic feature of metalliferous soils in Europe is the absence of woody and tree species. However, there are reports of the survival of trees in metal-contaminated environments. Their longevity means that the theories of the development of tolerance in populations in the classic sense have little relevance. There are increasing numbers of reports indicating that plants may be able to acclimatize to incidences of pollution and contamination. The concept that tolerance may be lost or induced is now being addressed, with increasing interest in the role of plasticity as a component of evolutionary change. Such an ability to acclimatize to fluctuating stresses may be particularly important in the survival of longer-lived species with a long generation time.

The physiological responses of plants to heavy metals are addressed. Methods of the quantification of plant metal tolerance are reviewed with respect to the measurement of the tolerance of larger and longer-lived plant species. The use of techniques such as pollen germination and tissue culture may be useful in the detection of tolerance traits in long-lived plants such as trees. Research has identified several aspects of plant tolerance; however, the complete mechanism of metal tolerance has yet to be described for any plant. Mechanisms of tolerance are reviewed, in particular the role of mycorrhizas in avoidance tolerance mechanisms and the importance of the production of metal binding compounds in inducible tolerance mechanisms. It is likely that there is a multiplicity of mechanisms determining metal tolerance.

Despite the considerable research to date into the responses of plants to heavy metals, there is still a lack of understanding of the precise mechanisms allowing survival in metal-contaminated environments. This lack of understanding is particularly evident with respect to the survival of longer-lived species. Further research is required into the evolutionary and physiological responses of plants to metal contamination.

Toxic Metals in Soil–Plant Systems. Edited by S. M. Ross

4.1 INTRODUCTION

Environments have been contaminated by heavy metals ever since the original magma of the earth solidified (Bradshaw, 1984). Without human impact excessive concentrations of some heavy metals in soils are the result of natural mineralisation caused by the presence of undisturbed ore bodies near the surface (Ernst, 1990). Many areas of metal contamination have arisen relatively recently. With industrial development over the last 150 years there has been an increased release of metals during human-related activities. The sources of anthropogenic inputs of metals into the environment are numerous and diverse and include mining, smelting and the combustion of fossil fuels (Friedland, 1990); pollution from car engines (Ernst, 1976); sewage sludge disposal (Giller and McGrath, 1988); power lines (Al-Hiyaly et al., 1988) and from the use of pesticides and fungicides (Dickinson et al., 1984, 1988a). This has led to an ever-increasing awareness of the potency of heavy metals as environmental pollutants (Baker and Walker, 1989).

In all there are about 38 heavy metals (Passow et al., 1961) although the term "heavy metal" is ill-defined and the number of metals defined by the term is open to debate. They have a variety of roles in biological systems, ranging from regulators of biological processes to being important structural components in proteins (Borovik, 1990); however, a common biological feature is that in large quantities they are toxic to most plants. Such toxicity of heavy metals to plants is well documented. Ochaia (1987) divided the mechanisms of metal toxicity into five groups: (1) the displacement of essential metal ions from biomolecules and other biologically functional units; (2) blocking essential functional groups of biomolecules, including enzymes and polynucleotides; (3) modifying the active conformation of biomolecules especially enzymes and polynucleotides; (4) disrupting the integrity of biomolecules; and (5) modifying other biologically active agents.

Metal concentrations are often particularly high in mining wastes, causing severe phytotoxic effects. Concentrations of lead and zinc in excess of 20 000 and 100 000 mg kg^{-1} are not uncommon (Smith and Bradshaw, 1979), for example, in many mine spoil sites in North Wales. High soil metal concentrations that originate from active smelters and refineries are also well known. At Prescot, Merseyside, concentrations of up to 10 000 mg kg^{-1} copper have been recorded in the surface soil of an area subject to airborne deposition from a copper refinery (Bradshaw and McNeilly, 1981). Such levels of contamination have acute phytotoxic effects on plants whilst lower but continuous concentrations of atmospheric contaminants may have chronic effects on long-lived plants (Pitelka, 1988; Schulze, 1989). It is rare that the soils are completely devoid of plants even in the most metal-contaminated environments. As well as bringing about severe phytotoxic action it can also act as a powerful selection force on plant populations, leading to the selection of tolerant genotypes (Bradshaw, 1984). The selection of tolerant genotypes is an evolutionary response. In order to achieve such an evolutionary response a multiplicity of physiological responses is required. An orchestrated response of physiological mechanisms enables a plant to survive the effect of the metal contamination or stress. Such responses can be divided into two strategies (Levitt, 1980): stress avoidance, where excessive metal uptake is avoided, or stress tolerance whereby a plant tolerates internally accumulated metals.

The purpose of this chapter is to review the evolutionary and physiological responses of plants to heavy metals. The subject covers a large amount of literature and has been extensively reviewed elsewhere (Antonovics *et al.*, 1971; Macnair, 1981; Bradshaw, 1984; Baker, 1987; Shaw, 1990; Cumming and Tomsett, 1992). The aim is not to duplicate other reviews but to highlight the areas where there is a particular need for further research and where there is particular current interest. The chapter is divided into two main parts: evolutionary and physiological responses; however, it is clear that one cannot be considered without the other.

4.2 EVOLUTIONARY RESPONSES OF PLANTS TO HEAVY METALS

4.2.1 OVERVIEW

Heavy metal tolerance of plants has been studied extensively and is the best documented example of evolution (Macnair, 1981; Bradshaw *et al.*, 1990). Prat (1934) first investigated the phenomenon of metal tolerance, discovering that *Silene dioica* seedlings derived from copper mine waste could grow on soil treated with copper carbonate whereas seedlings from other populations could not. In recent years, a great deal of work has been carried out on the evolution of herbaceous plant species to heavy metal pollution. One of the basic conditions for the colonisation of metalliferous soils is the ability of plant species to evolve metal resistance (Ernst, 1990). Most of the evidence produced so far suggests that only plant species containing the natural genetic variation for tolerance in their natural populations are able to develop tolerant populations; *de novo* mutation for a tolerant genotype induced by the contaminant has not been demonstrated (Bradshaw, 1984). Often the areas surrounding contaminated soil support a diverse flora; however, relatively few species appear to have the ability to colonise metalliferous soils. Whilst metal tolerance has been demonstrated in many families of vascular plants throughout the world, the majority of widespread British metal-tolerant taxa occur only in three families—the Poaceae, Brassicaceae and Caryophyllaceae —whilst other large families are very poorly represented (Baker and Proctor, 1990). Baker and Proctor suggest that species which evolve metal tolerance have the required genetic variability and do so from constitutional properties at the generic or even family level. Differences in metal sensitivity may therefore be inherent between taxa. The result of the inability of many species to survive on metal-contaminated soils is an extremely sparse flora with a low species richness (Ernst, 1990; Bradshaw, 1991). Ernst (1990) refers to a general incapacity of trees to evolve a resistance to high concentrations of heavy metals, with metalliferous soils being one of the few environments in Europe where trees are absent. Additionally, when an originally uncontaminated ecosystem is subjected to anthropogenic metal contamination a change in the species composition may take place.

4.2.2 THE DEVELOPMENT OF TOLERANCE IN PLANT POPULATIONS

Much of our understanding of the development of tolerance in plant populations is from studies of the early colonisation of newly created and metal-contaminated mine spoils (Macnair, 1987). Tolerant species found growing on metalliferous soils possess the genetic variability for tolerance which is also found in a small proportion of individuals in populations growing on uncontaminated soil. The explanation for the development of tolerant populations is that high levels of contamination exert a selection pressure favouring the tolerant genotype. Natural selection can only operate if the genetic variation for a character already exists in the population subject to the selection pressure. Such variation may arise *de novo* by mutation after environmental change or else may be pre-existing (Macnair, 1981). The probability that a population possesses the genetic variation for tolerance is affected by several factors, particularly the mutation rate, the selection pressure against tolerance in normal populations and the population size (Macnair, 1987).

Most of the research carried out to date has been concerned with short-lived herbaceous species and there are very few examples in the literature of tolerance in longer-lived species (Dickinson *et al.*, 1991*a*). The development of tolerance in short-lived plants by natural selection of tolerant genotypes is well understood. For example, when normal populations of *Agrostis tenuis* are screened for metal tolerance, a very low frequency of at least partially tolerant individuals is found on which natural selection can act (Gartside and McNeilly, 1974). These tolerant survivors proliferate, producing a population tolerant to the metal contamination. Ingram (cited in Bradshaw and McNeilly, 1981) showed that for 15 species of grasses screened for metal tolerance there was no case of a species that evolved tolerance but did not possess variability for tolerance in its normal population. There is now evidence, though, that plants growing in a contaminated environment may not possess a greater tolerance than populations growing in an uncontaminated environment; that is, the normal population has an inherent or constitutional tolerance. For example, the entire genus of *Calochortus* may possess a constitutional tolerance to Ni, Co and Cu (Fiedler, 1985). There are many other examples of such constitutional tolerance which demonstrate that the assumption that tolerant populations are differentiated from "normal" populations is not always the case (see Baker and Walker, 1989).

There are many examples of the occurrence of endemic plant communities on metalliferous soils throughout the world. Serpentine areas contain the most obvious examples of specialised metal-tolerant plant communities with a high degree of endemism (Brooks and Malaise, 1985). The degree of nickel tolerance possessed by populations is variable (see Chapter 12, this volume). Certain genera and families are particularly well represented on contaminated soils (Kuboi *et al.*, 1987); for example, 45 species of *Alyssum* and 12 species of *Thalspi* spp. tolerate and hyperaccumulate Zn on serpentine soils (Brooks, 1987). Tolerance to other factors as well as the high soil nickel content is also required. High magnesium content inhibits plant calcium uptake, and concentrations of nitrogen, potassium and phosphorus are low; however, cobalt and chromium concentrations may be very high. There are many books and reviews (e.g. Brooks and Malaise, 1985; Brooks, 1987) on the flora of serpentine areas and it is not the purpose of this chapter to review the literature.

A consequence of the low frequency of tolerance in normal populations is that fitness is usually reduced when tolerant plants are grown in competition with other genotypes or species on clean soils (Hickey and McNeilly, 1975; Nicholls and McNeilly, 1985; Bastow-Wilson, 1988). However, it is not clear whether this lowered competitive ability is due to the direct consequences of metal tolerance or to the presence of co-selected characters resulting from the colonisation of contaminated environments such as mines where the substrate may be of a low nutrient status. Cook *et al.* (1972) found that zinc and lead tolerant ecotypes of *Agrostis tenuis*, *Anthoxanthum odoratum* and *Plantago lanceolata* occupying a mine habitat were less fit than those occupying normal pasture habitats when grown in competition on normal soils. This difference in fitness was considered to be more than adequate to have prevented the spread of individuals, though not necessarily genes, of tolerant populations on to normal soils closely adjacent to contaminated mine soils. Hickey and McNeilly (1975) also found differences in the competitive ability of tolerant and non-tolerant populations growing on normal soil.

With selection pressures acting against tolerant individuals in a normal population it is perhaps surprising that even a low frequency of tolerance is maintained in the absence of metal contaminants, but ecological research on plant populations has shown that several distinct genotypes (genets) may exist within a single population, and that these tend to be preserved by natural selection (Harper, 1981).

4.2.3 THE DEVELOPMENT OF TOLERANCE IN ESTABLISHED PLANT COMMUNITIES

The development of tolerance in plant communities presents a different situation to the primary colonisation of a spoil heap (Baker, 1987). In the latter case, a strong selective force is exerted (Baker and Walker, 1989). Table 4.1 summarises these differences in the selection pressure exerted by relatively recent episodes of aerial contamination and mineralised soil or that which is contaminated from surface ore deposits.

Invasion by propagules takes place on spoil heaps, followed by the selection of tolerant genotypes and colonisation. Baker and Walker (1989) consider that when pollution incidence is gradual, such as aerial deposition from a smelter, selection forces will only be strong enough to bring about genotypic changes in the population when the levels have reached phytotoxic levels. Clearly, when this level is reached within a plant community, tolerant species or biotypes will be favoured over non-tolerant species. In response to increasing levels of pollution, a shift in community composition would be expected towards tolerant species or biotypes that possess the ability to tolerate the pollutant, or else there would be the invasion and establishment of propagules in gaps in the vegetation created by disturbances (Grubb *et al.*, 1982). The time required for these changes in plant communities may help to explain a gradual increase of copper tolerance in established lawns in relation to the accumulation of copper over a period of 70 years, compared to the rapid development of tolerance in invading populations of grasses on copper mines (Macnair, 1987).

A number of cases have been described in which air pollutants have altered the composition and stability of plant communities, such as ecosystems surrounding smelters, and lichen and bryophyte communities in urban conurbations (Hutchinson, 1984). This is illustrated around the copper refinery at Prescot, Merseyside. Before the refinery was established a diverse

Table 4.1. Some features of mineralised and aerially contaminated soils in relation to the selection forces for metal tolerance

Mine wastes and mineralised soils	Aerially polluted soils
May have been present for 200 to 2000 years or more	Pollution recent
Spoil heaps initially devoid of plants	Plants established before pollution episode
Metal concentrations very high	Metal concentrations initially low
Metal concentrations unchanging with time	Episodic pollution leads to a cumulative rise in concentrations
Uniform distribution of metals throughout the profile	Metal accumulation in the surface horizons
Soil is deficient in nutrients and organic matter; often poor physical structure	"Normal" soil structure, horizons and nutrients status
Selective forces strong and stable	Selective forces weak and changing with time

Source: Adapted from Baker (1987).

unpolluted neutral grassland would have been present in the area. After 70 years only five species remain growing on the most polluted soils, the only change having been the aerial fallout of metal compounds (Bradshaw, 1984). In examining the tolerance of *Agrostis stolonifera* in different-aged lawns around the refinery in Prescot a gradual increase in copper tolerance was identified with the increasing age of the lawns (Wu *et al.*, 1975). Other examples exist such as the area around the Sudbury smelter in Ontario (Hutchinson and Whitby, 1974). Greszta *et al.* (1979) found that tree species varied in their response to dusts containing different levels of heavy metals.

The gradual increase in tolerance identified in *A. stolonifera* in Prescot lawns (Wu *et al.*, 1975) contrasts with the rapid development of tolerance in invading populations of grasses on copper and lead mines (Macnair, 1987). When forest ecosystems are subjected to air pollution a change in the genetic structure occurs (Karnosky *et al.*, 1989). Natural selection takes place against sensitive species and individual genotypes within species. A consequent loss of germplasm may occur with a corresponding loss of genetic diversity. Similar changes in the genetic structure of forests when subjected to air pollutants have been reported by Bergman and Scholz (1989). Gregorius (1989) suggests that air pollution impairs the adaptation of tree populations in almost all aspects and fundamentally endangers the natural bases for the preservation of adaptability.

Although not related to metal-rich sites, the experiments of Turkington and Harper (1979) and Turkington (1989) showed that clones of *Trifolium repens* were strongly adapted to their local environment. Very localised adaptations of *Festuca ovina* are also known to occur in relation to As levels in soil (Macnair, 1981). In the development of tolerance in plant communities the effect of the pollutant will be relative to its concentration, availability and stability. The response of the community will depend on the composition of its constituent species and genotypes. Indeed, the ability of plant populations to evolve metal tolerance is one of the most important characteristics determining the structure, density and development of vegetation in metalliferous areas (Simon, 1978).

4.2.4 INVESTIGATIONS OF THE RESPONSES OF TREES TO POLLUTION

Trees are often absent from metal-contaminated sites (Ernst, 1990). There is little evidence of metal tolerance occurring in populations of long-lived trees species and as yet, no evidence of tolerance evolving in the classic sense. Little work has been carried out on the tolerance of trees to metal pollution; the size and longevity of most tree species making laboratory tests impractical (Turner *et al.*, 1991). There are numerous examples where forests have been devastated around metal-processing plants, such as smelters. Smith (1974) listed many of these examples. In many cases effects of emissions on tree growth have been attributed to the presence of gaseous pollutants in addition to metal particulates (Freedman and Hutchinson, 1980).

However, trees do survive in metal-polluted environments, either as individual specimens or in established communities present before and during recent pollution episodes. One example is at Prescot, Merseyside where mature stands of trees exist which were present before the emission of large amounts of airborne copper from a copper refinery over the last 70 years (Turner, 1991). A second example is a woodland adjacent to a lead–zinc–cadmium smelter and sulphuric acid plant near Avonmouth which has been subjected to high doses of metal pollution (Coughtrey et al., 1979). Another example is that of Platanus hybrida (plane trees) which thrive in polluted city environments. It is possible that such trees possess an innate or constitutive tolerance but, nevertheless, tree species which are not especially selected for metal tolerance are not totally excluded from contaminated areas (Borgegård and Rydin, 1989).

Forests are efficient sinks for atmospheric pollutants and the potential for large rates of deposition is provided by the turbulent structure of air above and within forest canopies (Fowler et al., 1989). Most work concerning the effect of pollution on trees has concentrated on the effect of gaseous pollutants such as SO_2 on forest decline (Anderson, 1989; Bergman and Scholz, 1989). Relatively few studies have been carried out on the adaptation of trees to metal pollution. In temperate zones, most species that colonise metalliferous spoils are herbaceous and many of the physiological investigations of the mechanisms of tolerance have been carried out on such plants. Some short-lived species have been studied, however. Populations of Betula pendula and B. pubescens have been shown to evolve tolerance to zinc on mine spoils. These species of birch are often found growing on metalliferous mine tailings (Denny and Wilkins, 1987a). At Minera, North Wales, a site used in past tolerance studies and comprising large areas of lead–zinc contaminated mine tailings, there are stands of birch and Salix spp. growing on mine waste containing up to 30 000 $mg\,kg^{-1}$ lead and 100 000 $mg\,kg^{-1}$ zinc (Turner and Dickinson, 1993b). Denny and Wilkins found that tolerant genotypes of Betula spp. are able to exclude zinc, thereby maintaining a constant tissue concentration of the metal over a range of increasing soil concentrations. This metal exclusion was largely due to the mycorrhizal associations with the roots. Flowering in birch may commence within the first 5–10 years of growth, producing winged, wind-dispersed fruits in large quantities (Harding, 1981). The development of tolerance in birch populations may therefore be explained by natural selection in the classic sense. Longer-lived trees tolerant to naturally occurring toxic metal levels in serpentine soils have also been described (Brooks, 1987). In this case it seems that a long period of time (since the Mesozoic era) has elapsed for adaptation and evolution to occur.

Evidence does exist for constitutive tolerance of longer-lived trees to their environment. Schaedle et al. (1989) reviewed the effect of aluminium on tree seedlings and reported that aluminium stress resistance is a common phenomenon in trees. Numerous examples of resistant broad-leaved and coniferous tree species were cited. In some cases differential Al sensitivity was reported in the same species (e.g. red oak, red spruce and Norway spruce), which may have been related to genetic differences in the source of seed or seedlings used. Hutchinson (1984) reported a number of cases in which air pollutants have altered the composition of forests. Where the dominant tree species for an area had declined in numbers due to a pollutant, other tree species increased in numbers. Work by Greszta et al. (1979) showed varying degrees of resistance of coniferous and broad-leaved tree species to dusts containing varying combinations of heavy metals. Bergman and Scholz (1989) found "remarkable" genetic differences in Norway spruce between tolerant and susceptible clones to various air pollutants (SO_2, HF, O_3). Tolerant clones originated from forest tree stands located in polluted areas. Their results demonstrated that genetic changes occur in tree stands if sensitive trees die prior to reproduction or if the reproductive rate of sensitive trees is decreasing. Tree populations not destroyed by pollution may suffer from the loss of individuals due to random or selection pressures. As a consequence of selection pressures particular genes will decrease in frequency.

Clearly the genetic variation for tolerance to certain pollutants is present in the populations of some tree species. Berrang et al. (1989) found that ozone-sensitive clones of *Populus tremuloides* are less common in relatively polluted parts than they are in more polluted parts of the eastern USA. Gregorius (1989) considers that forest trees depend on large amounts of genetic diversity and heterozygosity to preserve their adaptability to changing environments. Air pollution acts to reduce this diversity. With air pollution constituting a rapid environmental change in comparison to the long generation time of forest trees, Gregorius considers that forests are unlikely to ever adapt to and survive large amounts of air pollution.

Smith (1974) reviewed in detail the effects of air pollutants on trees. He considered that the influence of air pollution on trees should be grouped into three conceptual classes. Under conditions of low pollution load, class 1, the response of the individual tree is undetectable and may involve no, or minimal, physiological alteration because the forest acts as a sink for air contaminants. If the air pollution load is increased to an intermediate level and when some threshold of tolerance, varying with different species and pollutants, is exceeded, then class 2 responses occur. This causes significant direct or indirect physiological impairment to individual members of the biota with reduced growth, reduced reproduction or increased mortality, and changes in species composition may also occur. High air pollution loads elicit class 3 reactions which may include acute morbidity or mortality on the part of individual trees and simplification at the ecosystem level.

Dendroecological studies have recently been used to examine the effect of air pollution on tree growth. Air pollution is generally involved in the suppression of tree growth and tree-ring analysis provides a method of examining hypothesised air pollution effects on forests (Cook, 1987a,b). There have been a number of studies on the effect of various pollutants on tree growth using tree-ring width studies. Such studies have included the effect of acidic deposition (Johnson and Wood, 1987; LeBlanc et al., 1987a,b); rising emission levels and the increasing proximity of factories or smelters (Fox, 1980; Heikkinen and Tikkanen, 1981; Fox, et al., 1986). Changes in the growth pattern of sycamore trees close to the copper refinery at Prescot, Merseyside have been identified (Turner, 1991). Reductions in ring-width were recorded during the period of maximum emissions. It appeared that the rate of growth increased in recent years with the introduction of controls on the emissions from the factory.

When considering innate or constitutive tolerance, mature trees must be viewed differently from seedlings. Large trees differ from seedlings in a number of ways, including carbon allocation and canopy structure (Pye, 1988). The ratio of metabolic to catabolic tissue changes as a tree grows and the fraction of tissue that is photosynthetically inactive increases with age. Therefore the consideration of constitutional tolerance in mature trees may not be applicable to seedlings, which may be sensitive.

The existing literature on the evolution of metal tolerance in plant populations and communities is applicable only to short-lived herbaceous species where rapid selection of tolerant genotypes can occur. Such models do not describe a system in which plants could respond rapidly to episodic pollution. Rapid genetic change of long-lived species such as trees would clearly not be expected in response to episodes of pollution. The survival of these species may therefore be dependent on the adaptability of individual plants.

4.2.5 INDIVIDUAL PLANT RESPONSES TO HEAVY METALS

The longevity of some species means that mechanisms of the development of tolerance in populations and communities previously described have little significance. When grasses are considered, the generation time from germination to the production of seed may be only a few weeks; in contrast, a tree such as sycamore (*Acer pseudoplatanus* L.) may not start to bear fertile seed until it is 20–30 years of age (Jones, 1945). Thus a mature tree may be subjected to episodes of pollution that the seedlings and parent tree may not have been subjected to. A characteristic feature of metalliferous soils in Europe is the absence of woody and tree species (Ernst, 1990).

The concept that plants may be able to acclimate to incidences of pollution and contamination is now receiving greater attention. Such acclimation to pollution may involve the induction of tolerance. Bradshaw (1984) stated that although detailed testing has not been done, there is no evidence of a Lamarkian origin of tolerance by induction; an individual plant cannot be trained to tolerate metals, and tolerance is not lost in culture in non-toxic conditions. Nevertheless, there has been increasing interest in the importance of plasticity as a component of evolutionary change (Thompson, 1991). Whilst "true" tolerance may be the result of genetically inherited physiological mechanisms, there is evidence that tolerance can be lost (Baker and Walker, 1989) or induced by pretreatment (Brown and Martin, 1981). The ability of a plant to respond phenotypically to a stress may therefore be an important mechanism in the plant's survival. Phenotypic variation in plants, whether related to genotypic variation or not, can affect ecological, morphological, anatomical, physiological, biochemical and molecular characters (Schmid, 1992). Schmid (1992) cites many examples of phenotypic variation such as growth habit under varying physical conditions. Such variation can significantly alter the plant's appearance.

There is now evidence, albeit not conclusive, linking plasticity in tolerance to the nature of contamination (Baker and Proctor, 1990). In his review Thompson (1991) concludes that plasticity is a heritable trait; however, for plasticity to be adaptive it must combine both a physiological buffering to poor environmental conditions and an improved response to favourable conditions. Schmid (1992) describes the relationship between plant organisation, variation and evolution. In the review the difficulties of identifying phenotypic variation experimentally are discussed, but Thompson concludes that environment-induced phenotypic effects can be persistent and can be transmitted to clonal or even sexual progeny. Cumming and Tomsett (1992) consider that acquired tolerance has at its basis one or more mechanisms that detoxify metals, either in the environment or in the cytoplasm, that can be evoked by the appropriate physiological or biochemical mechanism. However, when considering phenotypic responses to contamination we must also include avoidance mechanisms. For example, atmospherically deposited pollutants may be heterogeneously distributed throughout the soil horizons, with highest concentrations of the metals bound in the upper layers (Dickinson *et al.*, 1984; 1988*a*). The ability of the roots of trees to forage uncontaminated soil zones may affect the survival of individual plants.

Inducible tolerance by pretreatment has been demonstrated by a number of workers. Baker (1984) reported on environmentally induced Cd tolerance in populations of *Holcus lanatus* collected from Cd-contaminated soils close to the smelter at Avonmouth, UK. When initially collected and tested for Cd tolerance, the population had a significantly higher Cd tolerance than *H. lanatus* from an uncontaminated area (Coughtrey and Martin, 1978). After only one year of growth on uncontaminated soil a significant loss of tolerance had occurred (McGrath *et al.*, 1980) such that after 5 years the tolerance index was less than that for the population from the uncontaminated site. Baker (1984) speculated that the tolerance identified in the original experiments was an environmentally induced plastic response and was not genetically determined. The tolerance was identified since Coughtrey and Martin (1978) cultivated plants in original soils after collection and prior to use in tolerance tests. Brown and Martin (1981) found that pretreatment of *Holcus lanatus* with low levels of Cd resulted in increased tolerance of both tolerant and non-tolerant populations. Further studies by Baker *et al.* (1986) showed that for *H. lanatus* derived from a metalliferous soil with a greater Cd content and subsequent strong stable selection force, Cd tolerance was stable on transplanting to uncontaminated soil. The difference in these responses was explained by the differences in the selection forces. The degree of plasticity in the response of *H. lanatus* was found to be much greater than other grass species collected from the same site. Baker *et al.* (1986) suggest that such phenotypic adjustment may allow plants to survive in habitats to which they are not well suited and, in the course of time, by mutation, genotypes may arise that are well adapted. In their discussion Baker *et al.* refer to the tolerance possessed by the *H. lanatus* populations on metalliferous soils as "genetically based" and that of the clones from aerially contaminated environments as "environmentally induced". Surely though, such plasticity is also genetically based with the degree of plasticity both species-dependent and presumably population-dependent.

Shaw (1987) demonstrated that pretreatment of the moss *Funaria hygrometrica* had an effect on the subsequent tolerance to copper and zinc, although the genetic differences between populations accounted for far more variation in tolerance. He suggested that a potentially significant expression of genetic variation among individuals colonising metal-contaminated habitats is the ability to acclimate to gradually increasing levels of contamination. Cumming *et al.* (1992) found aluminium tolerance in *Phaseolus vulgaris* L. to be an inducible trait. Two cultivars of *P. vulgaris* were tested for Al tolerance in aerated nutrient solutions. Initially the roots of both cultivars were sensitive to Al in solution; however, after a 24–48 hour period of exposure to Al, root elongation of the tolerant cultivar showed an apparent recovery. This tolerance was only apparent when longer term growth (over 21 days) of the two cultivars was compared and not when total root lengths of the cultivars over a 72 hour period were compared. In one study on trees, although not concerned with heavy metals, Pye (1988) described differential susceptibility of sun and shade leaves of trees to ozone gradients within a forest canopy as the result of phenotypic plasticity.

Using cell culture techniques, Cu-tolerance traits have been identified in cultures derived from mature sycamore trees growing in a woodland subjected to aerial deposition of copper (Turner and Dickinson, 1993*a*). Phytotoxic concentrations of copper and cadmium were present in the surface soil layers. In cell suspension cultures originating from uncontaminated sites, growth was inhibited at 12.5 and 15.0 mg litre^{-1} Cu, but cultures originating from trees at the metal-contaminated site were not affected by these concentrations (Figure 4.1(a)). The tolerance trait was stable for a period of more than 12 months. The trait was not present in a culture derived from a non-tolerant seedling at the metal-contaminated site. However, by repeated exposure to

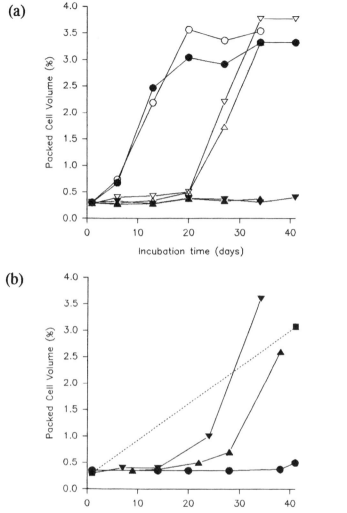

Figure 4.1. Copper tolerance traits in cell suspension cultures of *Acer pseudoplatanus* L. (Sycamore). (a) Growth of cultures (determined by measurement of packed cell volume) originating from trees growing at either a metal-contaminated site (open symbols) or an uncontaminated site (solid symbols) and previously unexposed to Cu in media. Growth is in media containing either 0 μg ml^{-1} (•,○), 12.5 μg ml^{-1} (▲ , △) or 15.0 μg ml^{-1} (▼ , ▽) Cu. No growth of the cultures derived from the uncontaminated site is evident in Cu-containing media. (b) Induction of increased Cu tolerance, shown by a progressively shortened growth cycle, in media containing 15 mg litre^{-1}, of an originally non-tolerant cell line, with gradually increased exposure to copper in successive subcultures. (Culture previously unexposed to Cu (•); after one subculture from media containing 12.5 μg ml^{-1} media (▲); after two subcultures from Cu-containing media (▼); after three subcultures from Cu-containing media (■)). Packed cell volume of subculture three was measured at 0 and 41 days only (experimental details in Turner and Dickinson, 1993*a*)

high copper concentrations the trait could be induced in a culture derived from a non-contaminated site (Figure 4.1(b)). Further work by Dickinson *et al.* (1992) showed that the cultures were also tolerant to cadmium in the medium. Metal uptake studies showed that the removal of metal from the medium was reduced in the metal-tolerant cultures compared to the non-tolerant cultures. These results have important implications in the survival of trees growing in environments subjected to episodic metal pollution. They demonstrate the occurrence of an alteration of gene expression in response to pollution stress *in vivo* and additionally the induction of tolerance in originally non-tolerant cultures *in vitro*. The work represents the first example of non-mycorrhizal adaptation to metal toxicity in woody plants.

Whilst there is increasing evidence of the importance of acclimation to contamination, there is little evidence to indicate whether when such acclimation takes place it is passed on to the offspring. In the work of Turner and Dickinson (1993 *a*,*b*), no evidence was found that the copper-tolerance traits identified in the cell cultures derived from mature trees was passed on to the offspring. Outridge and Hutchinson (1991) demonstrated the induction of Cd tolerance in the clonal fern *Salvinia minima* Baker. "Parent" ferns were acclimated to 0, 10, 25 or 50 μg litre^{-1} Cd and subsequently produced daughter ramets in a Cd-free medium. Parent acclimation to 25 μg litre^{-1} Cd increased the daughter ramets' tolerance index by 13–17% while a lower Cd concentration had no effect. Acclimation to 50 μg litre^{-1} produced "carry-over" toxicity effects. The implications of these results were that in patchily metal-contaminated environments rhizomatous clonal plants may be at an advantage over non-integrated clones of annual plants. Daughter ramets connected to acclimated parents may be more likely to survive in other contaminated patches, increasing the possibility that the clone will grow into contaminated areas. Turkington (1989) showed evidence that phenotypic traits of clones of *Trifolium repens* were maintained through repeated vegetative propagation of plants for up to 12 months. Karban (1990) cites a number of cases where phenotypic characteristics in trees have been retained. For example, Libby *et al.* (1972) showed that cuttings from hedged trees retain more juvenile characteristics than cuttings from unhedged trees of the same chronological age. Zagory and Libby (1985) found that resistance against rust in *Pinus radiata* was stable to propagation.

Clearly, phenotypic variation induced by environmental stress may play a role in the survival of individuals. Baker *et al.* (1986) suggested that such phenotypic adjustment may allow plants to survive in habitats to which they are not well adapted, and, in the course of time, by random mutation, genotypes may arise that are well adapted. Bradshaw and Hardwick (1989) consider that facultative adaptations to different stresses acting in a temporary or fluctuating manner, caused by phenotypic plasticity, can be found in plants in relation to different fluctuating stresses and operate over a wide range of time scales. However, they consider that such systems of response are under normal genetic control.

Gill (1986) proposed the idea of individual plants as mosaics of heritable genotypes. Central to this hypothesis is the argument that large plants with extensively branching architecture are actually colonies of heritable genotypes. This applies not only to the large canopy of trees but also to planar clones of grasses and other highly branched plants. A plant may consist of almost one million meristems, each with the ability to produce every kind of plant tissue, and because meristematic tissue is highly mutagenic, the number of developmental mutations should expand logistically as the plant grows. Central to the hypothesis is the notion that mutations may provide branches better suited to a changed environment. Meristems are totipotent and can give rise to vegetative and reproductive organs; therefore developmental mutations during vegetative growth have the dual potential of either being expressed somatically in the vegetative tissue or else developing into reproductive tissue and being inherited through the gametophytes. The hypothesis was developed to explain the responses of trees to insect attack, particularly in relation to variations in secondary metabolite production between different branches within plants. In a growing season when a tree is subjected to insect attack, susceptible branches would be expected to experience greater damage than resistant branches. Consequently, growth of the resistant branches will be greater than that of the susceptible branches which compete with each other for space in the tree canopy. A greater proportion of the tree will therefore be resistant in the successive growing seasons, but the situation may reverse as the tree grows if insect predation is reversed. A continuous process of selection and evolution within the tree therefore occurs with time.

If a tree experiences episodes of atmospheric pollution, the effect may be considered analogous to the damage caused by insect grazing. The genetic mosaicism hypothesis can therefore be extrapolated to the situation of a tree subject to pollution. The most appropriate canopy is selected each growing season as the tree increases in size. This may be a way of explaining how individual trees survive prolonged exposure to pollutants. If this were the case then the offspring of such trees would be expected to contain a greater percentage of tolerant individuals than trees not subjected to such selection pressures. However, it should be noted that there is a lack of experimental evidence to support the theory.

4.2.6 LIMITS OF TOLERANCE

Clearly there are limits to the degree of tolerance that plants can obtain. This is clearly shown by the relatively sparse colonisation of metal-contaminated environments even though adjacent non-contaminated areas support a normal diverse flora. However, there is a great lack of knowledge defining the precise quantitative limits of tolerance, i.e. the actual dosage level at which a chemical is toxic and the point beyond which no further adaptation can be achieved by a species (Dickinson et al., 1991b). This is due to either genetic variability or to the interactive effects of combinations of contaminants (Dueck et al.,

1987*a*; Hutchinson, 1984). It is generally accepted that only those plant species possessing the required genetic variation can develop tolerance in their populations (Gartside and McNeilly, 1974; Bradshaw, 1984; Baker and Proctor, 1990; Bradshaw, 1991). The majority of species lack the appropriate variation. Bradshaw (1991) proposed the term "genostasis" to describe the lack of appropriate variability possessed by most species.

Toxicity responses may also be determined by a number of physical factors which govern both availability and the relative toxicity of the metal contaminants such as the pH, clay content, organic matter content and the nutritional status (Clark and Clark, 1981; Dueck *et al.*, 1987*b*; Dickinson *et al.*, 1988*a,b*). Very little is known regarding the proportion of the total metal content in the soil and that which is available to plants. Many attempts have been made to estimate this fraction using various chemical extractants but in general an overestimate of metal bioavailability is obtained. Even methods where fresh soil is simply shaken with water appear to give an overestimate of the available metal in the soil (Turner, 1991). Advances in our understanding in this area are likely to come through the development of methods for determining the metal concentrations in the soil solution and then the determination of the metal speciation. Variation in the degree of tolerance between individuals in a population is evident when relative tolerance is determined through laboratory testing such as measuring root elongation in metal-containing solutions (Wilkins, 1957, 1978). Generally plants which survive in contaminated environments possess neither too much nor too little tolerance to allow optimal growth (Baker, 1987) and it seems that a plant provides an orchestrated response to a range of environmental variables within its immediate vicinity using a number of physiological mechanisms.

4.3 PHYSIOLOGICAL RESPONSES OF PLANTS TO HEAVY METALS

4.3.1 QUANTIFICATION OF PLANT METAL TOLERANCE

Most of the work quantifying and identifying tolerance in short-lived herbaceous plant species has been based on the use of the root growth technique developed by Wilkins (1957). This involves comparative measurements of the rates of root growth of test plants in control and metal-containing solutions. Although originally developed for grasses where the root system is fibrous, the test has been applied in studies on plants with a diverse range of root morphology such as *Mimulus guttatus*, *Armeria maritima*, *Plantago lanceolata* and *Leucanthemum vulgare* (Baker, 1987). A tolerance index (TI) on the basis of the relative root growth rates can then be calculated:

$$\text{T.I.}(\%) = \frac{\text{Root growth in solution with metal}}{\text{Root growth in solution without metal}} \times 100$$

A number of variations in calculating this index have been used (e.g. Jowett, 1958). The use of a range of metal concentrations allows the use of regression or probit analysis which is often a preferred method of analysis. Many variations on the methodology have been employed such as using sequential controls and measurement of total root length (see Wilkins (1978) and Baker (1987) for reviews). Single concentrations of metals have been used primarily because this enables a large number of populations to be screened without requiring large numbers of plants. Ideally a concentration should be used which slightly reduces the growth of the tolerant plants but which does not completely stop the growth of the non-tolerant plant (Wilkins, 1978). Commonly used metal concentrations in tolerance tests are $0.5\,\mu\text{g ml}^{-1}$, copper; $7.5\,\mu\text{g ml}^{-1}$, zinc; $12.0\,\mu\text{g ml}^{-1}$, lead and $2.0\,\mu\text{g ml}^{-1}$, nickel (Bradshaw and McNeilly, 1981). The use of single metals in tolerance tests may not give a reflection of the true nature of tolerance since sites are often contaminated with a number of metals and there is evidence of interaction between metals (e.g. Coughtrey and Martin, 1978).

The background solution in which the metal is supplied determines the toxicity of the test metal to the plants and there is much evidence for inter-actions between the toxic ions and other constituents of the solution. Traditionally, calcium nitrate has been used as the background solution (Wilkins, 1978); however, there have been warnings of the indiscriminate use of calcium nitrate (Davies and Snaydon, 1973). Many other background solutions have been used, sometimes in an attempt to mimic field conditions (Coughtrey and Martin, 1978, 1979; Symeonidis et al., 1985; Turner and Dickinson, 1993b).

The nature of the background solution can have a profound effect on the indices measured and the conclusions drawn. Humphrey and Nicholls (1984 warn that correlations between tolerance to different metals may arise through different characteristics of plants from different sites—known as vigour effects. These can have a genetic and/or an environmental basis which is independent of true tolerance and therefore may be misleading. Such vigour effects can be affected by changes in the composition of the background culture solution which forms a major part of the environment of the experiment. The use of tolerance indices to represent the data can be highly misleading, especially if a single metal solution is employed over one fixed time period (Baker, 1987; Macnair, 1990). Macnair (1990) considered that although the technique may be satisfactory for distinguishing large tolerance differences, it suffers from a serious flaw when more subtle differences are being explored. He points out that in using tolerance tests the genotype of the individual and the environmental effect are interacting and it may not be possible to separate the genotype × environment interaction of the root growth genes from the genotype × environment interaction of the tolerance genes. In many cases it is the inherent growth rate in the uncontaminated solution that is the major determining factor in the value of the tolerance index. Other methods have been developed to quantify tolerance, particularly where the root elongation test is not a practical method, as in the measurement of tree tolerance.

Alternative methods include the use of pollen germination and tissue culture techniques. Pollen germination and pollen tube growth have previously been used as indicators of tolerance. Pollen germination and pollen tube growth are affected by air pollutants. The stimulation and inhibition of these pollen characteristics depend on the pollen species as well as on the pollutant, its concentration, and a number of other factors such as relative humidity (Wolters and Martens, 1987). The response of pollen to a pollutant provides a parallel expression of tolerance to the parent plant. Estimates based on studies on tomato indicate that the genes expressed in pollen may represent up to 60% of genes expressed in the sporophytic stage (Tanksley *et al.*, 1981). Searcy and Mulcahy (1985) demonstrated the parallel expression of metal tolerance in pollen and sporophytes of *Silene dioica* and *Mimulus guttatus*. Tolerance of pollen was assessed by the determination of pollen germination in solutions containing copper or zinc salts. These results were linked to tolerance tests on the pollen sources.

Holub and Zelenâkova (1986) tested the tolerance to lead of various tree species from areas of varying soil levels of lead. Pollen germination and pollen tube growth were least influenced in trees from areas with the highest lead contents in the substrate. Lepp and Dickinson (1986) showed that pollen collected from stands of different-aged stands of *Coffea* spp. had an increased copper tolerance. Older stands had received considerable doses of copper-based fungicides over a period of more than 60 years. Germination of pollen from older stands occurred in much higher concentrations of copper than pollen from younger stands. Chaney and Strickland (1984) looked at the effects of various concentrations of various metals on Red Pine (*Pinus resinosa* Ait.) pollen germination and germ tube elongation. Large differences among metals in inhibition of the two growth parameters were found. The decreasing order of toxicity to germination was $Cd > Cu > Hg > Pb > Zn > Ba$. Cox (1988) assayed pollen from various coniferous and broadleaved trees to combinations of acidity and trace metals. Pollen used as an indicator of plant metal tolerance can prove a useful tool, particularly in large trees.

Tissue culture techniques offer a method of determining indices of metal tolerance representative of the whole plant (Wu and Antonovics, 1978; Qureshi *et al.*, 1981). Plant tissue culture has been widely used to identify a range of traits that are often maintained following the regeneration of whole plants (Smith *et al.*, 1989; Debergh and Zimmerman, 1991). Examples of the use of tissue culture techniques for the selection of tolerant plants include selection of tolerance to salt stress (Handa, *et al.*, 1982; Kurtz, 1982); aluminium tolerance in carrots (Ojima and Ohira, 1982); cold tolerance (Chen *et al.*, 1982) and heavy metals (Ten Hoopen *et al.*, 1985; Huang *et al.*, 1987).

Wu and Antonovics (1978) propagated callus material from meristematic tissue of clones of *Agrostis stolonifera* tolerant to zinc and copper. Growth of the callus in media containing metals demonstrated that the tolerance of the parent plants was reflected in the callus material. When the material was regenerated, propagated plants had the same degree of tolerance as the original plants and the tissue culture clones. Copper and cadmium tolerance traits have been identified in cell cultures derived from mature sycamore trees subjected to episodic metal pollution from a nearby refinery (Dickinson *et al.*, 1992; Turner and Dickinson, 1993*a*). In this case it was not possible to ascertain whether such traits were a parallel expression of the parent plant.

4.3.2 THE PHYSIOLOGICAL NATURE OF TOLERANCE

Research has identified several important aspects of metal tolerance but there is no model species in which the entire mechanism of tolerance has been determined (Jackson *et al.*, 1990). Plant survival in environments contaminated with potentially toxic levels of heavy metals can be achieved either by avoidance mechanisms where a plant is protected from the influence of heavy metal stress or by true tolerance mechanisms whereby a plant survives the effect of an internal stress (Macnair, 1981; Baker, 1987). The mechanisms are not mutually exclusive (Macnair, 1981; Tomsett and Thurman, 1988) and plant survival is likely to be dependent upon an orchestrated or integrated response involving a number of mechanisms (Baker and Walker, 1990). The different mechanisms of tolerance determine the pattern of metal uptake (Baker, 1981). In his review, Baker (1981) identified three basic strategies: (1) the excluder strategy; (2) the indicator strategy (where internal concentrations reflect the external metal concentration); and (3) the accumulator strategy, where metals are actively concentrated within plant tissues over the full range of metal concentrations.

Avoidance can be defined as an organism's ability to prevent excessive metal uptake into its body (Levitt, 1980). Such avoidance or metal-exclusion mechanisms include mycorrhiza formation, alteration of membrane permeability, proliferation of roots in uncontaminated horizons, alteration of membrane permeability, changes in the metal-binding capacity of the cell wall or increased exudation of metal-chelating substances (Verkleij and Schat, 1990). Verkleij and Schat (1990) consider a metal as having been taken up into the plant body once it has passed the plasmalemma of a root cell, and as such, tolerance is an organism's ability to cope with metals excessively accumulated, with mechanisms including production of metal-binding compounds, alteration of metal compartmentation patterns, alteration of cellular metabolism and alterations of membrane structure. There is a large amount of literature concerning the mechanisms of metal tolerance and the patterns of metal uptake. Detailed reviews of tolerance mechanisms and metal uptake by Jackson *et al.* (1990), Baker and Walker (1990), Verkleij and Schat (1990), and Cumming and Tomsett (1992) should be referred to. Two possible tolerance mechanisms that will be considered here are the role of mycorrhizas in avoidance tolerance mechanisms and the importance of the production of metal-binding compounds in inducible tolerance mechanisms.

4.3.3 INDUCIBLE TOLERANCE MECHANISMS: PRODUCTION OF METAL-BINDING COMPOUNDS

Recently, attention given to identifying mechanisms of tolerance has focused on the role of metal-binding polypeptides in the detoxification of metals (Robinson and Jackson, 1986; Robinson and Thurman, 1986; Tomsett and Thurman, 1988; Steffens, 1990). Metal-binding polypeptides reduce the concentration of cytotoxic free metal ions by chelation in the cytoplasm (Robinson, 1990). Early work in the field reported the production of metallothionein-like proteins in plants (Robinson and Jackson, 1986); however, there is now evidence and it is generally accepted that phytochelatins are the major complexing compounds in higher plants (Thurman, *et al.*, 1989; Davies *et al.*, 1991). The nomenclature concerning metallothioneins (MTs) is confusing (Robinson, 1990). There are three classes of metallothioneins. Class I and class II MTs have been isolated from animals, cyanobacteria and fungi (Evans *et al.*, 1992; Turner *et al.*, 1993) and are encoded for by structural genes. Metal-binding polypeptides produced by higher plants are termed class III MTs. They are known by a variety of names including cadystin, phytochelatin, Poly(γ-glutamylcysteinyl)glycine [$(\gamma EC)_n G$], phytometallothionein, γ-glutamyl metal-binding peptide and metallothiopeptide (Robinson, 1990; Tomsett *et al.*, 1989). Phytochelatins are secondary metabolites and are not synthesised by a mRNA template but by the enzyme γ-glutamylcysteinyl dipeptidyl transpeptidase (De Framond, 1991). Whether plant cells synthesise class I MTs has long been a controversial subject; however, recently Evans *et al.* (1990), De Framond (1991) and De Miranda *et al.* (1990) have reported the cloning and sequencing of plant genes encoding for proteins with similarity to class I MTs.

Phytochelatins are inducible in the whole plant kingdom, ranging from the phylogenetically simple algae to the highly advanced orchids (Grill *et al.*, 1988), and there are many examples of their production in the literature. In hydroponic culture of spinach in elevated copper concentrations, copper complexing proteins were identified in chloroplasts (Tukendorf, 1989). Robinson and Thurman (1986) conducted experiments on the formation of copper-binding proteins in the roots of *Mimulus guttatus*. After adding copper to the growth medium, considerable amounts of the precursor of heavy-metal-binding proteins were accumulated. Phytochelatins are induced upon exposure of plant suspension cultures to a wide range of heavy metals and much of the evidence for their production is based on the use of such techniques. Huang *et al.* (1987) induced cadmium tolerance in cell cultures of tomato. Tolerance was retained for a period of up to 12 months, Cd uptake was rapid and intracellular accumulation of Cd took place. It was found that Cd induced the synthesis and accumulation of phytochelatins.

There are few reports of the occurrence of phytochelatins in plants grown in the ecosystem, although Grill *et al.* (1988) examined the occurrence of phytochelatins in metal-sensitive *Acer pseudoplatanus* and resistant *Silene cucubalus* growing in a zinc-contaminated mining area. Phytochelatins were identified in root extracts of both plants, but when collected from a metal-uncontaminated stand neither species contained phytochelatins. The authors concluded that metal-binding phytochelatins are specifically induced in plants of heavy-metal-enriched ecosystems.

The phytochelatin response or synthesis of such compounds is one of few examples in which it can be readily demonstrated that the stress response (phytochelatin synthesis) is a truly adaptive response. Nevertheless, the extent to which this response accounts for the differential tolerance of metal-tolerant plants from metal-contaminated sites or cell cultures selected for heavy metals is far from clear (Steffens, 1990). Although phytochelatins are present in tissues of metal-tolerant ecotypes, the basis of evolved metal tolerance does not appear to involve mechanisms as simple as phytochelatin overproduction. This is owing to the energy required for sulphate reduction to support phytochelatin synthesis. Also the sporadic occurrence of toxic levels of metal ions in the biosphere seems unlikely to have exerted selection of a heavy metal detoxification system in plants (Steffens, 1990).

Tolerant genotypes have a higher relative growth rate than non-tolerant genotypes on heavy-metal-contaminated soils; in fact the latter may quickly die (Bradshaw and McNeilly, 1981). The frequency of occurrence of genotypes containing tolerance traits is extremely low in normal soil populations, and populations on heavy-metal-contaminated soils usually differ in tolerance from normal populations (Bastow-Wilson, 1988). Schultz and Hutchinson (1988) provided evidence against a key role for metal-binding proteins in the copper tolerance mechanism of *Deschampsia cespitosa* (L.) Beauv. The deliberate induction of sulphur deficiency limited the production of copper-inducible thiol-rich protein without causing a concomitant decrease in the tolerance of the copper-tolerant clones when exposed to elevated copper. Thiol-rich protein did not increase with duration of exposure to elevated copper despite a 43% increase in copper in the cell-free extract of one of the tolerant clones between a 4 and 30 day exposure to elevated Cu. In addition, the tolerant clones produced less copper-inducible thiol-rich protein than the non-tolerant ones. Schultz and Hutchinson concluded from the work that it was unlikely that any sulphur compounds constitute a major tolerance mechanism. However, whilst metal-binding proteins may not be the sole mechanism conferring the property of metal tolerance on a plant, they may be part of a range of orchestrated responses to metals. Cumming and Tomsett (1992) cite examples supporting the role of phytochelatins in metal detoxification. It appears that the production of metal-binding compounds may have a role in the metal tolerance of plants; however, the degree of importance in conferring tolerance on a plant is far from clear. Such a role is likely to be just one of a multiplicity of responses to elevated metal concentrations.

4.3.4 AVOIDANCE TOLERANCE MECHANISMS: THE ROLE OF MYCORRHIZAS

The effect of mycorrhiza in increasing plant growth has been well documented by many workers for many plant species. The importance of mycorrhizas in plant nutrition is well known, especially in the uptake of phosphorus (Bolan, 1991). Vesicular–arbuscular mycorrhizal infection can increase the rate of uptake of phosphorus, calcium, zinc, sulphur, bromide and chlorine (Gildon and Tinker, 1983a). The uptake of trace amounts of micronutrients such as Cu and Zn, present in soil solution at low concentrations, can also be increased by mycorrhizal infection (Gildon and Tinker, 1983a; Li et al., 1991). Additionally, the role of ectomycorrhiza is also well known, especially in the importance of tree nutrition (Wilkins, 1991). Recently it has been found that the heavy metal tolerance of certain plants may vary with mycorrhizal status. Many investigators have found an ameliorating effect of mycorrhizal infection on the growth of host plants cultivated at high metal concentrations (Colpaert and Van Assche, 1992a). Such effects have been mainly restricted to ecto-mycorrhizas (Wilkins, 1991) and ericoid mycorrhizas (Bradley et al., 1981, 1982).

There is little evidence of a role by vesicular–arbuscular mycorrhizas in ameliorating the toxicity of metals to their host plant, although recent studies have shown their importance in preventing unrestricted transport of manganese to shoots of mung bean (Stribley, 1988, cited in Baker and Walker, 1990). There is evidence that vesicular arbuscular (VA) endophytes can adapt to contaminated environments. Zinc- and cadmium-tolerant strains of *Glomus mossae* have been isolated from highly contaminated sites (Gildon and Tinker, 1983b). Considerable infection in white clover growing at the sites was found. The authors suggested that VA-mycorrhizal infection could protect plants against the effects of heavy metal additions although there is no strong evidence to support this yet. Ietswaart *et al.* (1992) found that infection of *Agrostis capillaris* by vesicular arbuscular mycorrhizal (VAM) fungi was reduced at a smelter site but there was high infection at sites with naturally high zinc concentrations. It was concluded that coevolution of *Agrostis* and *Glomus* towards a high degree of zinc tolerance had taken place. However, results indicated that VAM fungi do not operate as a sink for metals in high concentrations, with the concentration of mineral nutrients for three *Agrostis* populations related to soil concentrations but hardly modified by the degree of VAM infection.

Many areas contaminated with heavy metals are dominated by ericaceous plants (Allen, 1991). The role of mycorrhizal infection in increasing the nitrogen and phosphate uptake in ericaceous plants is well known. Bradley *et al.* (1981) first reported the beneficial effects on the growth and survival of mycorrhizal infection of *Calluna vulgaris* in metal-contaminated environments. In addition to *C. vulgaris*, Bradley *et al.* (1982) investigated the role of mycorrhizal infection in the metal tolerance of the ericaceous species *Vaccinium macrocarpon* and *Rhododendron ponticum*. Both mycorrhizal and non-mycorrhizal plants were grown in sand containing dilute nutrient solution and supplemented with Cu and Zn. In all cases mycorrhizal plants showed some growth in all metal treatments but non-mycorrhizal plants failed to grow in all but the lowest metal levels. Concentrations of metals were higher in the roots of mycorrhizal plants than in the roots of non-mycorrhizal plants. It was proposed that the hyphal complexes of the endophyte provided adsorptive surfaces within the cortical cells of the host roots thus restricting movement of metals to the shoots.

The role of ectomycorrhiza in the protection of trees from the effect of high external concentrations of heavy metals is now well established, with most of the work involving zinc but with some references to copper, nickel and aluminium (Wilkins, 1991). Brown and Wilkins (1985) found that mycorrhizas increased the tolerance of Zn to both tolerant and non-tolerant *Betula*. This was coupled with a reduction in the translocation of zinc to the shoots but with an accumulation in the mycorrhizas. In a series of papers, Denny and Wilkins (1987*a–d*) studied the zinc tolerance of *Betula*. Clones of tolerant and non-tolerant *B. pendular* Rot and *B. pubescens* Ehrg., collected as seed from zinc-contaminated mine tailings, were grown aseptically in a liquid-perlite medium in a range of zinc concentrations (Denny and Wilkins, 1987*a*). All the genotypes showed a limited initial rise in concentrations of zinc in tissue with increasing external concentrations; however, above a certain external zinc concentration a sudden increase in internal concentrations occurred. The point at which this inundation occurred was dependent upon the degree of zinc tolerance of the *Betula* clone but all the clones followed a similar pattern. Using energy dispersive x-ray microanalysis (EDAX), the location, concentration and form of the zinc in the root tissues of the genotype was studied (Denny and Wilkins, 1987*b*). At zinc concentrations above the lethal threshold, zinc accumulated in the endodermis and adjacent cortex as electron-dense granules. The granules represented zinc-induced cellular damage. No evidence was found of a wall-binding mechanism but it was speculated that zinc tolerance, control of zinc uptake and zinc-induced effects on cell extension were all linked by common events centred on the plasma membrane. An ameliorating effect of ectomycorrhiza formed by *Paxillus involutus* on zinc toxicity to *Betula* was found (Denny and Wilkins, 1987*c*). Interestingly, there was no indication of fungal adaptation to zinc at either an inter- or an intraspecific level. The ability of *P. involutus* strains isolated from both contaminated and uncontaminated soils to grow in agar, form ectomycorrhiza and produce a beneficial growth and zinc uptake with the plant was not related to the zinc status of the soil of fungus origin. The ameliorating effect of the fungus was found to be positively linked to the degree of compatibility between fungal strain and higher plant. The mechanism of mycorrhizal amelioration of zinc toxicity was investigated using X-ray microanalysis (Denny and Wilkins, 1987d). Ectomycorrhizal *Betula* spp. had lower concentrations of zinc in their root tissues than non-mycorrhizal plants. Zinc was adsorbed to the surface of extramatrical hyphae thereby reducing the zinc in the soil solution surrounding the roots. Less zinc is therefore taken up, allowing better growth than in non-mycorrhizal plants. The zinc may have been bound to electronegative sites in the hyphal cells and polysaccharide slime. In the series of papers it was not made clear how important the symbiotic relationship is in further increasing the zinc tolerance and therefore increasing fitness of mycorrhizal *Betula* spp. above the inherent tolerance level of the non-mycorrhizal plant. Kumpfer and Heyser (1989) provided further evidence of the role of ectomycorrhizas in tree tolerance using EDAX analysis to show that zinc is accumulated in the outer sheath of beech mycorrhizas with little detectable zinc in the inner sheath.

Nickel toxicity in mycorrhizal birch seedlings infected with *Lactarius rufus* or *Scleroderma flavidum* has been examined in other studies (Jones *et al.*, 1988; Jones and Hutchinson, 1988a). Infection by *S. flavidum* provided increased nickel tolerance; however, *L. rufus* appeared only to enhance tolerance to Ni in the seedlings at the early stages or not at all. Ni was concentrated in the roots with highest concentrations in the roots of plants infected with *S. flavidum*. The concentration of P was correlated with that of Ni in the stems and roots suggesting that high P concentrations in the roots may be important in the tolerance of mycorrhizal plants to Ni (Jones and Hutchinson, 1988b). Experiments examining the uptake of [63]Ni by birch using dinitrophenol as a metabolic inhibitor showed that *S. flavidum* birch mycorrhiza do not require metabolic energy to reduce Ni translocation from the roots to shoots (Jones *et al.*, 1988).

Many mycorrhizal species show a marked sensitivity to metals in the soil but there is some evidence of evolution of metal-tolerant mycorrhizal fungi in contaminated soils (Lepp, 1992). Morselt *et al.* (1986) examined the heavy metal tolerance of the ectomycorrhizal fungi *Pisolithus tinctorius* and *Cenococcum geophilum* using a specific stain to identify protein-bound disulphides and metal-thiolate clusters. The two fungi are known to have a relatively high and low tolerance for heavy metals respectively. Cultures of the two fungi were grown on agar containing sublethal concentrations of copper, cadmium or zinc. The existence of protein-bound disulphides and metal-thiolate clusters was demonstrated in the metal-tolerant *P. tintorius* but not in the non-tolerant *C. geophilum*. There are, however, many examples where metal-binding proteins are produced but they do not confer an increase in metal tolerance of the plant or fungus. Colpaert and Van Assche (1992b) examined the effects of cadmium and the cadmium–zinc interaction on the axenic growth of 11 strains of ectomycorrhizal fungi. Although a wide differential response to Cd was obtained between individual species, no clear relation between Cd tolerance and the site origin of the isolates was found, though such a relationship did exist between species. An interaction of Cd toxicity with zinc was identified where the addition of zinc to a medium resulted in a reduction of the toxic effect of Cd.

Clearly ectomycorrhizal fungi have a role in the survival of plants (particularly tree and ericaceous species) in metal-contaminated soils by enabling avoidance of the metal stress. Further research is required into the adaptations of the fungi to metal-contaminated environments and the role of the symbiosis on their survival. In addition, there is a need for research into the toxicity of heavy metals to VA mycorrhiza and their role in ameliorating plant metal uptake in contaminated environments.

4.4 CONCLUSIONS

There is a vast amount of literature concerning plant responses to heavy metals in the soil. With regard to the evolution of plant metal tolerance, studies have concentrated on short-lived species and considered evolution in the classic sense of natural selection of pre-existing tolerant genotypes already present at a low frequency in populations. The responses of trees and long-lived plants has largely been ignored, although there is now more research being carried out. It has long been assumed that plant populations are able to rapidly evolve tolerant populations on contaminated soils. However, it is only relatively recently that the questions have been raised as to why so few species are able to colonise metal-contaminated sites and why trees are absent. The concept of tolerance without the evolution of "tolerant" genotypes and the degree of plasticity of tolerance is now being addressed as evidence is produced that "tolerance" may be lost or induced. Of particular interest is the ability of plants to acclimate to heavy metals. Such an ability to acclimate to fluctuating stresses may be particularly important in the survival of longer-lived plant species with a long generation time. The evolutionary and physiological mechanisms achieving such acclimation are far from clear. Inducible physiological responses such as the synthesis of metal-binding polypeptides (phytochelatins) are receiving much attention at the moment. They appear to be present in most species of plants and fungi examined so far, are rapidly produced in response to metal stress, and appear to sequestrate metals from the critical metabolic systems in the cytoplasm. Their role in the conference of plant metal tolerance, however, is far from clear. Steffens (1990) argued against the role of overproduction of phytochelatins in plant tolerance due to the extra energy requirement for their synthesis. It does seem logical to pose the question that if phytochelatins are a major mechanism in the conference of metal tolerance, why then is tolerance restricted to relatively few species, and why are metalliferous soils so sparsely populated? Cumming and Tomsett

(1992) conclude in their review that phytochelatin synthesis is likely to be only one component of a complex mechanism required by metal-tolerant plant cells. However, the research to date in this area has concentrated on the production of phytochelatins *in vitro*, and their role *in vivo* is far from clear.

Information is required on the precise limits of tolerance of plants. The role of chronic sublethal doses of contamination in bringing about facultative adaptations in plants (in particular, long-lived plants) requires research. The concentrations of metals added to ecosystems that can bring about changes in species composition require definition if accurate risk assessment of land metal addition via air pollution (e.g. from smelters) or waste disposal is to be made. Such research is also required in view of the fact that alternative sources of sewage sludge disposal are now being sought in response to the imminent ban of disposal to sea in 1998. Sewage slude is often contaminated with heavy metals and a possible alternative disposal route is land application including disposal on agricultural (Smith and Giller, 1992) and forest land (Wolstenholme *et al.*, 1991). This gives cause for concern especially when the potentially long residence times of the metals in soil are considered (McGrath, 1987).

In conclusion, there is still much more research required into the evolutionary and physiological responses of plants to metal contamination. Information on the responses of long-lived species is particularly required. Conventional theories of evolution of metal tolerance do not explain the survival of longer-lived species when subjected to episodic pollution. The importance of phenotypic plasticity in determining survival needs further research. It is clear that there is a multiplicity of mechanisms determining plant tolerance.

REFERENCES

Al-Hiyaly, B. A., McNeilly, T. and Bradshaw, A. D. (1988) The effects of zinc contamination from electricity pylons – evolution in a replicated situation. *New Phytologist*, **110**, 571–580.

Allen, M. F. (1991) *The Ecology of Mycorrhizae*. Cambridge University Press, Cambridge.

Anderson, F. O. (1989) Air pollution impact on Swedish Forests—present evidence and future development. *Environmental Monitoring and Assessment*, **12**(1), 29–38.

Antonovics, J., Bradshaw, A. D. and Turner, G. (1971) Heavy metals tolerance in plants. *Advances in Ecological Research*, **7**, 1–75.

Baker, A. J. M. (1981) Accumulators and excluders. *Journal of Plant Nutrition*, **3**, 643–654.

Baker, A. J. M. (1984) Environmentally-induced cadmium tolerance in the grass *Holus lanatus*. *Chemosphere*, **13**(4), 585–589.

Baker, A. J. M. (1987) Metal tolerance. *New Phytologist*, **106** (suppl.), 93–111.

Baker, A. J. M. and Proctor, J. (1990) The influence of cadmium, copper, lead and zinc on the distribution and evolution of metallophytes in the British Isles. *Plant Systematics and Evolution*, **173**, 91–108.

Baker, A. J. M. and Walker, P. L. (1989) Physiological responses of plants to heavy metals and the quantification of tolerance and toxicity. *Chemical Speciation and Bioavailability*, **1**, 7–17.

Baker, A. J. M. and Walker, P. L. (1990) Ecophysiology of metal uptake by tolerant plants. In: Shaw, A. J. (Ed.) *Heavy Metal Tolerance in Plants: Evolutionary Aspects*, pp. 156–177. CRC Press, Boca Raton, Florida.

Baker, A. J. M., Grant, C.J., Martin, M. J., Shaw, S. C. and Whitebrook, J. (1986) Induction and loss of cadmium tolerance in *Holcus lanatus* L. and other grasses. *New Phytologist*, **102**, 575–587.

Bastow-Wilson, J. (1988) The cost of heavy metal tolerance: An example. *Evolution*, **42**(2), 408–413.

Bergman, F. and Scholz, F. (1989) Selection effects of air pollution in Norway Spruce. In: Scholz, F., Gregorius, H. R. and Rudin, D. (Eds) *Genetic Effects of Air Pollutants in Forest Tree Populations*, pp. 143–163. Springer Verlag, Berlin.

Berrang, P., Karnosky, D. F. and Bennett, J. P. (1989) Natural selection for ozone tolerance in *Populus tremeloides*: field verification. *Canadian Journal of Forest Research*, **19**, 519–522.

Bolan, N. S. (1991) A critical review on the role of mycorrhizal fungi in the uptake of phosphorus by plants. *Plant and Soil*, **134**, 189–207.

Borgegård, S. O. and Rydin, H. (1989). Biomass, root penetration and heavy metal uptake in birch, in a soil cover over copper tailings. *Journal of Applied Ecology*, **26**, 585–595.

Borovik, A. J. (1990) Characteristics of metals in biological systems. In: Shaw, A. J. (Ed.) *Heavy Metal Tolerance in Plants: Evolutionary Aspects* pp. 3–6. CRC Press, Boca Raton, Florida.

Bradley, R., Burt, A. J. and Read, D. J. (1981) Mycorrhizal infection and resistance to heavy metal toxicity in *Calluna vulgaris*. *Nature*, **292**, 335–336.

Bradley, R., Burt, A. J. and Read, D. J. (1982) The biology of mycorrhiza in the ericacae. VIII. The role of mycorrhizal infection in heavy metal resistance. *New Phytologist*, **91**, 197–209.

Bradshaw, A. D. (1984). Adaptations of plants to soils containing toxic metals—a test for conceit. In: Evered, D. and Collins, G. M. (Eds) *Origins and Development of Adaptation. Ciba Foundation Symposium 102*, pp. 4–19. Pitman, London.

Bradshaw, A. D. (1991) The Croonian lecture, 1991—genostasis and the limits to evolution. *Philosophical Transactions of The Royal Society of London, Series B*, **333**, 289–305.

Bradshaw, A. D. and McNeilly, T. (1981) *Evolution and Pollution*, Edward Arnold, London.

Bradshaw, A. D. and Hardwick, K. (1989) Evolution and stress—genotypic and phenotypic components. *Biological Journal of the Linnean Society*, **37**, 137–155.

Bradshaw, A. D., McNeilly, T. and Putwain, P. D. (1990) The essential qualities. In: Shaw, A. J. (Ed.) *Heavy Metal Tolerance in Plants: Evolutionary Apects*, pp. 323–333. CRC Press, Boca Raton, Florida.

Brooks, R. R. (1987) *Serpentine and its Vegetation—A multidisciplinary approach*. Croomhelm, London.

Brooks, R. R. and Malaise, F. (1985) *The heavy metal-tolerant flora of South Central Africa*. Balkema, Rotterdam.

Brown, H. and Martin, M. H. (1981) Pretreatment effects of cadmium on the root growth of *Holcus lanatus* L. *New Phytologist*, **89**, 621–629.

Brown, M. T. and Wilkins, D. A. (1985) Zinc tolerance in mycorrhizal *Betula*. *New Phytologist*, **99**, 101–106.

Chaney, W. R. and Strickland, R. C. (1984) Relative toxicity of heavy metals to red pine pollen germination and germ tube elongation. *Journal of Environmental Quality*, **13**(3), 391–394.

Chen, W. H., Cockburn, W. and Street, H. E. (1982) Cell plating and selection of cold tolerant cell lines in sugar cane. In: A. Fujiwaro (Ed.) *Proceedings 5th International Conference on Plant Tissue and Cell culture*, pp. 485–486. Japanese Association of Plant Tissue Culture, Tokyo.

Clark, R. K. and Clark, S. C. (1981) Floristic diversity in relation to soil characteristics in a lead mining complex in the Pennines, England. *New Phytologist*, **87**, 799–815.

Colpaert, J. V. and Van Assche, J. A. (1992*b*) The effects of cadmium and the cadmium–zinc interaction on the axenic growth of ectomycorrhizal fungi. *Plant and Soil*, **145**, 237–243.

Colpaert, J. V. and Van Assche, J. A. (1992*a*). Zinc toxicity in ectomycorrhizal *Pinus sylvestris*. *Plant and Soil*, **143**, 201–211.

Cook, E. R. (1987*a*) The decomposition of tree-ring series for environmental studies. *Tree Ring Bulletin*, **47**, 37–39.

Cook, E. R. (1987*b*). The use and limitations of dendroecology in studying the effects of air pollution on forests. In: Hutchinson, T. C. and Meema, K. M. (Eds) *Effects of Atmospheric Pollutants on Forests, Wetlands and Agricultural Ecosystems*, pp. 277–290. Springer-Verlag, Berlin.

Cook, S. C. A., Lefebure, C. and McNeilly, T. (1972) Competition between metal tolerant and normal plant populations on normal soil. *Evolution*, **26**, 366–372.

Coughtrey, P. J. and Martin, M. H. (1978) Tolerance of *Holcus lanatus* to lead, zinc and cadmium in factorial combination. *New Phytologist*, **81**, 147–154.

Coughtrey, P. J. and Martin, M. H. (1979) Cadmium, lead and zinc interactions and tolerance in two populations of *Holcus lanatus* L. grown in solution culture. *Environmental and Experimental Botany*, **19**, 285–290.

Coughtrey, P. J., Jones, C. H., Martin, M. H. and Shales, S. W. (1979) Litter accumulation in woodlands contaminated by lead, zinc, cadmium and copper. *Oikos*, **39**, 51–60.

Cox, R. M. (1988) Sensitivity of forest plant reproduction to long range transported air pollutants. *New Phytologist*, **110**, 33–38.

Cumming, J. R. and Tomsett, A. B. (1992) Metal tolerance in plants: Signal transduction and acclimation mechanisms. In: Adriano, D. C. (Ed.) *Biogeochemistry of Trace Metals*, pp. 329–364. Lewis, Boca Raton.

Cumming, J. R., Cumming, A. B. and Taylor, G. J. (1992) Patterns of root respiration associated with the induction of aluminium tolerance in *Phaseolus vulgaris* L. *Journal of Experimental Botany*, **43**(253), 1075–1081.

Davies, K. L., Davies, M. S. and Francis, D. (1991) The influence of an inhibitor of phytochelatin synthesis on root growth and root meristematic activity in *Festuca rubra* L. in response to zinc. *New Phytologist*, **118**, 565–570.

Davies, M. S. and Snaydon, R. W. (1973) Physiological differences among populations of *Anthoxatum odoratum* collected from the park grass experiments, Rothamsted. II. Response to aluminium. *Journal of Applied Ecology*, **10**, 47–55.

De Framond, A. J. (1991) A metallothionein-like gene from maize (*Zea mays*). *FEBS Letters*, **290**, 103–106.

De Miranda, J. R., Thomas, M. A., Thurman, D. A. and Thomsett, A. B. (1990) Metallothionein genes from the flowering plant *Mimulus guttatus*. *FEBS letters*, **260**, 227–280.

Debergh, P. C. and Zimmerman, R. H. (1991) *Micropropagation: Technology and Application*. Kluwer, Netherlands.

Denny, H. J. and Wilkins, D. A. (1987*a*). Zinc tolerance in *Betula* spp. I. Effect of external concentration of zinc on growth and uptake. *New Phytologist*, **106**, 527–524.

Denny, H. J. and Wikins, D. A. (1987*b*). Zinc tolerance in *Betula* spp. II. Microanalytical studies of zinc uptake into root tissues. *New Phytologist*, **106**, 525–534.

Denny, H. J. and Wilkins, D. A. (1987*c*). Zinc tolerance in *Betula* spp. III. Variation in response to zinc among ectomycorrhizal associates. *New Phytologist*, **106**, 535–544.

Denny, H. J. and Wilkins, D. A. (1987*d*) Zinc tolerance in *Betula* spp. IV. The mechanism of ectomycorrhizal amelioration of zinc toxicity. *New Phytologist*, **106**, 545–553.

Dickinson, N. M., Lepp, N. W. and Ormand, K. L. (1984) Copper contamination of a 68 year old *Coffea arabica* L. plantation. *Environmental Pollution (Series B)*, **7**, 223–231.

Dickinson, N. M., Lepp, N. W. and Surtan, G. T. K. (1988*a*). Lead and potential health risks from subsistence food crops in urban Kenya. *Environmental Geochemistry and Health*, **9**(2), 37–42.

Dickinson, N. M., Lepp, N. W. and Surtan, G. T. K. (1988*b*). Further studies on copper accumulation in Kenyan *Coffea arabica* plantations. *Agriculture Ecosystems and Environment*, **21**, 181–190.

Dickinson, N. M., Turner, A. P. and Lepp, N. W. (1991*a*). How do trees and other long-lived plants survive in polluted environments? *Functional Ecology*, **5**, 5–11.

Dickinson, N. M., Turner, A. P. and Lepp, N. W. (1991*b*). Survival of trees in a metal-contaminated environment. *Water Air And Soil Pollution*, **57–58**, 627–633.

Dickinson, N. M., Turner, A. P., Watmough, S. A. and Lepp, N. W. (1992) Acclimation of trees to pollution stress: Cellular metal tolerance traits. *Annals of Botany*, **70**, 569–572.

Dueck, T. A., Wotting, H. G., Moet, D. R. and Pasman, F. J. M. (1987*a*). Growth and reproduction of *Silene cucubalis* WIB intermittently exposed to low concentrations of air pollutants, zinc and copper. *New Phytologist*, **105**, 633–645.

Dueck, T. A., Tensen, D. and Duijff, B. J. (1987*b*). N-Nutrient fertilization, copper toxicity and growth in three grass species in the Netherlands. *Journal of Applied Ecology*, **24**, 1001–1010.

Ernst, W. H. O. (1976) Physiological and biochemical aspects of metal pollution. In: Mansfield, T. A. (Ed.) *Effects of Air Pollutants on Plants*, pp. 115–133. Cambridge University Press, Cambridge.

Ernst, W. H. O. (1990) Mine vegetation in Europe. In: Shaw, A. J. (Ed.) *Heavy Metal Tolerance in Plants: Evolutionary Aspects*, pp. 22–32, CRC Press, Boca Raton, Florida.

Evans, I. M., Gatehouse, L. N., Gatehouse, J. A., Robinson, N. J. and Croy, R. C. D. (1990) A gene from pea (*Pisum sativum* L.) with homology to metallothionein genes. *FEBS Letters*, **262**(1), 29–32.

Evans, K. M., Gatehouse, J. A., Lindsay, W. P., Shi, J., Tommey, A. M. and Robinson, N. J. (1992) Expression of the pea metallothionein-like gene *PsMT*$_A$ in *Escherichia coli* and *Arabidopsis thaliana* and analysis of trace metal ion accumulation: Implications for *PsMT*$_A$ function. *Plant Molecular Biology*, **20**, 1019–1028.

Fiedler, P. L. (1985) Heavy metal accumulation and the nature of edaphic endemism in the genus *Calochortus* (Liliaceae). *American Journal of Botany*, **72**, 1712–1718.

Fowler, D. P., Cape, J. N. and Unsworth, M. H. (1989) Deposition of atmospheric pollutants on forests. *Philosophical Transactions of The Royal Society of London, Series B*, **324**, 247–265.

Fox, C.A. (1980) The effect of air pollution on western larch as detected by tree-ring analysis. PhD, Arizona State University.

Fox, C. A., Kincaid, T. H., Nash III, T. H., Young, D. L. and Fritts, H. C. (1986) Tree-ring variation in western larch (*Larix occidentalis*) exposed to sulphur dioxide emissions. *Canadian Journal of Forest Research*, **16**, 283–292.

Freedman, B. and Hutchinson, T. C. (1980) Effects of smelter pollutants on forest leaf litter decomposition near a nickel–copper smelter at Sudbury, Ontario. *Canadian Journal of Botany*, **58**, 1722–1736.

Friedland, A. J. (1990) Movement of metals through soils and ecosystems. In: Shaw, A. J. (Ed.) *Metal Tolerance In Plants: Evolutionary Aspects*, pp. 22–32. CRC Press, Boca Raton, Florida.

Gartside, D. W. and McNeilly, T. B. (1974) The potential for evolution of heavy metal tolerance in plants. II: Copper tolerance in normal populations of different plant species. *Heredity*, **32**, 335–348.

Gildon, A. and Tinker, P. B. (1983*a*). Interactions of vesicular–arbuscular mycorrhizal infections and heavy metals in plants II. The effects of infection on uptake of copper. *New Phytologist*, **95**, 263–268.

Gildon, A. and Tinker, P. B. (1983*b*). Interactions of vesicular–arbuscular mycorrhizal infections and heavy metals in plants I. The effects of heavy metals on the development of vesicular-arbuscular mycorrhizas. *New Phytologist*, **95**, 247–261.

Gill, D. E. (1986) Individual plants as genetic mosaics: Ecological organisms versus evolutionary individuals. In: Crawley, M. J. (Ed.) *Plant Ecology*, pp. 321–343, Blackwell, Oxford.

Giller, K. E. and McGrath, S. P. (1988) Pollution by toxic metals on agricultural soils. *Nature*, **335**, 676.

Gregorius, H.-R. (1989) The importance of genetic multiplicity for tolerance of atmospheric pollution. In: Scholz, F., Gregorius, H. and Rudin, D. (Eds) *Genetic Effects of Air Pollutants on Forest Tree Populations*, pp. 163–172. Springer Verlag, Berlin.

Greszta, J., Braniewski, S., Marczynska, K., Alkowska, G. and Nose, A. (1979) The effect of dusts emitted by non-ferrous metal smelters on the soil, soil microflora and selected tree species. *Ekologia Polska*, **27**(3), 397–426.

Grill, E., Winnacker, E. L. and Zenk, M. H. (1988) Occurrence of heavy metal binding phytochelatins in plants growing in a mining refuse area. *Experimentia*, **44**, 539–540.

Grubb, P. J., Kelly, D. and Mitchley, J. (1982) The control of relative abundance in communities of herbaceous plants. In: Newman, E. I. (Ed.) *The Plant Community as a Working Mechanism*, pp. 79–97, Blackwell Scientific, Oxford.

Handa, A. V., Bresson, R. A., Handa, B. and Hasegawa, P. M. (1982) Tolerance to water and salt stress in cultured cells. In: Fujiwaro, A. (Ed.) *5th Int. Congress on Plant Tissue and Cell Culture*, pp. 471–475. Japanese Association of Plant Tissue Culture, Tokyo.

Harding, J. S. (1981) Regeneration of birch (*Betula pendula* Ehrh. and *Betula pubescens* Roth.). In: Newbold, A. N., Goldsmith, F. B. and Harding, J. S. (Eds) *Discussion Papers in Conservation no. 33*, pp. 83–112. University College of London, London.

Harper, J. (1981) *Population Biology of Plants*. Academic Press, New York.

Heikkinen, O. and Tikkanen, M. (1981) The effect of air pollution on growth in conifers: An example from the surroundings of the Skoldvik oil refinery. *Terra*, **39**, 134–144.

Hickey, D. A. and McNeilly, T. (1975) Competition between metal tolerant and normal plant populations—a field experiment on normal soil. *Evolution*, **29**, 458–464.

Holub, Z. and Zelenâkova, E. (1986) Tolerance of the reproductive processes of woods to the influence of heavy metals. *Ekologia*, **5**(1), 81–90.

Huang, B., Hatch, E. and Goldsbrough, P. B. (1987) Selection and characterisation of cadmium tolerant cells in tomato. *Journal of Plant Science*, **52**, 211–221.

Humphreys, M. O. and Nicholls, M. K. (1984) Relationships between tolerance to heavy metals in *Agrostis capillaris* (*A. tenuis* Sibth). *New Phytologist*, **98**, 177–190.

Hutchinson, T. C. (1984) Adaptation of plants to atmospheric pollutants. In: Evered, D. and Collins, G. M. (Eds) *Origins and Development of Adaptation*, pp. 52–72. Pitman, London.

Hutchinson, T. C. and Whitby, L. M. (1974) Heavy metal pollution in the Sudbury mining and smelting region of Canada. I. Soil and vegetation contaminated by nickel, copper and other metals. *Environmental Conservation*, **1**, 123–132.

Ietswaart, J. H., Griffioen, W. A. J. and Ernst, W. H. O. (1992) Seasonality of VAM infection in three populations of *Agrostis capillaris* (Gramineae) on soil with or without heavy metal enrichment. *Plant and Soil*, **139**, 67–73.

Jackson, P. J., Unkefer, P. J., Delhaize, E. and Robinson, P. J. (1990) Mechanism of trace metal tolerance in plants. In: Katterman, F. (Ed.) *Environmental Injury to Plants*, pp. 231–255. Academic Press, London.

Johnson, A. C. and Wood, M. (1987) Deionised distilled water as a medium for aluminium toxicity studies of *Rhizobium*. *Letters in Applied Microbiology*, **4**, 137–139.

Jones, E. W. (1945) Biological flora of *Acer pseudoplataus* L. *Journal of Ecology*, **32**, 220–238.

Jones, M. D. and Hutchinson, T. C. (1988*a*) Nickel toxicity in mycorrhizal birch seedlings infected with *Lactarius rufus* or *Scleroderma flavidum* I. Effects on growth, photosynthesis, respiration and transpiration. *New Phytologist*, **108**, 451–459.

Jones, M. D. and Hutchinson, T. C. (1988*b*) Nickel toxicity in mycorrhizal birch seedlings infected with *Lactarius rufus* or *Scleroderma flavidum* II. Uptake of nickel, calcium, magnesium, phosphorus and iron. *New Phytologist*, **108**, 461–470.

Jones, M. D., Dainty, J. and Hutchinson, T. C. (1988). The effect of infection by *Lactarius rufus* or *Scleroderma flavidum* on the uptake of ^{63}N by paper birch. *Canadian Journal of Botany*, **66**, 634–640.

Jowett, D. (1958) Populations of *Agrostis* spp. tolerant of heavy metals. *Nature*, **182**, 816–817.

Karban, R. (1990) Herbivore outbreaks on only young trees: testing hypotheses about aging and induced resistance. *Oikos*, **59**, 27–32.

Karnosky, D. F., Berrang, P. C., Scholz, F. and Bennett, J. P. (1989) Variation in and selection for air pollution tolerances in trees. In: Scholz, F., Gregorius, H. R. and Rudin, D. (Eds) *Genetic Effects of Air Pollutants in Forest Tree Populations*, pp. 29–37. Springer Verlag, Berlin.

Kuboi, T., Noguchi, A. and Yazaki, J. (1987) Relationship between tolerance and accumulation characteristics of cadmium in higher plants. *Plant and Soil*, **104**, 275–280.

Kumpfer, W. and Heyser, W. (1989) Zinc accumulation in beech mycorrhizae—A mechanism of zinc tolerance? *Agriculture Ecosystems and Environment*, **28**, 279–283.

Kurtz, S. M. (1982) *In-vitro* response of *Lycopersicum esculentum* to sodium chloride. In: Fujiwaro, A. (Ed.) *5th Int. Congress on Plant Tissue and Cell Culture*, pp. 479–480. Japanese Association of Plant Tissue Culture, Tokyo.

LeBlanc, D. C., Raynal, D. J. and White, E. H. (1987a). Acidic deposition and tree growth. I. The use of stem analysis to study historical growth patterns. *Journal of Environmental Quality*, **16**, 325–333.

LeBlanc, D. C., Raynal, D. J. and White, E. H. (1987b) Acidic deposition and tree growth. II. Assessing the role of climate in recent growth declines. *Journal of Environmental Quality*, **16**, 334–340.

Lepp, N. W. (1992) Uptake and accumulation of metals in bacteria and fungi. In: Adriano, D. C. (Ed.) *Biogeochemistry of trace metals*, pp. 277–298. Lewis Publishers, Boca Raton, Florida.

Lepp, N. W. and Dickinson, N. M. (1986) The effects of copper on germination of *Coffea* pollen: possible induction of copper tolerance. In *Proc. 2nd International Conference on Environmetal Contamination*, pp. 33–35. CEP Consultants, Amsterdam.

Levitt, J. (1980) *Responses of Plants to Environmental Stresses*, (2nd edn) Academic Press, New York.

Li, X.-L., Marschner, H. and George, E. (1991) Acquisition of phosphorus and copper by VA-mycorrhizal hyphae and root-to-shoot transport in white clover. *Plant and Soil*, **136**, 49–57.

Libby, W. J., Brown, A. G. and Fielding, J. M. (1972).Effects of hedging radiata pine on production and early growth of cuttings. *New Zealand Journal of Forest Science*, **2**, 263–283.

Macnair, M. R. (1981) Tolerance of higher plants to toxic materials. In: Bishop, J. A. and Cook, L. M. (Eds) *Genetic Consequences of Man-Made Change*, pp. 177–207. Academic Press, London and New York.

Macnair, M. R. (1987) Heavy metal tolerance in plants: A model evolutionary system. *Trends in Ecology and Evolution*, **2**(12), 354–359.

Macnair, M. R. (1990) The genetics of metal tolerance in natural populations. In: Shaw, A. J. (Ed.) *Heavy Metal Tolerance in Plants: Evolutionary Aspects*, pp. 235–255. CRC Press, Boca Raton, Florida.

McGrath, S. P. (1987) Long-term metal transfers following applications of sewage sludge. In: Coughtrey, P. J., Martin, M. H. and Unsworth, M. H. (Eds) *Pollutant Transport and Fate in Ecosystems*, pp. 301–317. Blackwell Scientific, Oxford.

McGrath, S. P., Baker, A. J. M., Morgan, A. N., Salmon, W. J. and Williams, M. (1980) The effects of interactions between cadmium and aluminium on the growth of two metal-tolerant races of *Holcus lanatus* L. *Environmental Pollution (Series A)*, **23**, 267–277.

Morselt, A. F. W., Smits, W. T. M. and Limonard, T. (1986) Histochemical demonstration of heavy metal tolerance in ectomycorrhizal fungi. *Plant and Soil*, **96**, 417–420.

Nicholls, M. K. and McNeilly, T. (1985) The performance of *Agrostis capillaris* L. genotypes, differing in copper tolerance in rye grass swards on normal soil. *New Phytologist*, **101**, 207–217.

Ochaia, E. I. (1987) *General Principles of Biochemisty of the Elements*. Plenum Press, New York.

Ojima, K. and Ohira, K. (1982). Characterisation and regeneration of an aluminium tolerant variant from carrot cell culture. In: Fujiwaro, A. (Ed.) *5th Int. Congress on Plant Tissue and Cell Culture*, pp. 475–476. Tokyo.

Outridge, P. M. and Hutchinson, T. C. (1991) Induction of cadmium tolerance by acclimation transferred between ramets of the clonal fern *Salvinia minima* Baker. *New Phytologist*, **117**, 597–605.

Passow, H., Rothstein, A. and Clarkson, T. W. (1961) The general pharmacology of heavy metals. *Pharmacological Review*, **13**, 185–224.

Pitelka, L. F. (1988) Evolutionary responses of plants to anthropogenic pollutants. *Trends in Ecology and Evolution*, 3(9), 233–236.

Prat, S. (1934) Die ehrlichkeit der resistenz gegen küpfer. *Berichte der Deutschen Botanischen Gesellschaft*, 52, 65–67.

Pye, J. M. (1988). Impact of ozone on the growth and yield of trees: A review. *Journal of Environmental Quality*, 17(3), 347–360.

Qureshi, J. A., Colin, H. A., Hardwick, K. and Thurman, D. A. (1981) Metal tolerance in tissue cultures of *Anthoxanthum odoratum*. *Plant Cell Reports*, 1, 80–82.

Robinson, N. J. (1990) Metal binding polypeptides in plants. In: Shaw, A. J. (Ed.) *Metal Tolerance in Plants: Evolutionary Aspects*, pp. 195–215. CRC Press, Boca Raton, Florida.

Robinson, N. J. and Jackson, P. J. (1986). "Metallothionein-like" metal complexes in Angiosperms; their structure and function. *Physiologia Plantarum*, 67, 499–506.

Robinson, N. J. and Thurman, D. A. (1986) Involvement of metallothionein-like copper complex in the mechanism of copper tolerance in *Mimulus guttatus*. *Proceedings of the Royal Society, Lond. (B)*, 227, 493–501.

Schaedle, M., Thornton, F. C., Raynal, D. J. and Tepper, H. B. (1989) Response of tree seedlings to aluminium. *Tree Physiology*, 5, 337–356.

Schmid, B. (1992) Phenotypic variation in plants. *Evolutionary Trends in Plants*, 6(1), 45–60.

Schultz, C. L. and Hutchinson, T. C. (1988) Evidence against a key role for metallothionine-like protein in the copper tolerance mechanism of *Deschampsia caespitosa* (L.) Beauv. *New Phytologist*, 110, 163–171.

Schulze, E. D. (1989) Air pollution and forest decline in a spruce (*Picea abies*) forest. *Science*, 244, 776–783.

Searcy, K. B. and Mulcahy, D. L. (1985) The parallel expression of metal tolerance in pollen and sporophytes of *Silene dioica* (L.) Clairv., *S. Alba* (Mill) Krouse and *Mimulus guttatus* (D.C.). *Theoretical and Applied Genetics*, 69, 597–602.

Shaw, A. J. (1990) *Heavy Metal Tolerance in Plants: Evolutionary Aspects*. CRC Press, Boca Raton, Florida.

Shaw, J. (1987) Effect of environmental pretreatment on tolerance to copper and zinc in the moss *Funaria hygrometrica*. *American Journal of Botany*, 74(10), 1466–1475.

Simon, E. (1978) Heavy metals in soils, vegetation development and heavy metal tolerance in plant populations from metalliferous areas. *New Phytologist*, 81, 175–188.

Smith, R. A. H. and Bradshaw, A. D. (1979) The use of metal tolerant plant populations for the reclamation of metalliferous wastes. *Journal of Applied Ecology*, 16, 595–612.

Smith, S. E., Meiners, M., Putzar, N. and Dobrenz, A. K. (1989) Potential use of measurements of callus and male gametophyte NaCl tolerance in lucerne breeding. *Euphyta*, 43, 245–251.

Smith, S. R. and Giller, K. E. (1992) Effective *Rhizobium leguminosarum* biovar *Trifolii* present in five soils contaminated with heavy metals from long-term applications of sewage sludge or metal mine spoil. *Soil Biology and Biochemistry*, 24(8), 781–788.

Smith, W. H. (1974) Air pollution-effects on the structure and function of the temperate forest ecosystem. *Environmental Pollution*, 6, 111–129.

Steffens, J. C. (1990) The heavy metal binding peptides of plants. *Annual Review of Plant Physiology and Plant Molecular Biology*, 41, 553–575.

Symeonidis, L., McNeilly, T. and Bradshaw, A. D. (1985) Differential tolerance of three cultivars of *Agrostis capillaris* L. to cadmium, copper, lead, nickel and zinc. *New Phytologist*, 101, 309–315.

Tanksley, S. D., Zamir, D. and Rick, C. M. (1981) Evidence for extensive overlap of sporophytic and gametophytic gene expression in *Lycopersicon esculentum*. *Science*, **213**, 453–455.

Ten Hoopen, H. J. G., Nobel, P. J., Schapp, A., Fuchs, A. and Roels, J. A. (1985) Effects of temperature on cadmium toxicity to the green alga *Scenedesmus acta*. I: Development of cadmium tolerance in batch cultures. *Antonie van Leeuwenhoek*, **51**(3), 344–346.

Thompson, J. D. (1991) Phenotypic plasticity as a component in evolutionary change. *Trends in Ecology and Evolution*, **6**, 246–249.

Thurman, D. A., Salt, D. E. and Tomsett, A. B. (1989) Copper phytochelatins of *Mimulus guttatus*. In: Hamer, D. H. and Winge, D. R. (Eds) *Metal Ion Homeostasis: Molecular Biology and Chemistry*, pp. 328–347, Alan R. Liss, New York.

Tomsett, A. B. and Thurman, D. A. (1988) Molecular biology of metal tolerance in plants. *Plant Cell and Environment*, **11**, 383–394.

Tomsett, A. B., Salt, D. E., De Miranda, J. R. and Thurman, D. A. (1989) Metallothioneins and metal tolerance. *Aspects of Applied Biology*, **22**, 365–372.

Tukendorf, A. (1989) Characteristics of Cu-binding proteins in chloroplasts of spinach tolerant to excess copper. *Journal of Plant Physiology*, **135**, 280–284.

Turkington, R. A. (1989) The growth, distribution and neighbour relationships of *Trifolium repens* in permanent pasture. VI. Conditioning effects by neighbours. *Journal of Ecology*, **77**, 734–746.

Turkington, R. A. and Harper, J. L. (1979) The growth and distribution and neighbour relationships of *Trifolium repens* in permanent pasture. IV. Fine-scale biotic differentiation. *Journal of Ecology*, **67**, 245–254.

Turner, A. P. (1991) *The Tolerance of Trees to Heavy Metals*. PhD, CNAA Liverpool Polytechnic.

Turner, A.P. and Dickinson, N. M. (1993*a*) Copper tolerance of *Acer pseudoplatanus* L. (Sycamore) in tissue culture. *New Phytologist*, **123**, 523–530.

Turner, A. P. and Dickinson, N. M. (1993*b*) Survival of *Acer pseudoplatanus* L. (Sycamore) seedlings on metalliferous soils. *New Phytologist*, **123**, 509–521.

Turner, A. P., Dickinson, N. M. and Lepp, N. W. (1991) Indexes of metal tolerance in trees. *Water Air And Soil Pollution*, **57–58**, 617–625.

Turner, J. S., Morby, A. P., Whitton, B. A., Gupta, A. and Robinson, N. J. (1993) Construction of Zn^{2+}/Cd^{2+} hypersensitive cyanobacterial mutants lacking a functional metallothionein locus. *The Journal of Biological Chemistry*, **268**(6), 4494–4498.

Verkleij, J. A. C. and Schat, H. (1990) Mechanisms of metal tolerance in higher plants. In: Shaw, A. J. (Ed.) *Heavy Metal Tolerance in Plants: Evolutionary Aspects*, pp. 179–193, CRC Press, Boca Raton, Florida.

Wilkins, D. A. (1957) A technique for the measurement of lead tolerance in plants. *Nature*, **180**, 37–38.

Wilkins, D. A. (1978) The measurement of tolerance to edaphic factors by means of root growth. *New Phytologist*, **80**, 623–633.

Wilkins, D. A. (1991) The influence of sheathing (ecto-) mycorrhizas of trees on the uptake and toxicity of metals. *Agriculture Ecosystems and Environment*, **35**, 245–260.

Wolstenholme, R., Dutch, J. D., Ferrier, R. C. and Edwards, A. C. (1991) Nutrient and metal mobility in forest soils subject to organic fertilisation with sewage sludge. Presented at *Soil Pollution* conference organised by British Society of Soil Science and Society of the Chemical Industry, London, 11–12 September 1991. Unpublished.

Wolters, J. H. B. and Martens, M. J. M. (1987). Effects of air pollutants on pollen. *The Botanical Review*, **53**(3), 373–409.

Wu, L. and Antonovics, J. (1978) Zinc and copper tolerance of *Agrostis stolonifera* in tissue culture. *American Journal of Botany*, **65**(3), 268–271.

Wu, L., Bradshaw, A. D. and Thurman, D. A. (1975) The potential for evolution of heavy metal tolerance in plants. III. The rapid evolution of copper tolerance in *Agrostis stolonifera*. *Heredity*, **34**, 165–187.

Zagory, D. and Libby, W. J. (1985). Maturation-related resistance of *Pinus radiata* to western gall rust. *Phytopathology*, **75**, 1443–1447.

5 Toxic Metals: Fate and Distribution in Contaminated Ecosystems

SHEILA M. ROSS
University of Bristol, UK

ABSTRACT

The complexity of ecosystems poses many problems for scientists interested in measuring the fate and distribution of contaminating metals. Two main types of approaches have been used in a range of different habitats to characterise the input, output and within-system distribution and transfers of metals: the "ecosystem audit" method and the "flux" method. In the main, flux approaches have been used by environmental scientists, especially hydrologists and hydrochemists, who measure rainfall, streamflow and ecosystem hydrological processes to chart the inputs, outputs and within-system hydrological transfers. Compartment audit approaches have been used by biologists who identify the compartments of interest in the ecosystem: leaves, fruits, wood, roots, soil organisms, etc., and measure their biomass amounts, metal concentrations and hence metal standing stocks. Methodological problems associated with these approaches are discussed. These include: difficulties in obtaining statistically representative samples, and sampling difficulties due to seasonal variability in compartment metal concentrations.

Factors influencing metal fate and distribution in ecosystems are: (i) metal input characteristics, i.e. whether from aerial deposition, sludge application to the soil surface, or mining and mineral weathering; (ii) vegetation characteristics; and (iii) soil and litter characteristics. Influencing vegetation characteristics include canopy geometry and density, leaf characteristics, rhizosphere effects, and differential metal uptake by different plant species, genera and families.

Relative topsoil enrichment (RTE) values, relating topsoil metal concentrations to subsoil metal concentrations, and concentration factors (CF), relating plant metal concentrations to soil metal concentrations, can be used to compare contaminated sites. In metal-impacted woodlands, forests, agriculture and wetlands, soils and sediments act as sinks for metals. Calculations based on metal influx and accumulation rates in forestry and agriculture indicate soil metal residence times of hundreds to thousands of years.

Toxic Metals in Soil–Plant Systems. Edited by S. M. Ross
© 1994 John Wiley & Sons Ltd

5.1 INTRODUCTION

Ecosystems, even man-made agricultural or forestry systems, are highly complex organisations of organisms, living, growing, reproducing and dying in response to the surrounding environment. Key characteristics of all ecosystems are *interactions* between individual organisms, between groups of organisms and between organisms and different environmental factors. Since both essential and non-essential metals may become toxic when their concentrations exceed certain levels, metal inputs can be considered as a stress and a perturbation to the system, potentially affecting species diversity, organism productivity and perhaps disrupting, or at least altering the overall functioning of the system in terms of nutrient cycling, water use or energy balance. There are several problems in attempting to examine how metals affect whole ecosystems. First, in order to understand how a perturbation influences ecosystem diversity and function, great detail about the system *prior* to perturbation should exist. This is rarely possible, particularly with metal pollution, and *a posteriori* comparisons must be made with other "control" systems. Secondly, the processes of metal transfer in ecosystems are rarely studied directly, but are calculated from changes in metal concentration and quantity in measured compartments of the system over time. Omission of a key compartment, considered *a priori* to be unimportant in comparison to other, larger, or more easily measured, compartments, may mean that while metal budgets can be calculated, *interpretation* of metal fluxes may be difficult or impossible. Thirdly, the vast majority of studies are short term, or even single sampling date analyses. Relatively few long-term (> 10 year) whole ecosystem studies have been published. A fourth problem also relates to the time element of ecosystem processes. A seasonality of soil metal availability and of foliar metal concentrations has been indicated in several studies, but long-term studies, by necessity, tend to measure metal budgets annually, potentially losing any seasonal dynamic pattern.

From this list of initial problems in studying the fate and effect of metals on ecosystems, it would appear that understanding the fate and distribution of metals in ecosystems may involve a series of different, but complimentary approaches, perhaps carried out at different spatial and temporal scales.

Smith (1974) classified the effects of air pollution on ecosystems into three categories:

— Class I: *Low level contamination*—ecosystem acts as an unaffected sink
— Class II: *Medium contamination*—sublethal adverse effects on individuals or populations
— Class III: *High contamination*—lethal effects on individuals or populations

Examples of class III metal contamination exist around smelters or in the proximity of metallurgical industries. Other severe contamination effects are usually caused by factors other than metals, but may be exacerbated by metal pollution. Such examples exist in parts of north-west Europe, affected by acid precipitation. In the main, researchers have concentrated their efforts on understanding the processes operating at lower level contamination class I and II sites and how metal accumulation may cause class I sites to pass into the class II category. It is with these sublethal metal conditions that this chapter is concerned.

A very wide range of experiences are reported in this text concerning the fate of metals in soil—plant systems. Is it possible to categorise the fate and effects of potentially toxic metals according to metal type, their origin, the system on which they impinge, the soil type or the vegetation type? The main aims of this chapter are thus threefold:

(i) to review *methods* of comparing relative accumulation and distribution of metals in different contaminated ecosystems, and to discuss their associated problems;

(ii) to examine the *factors* which influence distribution of metals in contaminated ecosystems; and

(iii) to compare the *distribution* of metals in selected different ecosystem types, including forests, grasslands, wetlands and agricultural systems.

While an attempt is made to draw examples from a range of different habitats, by far the bulk of published literature is on the fate of metals in forests and agricultural crops.

5.2 METHODOLOGICAL PROBLEMS IN STUDYING THE FATE AND EFFECTS OF METALS IN ECOSYSTEMS

In the main, approaches to characterising the trace metal budgets of contaminated ecosystems have examined the *concentrations* and have calculated *quantities* of metals in different above- and below-ground compartments of the system. Hughes (1981) has called this the "ecosystem audit" approach to studying metal cycling in ecosystems. Some long-term studies (>10 years) have now been reported and are beginning to indicate the patterns of change in *quantities* of metals in different ecosystem compartments over time. Changes in compartments over time are used to calculate metal *fluxes* between system compartments. Apart from hydrochemical processes, such as through-fall (e.g. Heinrichs and Mayer, 1977, 1980a) and soil leaching (e.g. Tyler, 1981; Bergkvist, 1987), extremely few studies have attempted to quantify ecosystem metal fluxes *directly*. This is mainly because most of the internal, ecosystem processes would be extremely difficult, if not impossible, to measure. Illustrations of the "compartment audit" and "flux" approaches to characterising the fate of metals in ecosystems are given in Figure 5.1. An example of the compartment audit approach for Cd, Pb, Zn and Cu in a contaminated woodland is presented by in Figure 9.2. In the following sections, the compartment audit approach will be discussed and problems associated with calculating fluxes, or of measuring them, will be assessed.

5.2.1   COMPARTMENT STUDIES OR ECOLOGICAL "AUDITS"

At the largest scale, whole watershed input–output studies of heavy metals in ecosystems (e.g. Swanson and Johnson, 1980) give a good idea of the gross retention, accumulation or mobilisation of metals by plants and soils. Superimposed on these studies are smaller scale, more detailed, metal budget studies, usually of representative plot- or field-sized, intact ecosystems (e.g.

'COMPARTMENT' APPROACH

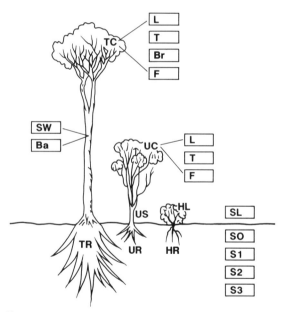

Key :

TC - tree canopy	**UC** - understorey canopy
L - leaves	**US** - understorey stems
T - twigs	**UR** - understorey roots
Br - branches	**HL** - herbaceous leaves
F - fruits	**HR** - herbaceous roots
SW - stemwood	**SL** - soil litter layer
Ba - bark	**SO** - soil organic horizon(s)
TR - tree roots	S_1
	S_2 -- 3 depths of soil horizons
	S_3

Parker *et al.*, 1978; Martin and Coughtrey, 1987; Levine *et al.*, 1989). At an even smaller scale, individual parts of the system may be studied in detail *in situ*, in intact lysimeters (e.g. Sheppard and Sheppard, 1991) or, usually only for agricultural systems, in field plots (e.g. Latterell *et al.*, 1978; Rappaport *et al.*, 1988), glasshouse pots (e.g. Haq *et al.*, 1980; LeClaire *et al.*, 1984; Tadesse *et al.*, 1991) or tubs (e.g. Jackson and Alloway, 1991) in which experiments can be designed to examine individual processes in detail. Pot experiments are an attractive option for studying plant metal uptake under controlled conditions, but suffer from over-simplification and difficulties in relating results to real field conditions.

In a review of aerial heavy metal pollution on terrestrial ecosystems, Hughes *et al.* (1980) divide ecosystems up into four types of "trapping zones": leaves, stems, epiphytes and litter + soil. For most purposes and most ecosystems, this system compartmentalisation is too crude to be useful for charting the fate of

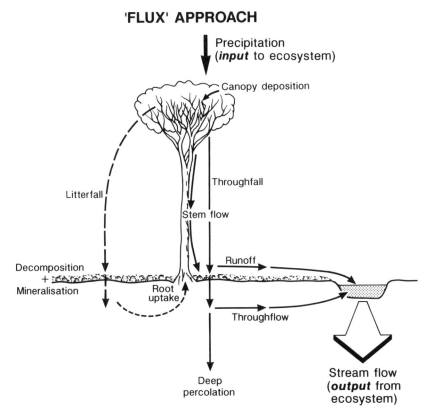

Figure 5.1. Diagrammatic comparison of the "compartments audit" and "flux" approaches to studying distribution and fate of metals in ecosystems

heavy metals. Most commonly, forestry studies, for example, divide the ecosystem into various plant and soil standing pools (see Figure 5.1(a)):

— *canopy components*: leaves, twigs, branches, fruits
— *litterfall*
— *stem*: bark, stemwood
— *soil*: litter layer, organic horizon(s), topsoil, subsoil

Two such compartment studies in forestry are in the Black Forest (e.g. Zottl, 1985) and in the Solling region (e.g. Heinrichs and Mayer, 1977, 1980*a*) of Germany. In herbaceous ecosystems and in agriculture, compartment audits are usually much simpler, at best identifying: leaves, fruits, stems, roots and topsoil. Examples of such compartment studies include metal contamination near the Cu–Ni smelter at Coniston, Sudbury, Ontario (Hutchinson and Whitby, 1974), Pb budgetting in grassland (Brabec *et al.*, 1983) and Cr budgetting in fescue grass (Taylor, 1983). Only in a few studies has detailed quantification of metals in soil organism compartments (e.g. Chapter 9, this volume; Hunter *et al.*, 1987*a*; Roth, 1992) or animal compartments (e.g. Munshower and Neuman, 1979; Hunter *et al.*, 1987*b*) been given.

There are several difficulties in attempting to compare different ecosystem metal audits. Hughes (1981) identifies four main problems:

1. rarely is the ecosystem divided up into similar biological or soil compartments,
2. rarely is the same element analysed,
3. different sampling strategies are frequently adopted, and
4. different analytical techniques are frequently adopted.

For these reasons, Hughes suggested that only very *general* conclusions can be drawn from comparisons of ecosystem metal audits. Compartment studies may be potentially useful in examining different responses by tolerant and non-tolerant systems. Long-term compartment studies are potentially useful in identifying how ecosystems of differing sensitivity have responded to metal inputs, for example, through air pollution. Using retrospective studies may help in *predicting* how systems may respond in the future to similar metal stresses and perturbations, or how the same ecosystems may respond to *continued* metal stress. Where it is possible to combine the compartmental and time series approaches, very useful information can be generated about ecosystem fluxes, time responses to perturbation and tolerance of whole systems or system components.

There are two key problems associated with using compartment differences over time to calculate metal fluxes. First, the time interval between measurements is frequently at least one year. Annual differences represent *gross* effects and may mask dynamic detail related to seasonal or other environmental or biological influences on the ecosystem. Secondly, the difference approach places heavy reliance on representative sampling and accurate analyses of component parts. It is more than likely that many published results fall seriously short of statistical requirements for representative sampling. To obtain a statistically representative sample for determining concentration of total soil Pb, Zn, Ni, Cu and Cd in 1 ha, Wopereis *et al.* (1988) calculated that the required number of soil samples would be 4, 7, 7, 16 and 121 respectively. Existing data indicate that the concentrations of trace metals in biological tissues, such as plant material (e.g. Garten *et al.*, 1977) and in soil organisms and animals in food chains (e.g. Hopkin, 1993), may not show a normal distribution. If this is the case, and statistical calculations (e.g. means and standard deviations) or statistical analyses (e.g. simple types of analysis of variance) are based on populations of data showing skewed metal concentration distributions, the results could be seriously biased. Similar criticisms could be aimed at models which use mean concentrations or fluxes based on populations of data with skewed concentration distributions.

5.2.2 MEASURING SYSTEM METAL FLUXES

Evidence of the strength of metal retention in contaminated soils (Chapter 3) indicates that metal accumulation in the food chain, with effects on vegetation, including crops, and on animals, including humans, is a higher probable risk than is metal leaching through soils and contaminating groundwaters. To assess the likelihood of these risks, information is needed on potential transfers of metals from soils to plants and from plants to animals. This is not a simple task. We have already seen that just one part of the story—transfers of metals from soils to plants—involves a large number of "unknowns", including: adsorption and exchange in the soil, mass flow and diffusion of metal ions, precipitation and dissolution of solid phases, root biomass and distribution in the soil in relation to soil metal sources, passive and active uptake processes at the root cellular level and microbial relations in the rhizosphere. Some of these characteristics and processes which relate to soil metals have not been properly quantified in isolation, far less as components in a complete system, linking soil metal to metal species uptake by roots. Metal transfers in ecosystems are thus highly complex. In general, methods to characterise metal fluxes in ecosystems have either concentrated on a single, small aspect of the system (e.g. modelling the root zone metal transfer by Youssef and Chino (1991)), or have quantified system compartments over time and implied fluxes from measured compartment differences (e.g. Martin and Coughtrey (1987), for downprofile soil compartments).

5.2.2.1 Seasonality of metal fluxes

Soil–plant relationships and the seasonality of soil chemistry, element bioavailability and plant physiological responses involve numerous processes that are only just beginning to be examined. There is now good evidence of

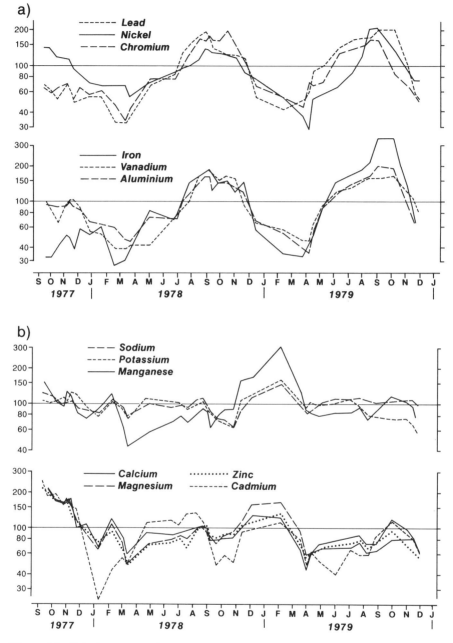

Figure 5.2. Relative seasonal variation of metal concentrations in soil lysimeter percolates under spruce in south Sweden (mean of the whole study period = 100 for each metal). Reproduced from Tyler (1981) by permission of Kluwer Academic Publishers

seasonal patterns in metal concentrations at different stages in the food chain: incoming rainfall, soil, plant foliage, invertebrates and higher animals. Seasonal variations in the influx of trace metals in precipitation were observed by Ross (1987) in the rainfall monitored at six locations in Sweden. Cd, Pb and Zn showed peaks in concentration from February to April and lower concentrations in summer from July to September. Cu, Fe and Mn showed peaks a little later, from April to June, and low concentrations from August to January. The differences in metal concentration were related both to rainfall characteristics, such as amount and intensity, and to air pollution and climatic characteristics such as season and incidence of cyclonic or anticyclonic conditions.

Several lysimeter studies report seasonal leaching patterns of soil metals (e.g. Tyler, 1981; Berkvist, 1987). Tyler (1981) used rather shallow lysimeters, of 29 cm depth. Soil water percolates in the top 29 cm represent patterns of mobility in the rooting zone, and solutes percolating below 29 cm may represent mobilisation in excess of root uptake. Tyler (1981) identified two types of seasonal metal mobilisation patterns (Figure 5.2). One group of metals, comprising Fe, Al, Pb, Ni, Cr and V, all showed distinct peaks in leachate concentration in late summer and autumn, with clear minima in winter and early spring. This is the same pattern as seen in soluble organic matter in soil leachates. A second group of metals, comprising the alkali metals: K, Na, Ca, Mg and the trace metals Mn, Zn and Cd, showed concentration maxima in winter. The mobility of these elements is negatively related to pH. Although Tyler (1981) found that late summer leachates were only 0.1–0.2 pH units higher than in the winter, this appeared sufficient to imbalance cation exchange processes and mobilise the second group of metals. The metal mobilising effects of low soil pH and soluble organic matter have been reported in many studies but seasonal variation in soil metal availability *per se* has not been confirmed.

There are now many examples of plants growing in contaminated sites whose foliar metal concentrations show seasonal variations. Grassland species and herbs show foliar metal concentration maxima in winter and troughs in summer (e.g. Crump *et al.*, 1980; Matthews and Thornton, 1980; Hunter *et al.*, 1987c). These authors suggested that aerial deposition was a major factor affecting foliar metal variability, but that soil uptake was also important. The seasonal pattern of Cd and Cu concentrations in the foliage of two grasses and the horsetail *Equisetum arvense* in the vicinity of a Cu refinery on Merseyside in northern England are illustrated in Figure 5.3(a) and (b). There is a more obvious lowering in summer Cu concentrations in the foliage of all three plants than there is for Cd. Hunter *et al.* (1987a,b,c) studied the transfer of Cd and Cu through the grassland food chain. Cd and Cu concentrations in invertebrates from the same Merseyside site showed very clear peaks in metal concentrations in the summer months, with lowest concentrations in winter months. Seasonal metal concentration patterns in two

198

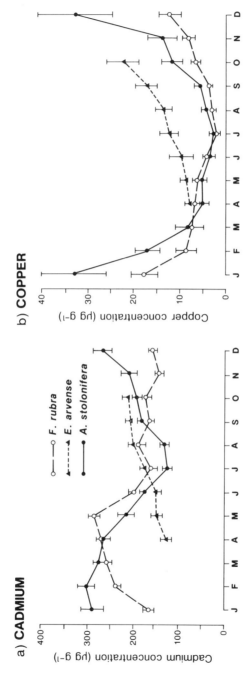

Figure 5.3. Seasonal variation in the (a) cadmium and (b) copper concentration (μg g^{-1} dry wt $\pm$ SE) of *in situ* vegetation near a Cu refinery. ($\bullet$) *Agrostis stolonifera*, ($\circ$) *Festuca rubra*, ($\blacktriangle$) *Equisetum arvense*. Reproduced by permission of Blackwell Scientific Publications from Hunter *et al.* (1987c)

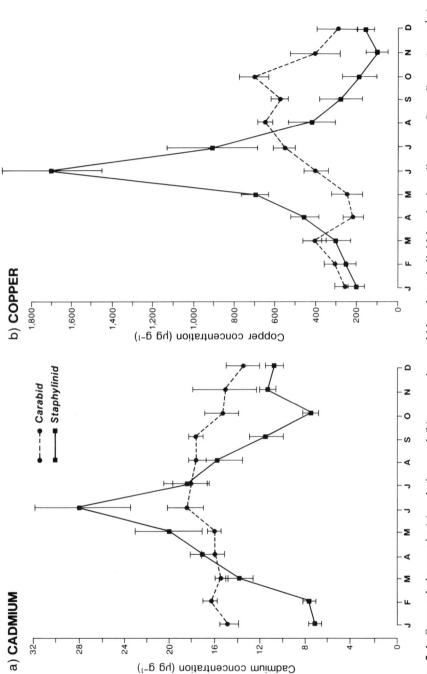

Figure 5.4. Seasonal changes in (a) cadmium and (b) copper in carabid and staphylinid beetles in soil near a Cu refinery (μg g^{-1} dry weight $\pm$ SE). Reproduced by permission of Blackwell Scientific Publications from Hunter et al. (1987a)

types of beetle are illustrated in Figure 5.4(a) and (b). Hunter *et al.* (1987*a*) went on to show that small mammals feeding on the herbage and invertebrates experienced seasonal fluctuations in their diet. In field vole, wood mouse and common shrew, the body tissue Cd concentrations closely paralleled diet Cd concentrations.

Foliar metal concentrations in woody plants have also been shown to vary seasonally. In a study of the trace metal concentrations in 18 different deciduous tree species, Guha (1961) found clear evidence of accumulation in actively growing tissues, such as shoots and young leaves. Many of the species, such as sycamore, beech and horse chestnut showed accumulation of Fe and Zn in leaves at the end of the growing season. Other studies have confirmed this concentration of metals in leaves of deciduous trees prior to shedding. Martin and Coughtrey (1982), for example, found generally higher Zn and Pb in hazel leaves and higher Cd and Pb in leaves of field maple at the end of the growing season in October, compared to earlier in the year.

Evidence of seasonality and variability in foliar metal concentrations of a range of different vegetation types suggests that either soil metal availability shows a seasonal pattern, or that growth dilution or metal shunting occurs in plant tissues. To characterise such complex processes requires careful sampling scheme selection in order to obtain representative tissue material. Seasonal variations complicate the interpretation of simple, short-term studies and of model laboratory or greenhouse studies performed under controlled conditions using test plants and simple forms of metals which are frequently different from those forms found in natural environments. For realistic assessments of environmental toxicity and metal retention or accumulation, we require longer-term studies of the fate of toxic metals in real ecosystems. There are now examples of a few such studies, of which other chapters in this volume by McGrath (Chapter 6), Brewer, Benninger-Truax and Barrett (Chapter 7), and Martin and Bullock (Chapter 9) are examples.

5.3 FACTORS INFLUENCING DISTRIBUTION OF METALS IN CONTAMINATED SOIL–PLANT SYSTEMS

Key influences on the eventual distribution of metal in an ecosystem include (i) a series of metal input characteristics, (ii) a series of vegetation characteristics, and (iii) a series of litter and soil profile characteristics.

5.3.1 METAL INPUT INFLUENCING METAL DISTRIBUTION

There are three main routes of metal input to ecosystems:

(i) aerially, involving soluble forms of metals, metal aerosols, and metal particles from industrial pollution, domestic and vehicular fuel combustion, mine wastes and smelter;

(ii) to the soil surface, involving industrial and domestic waste sludges, irrigation waters; and

(iii) to the subsoil, via weathering of parent rocks, mineworking, minewastes or waste dumps.

The role of the first two routes of metal input to plant uptake and the fate of metals from different sources will be only briefly assessed, since various reviews of aerial metal inputs (e.g. Hughes *et al.*, 1980; Nriagu and Pacyna, 1988; Bergkvist *et al.*, 1989) and sludge metal inputs (e.g. McGrath, 1987; Henry and Harrison, 1992; Juste and Mench, 1992) to ecosystems have already been published.

5.3.1.1 Aerial metal inputs

A number of factors determine the form and quantities of atmospheric metal input and fate in the receiving plant, soil or water system, including: (i) particle size, (ii) solubility, (iii) distance of receiving system from metal source, and (iv) acidity of rainfall. Atmospheric metal aerosols are present in a range of particle diameters, from $0.01-1.0 \mu m$ (car exhaust lead, oil smoke, metallurgical fumes), and $1.0-100 \mu m$ (pulverised fuel ash, metallurgical dust) to $10-80 \mu m$ (stoker fly ash) (summarised by Livett, 1988). The largest metal pollution particles fall as dry deposition close to their source, while smaller particles can be transported long distances, particularly if they achieve sufficient altitude at their point of origin. Particles $< 1 \mu m$ diameter act as focii for condensation. "Washout" or "rainout" transfer soluble metal components from the atmosphere as wet deposition. Wet deposition is the dominant process of metal influx to ecosystems at further distances from the metal source, particularly when aerosols reach rain-forming altitudes. Pb and Cd are associated with particles $< 1 \mu m$ in diameter, which were found to have atmosphere residence times of around 7 days: estimated to be sufficient for transport over thousands of kilometres (Pacyna, 1987). Evidence of such long-distance movement of metals in the atmosphere comes from the discovery of highly elevated metal concentrations in polar ice cores (reviewed by Peel, 1989). This atmospheric pollution is predominantly a Northern Hemisphere problem. Shaw (1989), for example, quotes a 400-fold increase in Pb deposition in contemporary Greenland ice compared to that of Antarctica.

Techniques used to monitor atmospheric metal inputs to ecosystems have been reviewed elsewhere (e.g. Hughes *et al.*, 1980). The main types of collectors used for atmospheric aerosols are: total deposit gauges, dry deposit gauges and moss bags. The latter were designed to simulate deposition on naturally growing epiphytic mosses, but now more widely used for air pollution assessment. Quantities of trace metals input in precipitation and dry deposition vary according to distance from polluting sources. Table 5.1 summarises the results of trace metal concentration in precipitation from different parts of the world. The extremely low minimum values reported for

Table 5.1. Metal concentrations in precipitation

Location	Metal concentration range (μg litre^{-1})					Source
	Pb	Cd	Cu	Zn	Mn	
North-east Scotland	0.6–29.0	0.1–1.52	0.2–13.0	2.5–95.0	0.8–21.0	Balls (1987)
Southern New Jersey	4–118	<0.1–5.1	<1–16.0	—	1.0–112	Swanson and Johnson (1980)
Northern Germany (Solling)	11–14	0.19–0.35	2.3–2.5	320	—	Schultz (1987)
Southern Sweden	7.9–8.5	0.13–0.16	1.3–2.0	25–37	—	Bergkvist et al. (1989)
Sweden (ranges of six sites)	0.03–45	0.01–0.77	0.13–6.3	0.9–66	0.08–36	Ross (1987)

Sweden (Ross, 1987) are for a northern, remote monitoring station. The pattern of higher concentrations of Zn, Pb and Mn in precipitation compared to Cd and Cu are typical of all reported results.

The acidity of rainfall has been shown to influence solubilities of metals in the atmosphere. Zimmerman (1986) found that the solubility of Pb in deposition increased significantly when rain pH decreased from 6.4 to 3.4. The solubility of Fe and Mn were unaffected by rain pH. Solubility of metal aerosols determines whether they are predominantly deposited in wet or dry form. Grosch (1986) observed that Pb and Cd were mainly found in wet deposition, while Mn and Fe were mainly found in dry deposition. Insoluble particles in dry deposition can also act as adsorbing surfaces to remove solubilised metals from solution in washout. Rohbock (1986) measured the proportions of soluble Pb, Cd, Mn and Fe in precipitation and in dry deposition (Table 5.2). These data confirm that Pb and Cd are highly soluble in precipitation, but both, particularly Pb, are much less soluble in dry deposition. The lower solubilities of metals in dry deposition will cause a leaf surface accumulation until rainfall physically washes off the deposits.

Bergkvist (1987) calculated that bulk atmospheric deposition accounted for around 50% of the total input (including litterfall) of Pb and Cd to the forest floor of a spruce forest in south-west Sweden. The corresponding value for Cu, Zn, Cr and Ni was lower, at 10–30%. At altitude, fog and other forms of occult deposition are frequently more important than bulk deposition. The proportion that fog can contribute to the metal content of the forest floor can be high. Kazda (1986) reported the fog contribution to the forest floor of a beech forest in Austria. The Cd contribution was highest, at 43%, with 24% for Pb, 24% for Ni and 15% for Cu.

Plants can take up metals both from the atmosphere via leaf surfaces and from the soil via roots. Washout landing on leaf surfaces can contain high proportions of soluble metals (Table 5.2), especially if rainfall is acidic. Cd and Mn in dry deposition may also be relatively soluble. Little (1973) identified three fractions of heavy metal cover on leaf surfaces: a water-soluble fraction (removable with distilled water), a physically attached fraction (removable

Table 5.2. Average percentages of soluble Pb, Cd, Mn and Fe in precipitation and in dry deposition, measured at 11 sites in West Germany

Metal	Precipitation (%)	Dry deposition (%)
Pb	90	10
Cd	95	55
Mn	85	50–55
Fe	55–65	5

Source: Rohbock (1986).

with 2% detergent solution) and an exchangeable fraction (removable with weak acid). By far the largest metal fraction reported for Cd, Zn and Pb deposits on the leaves of a number of tree species is the exchangeable component (Little, 1973; Lindberg and Harriss, 1981; Hagemeyer et al., 1986). Metal ions present in solution at the leaf surface can diffuse into leaf cells or may bind to the cuticle or cell wall. There is no endodermis in leaf tissues to regulate metal diffusion from the cell apoplast to vascular tissues (see Chapter 2, Section 2.2.2, this volume). Proportions of metals diffusing directly into leaf tissue were not thought to be high enough to contribute greatly to the metal fund of plants. However, Harrison and Johnston (1987) and Harrison and Chirgawi (1989a) have shown that aerial sources of metals can contribute > 90% of the Pb present in the leaves of various food plants and > 80% of the Cr and Zn. Lepp and Dollard (1974) have also shown that aerially derived metals can also be taken up from the bark. This is an important finding since metals concentrations in bark of contaminated forest systems is commonly high (see Section 5.5.1).

Comparisons of metal concentrations in incident precipitation with those in subsequent throughfall have indicated that several metals can become concentrated in the canopy (Lindberg and Harriss, 1981; Grosch, 1986), but this depends on tree and canopy type (see Section 5.3.2). Dry deposition and insoluble particles are trapped and accumulate on leaf surfaces until rainfall is heavy enough to wash metals off as throughfall, to become secondary deposits on understorey plants, litter and the soil. An indication of how mobile some leaf deposits can be was indicated by distilled-water leaf washing experiments, which removed up to 80% of the total leaf Pb content (Schuck and Locke, 1970; Buchauer, 1973).

5.3.1.2 Sludges and wastes added to soil

A range of different industrial and domestic waste sludges has been added to soils, both as a means of disposal and as a fertiliser and conditioner of the soil, since they contain useful amounts of plant nutrients and organic matter. Many waste sludges contain high levels of potentially toxic metals, sometimes used as catalysts in industrial processes, or, in the case of animal manures, from feedstuffs. Sewage sludges and composted refuse vary widely in composition (Table 1.6) but can be particularly high in Cr, Cu, Cd and Zn. Metals are present in sludges in various forms, including free ions, carbonates and both soluble and insoluble organic chelates (see Chapter 3, Section 3.4.1). Stover et al.'s (1976) fractionation of sewage sludge showed that carbonates were the main fraction for Pb, Cd and Ni (at 61%, 49% and 32% respectively), Zn was mainly present in organic forms (35%), while Cu was mainly present as sulphides (35%). Few of these forms are initially mobile in soil or available for plant or microbial immobilisation. The relatively mobilisable fractions in sewage sludge—the adsorbed and exchangeable fractions—made up less than

17% of the Cd, Pb, Zn and Cu content of Stover *et al.*'s (1976) sludge. Although researchers have speculated that the processes of soil organic matter decomposition and soil acidification could alter the fractionation of metals in sludge-amended soils, rendering metals more mobile and bioavailable, there is little evidence of this in published results from agricultural soils, with most authors reporting significant metal accumulation in sludge-treated topsoils.

A large number of studies have examined the fate of sludge metals in agricultural soils and their uptake by crops. The fate of sludge metals in soils, particularly their potential mobility, is discussed in Chapter 3, Section 3.3.5. In general, sludge metals are unlikely to leach through soils and into ground-waters, but acidity and the presence of soluble organic matter may increase metal mobility in soil. Much evidence indicates that metals input in sewage sludge accumulate in topsoils and, even over the longer term, measured in years, do not become bioavailable for crop uptake. McGrath (1987), for example, estimated that <0.5% of sewage sludge-applied metals was taken up by crops over a 20 year period from a field receiving annual sludge treatments. The situation may be a little different for soil organisms. Inhibition of soil microbial activities by metals from sewage sludge has been reported (e.g. Giller *et al.*, 1989). These issues are discussed in more detail in Section 5.5.2.

5.3.2 VEGETATION CHARACTERISTICS INFLUENCING METAL DISTRIBUTION

Vegetation can influence the fate of metals in ecosystems in two main ways. First, the canopy characteristics of plants influence the way that metal aerosols are trapped and the shape and geometric configuration of leaves influences how wet and dry metal deposition is accumulated. Secondly, plant roots alter their immediate environment through exudation and small changes in soil pH and organic matter status in the rhizosphere. In addition to these effects, there is clear evidence that different species, genera and families of plants show differential metal uptake. These three influences will be briefly examined.

5.3.2.1 Canopy characteristics

Two types of foliage "metal-capture" effects should be differentiated: trapping of particles and aerosols by leaves and interception of wet and dry deposition. Several characteristics of vegetation canopies, leaves and stems will influence the way that they trap incoming metal aerosols. Conifer canopies are better aerosol trapping structures because their fine needle leaves, dense canopy branching and year-round foliage provide a very large surface area. Similarly, dry deposition to a forest canopy is much higher than to an open pasture because of the greater leaf surface area. This effect was seen by Little and Martin (1972) who found that woodland soils near a Zn smelting complex had significantly higher Zn, Pb and Cd concentrations than soils of adjacent open agricultural fields. Leaf surface characteristics also play an important role in capturing and trapping metal aerosols. Wedding *et al.* (1975) showed that hairy *Helianthus* leaves trapped more aerosol particles than *Liriodendron* leaves which have smooth surfaces. Little and Wiffen (1977) also observed more particle trapping by hairy-leaved roadside plants than by smooth-leaved.

Proportions of incident rainfall intercepted by tree canopies varies widely, with an average figure of around 25% reported by Bergkvist *et al.* (1989). However, average figures for interception and throughfall are unreliable since there are so many influencing factors, including rainfall intensity, antecedent meteorological conditions, whether the tree is deciduous or not, leaf shape, branch angle, and canopy geometry and density. In the Solling region of Germany, Heinrichs and Mayer (1980*a*) measured canopy throughfall in beech and spruce stands to be 72.5 and 70.8% of annual rainfall respectively. The remaining 27.5 and 29.2% are accounted for as interception, direct evaporation from the canopy during and after the storm, and stemflow. Temperature and other microclimatological conditions, including wind, will determine the proportion of intercepted precipitation that is evaporated back to the atmosphere directly from leaf surfaces. It is likely that in some measurements of dry deposition, a portion should be attributed to the precipitation on leaf surfaces of insoluble materials after wet deposition has been evaporated. This portion will be dissolved at the next occurrence of wet deposition.

Grosch (1986) used a canopy input–output calculation to show how metals were concentrated by the canopies of three spruce forests in Germany. There are two components of canopy concentration: (i) accumulation of leaf deposition, followed by washing, and (ii) leaching of ions from the leaf. Compared to above canopy concentrations in bulk rainfall, Grosch (1986) found that Pb, Cd, Cu and Fe were all concentrated below the canopy, in throughfall, by a factor of around 2.5–5. Mn and Cr above:below canopy concentration factors were higher, at 8–10 and 15 respectively. These patterns of canopy metal concentration have been confirmed elsewhere. In the Solling region in Germany, both Zottl (1985) and Heinrichs and Mayer (1980*a*) report very big concentration effects for Mn in throughfall under spruce canopies. Heinrichs and Mayer found a 300 times increase in Mn concentration in throughfall compared to rainfall (Table 5.3). Heinrichs and Mayer (1980*a*) compare the throughfall metal concentration associated with beech and spruce in the same region. While spruce canopies show increased concentrations of Cr, Mn, Fe, Ni, Zn and Pb in throughfall, only Mn and possibly Ni are elevated in concentration in throughfall under beech. Differences in canopy concentration effects of deciduous (hickory and oak) and coniferous (loblolly pine) forest were studied by Capellato *et al.* (1993) in Georgia, USA. These authors have used their data to propose that leaching of base cations from canopies by H^+ exchange was one of the most important processes

Table 5.3. Concentrations of trace metals in bulk precipitation, throughfall and stemflow in Solling, Germany (all values in μg litre^{-1})

Trace metal concentration (μg litre^{-1})

Hydrological flux		Mn Beech	Mn Spruce	Fe Beech	Fe Spruce	Cu Beech	Cu Spruce	Zn Beech	Zn Spruce	Cd Beech	Cd Spruce	Pb Beech	Pb Spruce
Open field bulk precipitation	a		35		74		33		180		3.4		29
	w		31		110		53		320		5.3		38
	s		39		43		14		44		1.5		21
Throughfall	a	610	970	207	80	30	17	350	120	1.8	3.2	33	70
	w	270	1000	250	110	35	19	480	170	3.0	4.6	55	90
	s	950	900	170	55	25	14	217	67	0.6	1.9	10	50
Stemflow (beech only)	a	1050		240		18		1720		2.0		78	
	w	950		380		20		1900		2.6		120	
	s	1200		110		15		1500		1.4		37	
Seepage below root zone (>80 cm depth)	a	1000	2000	36	36	27	18	570	190	2.9	5.1	51	63

Source: Heinrichs and Mayer (1980).

a = annual averages; w = winter averages (November–April); s = summer averages (May–October).

contributing Ca^{2+}, Mg^{2+} and K^+ to forest floors. A similar process may be very important in metal leaching from contaminated forest canopies.

Throughfall *concentrations* indicate the effects taking place in single rainstorms. Multiplying concentrations by throughfall amounts gives an indication of throughfall metal quantities compared to rainfall metal inputs. Compared to rainfall, Zottl (1985) reported a tenfold increase in the quantity of Mn in throughfall under spruce and a threefold increase in quantity of Cd over a one-year period. Equivalent figures in the Heinrichs and Mayer (1980*a*) study are a 15.5 and 19.3 times increase in quantity of Mn in throughfall under spruce and beech respectively. The only other trace metals to show any increased input in throughfall were Zn, with an input around 1.5 times higher than rainfall in both spruce and beech, and Fe, with a doubling of input under spruce. Both Cd and Cu showed small decreases in thoughfall input compared to rainfall, perhaps indicating deposition, adsorption or absorption in the canopy. Throughfall under conifers frequently shows higher acidity than under deciduous trees (e.g. Nihlgard, 1970; Miller and Miller, 1980) and this may be the reason for enhanced throughfall concentrations of trace metals under spruce canopies. The effect of more acidic throughfall in conifers may also leach elements, including trace metals, from leaves (cf. Little, 1973).

Although stemflow usually contributes only a small proportion of the water reaching the soil, i.e. around 5–10% (9.9% for beech in the Heinrichs and Mayer (1980*a*) study), stemflow metal concentrations are nearly always higher than in rainfall (Bergkvist *et al.*, 1989). Quantities of stemflow and its chemistry depend very much on bark configuration, particularly bark smoothness, presence of epiphytic lichens and mosses and, in the case of deciduous species, geometry of the branches, including branch angle, which influences stemflow more strongly in winter. Large-crowned emergent trees with smooth bark and raised branches produce the most stemflow (Parker, 1983). In beech stemflow (Table 5.3), we can see that Mn, Fe, Zn and Pb are all more concentrated in stemflow than in rainfall. Zn in stemflow shows particularly high concentration effects, at values around 10 times higher than in rainfall. Bergkvist *et al.* (1989) provide evidence to show that metals in stemflow can be more highly concentrated in conifers than in deciduous broadleaves. Stemflow episodes can lead to metal enrichment in the bole and root zone of individual trees and may adversely affect the growth of fine roots in this zone. An extensive review of available published data on hydrological fluxes of metals in temperate forest ecosystems is given by Bergkvist *et al.* (1989).

5.3.2.2 Rhizosphere characteristics

Fine roots of plants and trees influence the availability and uptake of metals from soil through small changes in soil conditions in the rhizosphere. Two main rhizosphere influences are slightly lower pH and increased quantities of different organic molecules in this zone of exudation and sloughage. Sarkar and Wyn Jones (1982) showed that small changes in pH in the rhizosphere could make trace metals more available for uptake. Xian and Shokohifard (1989) suggested that exudation of H_2CO_3 by roots may help to solubilise metal carbonates and make them more bioavailable. The influence of different organic compounds in the rhizosphere is complex. Mycorrhizal fungi and bacteria, as well as plant roots, all contribute organic molecules to the rhizosphere. Several authors have suggested that exuded organic molecules can bind trace metals in the rhizosphere and make them unavailable for uptake (e.g. Denny and Wilkins, 1987; Cumming and Weinstein, 1990). Siderophores, which are iron-binding organic molecules, can be found in mycorrhizal exudates (e.g. Morselt et al., 1986) and compounds secreted by rhizosphere bacteria (Raymond et al., 1984). These compounds bind with metals in the rhizosphere and make them unavailable for uptake (see Chapter 2, Section 2.2.2 and Chapter 4, Sections 4.3.3 and 4.3.4, this volume).

5.3.2.3   The effect of species on metal uptake

Mitchell et al. (1957) observed that legumes appeared to accumulate molybdenum and copper in a different manner to grasses, with both plants showing a wide range of tissue concentrations, but grasses taking up generally lower amounts under both toxic and less toxic soil conditions. In numerous experiments, food crops have been shown to take up different quantities of potentially toxic metals when grown under identical conditions (Table 5.4).

Table 5.4. Relative metal uptake by a range of different crop plants growing in contaminated soils

Metal	Relative metal uptake	Source
Cd	lettuce > radish > carrot > spinach > cauliflower > oats > pea	John (1973)
Cd	turnip, spinach > tomato, lettuce > swiss chard, radish, carrot > corn > fescue, wheat, field bean > rice	Page et al. (1981)
Cd	spinach > lettuce > beetroot > radish > sweet corn > spring greens > spring onions > leeks > cauliflower > potato > cabbage	Chumbley and Unwin (1982)
Cd	bermuda grass > tall fescue > alfalfa > white clover > sudan grass	Bingham et al. (1976)
Pb	spinach > radish > spring greens, lettuce > cauliflower > leeks > spring onions > beetroot > cabbage > potato > sweet corn	Chumbley and Unwin (1982)
Pb	lettuce > cabbage > carrots > radish	Alloway and Morgan (1986)
Zn	lettuce > kidney beans > peas > potatoes	Purves (1985)
Cu	kidney beans > lettuce > peas > potatoes	Purves (1985)
Ni	radish > lettuce > cabbage > carrot	Alloway and Morgan (1986)

Sauerbeck (1991), summarising the results of a number of long-term crop field experiments with sewage sludge treatment in Germany, found that many dicotyledonous crop plants, such as spinach or lettuce, absorbed more heavy metals than did monocotyledonous crop plants such as oats or wheat.

Kuboi *et al.* (1986) have subsequently examined Cd uptake in 34 plants species of nine different families. They found that different plant families responded differently and classified them into three groups:

1. *Low accumulation*: Leguminosae
2. *Moderate accumulation*: Gramineae, Liliaceae, Cucurbitaceae and Umbelliferaceae
3. *High accumulation*: Chenopodiaceae, Cruciferae, Solonaceae and Compositae

Differential metal uptake by plants is also indicated in the excluder, accumulator or hyperaccumulator strategies of characteristic floras of metalliferous lithologies.

5.3.3 LITTER AND SOIL PROFILE CHARACTERISTICS INFLUENCING METAL DISTRIBUTION

5.3.3.1 Litter characteristics

Hydrologically, litter at the soil surface provides a filter and controlling mechanisms for water infiltration into the mineral soil profile. Thick leaf litter layers absorb incoming rainfall and throughfall, slowing down water percolation into soil and providing time for metals in percolating water to be adsorbed by decaying litter and organic matter. Water absorption by dead leaves and litter thus reduces runoff and soil erosion. An intact layer at the soil surface, particularly in forest soils, provides soil protection and helps to prevent disruption of soil particles which is the start of soil erosion (e.g. Ross *et al.*, 1990).

The litter layer is also the repository for metals accumulated in leaves prior to shedding. Several authors have reported a build-up of litter on the forest floor in woodlands adjacent to metal pollution sources (e.g. Jackson and Watson, 1977; Strojan, 1978; Coughtrey *et al.*, 1979), caused by reduced rates of organic matter decomposition. Strojan (1978) observed that metal contamination more strongly affected the later stages of litter decomposition. In an experiment to study the decomposition of contaminated Scots pine needles near a brass works, Berg *et al.* (1991) found that lignin decay was more sensitive to metal contamination than was whole litter. As in acidic forest litter layers, the lignin component of contaminated forest floors accumulates relative to the non-lignin tissues and becomes the rate-limiting factor in litter decomposition. During the decomposition of leaf litter on the forest floor, it is possible to identify a sequence of events which influence the metal concentration of the litter:

— *stage 1:* the cuticle of a fallen leaf breaks down,
— *stage 2:* internal cell walls become exposed and provide sites for metal ion sorption, and
— *stage 3:* as decomposition proceeds, metal sorption continues and metal concentration in the litter increases.

Nilsson (1972) has thus shown that it is possible for fallen and aged litter to have higher metal concentrations than living leaves.

5.3.3.2   Soil conditions influencing distribution and fate of metals

There is apparently conflicting evidence concerning the mobility or retention of metals by soil organic matter, particularly surface organic horizons. Three main processes have been identified in the surface organic horizons which may make metals more mobile and leachable:

(i) production of *soluble organic matter* forms soluble chelates and mobilises metals such as Cd, Mn, Zn and Ni (Bergkvist *et al.*, 1989; Berggren, 1992*a*),

(ii) *organic matter decomposition* releases metals that were bound to organic molecules (Ng and Bloomfield, 1962), and

(iii) *microflora* at the surface of soil particles restrict metal ion exchange processes which leads to accelerated leaching of metals (Ausmus *et al.*, 1977).

Having identified organic matter processes which aid metal mobility, there is also evidence to show that mor surface organic horizons and peat are highly efficient absorbers of trace metals. Dumontet *et al.* (1990) and Wieder (1990) present evidence for the strong adsorption and retention of metals by peat (see, e.g. Figure 3.11(*b*)). Peat substrates have also been used as "cleaning" agents to remove metals from contaminated wastes (e.g. McClellan and Rock, 1988). Much of the forest soil metal contamination research indicates that metals are strongly retained and accumulate in the litter and more layers of forest soils; this is especially true for Pb and Cu (e.g. Andresen *et al.*, 1980; Friedland *et al.*, 1984).

Only soils developed from parent materials containing metal minerals could display inherently high metal concentrations throughout the profile, as a result of *in situ* rock weathering. Baseline data on metal distribution in generally uncontaminated soils in Scotland are provided by Swaine and Mitchell (1960) and Berrow and Reaves (1986). Databases of this type provide the reference against which the accumulation of metal pollutants can be measured. Billet *et al.* (1991), for example, measured the Cu, Pb and Zn concentrations in 19th and 20th century conifer plantation soils in northern Scotland and compared them with soil samples taken from exactly the same locations 40 years earlier. When the soils were first sampled, no significant differences were seen in trace metal concentrations between the 19th and 20th century plantations. All forest organic horizons contained significantly higher concentrations of metals than 40 years previously. Soils under 19th century forestry plantations showed a 40% increase in Pb concentration over the 40 year period, while soils under 20th century plantations showed a 12% increase. Over the same 40 year period, soil Zn concentrations showed the opposite effect, with 34% and 10% decreases in soil Zn concentration for 19th and 20th century plantations respectively. Soil Pb concentrations were positively correlated with organic matter content, while soil Zn concentrations were positively correlated with pH. These results illustrate (a) the better ability of the older forest to trap aerosol Pb and accumulate it in surface soil organic horizons and (b) the higher mobility and bioavailability of Zn under the more acid conditions associated with older plantations.

In mineral soil, the main factors influencing metal adsorption or mobility are pH, organic matter content, content of Fe and Mn oxides and percentage clay content (Chapter 3, Section 3.3.3). By quantifying these characteristics in field soils we can begin to predict how soils may respond to different metal inputs. King (1988) measured metal contents of 13 major soil groups in the USA and compared them to key soil characteristics. Over all soil types the relative retention of metals was found to follow the sequence:

$$Pb > Sb > Cu > Cr > Zn > Ni > Co > Cd$$

Sorbed and non-exchangeable Cd, Co, Cu, Ni and Zn were related mainly to soil pH or the content of crystalline Fe oxides. King (1988) concluded that although cation exchange capacity (CEC) is widely used in regulations governing metal loadings to agricultural land, Fe oxide and clay contents would be better parameters to use. A discussion of soil factors and processes influencing metal distribution, transformation, retention, mobility and availability in soil is given in Chapter 3.

Since rock weathering, air pollution and addition of metalliferous wastes and sludges introduce quite different forms of metals into soils, metals of different origins may also be expected to behave differently in soils. The following discussion deals mainly with soils receiving surface metal inputs.

5.3.3.3 Indices of soil contamination

Several authors have developed arithmetic indices to allow comparisons of soil contamination. Colbourn and Thornton (1978) calculated relative topsoil enrichment (RTE) values for soils in a lead mining region in the north of England. RTE values were calculated as:

$$RTE = \frac{\text{total metal in 0–15 cm zone of soil}}{\text{total metal in 30–45 cm zone of soil}}$$

In control (uncontaminated) soils, values averaged 1–2, while in contaminated soils near mineworkings and smelters, values ranged up to 20. Martin and Coughtrey (1982) also calculated similar topsoil "enrichment ratios" for a number of soils in the proximity of metal mining and smelter sites. They show that surface metal enrichment can occur in soils not only close to active sources of aerial metal contamination, but also distant from such sources. The data of Hutchinson and Whitby (1974) for soil metal contamination in the vicinity of a Cu–Ni smelter in Sudbury, Ontario is illustrated in Figure 5.5. The actual topsoil metal concentrations of Cu, Ni and Pb decline with distance from the

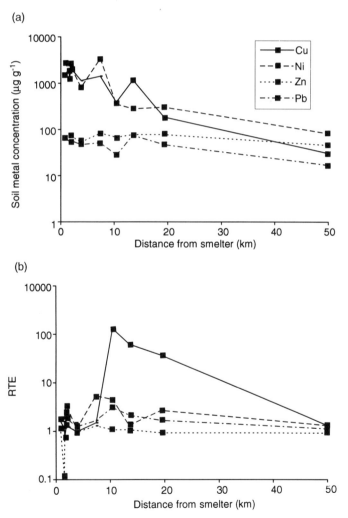

Figure 5.5. (a) Total soil metal concentrations and (b) relative topsoil enrichment (RTE) values with distance from a Ni–Cu smelter near Sudbury, Ontario. Calculated and drawn from the data of Hutchinson and Whitby (1974)

smelter, but the Zn concentrations remain high, even at a distance of 50 km. When the RTE values are calculated from available data (total metal concentration in superficial soil/total metal concentration at 10 cm depth) (Figure 5.5(b)) the ratios are not as high as in the Martin and Coughtrey (1982) study possibly because subsoil metal concentrations should have been taken at a greater soil depth. The RTE values close to a Cu smelter near Tacoma, Washington, show values ranging from 15.27 to 35.95 for copper, and from 2.86 to 15.0 for zinc (calculated from the data of Kuo et al., 1983). The distribution of downprofile soil total metal concentrations is shown in Figure 5.6. A sharp reduction of Cu, Zn and Cd contamination occurs at a soil depth of around 30 cm. Significantly higher metal contamination at the soil surface can be due to the combined effects of aerial input, strong surface soil adsorption and biological concentration processes, including root uptake, metal translocation within the plant and subsequent litterfall.

5.3.3.4 Comparison of different soil conditions

The importance of factors such as pH and organic matter content immediately indicate that the fate of potentially toxic metals could be very different in different soil types. In an attempt to compare different soil profile responses to toxic metal inputs, a literature search was made for data on downprofile metal contents in different soil types under differing vegetation types. In the main, such studies have been carried out in "type sites", such as the Black Forest in Germany, peatland of Finland or in agricultural experiments, usually where sewage sludges have been added. One major difficulty in comparing different studies is the wide range of analytical techniques used for metal analysis in soils. Techniques range from using ceramic suction samplers which extract a sample of the soil water under tension (e.g. Dreiss, 1986), to zero-tension lysimeters which collect only gravitational drainage waters (e.g. Berggren, 1992a,b), to soil sample extracts using a range of chemical extractants of varying strengths, such as magnesium chloride or ammonium oxalate (e.g. Shuman, 1979), and finally to determinations of total metals in soils using concentrated acid digestions (e.g. Dudka and Chlopecka, 1990; Davies et al., 1987) or ICP-AES (inductively coupled plasma-emission spectrometry (e.g. Jarvis and Higgs, 1987, in sediments) and ICP-MS (inductively coupled plasma-mass spectrometry) (Malmer, N., 1989, pers. comm., in peats).

Despite these differences, it has been possible to begin to compare different metal fractions in different soils (see Chapter 3, Section 3.4, this volume). Based on the review of soil characteristics influencing metal retention and mobility in soil, we might expect Cd and Zn retention and mobility to be mainly related to soil pH and that of Cu and Pb to be mainly related to soil organic matter.

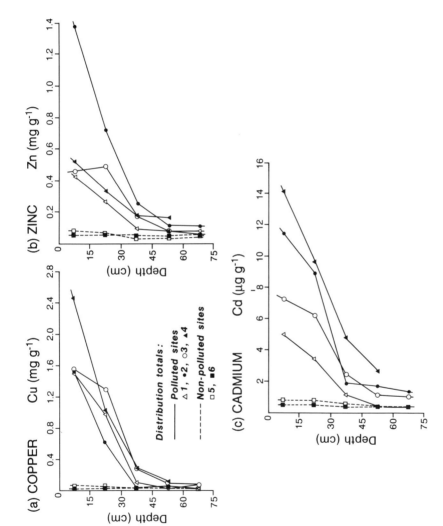

Figure 5.6. Distribution of total Cu, Zn and Cd with depth in four polluted soils near a copper smelter (sites 1,2,3,4 polluted; sites 5,6, non-polluted). Reproduced by permission of Williams & Wilkins from Kuo *et al*., 1983

216

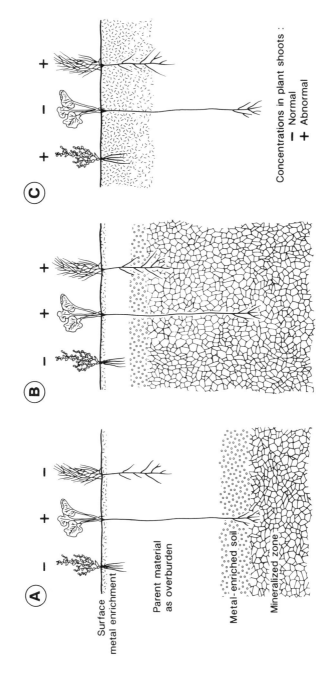

Figure 5.7. Relationship between plant rooting depth, location of soil metal enrichment and metal uptake into aerial parts of plants. Reproduced by permission of Chapman and Hall from Martin and Coughtrey (1982)

5.3.3.5   The relationship between soil metal concentrations and plant metal concentrations

Martin and Coughtrey (1982) give an excellent introduction to the influence both of geology on soil metal distribution and of plant rooting characteristics on metal uptake from the soil. As well as metal availability, the coincidence or otherwise of zones of metal enrichment in the soil with the depth of rooting and root distribution can be critical in determining how much metal can be taken up by the plant. Figure 5.7 illustrates these ideas for a range of rooting types and soil enrichment scenarios. This figure also illustrates how metals can be brought up by deep-rooting plants from metal mineralisation zones at depth, translocated through the plant to the leaves, and deposited in litterfall, to enrich the surface horizon. Shallow-rooted plants can then suffer from metal toxicities in *their* rooting zones which would not have occurred if adjacent plants had also been shallow rooted.

In Chapter 1 of this text, the mean concentrations of metals in soils and plants was reviewed (Table 1.4). These data indicate that soil metal concentrations are almost always higher than plant concentrations, with the possible exceptions of Zn, Cd, Hg, Ag and Sn. Many authors have calculated the ratio of metal concentration in plants to that in associated soil. Measurements of metals in plants suffer from several technical problems. A major difficulty is to adequately wash the roots clean of "contaminating" soil and of adequately accounting for aerial pollution on the leaves.

Although much effort has been spent testing different soil extraction techniques for characterising metal phytoavailability (Chapter 3, Section 3.4.2), Sharma and Shupe (1977) found surprisingly good correlations between total metal concentrations in the soil and total metal concentrations in plants. The best relationships were for Zn, Cd and Cu, with correlation coefficients of 0.884, 0.883 and 0.875 respectively. Pb showed a much poorer correlation of $r = 0.517$, although improved correlation coefficients of 0.889, 0.793 and 0.977 were obtained when individual soil depths of 0–5 cm, 5–10 cm and 10–15 cm were used with a $y = ax^b$ regression model. Statistical relationships between soil and plant metal concentrations, based on single sampling dates, may be fortuitous since there is now ample evidence that the trace metal concentrations in the tissues of many different plants show significant seasonal fluctuations (Section 5.2.2).

5.4   METHODS OF COMPARING ACCUMULATION AND DISTRIBUTION OF METALS IN DIFFERENT ECOSYSTEMS

Most earlier studies of metal uptake by plants have assumed that the prime source of metals is the soil. Chamberlain (1983) characterised the amount of

Table 5.5. Concentration factors (CF) for a range of different plants and soil types

Plant material	Cd	Cr	Ni	Zn	Pb	Source
Natural vegetation	0.6–4.0	0.002–0.05	0.05–0.1	0.05–3.5	—	Thorne and Coughtrey (1983)[a]
Pasture grass	0.4–2.0	0.002–0.05	0.07–0.9	0.08–1.7	—	
Edible plants	0.07–4.0	0.002–0.1	0.02–0.2	0.06–5.0	—	
Cereal grains	0.09–0.45	0.002–0.1	0.01–0.65	0.20–7.8	—	
Pea fruits	0.005–0.011	0.1–0.24	0.17–0.40	2.24–2.37	0.005–0.011	Harrison and Chirgawi (1989a)[b]
Pea (average in pods, leaves)	0.2–0.6	0.1–0.33	0.23–0.43	0.8–1.0	0.002–0.055	
Lettuce	0.7–2.1	0.02–0.2	0.15–0.58	0.4–1.2	0.006–0.02	
Radish roots	0.01–0.03	0.02–0.12	0.08–0.3	0.55–0.8	0.001–0.02	
Leaves	0.03–1.9	0.07–0.19	0.11–0.5	0.3–2.1	0.001–0.03	
Pea fruits	0.05	0.1	0.1	0.8–1.0	0.002–0.007	Harrison and Chirgawi (1989c)[c]
Pea (average in pods, leaves)	0.06–0.8	0.05–0.25	0.05–0.4	0.2–0.95	0.001–0.0095	
Lettuce	1.1–2.3	0.1	0.5–1.15	0.3–0.7	0.005–0.2	
Radish root	0.05	0.05	0.1–0.22	0.3–0.4	0.005	
Leaves	1.0–1.65	0.05–0.25	0.2–0.5	0.1–0.3	0.005–0.022	

[a] Calculated from published data.
[b] Estimated from bar graphics of Harrison and Chirgawi (1989a), for plants in air-filtered growth cabinets.
[c] Estimated from bar graphics of Harrison and Chirgawi (1989c) for plants in the field.

Pb that different plants could take up by calculating plant:soil Pb ratios called concentration factors (CF):

$$CF = \frac{\text{metal concentration in plant } (\mu g \ g^{-1} \ dry \ wt)}{\text{metal concentration in soil } (\mu g \ g^{-1} \ dry \ wt)}$$

Other authors have subsequently calculated similar ratios for Cd, Cr, Ni and Zn, for a range of different types of plants in different soils. Some examples are given in Table 5.5. Harrison and Chirgawi (1989a,b) report an order of CF values of Zn > Cd > Ni > Cr > Pb. The much higher CF_{Zn} and CF_{Cd} values are attributed to greater mobility and bioavailability of these metals in the soil. CF values were calculated from the data of Chumbley and Unwin (1982) who reported Cd and Pb concentrations in a range of crop plants grown in sewage sludge treated soil. CF_{Pb} values were again extremely low, ranging from 0.0026 for cabbage, to 0.0388 for beetroot. CF_{Cd} values ranged from 0.056 for potato, to 1.00 for spinach.

CF values calculated for the soil and vegetation contamination at Sudbury, Ontario (Hutchinson and Whitby, 1974) show much variability with distance from the Sudbury Cu–Ni smelter (Figure 5.8). CF_{Zn} are generally higher than CF_{Ni} and CF_{Cu}. The most variable CF values are in *Dechampsia flexuosa*, especially for Cu and Ni, which may have been most variable in atmospheric deposition close to the smelter. *Acer* showed both lower and less variable CF valves for Cu and Ni, perhaps indicating deeper rooting into less contaminated subsoil horizons than *Deschampsia*. Rooting depth cannot be the only factor influencing tissue metal concentrations since the pattern of CF_{Cu} and CF_{Ni} for the shallow rooting *Vaccinium* and *Acer* are identical within a 16 km distance of the smelter. Sheppard and Sheppard (1991) report CF_{Pb} values for plants growing in acidic soils under boreal climatic conditions in northern Ontario, Canada. Under acidic and organic soil conditions, the authors expected to obtain higher CF_{Pb} than in other studies on neutral to alkaline mineral soils. In "natural" field vegetation, *Vaccinium* (blueberries) showed very low CF_{Pb} values at 0.0008–0.1. Soil ^{210}Pb application studies, using intact lysimeters, yielded CF_{Pb} values of 0.1 in blueberries and over 0.059 for all other crops studied. These results, together with Pb leaching data in the same study (Sheppard and Sheppard, 1991), indicate how Pb mobility and bioavailability acid and organic soils can be much higher than in many agricultural soils.

In an attempt to estimate the amount of metals taken up from soils compared to that taken up from air pollution, Harrison and Chirgawi (1989a,b) used Chamberlain's (1983) air accumulation factor (AAF) to estimate the proportional plant metal accumulation from the atmosphere, where:

$$AAF = \frac{\text{lead in foliage } (mg \ kg^{-1} \ dry \ wt)}{\text{lead in air } (\mu g \ m^{-3})}$$

From these calculations, Harrison and Chirgawi (1989a) estimated percentage contributions of atmospheric metals in different plants (Table 5.6). The

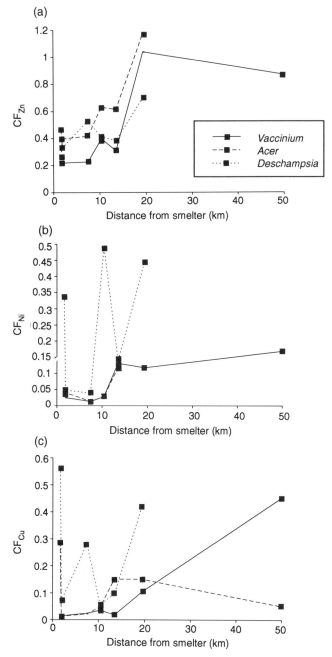

Figure 5.8. Concentration factors for Zn, Ni and Cu (CF_{Zn}, CF_{Ni} and CF_{Cu}) for plants in the vicinity of a Ni–Cu smelter near Sudbury, Ontario. Calculated and drawn from the data of Hutchinson and Whitby (1974)

Table 5.6. The range of % atmospheric contribution to plant metal uptake in a filtered air cabinet experiment[a]

Plant material	% Atmospheric contribution				
	Cd	Cr	Ni	Zn	Pb
Radish					
leaves	18–43	38–85	41–47	7–12	89–96
roots	25–47	25–41	22–31	3–39	46–88
Turnip					
leaves	29–40	13–21	36–46	3–34	81–90
roots	5–6	18–72	8–30	1–15	34–52
Carrot					
leaves	12–18	18–28	8–17	8–10	50–87
roots	4–8	33–67	8–17	2–7	32–44
Spinach					
leaves	23	20	35	7	85
stalk	0	0	24	1	0
Pea					
leaves	38–48	43–60	9–39	84–87	7–11
pods	6–30	33–43	28–29	68–79	5–7
peas	0	0	0–8	0	0–6
Lettuce	7–21	17–58	1–43	5–16	76–91

Dry deposition only. Reproduced by permission of Elsevier Science Publishers BV from Harrison and Chirgawi (1989a).

[a]Calculated as:

$$\frac{\text{Metal concentration in plant (unfiltered air)} - \text{metal concentration in plant (filtered air)}}{\text{Metal concentration in plant (unfiltered air)}} \times 100$$

exposed plant parts (the leaves) accumulate larger quantities of air-derived metals than do unexposed plant parts (e.g. roots and pea fruit). Again, the generally lower air contribution to plant Zn content reflects the ease of uptake of soil Zn. Although Pb is generally considered not to be easily translocated within plant tissues, Harrison and Chirgawi's (1989a) data indicate that atmospheric Pb contributed to the unexposed plant parts, especially the storage roots of radish, turnip and carrot. This may simply reflect relatively high air Pb concentrations and the low efficiency of soil Pb uptake compared to the other trace metals studied. There are rather few corroborative published data. Hovmand *et al.* (1983) reported that anything from 20 to 60% of Cd uptake by a range of different edible crops could be air-derived. The leaves of kale and the grain of barley showed highest Cd concentration derived from air pollution, at 50–60% and 40–60% respectively. The studies of Harrison and Johnston (1987) and Harrison and Chirgawi (1989a,b) indicate a relative ease of uptake of Zn and Cd from soil, but not for Pb. Their experiments also indicate that metals in atmospheric pollution can contribute substantially to the metal fund of the plant and be translocated to unexposed plant parts.

5.5 COMPARISON OF THE FATE OF METALS IN DIFFERENT ECOSYSTEMS

5.5.1 WOODLANDS AND FORESTRY

5.5.1.1 Forest soils

Many authors have reported strong retention and accumulation of trace metals in the forest floor of woodlands and forests. This concentration effect has been observed for a range of different metal contamination conditions, including woodlands close to sources of metal pollution (e.g. Jackson and Watson, 1977; Parker *et al.*, 1978; Martin and Coughtrey, 1987), by roadsides (Smith, 1976) and due to atmospheric deposition (e.g. Heinrichs and Meyer, 1977; Friedland *et al.*, 1984; Turner *et al.*, 1985). Extremely high quantities of metals have been measured in the forest floor of stands in the north-east United States. In a study of 51 forested sites from Virginia to Massachusetts, Andresen *et al.* (1980) reported mean forest-floor metal levels of 1234 ± 100 mg Pb m^{-2}, 170 ± 35 mg Cu m^{-2} and 1013 ± 298 mg Zn m^{-2}. Using calculations of Pb inputs and ^{210}Pb studies, Benninger *et al.* (1975) estimated that the mean retention time of Pb in the forest floor at Hubbard Brook, New Hampshire, is 5000 years.

Friedland *et al.* (1984) measured metal accumulation rates over a 14 year period in the forest floors of three forest types in Vermont: northern hardwood, transition and boreal (Table 5.7). Lead accumulation rates were highest, probably reflecting higher deposition rates over the period compared to the other metals. Rates of organic matter accumulation in the forest floor were lower than metal accumulation rates, indicating that factors other than reduced organic matter decomposition and organic matter accumulation were responsible for increases in trace metal content over the 14 year period. Friedland *et al.* found higher metal accumulation rates at higher elevations, probably reflecting higher precipitation rates at altitude.

Table 5.7. Annual metal accumulation rates in the forest floor of three forest types in Vermont, calculated over a 14 year period

Forest type	Annual metal accumulation rate (mg m^{-2} year^{-1})		
	Pb	Cu	Zn
Northern hardwood	15.5	2.14	17.8
Transition	59.9	6.52	40.8
Boreal	52.1	2.78	9.64

Source: Friedland *et al.* (1984).

Much evidence is available to show that Pb is retained strongly in forest soils, with very little atmospheric Pb being transported to streams. In the Hubbard Brook catchment, New Hampshire, Smith and Siccama (1981) measured around 2% of atmospheric Pb issuing in streamwater. Using lysimeters to collect water percolating to different depths in the soil profile, they calculated that this amount could be accounted for by direct precipitation on the stream channel. In a white pine forest in Massachusetts, Siccama *et al.* (1980) showed that 20% of incoming Pb leached through the surface soil organic layers and into the mineral soil below. Subsequently, Bergkvist (1987), also using lysimeters to collect percolating soil water, showed that Pb and other trace metals, were leached into the B and C horizons of podsols under spruce forest. There is clearly evidence that, although only very small amounts of trace metals are seen in stream outflow from forests, there may be down-profile movement through soil, associated with dissolved organic matter (Swanson and Johnson, 1980; Tyler, 1981; Bergkvist, 1987). Bergkvist (1987) found Fe, Cu, Pb and Cr to be leached in association with dissolved organic matter from surface horizons, but these metals were subsequently deposited in the soil B horizon.

5.5.1.2 Forest canopies: metal accumulation, hydrochemistry and litterfall

We have seen in Section 5.3.1 that forest canopies can modify both concentrations and quantities of trace metals reaching the forest floor in throughfall. A proportion of dry deposition and insoluble wet deposition can remain on leaf surfaces. Together with foliar metals, these provide the input to the forest floor in litterfall. Quantities of leaf litter biomass returned to the forest floor are more or less the same for evergreen conifers and deciduous broadleaves when averaged over several years (Bray and Goreham, 1964; Vogt *et al.*, 1986). Differences in litterfall input of metals between conifer and deciduous forests are thus likely to reflect concentration differences in foliage chemistry rather than differences in litterfall biomass.

A large database of data on metal fluxes in temperate forest ecosystems is presented by Bergkvist *et al.* (1989). Table 5.8 reproduces a selection of data on Zn, Pb and Cd inputs and outputs for a range of conifer and deciduous forest types. The Solling beech and spruce sites of Heinrichs and Mayer (1980*a*) would appear to be more heavily metal polluted than other sites selected. However, the patterns of metal flux through the forest ecosystems are similar. An augmentation of Zn input in throughfall compared to rainfall is seen in virtually all studies, for both conifer and deciduous trees. Quantities of Zn input to the forest floor in litterfall are generally smaller than through-fall Zn quantities, except for birch (Bergkvist *et al.*, 1989) and alder (Asche, 1985). The extremely large difference between throughfall and litterfall Zn inputs in Heinrichs and Mayer's study is out of line with other data and

Table 5.8. Fluxes of Pb, Zn and Cd through different forest systems (all values in g ha^{-1} year^{-1})

Author and forest type	Zn				Pb				Cd			
	IP[a]	T+S[b]	L[c]	SL[d]	IP	T+S	L	SL	IP	T+S	L	SL
Heinrichs and Mayer (1980a)												
Beech	1890	2720	260	1100	310	340	120	30	35	16	2.3	17
Spruce	1890	2700	260	2400	310	532	256	27	35	25	1.9	22
Bergkvist (1987)												
Spruce	100	320	210	560	87	77	74	6.4	1.2	2.3	0.7	5
Spruce + pine	318	283	230	262	64	49	9	2.4	1.9	2.3	0.5	4.2
Zöttl (1985)												
Spruce	ND	ND	ND	ND	128	74–80	111–172	12–18	4.3	10.5–12.3	0.62–1.52	3.73–5.93
Bergkvist et al. (1989)												
Spruce	190–220	285–460	110–210	650–1300	51–66	44–77	40–83	2.1–5.8	0.9–1.2	1.7–2.6	0.3–0.7	4.4–12
Beech	190–220	111–190	59–200	1200–1800	51–66	15–31	18–38	1.3–2.8	0.9–1.2	0.9–1.1	0.35–1.2	5.2–15
Birch	190–220	248–306	230–480	58–250	51–66	28–42	15–30	1.7–15	0.9–1.2	1.4–1.7	0.54–1.1	1.6–3.2
Schultz (1987)												
Pine	184	369	107	599	110	144	30	24.2	1.77	3.35	1.32	7.6
Oak	184	288	131	211	110	121	40	9–9	1.77	2.46	0.86	10.5
Asche (1985)												
Alder	241	484	528	ND	138	145	42	ND	2.2	2.5	1.0	ND
Oak-Hornbeam	241	434	262	ND	138	164	40	ND	2.2	3.7	1.3	ND

Source: data taken from review by Bergkvist *et al.* (1989).
[a]IP = Incident (bulk) precipitation.
[b]T + S = Throughflow and stemflow.
[c]L = Litterfall.
[d]SL = Soil leaching below the rooting zone.
ND = No data.

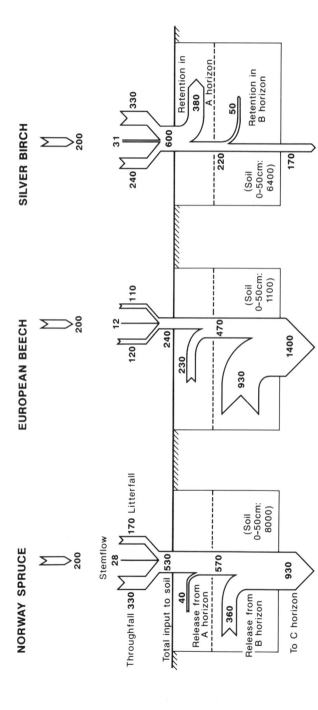

Figure 5.9. Annual flow of Zn through Norway spruce, European larch and Silver birch forest ecosystems in southern Sweden. Mean of five sites, from June 1984 to May 1987 (all values in g ha^{-1} year^{-1}). Reproduced by permission of Kluwer Academic Publishers from Bergkvist *et al.* (1989)

remains unexplained. Quite large quantities of Zn leach below the rooting zone of the soil. This may be associated with soil acidification under conifers and beech, since the birch soil shows significantly less Zn mobility. Throughfall quantities of Pb are generally lower than incident bulk precipitation apart from the Solling beech and spruce (Heinrichs and Mayer, 1980a) and the deciduous species of Asche (1985). Only small quantities of Pb are leached below the rooting zone in all studies. Of the three metals in Table 5.8, Cd inputs in precipitation and leaching in soil are the smallest. Litterfall consistently provides less Cd to the forest floor than throughfall in both conifer and deciduous forest. This very small selection of data on metal fluxes in forests provides some evidence that Pb and Cd are retained by forests and forest soils, while Zn is more mobile.

It is extremely difficult to make *direct* comparisons of metal fluxes in different forest types since rarely are equivalent data available. Many more studies report metal compartment audits. These will be discussed in the following section. For a Norway spruce forest in south-west Sweden, Bergkvist (1987) compares system fluxes for a range of elements, including Pb, Cu, Cr, Zn, Cd and Ni. His data illustrate net canopy and soil retention of Pb, canopy "releases" of Cu and Cd, and soil B horizon releases of Zn, Cd and Ni. Bergkvist *et al.* (1989) illustrate differences in Zn fluxes for three different forest types in southern Sweden (Figure 5.9). The relative mobilising effect on soil Zn, particularly of larch, but also of spruce, is clearly compared to the soil retention of Zn under birch.

5.5.1.3 Forest compartment audits

A large number of forest compartment metal audits have now been published for a range of different polluting conditions, although the majority are for general aerial pollution (e.g. Turner *et al.* (1985) and Friedland and Johnson (1985) for Pb; Heinrichs and Mayer (1980b) for Ni; Parker *et al.* (1978) for Zn, Cu, Cd and Pb) or pollution in the vicinity of metalliferous industry or smelters (e.g. Martin and Coughtrey, 1981; see Chapter 9, this volume).

In contaminated woodlands and forests, the largest pool of total metals is that accumulated in the soil. The soil, particularly in litter and surface organic horizons, also shows the highest concentrations of metals in forest ecosystems. Table 5.9 shows the metal concentrations for different compartments of beech and spruce forests in Solling, northern Germany. Under both beech and spruce, concentrations of all four metals—Pb, Zn, Cd and Cu—are by far the highest in the surface soil organic horizon. This is especially true for Pb and Zn. Vegetation compartments of the forestry system show highest concentrations of Pb and Zn in the bark and lowest concentrations in stem and branch wood. For Cd in beech, the highest concentrations are in leaves and litter. Concentrations of Cu in all compartments are very much less variable than for Pb, Zn and Cd. In spruce, the metal concentrations in different aged

Table 5.9. Metal concentrations in vegetation and soil compartments of spruce and beech forests at Solling, northern Germany

Compartment	Metal concentration ($mg\,kg^{-1}$)			
	Pb	Zn	Cd	Cu
Beech				
Stem wood	3.0	11.0	0.19	41.0
Stem bark	35.0	41.0	0.064	32.0
Branch wood	3.3	9.7	0.15	30.0
Branch bark	73.0	94.0	0.17	41.0
Green leaves	20.0	46.0	0.66	24.0
Litter	33.0	69.0	0.70	20.0
Soil:				
organic	375	205	1.55	72
mineral	41	44	0.058	21
Spruce				
Stem wood	7.0	21.0	0.36	46.0
Stem bark	19.0	178.0	1.45	32.0
Branches (average)	8.6–80.0	59–105	0.61–1.4	24–48
Green needles (average)	5.4–16.0	33–63	0.24–0.39	24–46
Litter	65.0	55.0	0.47	61.0
Soil:				
organic	405	163	0.82	74.0
mineral	62	34	0.07	32.0

Reproduced by permission of the American Society of Agronomy from Heinrichs and Mayer (1980*a*).

branches and needles was also studied by Heinrichs and Mayer (1980*a*). With increasing age of branches, from one year to five years, Pb, Zn and Cu concentrations generally decreased, while Cd concentration clearly increased. With increasing age of needles from one to five years, Pb and Zn concentrations increased slightly, Cu decreased slightly and Cd concentrations remained the same. Very few authors report metal concentrations for the same forest compartments or all the same metals. Friedland and Johnson (1985) measured the Pb concentration in compartments of spruce (*Picea rubens*) and birch (*Betula pubescens*) in Vermont. They found that twigs showed high concentrations, as did bark and roots, with the wood component again showing low concentrations. Turner *et al.* (1985) found highest concentrations of Pb in the fine roots of a mixed pine, oak, cedar and maple forest in New Jersey.

When compartment concentrations are converted into stocks in relation to compartment biomass (or soil mass/volume), a slightly different metal distribution pattern emerges. The largest above-ground stocks of Pb, Cd, Zn and Cu are in the woody tissues of both beech and spruce forests (Heinrichs and Mayer, 1980*a*). Major compartment stocks of Pb, Zn and Cd are given in

Table 5.10 for three different forest types. The magnitude of metals accumulation in soil is again very clear from these data. The proportional contribution that the soil metal compartment makes to the whole forest metal fund (Table 5.11) is very large indeed. Clearly, any environmental processes that are likely to influence soil metal mobility, such as acid rain causing soil acidification, must be monitored and studied with care, so that data and evidence are available to aid environmental decision-making and solutions to subsequent metal pollution may be found.

5.5.2   FIELD AGRICULTURAL SYSTEMS

Much process work on metal transfers in agricultural soils and crops is carried out in laboratory and glasshouse experiments, using simple forms of metals under controlled conditions. As far as possible, this section will focus on research that has attempted to chart the fate of metals under field conditions. Examples of field crop contamination from air pollution and from minespoil or minewaste are provided, for example, by Alloway and Davies (1971), Tunney et al. (1972), Davies (1983), Hovmand et al. (1983), and Davies et al. (1987). By far the largest literature on the fate of metals in field agricultural systems comes from studies of sewage-sludge-treated fields and crops (e.g. Soon et al., 1980; Chumbley and Unwin, 1982; McGrath 1987; Juste and Mench, 1992). It is on studies of metal transfers in sludge-treated agricultural

Table 5.10. Major metal standing stocks in three forest types (all units in $kg\,ha^{-1}$)

Compartment	Metal stock ($kg\,ha^{-1}$)			Source
	Pb	Zn	Cd	
Mixed deciduous				
Tree canopy	3.24	8.02	0.21	van Hook et al. (1977)
Stem	4.08	15.2	0.44	
Roots	5.49	17.77	0.28	
Total vegetation	12.81	40.99	0.93	
Litter	34.2	78.1	0.52	
Total soil	6400	14 170	30.0	
Beech				
Total vegetation	2.0	4.6	0.057	Heinrichs and Mayer (1980a)
Litter	16.7	8.5	0.063	
Soil	150	315	0.458	
Spruce				
Total vegetation	3.5	12.7	0.16	Heinrichs and Mayer (1980a)
Litter	25.0	9.8	0.05	
Soil	220	262	0.39	

Table 5.11. Percentage contribution of soil metal content to total metal stock of three forest ecosystems described in Table 5.10.

Forest type	Percentage soil contribution to total ecosystem metal fund		
	Pb	Zn	Cd
Mixed deciduous[a]	99.3	99.2	95.4
Beech[b]	88.9	96.0	79.2
Spruce[b]	88.5	92.1	65

Source: [a]van Hook et al. (1977); [b]Heinrichs and Mayer (1980a).

systems, particularly those receiving long-term applications, that much of the current research effort is focused. This will also be the basis of discussion in this section. To understand the fate and distribution of metals in contaminated agricultural systems, a combination of field and pot experiments can add more information than either type of study in isolation. In studies of the fate of metals applied in sludges, numerous examples of both approaches have been published. Only some of the major findings will be discussed briefly here.

The predominant effect of both sewage sludge application and aerial pollution, is the significant accumulation of high metal concentrations in surface soils (see Chapter 3, Section 3.3.5). These accumulations occur within the agriculturally cultivated zone, where root and microbial activity are highest. Both root and microbial metal toxicity effects have been recorded under these conditions (e.g. Chapter 6, this volume; Brookes and McGrath, 1984; Giller et al., 1989; Koomen et al., 1990). As soil metal concentrations increase during long-term sewage sludge application to agricultural soils, crops also take up increasing amounts of metals (e.g. Juste and Mench, 1992). Chang et al. (1987) established a regression model for cumulative sludge metal loading and metal uptake by swish chard and radish:

$$C_p = a + b(1 - e^{-cx})$$

where:

C_p = metal concentration in plant tissue,
x = total cumulative metal loading in sludge, and
a,b,c = experimentally derived parameters.

Crop responses to long-term metal contamination were studied in a series of cropping experiments carried out on long-term sewage-sludge-treated agricultural soils in Germany (Sauerbeck, 1991). Metal concentrations in different crop compartments were studied in a series of comparisons. In a comparison of metal concentrations in grains and vegetable storage organs versus that of leaves, stalks and straw of a large number of different crop plants, Cd and Pb were found to be more highly concentrated in storage organs, while Zn and Cu were more highly concentrated in foliage and stalks. For Zn, Cu and Pb, radish provided an exception to the above rules, perhaps indicating problems in its use as a test plant for assessing metal phytoavailability. In a second comparison, Sauerbeck (1991) reported the concentrations of metals in shoots versus roots of a number of crops. Pb and Cd concentrations were found to be higher in roots; Cu and Zn concentrations were highest in shoots. Juste and Mench (1992) review the results of many compartment studies on sewage sludge treated crops. Their main conclusions were that Ni and Zn are most bioavailable, while Cr and Pb uptake was insignificant. Results of crop experiments carried out on contaminated soils after sludge treatment was terminated, indicated a progressive decrease in metal uptake with time after sludge application ceased.

McGrath (1987) and McGrath and Lane (1989) calculated the metal budgets for long-term (40 year) sewage sludge treated agricultural systems at Woburn, England. The crops grown over the period included potatoes, leeks, carrots, beans, sugar beet and red beet. Their results indicate that extremely small quantities of sludge-applied metals are removed in harvested crops over a 20 year period (Table 5.12). The largest percentage metal removal was for Zn, at only 0.57% of that added in sludges over 20 years. If crop harvesting is assumed to be the only metal removal process from these fields, the Zn removal rates translate to a soil residence time of 3700 years if identical cropping regimes are assumed. Cr and Pb are removed in the smallest proportions in crops, resulting in extremely long predicted soil residence. Residence times will, of course, be shorter than this, since other soil processes also contribute to metal loss from sludge-treated fields. Although metal leaching is not an important loss in most soils (see Chapter 3, Section 3.3.5,

Table 5.12. Amounts of metals removed in harvested crops, related to amounts added in sewage sludge, and their calculated soil residence times

Metal	Metal offtake 1960–80 $(kg\,ha^{-1})$	Offtake as % of total metals added in sludges	Residence time (years)[a]
Zn	11.74	0.57	3700
Cu	1.33	0.16	13 100
Ni	0.45	0.37	5700
Cd	0.18	0.28	7500
Cr	0.22	0.03	70 000
Pb	0.41	0.06	35 000

Reproduced by permission of Blackwell Scientific Publications from McGrath (1987).
[a]Number of years of same cropping regime required to remove 100% of the added metals, calculated to the nearest 100 years.

this volume), McGrath and Lane (1989) and Juste and Mench (1992) identify lateral soil and particulate movement (the beginnings of soil erosion) as an important process removing sludge metals from their applied location.

5.5.3 OTHER ECOSYSTEMS

5.5.3.1 Plants typical of metalliferous sites

While understanding of the *exact* metal transfer processes between rock types, soils and plants is not well developed, it has been appreciated since Roman times that natural vegetation reflects to some extent its rock and soil substrate. Geological and geochemical prospecting relies on statistically tested relationships between rock types, soil conditions and plant communities and individuals (Brooks, 1983). Unique floras (*indicator* or *characteristic* floras) have been identified in many parts of the world, including limestone areas, where *calcium* is the main influencing element, but similarly, *selenium* floras in the western United States and Queensland, *serpentine* floras (on soils high in Cr, Co and Ni) in many parts of the world, *calamine* floras (on soils high in Zn) in eastern Belgium and western Germany, and Cu–Co floras in Zaire and Zambia (Brooks, 1983). The ecology and Cr–Ni nutrition of serpentine flora has received much attention (e.g. Chapter 12, this volume; Brooks, 1987; Proctor et al., 1992; Roberts and Proctor, 1992). Brooks et al. (1990) identified plants in Goiás State, Brazil, which are Ni hyperaccumulators, with foliar tissue Ni concentrations exceeding 3000 ppm. Many of the indicator plants of these rich metalliferous conditions are rare and have attracted much botanical attention over the years. Only relatively recently have their ecological and metal tolerance behaviour been studied. In the UK, the colonisation of metal-rich sites by plant species which possess metal avoidance or tolerance strategies has been reviewed by Baker and Proctor (1990). The fate and distribution of metals in such *adapted* species and communities is likely to be different from that of non-adapted metal-impacted ecosystems.

5.5.3.2 Metal-impacted wetland ecosystems

Freshwater and wetland habitats have recently received more attention than others in the study of metal fate and cycling. This is due to the potential value of wetland and freshwater plants in "cleaning up" the environmental metal contamination of various kinds of industrial and domestic wastes and dredged sediments. Emergent macrophytes such as *Cyperus* spp. have been shown to take up more metals from contaminated sediments which are well oxidised than from those which are reduced (Lee et al., 1983). This result reflects the lower mobility and bioavailability of many metals under waterlogged, anaerobic and reducing conditions, and may indicate that designs for wastewater treatment should include some consideration of the redox conditions of the rooting zone. Submerged freshwater macrophytes may also have the potential for "stripping" polluting metals from contaminated water bodies. Several studies have indicated that macrophyte metal concentrations are generally much higher than ambient water metal concentrations (e.g. Leland and McNurney, 1974; Reimer, 1989). Campbell and Tessier (1991) quote concentration factors for freshwater macrophytes of the order of 10^3 to 10^4. In the natural habitat, these concentration effects mark the beginning of metal transferal up the freshwater food chain from the contaminated water column.

Correlations have been shown between metal concentrations in sediments and waters and metal concentrations in freshwater plants (e.g. Welsh and Denny, 1976, 1979; Franzin and MacFarlane, 1980; Whitton *et al.*, 1981). Campbell and Tessier (1991) successfully predicted the Pb, Cu and Zn concentrations in rhizomes and roots of five floating and emergent freshwater macrophytes from the [Fe-OM/Fe-OH$_x$] ratio of the host sediments. Discovery of these relationships has led to the testing of botanical type materials such as the alga *Navicula* spp., aquatic mosses such as *Fontinalis antipyretica*, and floating macrophytes such as *Elodea canadensis* and *Potamogeton pectinatus*, for use as sensitive biological monitors of heavy metal pollution in freshwaters (Everard and Denny, 1985).

One of the few published metal budgets for a metal-impacted wetland is given by Finlayson (see Chapter 13, this volume). The seasonally inundated Magela floodplain of Northern Territory, Australia, acts as a net sink for the trace metals Cu, Pb, Zn and U, input to the catchment from uranium mining. As in the Magela study, and in salt marsh (DeLaune *et al.*, 1981), Silva *et al.* (1990) found that the sediments of a metal-impacted mangrove in Sepitiba Bay, Rio de Janeiro, acted as a sink for heavy metals input to the system from a large metallurgical processing complex located along the bay's coast. In all three habitats, inundated floodplain, salt marsh and mangrove swamp, metals accumulate in the sediment in a strongly bound fraction, which does not appear to cycle through the biomass to any great extent.

5.6 MODELLING THE FATE OF METALS IN ECOSYSTEMS

A range of modelling approaches has been used for understanding, predicting and simulating the fate and distribution of metals in soil and in plants, but only occasionally in whole soil–plant systems. Whole system approaches have invariably addressed the foodchain contamination problem, particularly the problems of metal toxicity or radionuclide ingestion by animals and humans (e.g. Thorne and Coughtrey, 1983; Jackson *et al.*, 1985). Rarely have modellers attempted to simulate metal transfers in ecosystems with any great species complexity. Harmsen (1992) simulated the long-term retention and mobility of metals in agricultural soil, but not in the plant compartments. Martin and Coughtrey (1987) used a time series of soil metal data to simulate and predict future metal retention and mobility in a woodland soil. Other simulation models are more environmentally based, driven by atmospheric and hydrological components (e.g. Wagenet *et al.*, 1979; Wagenet and Grenney, 1983). The following brief introduction to modelling aims to outline several approaches and to illustrate, where possible, examples of each. Problems and gaps in research will be introduced.

Four main types of models have been used in metal transfer studies in soil–plant systems: (i) hardware models, (ii) conceptual models, (iii) empirically based prediction models, and (iv) mathematical simulation models. If we consider lysimeters to be simple *hardware* models, these, and even pot studies, are by far the most common type of modelling study. They yield black-box input–output data which can be used in simple *predictive* models. We have seen in Chapter 3, Section 3.3.5, that many predictions of metal phytoavailability in soil are based on lysimeter or pot-plant bioassays, correlated against some soil chemical index, based on extraction of metal from the soil. The resultant simple regression equation is used to predict metal availability (e.g. Browne *et al.*, 1984). The illustration in Figure 3.16 (p. 136), showing prediction of Cd uptake by radish, using soil Cd extracted by NH_4NO_3 is a good example of this empirical, predictive approach.

Many authors have provided *conceptual* models of the fate of metals in different ecosystems. These models generally take the form of systems diagrams of various degrees of complexity and are necessary precursors to prediction or simulation modelling. Lepp (1979) presented a compartment system diagram for Cu cycling in a woodland system. A similar system was developed by Martin and Coughtrey (1982) for a metal-impacted deciduous woodland close to a Zn smelter.

A range of *simulation* models for plant nutrient uptake have been developed and some have been applied to the problem of metal uptake. Scales of model operation range from the plant root in contact with contaminated soil (e.g. Cushman, 1979*a*,*b*; 1984) to whole-plant compartment transfer models (e.g. Cowan *et al.*, 1983, for plutonium). Mullins *et al.* (1986) adapted the Cushman (1984) root nutrient uptake model to the *Zea mays* uptake of Cd and Zn from sewage-sludge-treated soils. They ran a series of model sensitivity analyses to identify which plant and soil characteristics were the main controls on metal uptake. Not surprisingly, these analyses identified metal concentration in the soil solution as the most important soil parameter and the proportion of short, fine roots as the most important plant parameter. Very similar results were obtained by Barber and Claassen (1977), simulating plant uptake of Zn, using soil and plant data gleaned from the literature.

Disappointingly few attempts have been made to simulate metal fate and distribution in anything other than simplified soil–plant systems. The models described by Cowan *et al.* (1983) for plutonium cycling, and by Jackson and

Smith (1987), for radionuclide transfers to grazing animals, include a series of soil and plant compartment flux coefficients which could simply be adapted for use in modelling metal transfers, if similar compartment flux coefficients were available, or could be derived, for the major polluting metals.

5.7 CONCLUSIONS

Briefly reviewing the compartmentalisation of metals in a range of different ecosystems has indicated that in most cases metals accumulate in soils and sediments, especially in organic horizons. This is especially true for Pb, Zn and Cu. From the data given in Table 5.11, forest soils can be seen to account for anything from 65% to 99.3% of the total ecosystem metal fund (Table 5.10). Relatively small proportions of metals added to ecosystems in air pollution, metalliferous wastes or contaminated waters, are taken up and recycled through the biomass. In terms of *quantity*, woody tissues of forest stands provide substantial *standing stocks* of metals, but metal concentrations in wood are generally low. In contaminated forests, older leaves and fallen leaves contain some of the highest concentrations of metals, particularly Pb, Zn and Cd, of the entire biomass. The concentrations of metals in fallen leaves can increase over time on the forest floor as metal ions exchange onto the cation exchange sites of the decaying litter.

While much is known about metal *input* to different ecosystems, and metal *output* from ecosystems, relatively little is known about metal transformation and translocation processes *within* ecosystems. This is partly due to (a) difficulties of *in situ* monitoring of ecosystem transfers and (b) problems in using difference calculations for ecosystem compartments to calculate system fluxes. There are also difficulties in comparing the fate and distribution of metals in different ecosystems. Three main types of problems arise:

(i)   lack of comparable biomass compartments measured;
(ii) lack of statistically representative sampling of both biomass and soil compartments—this is especially true for more complex ecosystems, but easier in simpler ecosystems, such as monospecific forestry stands, or in agriculture;
(iii) lack of sufficient long-term (>10 year) ecosystem studies, with regular enough monitoring of compartments to estimate metal fluxes.

Most of the approaches outlined earlier in this chapter are represented in the case studies which follow in the second section of this book. Ecosystem compartment studies are illustrated by the work of Martin and Bullock in deciduous woodland (Chapter 9), Lepp and Dickinson in tea, coffee and cacao plantations (Chapter 10) and Finlayson in inundated monsoonal wetlands (Chapter 13). Process-orientated studies are represented by the work of McGrath on soil microbial processes affected by sewage sludge applications

(Chapter 6) and Hopkin, on soil arthropods and decomposition processes (Chapter 8).

REFERENCES

Alloway, B. J. and Davies, B. E. (1971) Heavy metal content of plants growing on soils contaminated by lead mining. *Journal of Agricultural Science (Cambridge)*, **76**, 321–323.

Alloway, B. J. and Morgan, H. (1986) The behaviour and availability of Cd, Ni and Pb in polluted soils. In: Assink, J. W. and van den Brink, W. J. (Eds) *Contaminated Soil*, pp. 101–113. Martinus Nijhoff Publishers, Dordrecht.

Andresen, A. M., Johnson, A. H. and Siccama, T. G. (1980) Levels of lead, copper and zinc in the forest floor in the northeastern United States. *Journal of Environmental Quality*, **9**, 293–296.

Asche, N. (1985) Komponenten de Schwarmetallhaushalts von zwei Waldokosystemen. *VDI-Berichte*, **560**, (reference quoted by Bergkvist *et al.*, 1989).

Ausmus, B. S., Jackson, D. R. and Dodson, G. J. (1977) Assessment of microbial effects on Cd[109] movement through soil columns. *Pedobiologia*, **17**, 183–188.

Baker, A. J. M. and Proctor, J. (1990) The influence of cadmium, copper, lead, and zinc on the distribution and evolution of metallophytes in the British Isles. *Plant Systematics and Evolution*, **173**, 91–108.

Balls, P. W. (1989) Trace metal and major ion composition of precipitation at a North Sea coastal site. *Atmospheric Environment*, **23**, 2751–2759.

Barber, S. A. and Claassen, N. (1977) A mathematical model to simulate metal uptake by plants growing in soil. In: Drucker, H. and Wildung, R. E. (Eds) *Biological Implications of Metals in the Environment*, pp. 358–364. National Technology Information Service, US Department of Commerce, Springfield, VA.

Benninger, L. K., Lewis, D. M. and Turekian, K. K. (1975) The use of natural Pb-210 as a heavy metal tracer in the river estuarine system. In: Church, T. M. (Ed.) *Marine Chemistry and the Coastal Environment*, pp. 201–210. American Chemical Society Symposium Series No. 18.

Berg, B., Ekbohm, G., Soderstrom, B. and Staaf, H. (1991) Reduction of decomposition rates of Scots pine needle litter due to heavy-metal pollution. *Water, Air and Soil Pollution*, **59**, 165–177.

Berggren, D. (1992*a*) Speciation and mobilization of aluminium and cadmium in podzols and cambisols of S. Sweden. *Water, Air and Soil Pollution*, **62**, 125–156.

Berggren, D. (1992*b*) Speciation of copper in soil solutions from podzols and cambisols of S. Sweden. *Water, Air and Soil Pollution*, **62**, 111–123.

Bergkvist, B. (1987) Soil solution chemistry and metal budgets of spruce forest ecosystems in S. Sweden. *Water, Air and Soil Pollution*, **33**, 131–154.

Bergkvist, B., Folkeson, L. and Berggren, D. (1989) Fluxes of Cu, Zn, Pb, Cd, Cr and Ni in temperate forest ecosystems. *Water, Air and Soil Pollution*, **47**, 217–286.

Berrow, M. L. and Reaves, G. A. (1986) Total chromium and nickel contents of Scottish soils. *Geoderma*, **37**, 15–27.

Billett, M. F., Fitzpatrick, E. A. and Cresser, M. S. (1991) Long-term changes in the Cu, Pb and Zn content of forest soil organic horizons from north-east Scotland. *Water, Air and Soil Polution*, **59**, 179–191.

Bingham, F. T., Page, A. L., Mahler, R. J. and Ganje, T. J. (1976) Yield and cadmium accumulation of forage species in relation to cadmium content of sludge-amended soil. *Journal of Environmental Ouality*, **5**, 57–59.

Brabec, E., Cudlin, P., Rauch, O. and Skoda, M. (1983) Lead budget in a smelter-adjacent grass stand. In: *Heavy Metals in the Environment*, pp. 1150–1153. Proceedings of an International Conference, Heidelberg. CEP Consultants, Edinburgh.

Bray, J. R. and Goreham, E. (1964) Litter production in forests of the world. *Advances in Ecological Research*, **2**, 101–157.

Brookes, P. C. and McGrath, S. P. (1984) Effects of metal toxicity on the size of the soil microbial biomass. *Journal of Soil Science*, **35**, 341–346.

Brooks, R. R. (1983) *Biological Methods of Prospecting for Minerals*. Wiley Interscience Publication, Wiley, New York.

Brooks, R. R. (1987) *Serpentine and its Vegetation: A Multidisciplinary Approach*. Dioscorides Press, Portland, Oregon.

Brooks, R. R., Reeves, R. D., Baker, A. J. M., Rizzo, J. A. and Ferreira, H. D. (1990) The Brazilian serpentine plant expedition (BRASPEX), 1988. *National Geographic Research*, **6**, 205–219.

Browne, C. L., Wong, Y.-M. and Buhler, D. R. (1984) A predictive model for the accumulation of cadmium by container-grown plants. *Journal of Environmental Quality*, **13**, 184–188.

Buchauer, M. J. (1973) Contamination of soil and vegetation near a zinc smelter by zinc, cadmium, copper and lead. *Environmental Science and Technology*, **7**, 131–135.

Campbell, P. G. C. and Tessier, A. (1991) Biological availability of metals in sediments: Analytical approaches. In: Vernet, J.-P. (Ed.) *Heavy Metals in the Environment. Trace Metals in the Environment*, Vol. I, pp. 161–173. Elsevier, Amsterdam.

Cappellato, R., Peters, N. E. and Ragsdale, H. L. (1993) Acidic atmospheric deposition and canopy interactions of adjacent deciduous and coniferous forests in the Georgia Piedmont. *Canadian Journal of Forest Research*, **23**, 114–124.

Chamberlain, A. C. (1983) Fallout of lead and uptake by crops. *Atmospheric Environment*, **17**, 693–706.

Chang, A. C. Page, A. L. and Warneke, J. E. (1987) Long-term applications on cadmium and zinc accumulation in Swiss Chard and Radish. *Journal of Environmental Quality*, **16**, 217–221.

Chumbley, C. G. and Unwin, R. J. (1982) Cadmium and lead content of vegetable crops grown on land with a history of sewage sludge application. *Environmental Pollution (Series B)*, **4**, 231–237.

Colbourn, P. and Thornton, I. (1978) Lead pollution in agricultural soils. *Journal of Soil Science*, **29**, 513–526.

Coughtrey, P. J., Jones, C. H., Martin, P. J. and Shales, S. W. (1979) Litter accumulation in woodlands contaminated with Pb, Zn, Cd and Cu. *Oecologia (Berlin)*, **39**, 51–60.

Cowan, C. E., Jenne, E. A., Simpson, J. C. and Cataldo, D. A. (1983) Nutrient-contaminant (Pu) plant accumulation model. *The Science of the Total Environment*, **28**, 277–286.

Crump, D. R., Barlow, P. J. and Van Dest, D. J. (1980) Seasonal changes in the lead content of pasture grass growing near a motorway. *Agriculture and Environment*, **5**, 213–215.

Cumming, J. R. and Weinstein, L. H. (1990) Aluminium–mycorrhizal interactions in the physiology of pitch pine seedlings. *Plant and Soil*, **125**, 7–18.

Cushman, J. H. (1979*a*) An analytical solution to solute transport near root surfaces for low initial concentrations. I—Equations' development. *Soil Science Society of America*, **43**, 1087–1090.

Cushman, J. H. (1979*b*) An analytical solution to solute transport near root surfaces for low initial concentration. II—Applications. *Soil Science Society of America*, **45**, 1090–1095.

Cushman, J.H. (1984) Nutrient transport inside and outside the root rhizosphere: Generalized model. *Soil Science*, **138**, 164–171.

Davies, B. E. (1983) Heavy metal contamination from base metal mining and smelting: implications for man and his environment. In: Thornton, I. (Ed.) *Applied Environmental Geochemistry*, pp. 425–462.

Davies, B. E., Lear, J. M. and Lewis, N. J. (1987) Plant availability of heavy metals in soils. In: Coughtrey, P. J., Martin, M. H. and Unsworth, M. H. (Eds) *Pollutant Transport and Fate in Ecosystems*, pp. 267–275. Blackwell Scientific, Oxford.

DeLaune, R. D., Reddy, C. N. and Patrick, W. H. Jr., (1981) Accumulation of plant nutrients and heavy metals through sedimentation processes and accretion in a Louisiana salt marsh. *Estuaries*, **4**, 328–334.

Denny, H. J. and Wilkins, D. A. (1987) Zinc tolerance in *Betula* spp. IV. The mechanism of ectomycorrhizal amelioration of zinc toxicity. *New Phytologist* **106**, 545–553.

Dreiss, S. J. (1986) Chromium migration through sludge-treated soils. *Ground Water*, **24**, 312–321.

Dumontet, S., Levesque, M. and Mathur, S. P. (1990) Limited downward migration of pollutant metals (Cu, Zn, Ni and Pb) in acid virgin peat soils near a smelter. *Water, Air and Soil Pollution*, **49**, 329–342.

Everard, M. and Denny, P. (1985) Flux of lead in submerged plants and its relevance to a freshwater system. *Aquatic Botany*, **21**, 181–193.

Franzin, W. G. and MacFarlane, G. A. (1980) An analysis of the aquatic macrophyte, *Myriophyllum exalbescens*, as an indicator of metal contamination of aquatic ecosystems near a base metal smelter. *Bulletin of Environmental Contamination and Toxicology*, **24**, 597–605.

Friedland, A. J. and Johnson, A. H. (1985) Lead distribution and fluxes in a high-elevation forest in northern Vermont. *Journal of Environmental Quality*, **14**, 332–336.

Friedland, A. J., Johnson, A. H. and Siccama, T. G. (1984) Trace metal content of the forest floor in the Green Mountains of Vermont: spatial and temporal patterns. *Water, Air and Soil Pollution*, **21**, 161–170.

Garten, C. T., Gentry, J. B. and Sharitz, R. R. (1977) An analysis of elemental concentrations in vegetation bordering a southeastern United States coastal plain stream. *Ecology*, **58**, 979–992.

Giller, K. E., McGrath, S. P. and Hirsch, P. R. (1989) Absence of nitrogen fixation in clover grown on soil subject to long-term contamination with heavy metals, is due to survival of only ineffective *Rhizobium*. *Soil Biology and Biochemistry*, **21**, 841–848.

Grosch, S. (1986) Wet and dry deposition of atmospheric trace elements in forest areas. In: Georgii, H. W. (Ed.) *Atmospheric Pollutants in Forest Areas*, pp. 35–45. D. Reidel Publishing, Dordrecht.

Guha, M. (1961) A study of the trace element uptake of deciduous trees. PhD Thesis, Aberdeen University.

Hagemeyer, J., Kahle, H., Breckle, S.- W. and Waisel, Y. (1986) Cadmium in *Fagus sylvatica* L. trees and seedlings: leaching, uptake and interconnection with transpiration. *Water, Air and Soil Pollution*, **29**, 347–359.

Haq, A. U., Bates, T. E. and Soon, Y. K. (1980) Comparison of extractants for plant-available zinc, cadmium, nickel and copper in contaminated soils. *Soil Science Society of America*, **44**, 772–777.

Harmsen, K. (1992) Long-term behaviour of heavy metals in agricultural soils: a simple analytical model. In: Adriano, D. C. (Ed.) *Biogeochemistry of Trace Metals*, pp. 217–247. Lewis Publishers, Boca Raton.

Harrison, R. M. and Chirgawi, M. B. (1989*a*) The assessment of air and soil as contributors of some trace metals to vegetable plants. I—Use of a filtered air growth cabinet. *The Science of the Total Environment*, **83**, 13–34.

Harrison, R. M. and Chirgawi, M. B. (1989*b*) The assessment of air and soil as contributors of some trace metals to vegetable plants. III—Experiments with field-grown plants. *The Science of the Total Environment*, **83**, 47–63.

Harrison, R. M. and Johnston, W. R. (1987) Experimental investigations on the relative contribution of atmosphere and soils to the lead content of crops. In: Coughtrey, P. J., Martin, M. H. and Unsworth, M. H. (Eds) *Pollutant Transport and Fate in Ecosystems*. Special Publication of the British Ecological Society, No.6, pp. 277–287. Blackwell Scientific, Oxford.

Heinrichs, H. and Mayer, R. (1977) Distribution and cycling of major and trace elements in two Central European Forest Ecosystems. *Journal of Environmental Quality*, **6**, 402–407.

Heinrichs, H. and Mayer, R. (1980a) The role of forest vegetation in the biogeo-chemical cycle of heavy metals. *Journal of Environmental Quality*, **9**, 111–118.

Heinrichs, H. and Mayer, R. (1980b) Distribution and cycling of nickel in forest ecosystems. In: Nriagu, J. O. (1980) *Nickel in the Environment*, pp. 431–455. Wiley-Interscience, John Wiley, New York.

Henry, C. L. and Harrison, R. B. (1992) Fate of trace metals in sewage sludge compost. In: Adriano, D. C. (Ed.) *Biogeochemistry of Trace Metals*, pp. 195–216. Lewis Publishers, Boca Raton.

Hopkin, S. P. (1993) Ecological implications of "95% protection levels" for metals in soil. *Oikos*, **66**, 137–141.

Hovmand, M. F., Tjell, J. C. and Mosbaek, H. (1983) Plant uptake of airborne cadmium. *Environmental Pollution (Series A)*, **30**, 27–38.

Hughes, M. K. (1981) Cycling of trace metals in ecosystems. In: Lepp, N. W. (Ed.) *Effect of Heavy Metal Pollution on Plants. Vol. 2: Metals in the Environment*, pp. 95–118. Applied Science Publishers, London.

Hughes, M. K., Lepp, N. W. and Phipps, D. A. (1980) Aerial heavy metal pollution and terrestrial ecosystems. *Advances in Ecological Research*, **11**, 217–327.

Hunter, B. A., Johnson, M. S. and Thompson, D. J. (1987*a*) Ecotoxicology of copper and cadmium in a contaminated grassland ecosystem. II—Invertebrates. *Journal of Applied Ecology*, **24**, 587–599.

Hunter, B. A., Johnson, M. S. and Thompson, D. J. (1987*b*) Ecotoxicology of copper and cadmium in a contaminated grassland ecosystem. III—Small mammals. *Journal of Applied Ecology*, **24**, 601–614.

Hunter, B. A., Johnson, M. S. and Thompson, D. J. (1987*c*) Ecotoxicology of copper and cadmium in a contaminated grassland ecosystem. I—Soil and vegetation contamination. *Journal of Applied Ecology*, **24**, 573–586.

Hutchinson, T. C. and Whitby, L. M. (1974) Heavy-metal pollution in the Sudbury mining and smelting region of Canada. I—Soil and vegetation contamination by nickel, copper and other metals. *Environmental Conservation*, **1**, 123–132.

Jackson, A. P. and Alloway, B. J. (1991) The bioavailability of cadmium to lettuce and cabbage in soils previously treated with sewage sludges. *Plant and Soil*, **132**, 179–186.

Jackson, D. and Smith, A. D. (1987) Generalized models for the transfer and distribution of stable elements and their radionuclides in agricultural systems. In: Coughtrey, P. J., Martin, M. H. and Unsworth, M. H. (Eds) *Pollutant Transport and Fate in Ecosystems*, pp. 385–402. Special Publication of the British Ecological Society, No. 6. Blackwell Scientific, Oxford.

Jackson, D., Coughtrey, P. J. and Crabtree, D. F. (1985) Dynamic models for soil–plant–animal systems. *Nuclear Europe*, 5, 29–31.

Jackson, D. R. and Watson, A. P. (1977) Disruption of nutrient pools and transport of heavy metals in a forested watershed near a smelter. *Journal of Environmental Quality*, 6, 331–338.

Jarvis, I. and Higgs, N. (1987) Trace-element mobility during early diagenesis in distal turbidites: late Quaternary of the Madeira abyssal plain, N. Atlantic. In: Weaver, P. P. E. and Thomson, J. (Eds) *Geology and Geochemistry of Abyssal Plains*, pp. 179–213. Geological Society Special Publication No. 31.

John, M. K. (1973) Cadmium uptake by eight food crops as influenced by various soil levels of cadmium. *Environmental Pollution*, 4, 7–15.

Juste, C. and Mench, M. (1992) Long-term application of sewage sludge and its effects on metal uptake by crops. In: Adriano, D. C. (Ed.) *Biogeochemistry of Trace Metals*, pp. 159–193. Lewis Publishers, Boca Raton.

Kazda, M. (1986) Untersuchungen von Schwermetalldepositionsvorgangen und Analysen fractionell gesammelter Stammabflussproben und Jahresgang Del Schwermetalldeposition in einem Buchenwaldokosystem Del stadtnahen Wienerwaldes. Dissertation der Universitat fur Bodenkultur in Wein 27. VWGO. Vienna (reference quoted in Bergkvist *et al.*, 1989).

King. L. D. (1988) Retention of metals by several soils of the southeastern United States. *Journal of Environmental Quality*, 17, 239–246.

Koomen, I., McGrath, S. P. and Giller, K. E. (1990) Mycorrhizal infection of clover is delayed in soils contaminated with heavy metals from past sewage sludge applications. *Soil Biology and Biochemistry*, 22, 871–873.

Kuboi, T., Noguchi, A. and Yazaki, J. (1986) Family-dependent cadmium accumulation characteristics in higher plants. *Plant and Soil*, 92, 405–415.

Kuo, S., Heilman, P. E. and Baker, A. S. (1983) Distribution and forms of copper, zinc, cadmium, iron and manganese in soils near a copper smelter. *Soil Science*, 135, 101–109.

Latterell, J. J., Dowdy, R. H. and Larson, W. E. (1978) Correlation of extractable metals and metal uptake of snap beans grown on a soil amended with sewage sludge. *Journal of Environmental Quality*, 7, 435–440.

LeClaire, J. P., Chang, A. C., Levesque, C. S. and Sposito, G. (1984) Trace metal chemistry in arid-zone field soils amended with sewage sludge: IV—Correlations between zinc uptake and extracted soil zinc fractions. *Soil Society of America*, 48, 509–513.

Lee, C. R., Folsom, B. L. Jr. and Bates, D. J. (1983) Prediction of plant uptake of toxic metals using a modified DTPA soil extraction. *The Science of the Total Environment*, 28, 191–202.

Leland, H. V. and McNurney, J. M. (1974) Lead transport in a river ecosystem. In: *Proceedings of an International Conference of Transport of Persistent Chemicals in Aquatic Ecosystems*, Vol. III, pp. 17–23. Ottawa.

Lepp, N. (1979) Cycling of copper in a woodland ecosystem. In: Nriagu, J. O. (Ed.) *Copper in the Environment. Vol. 1 Ecological Cycling*, pp. 289–323. John Wiley, New York.

Lepp, N. W. and Dollard, G. J. (1974) Studies on lateral movement of ^{210}Pb in woody stems. Patterns observed in dormant and non-dormant stems. *Oecologia (Berlin)*, 7, 413–416.

Levine, M. B., Hall, A. T., Barrett, G. W. and Taylor, D. H. (1989) Heavy metal concentrations during ten years of sludge treatment to an old-field community. *Journal of Environmental Quality*, **18**, 411–418.

Lindberg, S. E. and Harriss, R. C. (1981) The role of atmospheric deposition in an Eastern US deciduous forest. *Water, Air and Soil Pollution*, **16**, 13–31.

Little, P. (1973) A study of heavy metal contamination of leaf surfaces. *Environmental Pollution*, **5**, 159–172.

Little, P. and Martin, M. H. (1972) A survey of zinc, lead and cadmium in soil and natural vegetation around a smelting complex. *Environmental Pollution*, **3**, 241–254.

Little, P. and Wiffen, R. D. (1977) Emission and deposition of petrol exhaust Pb. I—Deposition of exhaust Pb to plant and soil surfaces. *Atmospheric Environment*, 12, 1331–1343.

Livett, E. A. (1988) Geochemical monitoring of atmospheric heavy metal pollution: theory and practice. *Advances in Ecological Research*, 18, 65–177.

Martin, M. H. and Coughtrey, P. J. (1982) *Biological Monitoring of Heavy Metal Pollution*. Chapman and Hall, London.

Martin, M. H. and Coughtrey, P. J. (1987) Cycling and fate of heavy metals in a contaminated woodland ecosystem. In: Coughtrey, P. J., Martin, M. H. and Unsworth, M. H. (Eds) *Pollutant Transport and Fate in Ecosystems*, pp. 319–336. Special Publication of the British Ecological Society No. 6. Blackwell Scientific, Oxford.

Matthews, H. and Thornton, I. (1980) Agricultural implications of zinc and cadmium contaminated land at Shipham, Somerset. In: Hemphill, D. D. (Ed.) *Trace Substances in Environmental Health*, pp. 478–488. Vol. XIV. University of Missouri, Columbia.

McGrath, S. P. (1987) Long-term studies of metal transfers following application of sewage sludge. In: Coughtrey, P. J., Martin, M. H. and Unsworth, M. H. (Eds) *Pollutant Transport and Fate in Ecosystems*, pp. 301–317. Special Publication of the British Ecological Society No. 6. Blackwell Scientific, Oxford.

McGrath, S. P. and Lane, P. W. (1989) An explanation for the apparent losses of metals in a long-term field experiment with sewage sludge. *Environmental Pollution*, **60**, 235–256.

McClellan, J. K. and Rock, C. A. (1988) Pretreating landfill leachate with peat to remove metals. *Water, Air and Soil Pollution*, **37**, 203–215.

Miller, H. G. and Miller, J. D. (1980) Collection and retention of atmospheric pollutants by vegetation. In: Drablos, D. and Tollan, A. (Eds) *Ecological Impacts of Acid Precipitation*, pp. 33–40. SNSF Project, Oslo-As, Norway.

Mitchell, R. L., Reith, J. W. S. and Johnston, I. M. (1957) Trace element uptake in relation to soil content. *Journal of the Science of Food and Agriculture*, **8** (Supplement Issue), 51–58.

Morselt, A. F. W., Smits, W. T. M. and Limonard, T. (1986) Histochemical demonstration of heavy metal tolerance in ectomycorrhizal fungi. *Plant and Soil*, **96**, 417–420.

Mullins, G. L., Sommers, L. E. and Barber, S. A. (1986) Modeling the plant uptake of cadmium and zinc from soils treated with sewage sludge. *Soil Science Society of America*, **50**, 1245–1250.

Munshower, F. F. and Neuman, D. R. (1979) Pathways and distribution of some heavy metals in a grassland ecosystem. In: Perry, R. (Ed.) *Management and Control of Heavy Metals in the Environment*, pp. 206–209. Proceedings of an International Conference. CEP Consultants, Edinburgh.

Ng, S. K. and Bloomfield, C. (1962) The effect of flooding and aeration on the mobility of certain trace elements in soils. *Plant and Soil*, **16**, 108–135.

Nihlgard, B. (1970) Precipitation, its chemical composition and effects on soil and water in a beech and spruce forest in South Sweden. *Oikos*, **21**, 208–217.

Nilsson, I. (1972) Accumulation of metals in spruce and needle litter. *Oikos*, **23**, 132–136.

Nriagu, J. O. and Pacyna, J. M. (1988) Quantitative assessment of worldwide contamination of air, water and soils by trace metals. *Nature (London)*, **333**, 134–139.

Pacyna, J. M. (1987) Atmospheric emissions of arsenic, cadmium, lead and mercury from high temperature processes in power generation and industry. In: Hutchinson, T. C. and Meema, K. M. (Eds) *Lead, Mercury, Cadmium and Arsenic in the Environment*, pp. 69–87. SCOPE Vol. 31. John Wiley, New York.

Page, A. L., Bingham, F. T. and Chang, A. C. (1981) Cadmium. In: Lepp, N. W. (Ed.) *Effect of Trace Metals on Plant Function*, pp. 77–109. Applied Science Publishers, London.

Parker, G. G. (1983) Throughflow and stemflow in the forest nutrient cycle. *Advances in Ecological Research*, **13**, 57–133.

Parker, G. R., McFee, W. W. and Kelly, J. M. (1978) Metal distribution in forested ecosystems in urban and rural Northwestern Indiana. *Journal of Environmental Quality*, **7**, 337–342.

Peel, D. A. (1989) Trace metals and organic compounds in ice cores. In: Oeschger, H. and Langway, C. C. (Eds) *The Environmental Record in Glaciers and Ice Sheets*, pp. 207–223. Report of the Dahlem Workshop, Berlin, 1988. John Wiley, Chichester.

Proctor, J., Baker, A. J. M. and Reeves, R. D. (Eds) (1992) *Proceedings of the First International Conference on Serpentine Ecology*. University of California, Davis, June 1991. Intercept Ltd., Andover, Hants.

Purves, D. (1985) *Trace-Element Contamination of the Environment*, revised edition. Elsevier, Amsterdam.

Rappaport, B. D., Martens, D. C., Reneau, R. B. and Simpson, T. W. (1988) Metal availability in sludge-amended soils with elevated metal levels. *Journal of Environmental Quality*, **17**, 42–47.

Raymond, K. N., Muller, G. and Mayzanke, B. F. (1984) Complexation of iron by siderophores. A review of their solution and structural chemistry and biological function. *Topics in Current Chemistry*, **123**, 49–102.

Reimer, P. (1989) Concentration of lead in aquatic macrophytes from Shoal Lake, Manitoba, Canada. *Environmental Pollution*, **56**, 77–84.

Roberts, B. A. and Proctor, J. (Eds) (1992) *The Ecology of Areas with Serpentinized Rocks: A World View*. Kluwer Academic, Dordrecht.

Rohbock, E. (1986) Water solubility of heavy metals in deposition samples— interpretation and prediction of bioavailability. In: Georgii, H. W. (Ed.) *Atmospheric Pollutants in Forest Areas*, pp. 201–214. D. Reidel Publishing, Dordrecht.

Ross, H. B. (1987) Trace metals in precipitation in Sweden. *Water, Air and Soil Pollution*, **36**, 349–363.

Ross, S. M., Thornes, J. B. and Nortcliff, S. (1990) Soil hydrology, nutrient and erosional response to the clearance of terra firme forest, Maraca Island, Roraima, Northern Brazil. *The Geographical Journal*, **156**, 267–282.

Roth, M. (1992) Metals in invertebrate animals of a forest ecosystem. In: Adriano, D. C. (Ed.) *Biogeochemistry of Trace Metals*, pp. 299–328. Lewis Publishers, Boca Raton.

Sarker, A. N. and Wyn Jones, R. G. (1982) Effect of rhizosphere pH on the availability and uptake of Fe, Mn and Zn. *Plant and Soil*, **66**, 361–372.

Sauerbeck, D. R. (1991) Plant, element and soil properties governing uptake and availability of heavy metals derived from sewage sludge. *Water, Air and Soil Pollution*, **57–58**, 227–237.

Schuck, E. A. and Locke, J. K. (1970) Relationship of automative lead particulates to certain consumer crops. *Environmental Science and Technology*, **4**, 324–330.

Schultz, R. (1987) Vergleichende Betrachtung des Schwarmetallhaushalts verschiedener Waldokosysteme Norddeutschlands. *Ber. d. Forschungszentr. Waldokosysteme/ Waldsterben, Reihe A. Bd. 32*. Universitat Gottingen, Gottingen (reference quoted by Bergkvist *et al.*, 1989).

Sharma, R. P. and Shupe, J. L. (1977) Lead, cadmium and arsenic residues in animal tissues in relation to their surrounding habitat. *Science of the Total Environment*, **7**, 53–62.

Shaw, G. E. (1989) Aerosol transport from sources to ice sheets. In: Oeschger, H. and Langway, C. C. (Eds) *The Environmental Record in Glaciers and Ice Sheets*, pp. 13–27. Report of the Dahlem Workshop, Berlin, 1988. John Wiley, Chichester.

Sheppard, S. C. and Sheppard, M. I. (1991) Lead in boreal plants. *Water, Air and Soil Pollution*, **57–58**, 79–91.

Shuman, L. M. (1979) Zinc, manganese and copper in soil fractions. *Soil Science*, **127**, 10–17.

Siccama, T. G., Smith, W. H. and Mader, D. L. (1980) Changes in lead, zinc, copper, dry weight, and organic matter content in the forest floor of white pine stands in central Massachusetts over 16 years. *Environmental Science and Technology*, **14**, 54–56.

Silva, C. A. R., Lacerda, L. D. and Rezende, C. E. (1990) Metals reservoir in a red mangrove forest. *Biotropica*, **22**, 339–345.

Smith, W. H. (1974) Air pollution—effects on the structure and function of the temperate forest ecosystem. *Environmental Pollution*, **6**, 111–129.

Smith, W. H. (1976) Lead contamination of the roadside ecosystem. *Journal of the Air Pollution Control Association*, **26**, 753–766.

Smith, W. H. and Siccama, T. G. (1981) The Hubbard Brook ecosystem study: biogeochemistry of lead in the northern hardwood forest. *Journal of Environmental Quality*, **10**, 323–333.

Soon, Y. K., Bates, T. E. and Moyer, J. R. (1980) Land application of chemically treated sewage sludge: III—Effects of soil and plant heavy metal content. *Journal of Environmental Quality*, **9**, 497–505.

Stover, R. C., Sommers, L. E. and Silviera, D. J. (1976) Evaluation of metals in wastewater sludge. *Journal of the Water Pollution Control Federation*, **48**, 2165–2175.

Strojan, C. L. (1978) Forest leaf litter decomposition in the vicinity of a zinc smelter. *Oikos*, **31**, 41–46.

Swaine, D. J. and Mitchell, R. L. (1960) Trace-element distribution in soil profiles. *Journal of Soil Science*, **11**, 347–368.

Swanson, K. A. and Johnson, H. (1980) Trace metal budgets for a forested watershed in the New Jersey Pine Barrens. *Water Resources Research*, **16**, 373–376.

Tadesse, W., Shuford, J. W., Taylor, R. W., Adriano, D. C. and Sajwan, K. S. (1991) Comparative availability to wheat of metals from sewage sludge and inorganic salts. *Water, Air and Soil Pollution*, **55**, 397–408.

Taylor, F. G. (1983) Cycling and retention of hexavalent chromium in a plant–soil system. In: *Heavy Metals in the Environment*, Vol. 2, pp. 749–752. Proceedings of an International Conference, Heidelberg. CEP Consultants, Edinburgh.

Thorne, M. C. and Coughtrey, P. J. (1983) Dynamic models for radionuclide transport in soils, plants and domestic animals. In: Coughtrey, P. J., Bell, J. N. B. and

Roberts, T. M. (Eds) *Ecological Aspects of Radionuclide Release*, pp. 127–139. Blackwell Scientific, Oxford.

Tunney, A., Flemming, G. A., O'Sullivan, A. N. and Molley, J. P. (1972) Effects of lead-mine concentrates on the lead content of ryegrass and pasture herbage. *Irish Journal of Agricultural Research*, **11**, 144–147.

Turner, R. S., Johnson, A. H. and Wang, D. (1985) Biogeochemistry of lead in McDonalds Branch Watershed, New Jersey Pine Barrens. *Journal of Environmental Quality*, **14**, 305–314.

Tyler, G. (1981) Leaching of metals from the A-horizon of a spruce forest soil. *Water, Air and Soil Pollution*, **15**, 353–369.

van Hook, R. I., Harris, W. F. and Henderson, G. S. (1977) Cadmium, lead and zinc distributions and cycling in a mixed deciduous forest. *Ambio*, **6**, 281–286.

Vogt, K. A., Grier, C. C. and Vogt, D. J. (1986) Production, turnover and nutrient dynamics of above and below ground detritus of world forests. *Advances in Ecological Research*, **15**, 303–377.

Wagenet, R. J. and Grenney, W. J. (1983) Modelling the terrestrial fate of heavy metals. In: Jorgensen S. E. and Mitsch, W. J. (Eds) *Application of Ecological Modelling in Environmental Management*, Part B, 7–34.

Wagenet, R. J., Grenney, W. J., Wooldridge, G. L. and Jurinak, J. J. (1979) An atmospheric–terrestrial heavy metal transport model. I—Model theory. *Ecological Modelling*, **6**, 253–272.

Wedding, J. B., Carlson, R. W., Stukel, J. J. and Bazzazz, F. A. (1975) Aerosol deposition on plant leaves. *Environmental Science and Technology*, **9**, 151–153.

Welsh, R. P. H. and Denny, P. (1976) Waterplants and the recycling of heavy metals in an English lake. In: Hemphil, D. D. (Ed.) *Trace Substances in Environmental Health*, Vol. X, pp. 217–223. A Symposium. University of Missouri, Columbia.

Welsh, R. P. H. and Denny, P. (1979) The lead and copper in two submerged aquatic angiosperm species. *Journal of Experimental Botany*, **30**, 339–345.

Whitton, B. A., Say, P. J. and Wehr, J. D. (1981) Use of plants to monitor heavy metals in rivers. In: Say, P. J. and Whitton, B. A. (Ed.) *Heavy Metals in Northern England, Environmental and Biological Aspects*, pp. 135–145. University of Durham, Department of Botany.

Wieder, R. K. (1990) Metal cation binding to *Sphagnum* peat and sawdust: relation to wetland treatment of metal-polluted waters. *Water, Air and Soil Pollution*, **53**, 391–400.

Wopereis, M. C., Gascuel-Odoux, C., Bourrie, G. and Soignet, G. (1988) Spatial variability of heavy metals in soil on a one-hectare scale. *Soil Science*, **146**, 113–118.

Xian, X. and Shokohifard, G. (1989) Effect of pH on chemical forms and plant availability of cadmium, zinc and lead in polluted soils. *Water, Air and Soil Pollution*, **45**, 265–273.

Youssef, R. A. and Chino, M. (1991) Movement of metals from soil to plant roots. *Water, Air and Soil Pollution*, **57–58**, 249–258.

Zimmermann, R. (1986) Temporal variations of trace substances during individual rain events. In: Georgii, H. W. (Ed.) *Atmospheric Pollutants in Forest Areas*, 155–164. D. Reidel Publishing, Dordrecht.

Zöttl, H. W. (1985) Heavy metal levels and cycling in forest ecosystems. *Experiencia*, **41**, 1104–1113.

SECTION II

CASE STUDIES

6 Effects of Heavy Metals from Sewage Sludge on Soil Microbes in Agricultural Ecosystems

STEVE P. McGRATH

AFRC Institute of Arable Crops Research, Rothamsted Experimental Station, Harpenden, Hertfordshire, UK

ABSTRACT

This chapter deals with the toxicity of heavy metals to soil microorganisms: in particular, those derived from disposal of metal-contaminated sewage sludges on agricultural land. It is concerned with the long-term effects on microbial activity, rather than the short-term changes associated with the additions of nutrients and organic matter in the sludge. Soil respiration and soil microbial biomass are both affected by metals, but gross respiration reacts rather insensitively to metal pollution, decreasing only at very high metal loadings. Microbial biomass is much more sensitive, and the proportion of total soil carbon which exists as living biomass shows good potential as an indicator of metal pollution. Soil enzyme activities have been investigated in metal-contaminated soils, but these results are more difficult to interpret than measurements of organisms themselves or their activity. Some microbially mediated processes such as the mineralisation of carbon and nitrogen seem to be relatively insensitive at metal concentrations in soil which are relevant in this context, i.e. around the maxima set in the EC Directive on sewage sludge applications to land.

Fixation of atmospheric nitrogen by microorganisms is a more metal-sensitive process. Thus N_2-fixation by free-living bacteria and cyanobacteria decreases when soil metal concentrations are below EC limits, because of a decrease in the numbers of these organisms in soil. Symbiotic N_2-fixation by legumes with rhizobia is also inhibited, depending on the species from the Rhizobiaceae involved. Most of the evidence for this is for the clover–*Rhizobium leguminosarum* biovar *trifolii* association, where decreased clover yields and N_2-fixation are associated with reduced survival of the bacteria in contaminated soils.

Effects of metals in sludge-treated soils on mycorrhizal associations do occur, but the results are often contradictory, as they are confounded by factors such as the large amounts of available phosphate in such soils.

These results are discussed in relation to control of metal contamination in soils. Such a debate is hampered by the relative lack of measurements on the abundance of microorganisms and their activity made under realistic conditions, i.e. long-term sludge-treated soils. In particular, the importance of each heavy

Toxic Metals in Soil–Plant Systems. Edited by S. M. Ross
© 1994 John Wiley & Sons Ltd

metal or combinations of these is unclear because there are few experiments where these factors have been controlled. However, adverse effects on microbial activity at or below the upper EC limits argue for a precautionary approach in which maxima for soil protection are set below these upper limits. One final area of concern is that almost no studies on soil microbes included measurement of the species of metals present in the soils. This could be very important for future research, as it could explain the large differences observed between soils with contrasting physico-chemical properties such as texture and pH.

6.1   INTRODUCTION AND SCOPE

Sewage sludge contains larger concentrations of heavy metals than most soils, and once the metals have entered the soil they are not easily leached. Therefore, disposal of sludge to land results in an accumulation of these potentially toxic elements in topsoil. It seems likely that more sludge will be disposed on land in future, as a result of the decision to cease disposal of UK sewage sludge to sea before 1998.

Because of this, and in the light of new information on adverse effects of sludge-derived metals on soil microbial processes, it may be necessary to re-appraise the limits for metal concentrations in sludge-treated soils. Studies of the effects of metals on the yield and quality of plant produce from sludge-treated soils began earlier and a general consensus has emerged about the measures needed to protect against phytotoxicity and adverse effects on the food chain (e.g. Sauerbeck and Styperek, 1989); possible exceptions are cadmium (Cd) and chromium (Cr). This chapter focuses on the effects of heavy metals on soil microbes and their activity. Work on soil fauna is not included as there have been few studies done on sludge-treated soils, but they may be either of similar sensitivity to heavy metals as microorganisms (Ma, 1988; Witter, 1992) or possess greater resistance (Hopkin, 1989). Tolerance or adaptation of microbes to metals is not specifically addressed, because although increased frequencies of tolerant microorganisms have been found in severely polluted environments (Bååth, 1989), adverse effects on either amounts of microbes or microbially mediated processes are still apparent in more modestly polluted soils.

Sewage sludges usually contain much more zinc (Zn), copper (Cu), nickel (Ni), Cd, Cr and lead (Pb) than soils, so the emphasis will be on these metals. However, Cr and Pb are so insoluble in sludge-treated soils that they are unlikely to be bioavailable and cause adverse effects (McGrath, 1987). Lead is nevertheless a problem for grazing animals and for children due to direct ingestion of soil and its solubilisation in gastric juices. In addition to the above elements, the European Community Directive of 1986 (CEC, 1986; see also Table 6.1) also laid down maximum permitted concentrations for mercury (Hg). However, Hg is not present at large concentrations in most sludges, and,

Table 6.1. Comparison of the range of metal concentrations in plots of the Woburn Market Garden Experiment containing small microbial biomasses and decreased clover yields with CEC and UK (SI 1263) limits for soils receiving sewage sludge

Element	Woburn plots			CEC (1986) (mg kg^{-1})	UK (1989)[d] (mg kg^{-1})
	tss[a] (μg litre^{-1})	[M^{2+}][b] (μg litre^{-1})	total[c] (mg kg^{-1})		
Zinc	8–79	1.5–10	180–435	150–300	300
Copper	63–274	0.6–2.7[e]	70–150	50–140	135
Nickel	37–117	0.4–1.2[e]	22–33	30–75	75
Cadmium	1.3–2.7	0.3–0.4	6–13	1–3	3
Chromium	—	—	105–210	100–150[f]	400[g]
Lead	3–13	bd[h]	100–175	50–300	300

[a] Total concentrations in displaced soil solutions.
[b] Free metal ion concentrations in soil solutions.
[c] Aqua regia digest.
[d] For soils of pH 6–7.
[e] For these elements the concentrations of uncomplexed M^{2+} in solution were so small that they could not be measured accurately, but were certainly <2% of total in solution (tss); 1% free M^{2+} has been assumed for these elements.
[f] Under discussion.
[g] Provisional.
[h] Below detection.

once in soil, it is also very insoluble and little is bioavailable. The EC Directive was passed into UK law in 1989 (UK, 1989). This regulation limits the maximum total concentrations of metals allowed in UK arable soils receiving sewage sludge, which for Zn, Cu and Ni are varied according to the soil pH (Table 6.2).

Most emphasis will be put on studies in which metal-contaminated sewage sludge is added to soil rather than laboratory tests in artificial media (e.g. sand, solution culture, agar, etc.) as extrapolation of the latter results to the field is impossible. Soil is a very complex substrate and completely different levels of bioavailability occur once metals are added to it, compared with what might be expected from the tests with simple media. This is due to many factors which operate in soils, including sorption, chelation by organic matter or precipitation reactions. These factors are the cause of varying bioavailability in different soils, and total concentrations do not adequately characterise the risk of a given concentration in all soils, as is assumed in the CEC and UK regulations. Measurement of the amount of metals in the soil solution which bathes plant roots and soil organisms may be a better indicator of exposure. In Table 6.1, the total concentrations in soil are compared with the amounts in displaced soil solutions (Sanders *et al.*, 1987), and the free metal ion concentrations [M^{2+}] in these solutions measured by the ion exchange equilibrium technique and graphite-furnace atomic absorption

Table 6.2. Maximum permissible concentrations of potentially toxic elements in soil with arable crops after application of sewage sludge (samples taken to a depth of 25 cm)

Element	Maximum permissible concentration of element in soil (mg kg^{-1} dry solids)			
	pH 5.0–5.5	pH 5.5–6.0	pH 6.0–7.0	pH >7.0[a]
Zinc	200	250	300	450
Copper	80	100	135	200
Nickel	50	60	75	110
Cadmium[b]	3			
Lead[b]	300			
Mercury[b]	1			
Chromium[b]	400 (provisional)			

Source: UK SI 1263 (1989).
[a]Applies only to soils with >5% calcium carbonate. [b]For pH 5.0 and above.

spectrometry (Sanders, 1984; McGrath *et al.*, 1986). The total concentrations in soil solution are small (μg litre^{-1} or ppb range), and the free [M^{2+}] are even smaller (dropping into the ng litre^{-1} or ppt range) and not directly related to the total concentrations in soil solution in the case of metals like Cu, Ni and Pb which are strongly complexed with soluble organic matter. Because it is now widely believed that organisms respond to the free [M^{2+}] in solution, the immediate implications are that this is the parameter which should be measured. Such an approach may put many presently contradictory studies into a unified framework for interpretation. For example, there are numerous studies that show that the form in which heavy metals are added to soil affects the severity of toxic effects (Sommers *et al.*, 1987). Addition of soluble metal salts to soil results in many differences in comparison with the same amount of metals in sludge, including much greater activities of free metal ions, lack of the organic complexes derived from sludge, and no input of nutrients or of organic matter that provides more nutrients as it mineralises. Almost no studies have measured free ion activities, so the complete explanation for these differences is lacking at present. Therefore, results of laboratory studies will only be referred to when specified and where they provide useful background information.

   Numerous papers from laboratory experiments on the effects of heavy metals on microorganisms which did not use soil often involved the addition of unusually large amounts of metal-contaminated sewage sludges or metal salts so that adverse effects could be measured in a short time (often hours). But adding sewage sludge generally stimulates microbial activity because it is an organic substrate and a source of nutrients. Any negative effects must therefore be measured against this background, and may only be clear when the sludge has largely been mineralised. Also, the deleterious effects are thought to be due to a reduction in diversity of organisms present, with those not able to survive dying out over a period of time (McGrath and Hirsch, 1989).

Therefore, soil microbial processes that are carried out by a diverse collection of organisms, such as mineralisation of organic matter, or simply soil respiration, are unlikely to be greatly affected by moderate amounts of heavy metals. Conversely, processes that only few specialised organisms are able to perform, such as nitrogen fixation or nitrification, could be expected to be more easily adversely affected by metals. The elimination of the most sensitive species from the soil microbial community appears to occur at metal concentrations similar to those which have measurable effects on microbial processes (Witter, 1989).

6.2 SOIL RESPIRATION

Only at very high soil metal concentrations is carbon dioxide evolution from bulk soil reduced. For example, Tyler (1974, 1981) studied acid forest soils close to a brass smelter and respiration decreased with 1000 mg kg^{-1} or more each of Cu and Zn in the soil. Associated with such extreme cases of pollution, reduced decomposition of plant residues often occurs. But at lower levels, say close to concentrations of metals in soil given in the EC Directive, respiration is unaffected (Cornfield et al., 1976). As discussed below, respiration measurements alone can be misleading and are probably not a good indicator of metal contamination of agricultural soils.

6.3 SOIL MICROBIAL BIOMASS

The term "soil microbial biomass" is used to describe the soil microbial population as a whole. When soils are at near steady-state conditions with respect to total organic matter content, the soil microbial biomass comprises about 2–3% of the total organic carbon in a wide range of arable, grassland and woodland soils (Jenkinson and Ladd, 1981; Powlson and Jenkinson, 1981; Anderson and Domsch, 1989). In soils in which soil organic matter content is changing, e.g. due to inputs of organic manures such as sewage sludge, soil microbial biomass changes more rapidly than the total amount of organic carbon remaining in the soil, and so would be greater than the percentage expected under steady-state conditions (Powlson et al., 1987). Microbial biomass is therefore a useful indicator of changes in soil conditions, which is measurable before significant changes in total organic carbon can be observed.

Brookes and McGrath (1984) applied this concept to soils from plots of the Woburn Market Garden Experiment more than 20 years after the last additions of (i) inorganic fertilisers, (ii) farmyard manure (FYM), or (iii) sewage sludge contaminated with metals (McGrath, 1984). In these plots of sandy loam soil (10% clay), which are all close to pH 6.5, only the low-metal plots (inorganic or FYM treatments) showed the usual relationship between total organic carbon and microbial biomass content. In the sludge-treated plots, no obvious relationship was seen, and the total amount of biomass was roughly half that in the low-metal soils. The smaller biomass in the metal-contaminated soil showed a much higher rate of respiration per unit weight of biomass (Brookes and McGrath, 1984), and the adenylate energy charge, which is said to indicate the level of metabolic activity, was high in both soils (Brookes and McGrath, 1987), confirming that the smaller biomass in the contaminated soil was in a state of high metabolic activity. The "total" metal concentrations in the sludge-treated soils that had lower biomasses than expected are compared with the CEC (1986) and UK limits in Table 6.1. These were either comparable with, or lower than the range given in the Directive, apart from Cd which was greater.

Chander and Brookes (1991a) examined whether the same effects occur at two other sites in the UK with different soils. The sewage sludge experiments at Luddington and Lee Valley (both Experimental Husbandry Stations of the Agricultural Development and Advisory Service) started in 1968 and in the different treatments each received a single large dose of sewage sludge contaminated predominantly with Zn, Cu or Ni. In addition, there was a relatively uncontaminated sludge and a "control" treatment with inorganic fertilisers only. More than 20 years later, biomass carbon as a proportion of the total carbon in the soil was significantly smaller where Zn and Cu-enriched sludges had been added (Chander and Brookes, 1991a; see also Table 6.3). Both soils were under grass and had similar pH values (5.6–5.9). At Luddington, the Cu plots (low rate) contained about 1.5 times more Cu than is currently permitted for arable land receiving sludge in the UK (DoE, 1989;

Table 6.3. Biomass carbon as a percentage of total soil organic carbon in the Luddington and Lee Valley sludge experiments

Treatment	Luddington	Lee Valley
Control soil	1.47	1.96
Uncontaminated sludge	1.55	1.73
Low-Zn sludge	ND	1.76
High-Zn sludge	1.28	1.13
Low-Cu sludge	0.96	1.84
High-Cu sludge	0.65	0.95
High-Ni sludge	1.47	1.86

Source: After Chander and Brookes (1991a). Copyright 1991 Pergamon Press

see Table 6.2) and had about 18% less biomass C than the uncontaminated plots. However, the Cu plots (low rate) at Lee Valley contained 2.1 times the limit and biomass was unaffected, suggesting lower bioavailability of Cu in the heavier-textured soil at this site. After larger additions of Cu, giving about 3.7 times the permitted UK limits in soil, the biomass was decreased by 40% at both sites. This also increased accumulation of organic C and N by about 30% in the sandy loam and 13% in the silty loam. Zinc, at about 2.9 times the limit in the sandy loam and 3.4 times in the silty loam, decreased biomass by about 40% and 30% respectively, while soil organic matter increased by only 9–14%. Nickel, at 2.7 and 4.5 times the limit at Luddington and Lee Valley respectively, had no effect on the biomass or accumulation of soil carbon. The differences in response between the sites were ascribed by the authors to the effects of soil texture: Luddington has less clay and 3% organic matter, but the Lee Valley soil is a heavier silt loam with 7.8% organic matter. Heavier soils, and those with more organic matter, are known to bind heavy metals and make them less available to microorganisms (Bååth, 1989). Cadmium was present as a contaminant in the "uncontaminated" sludge used on the silty loam (Lee Valley), and this had no effect even at 2–2.5 times the limit for sludged soils.

Work by Chander (1991) and Chander and Brookes (1991 a,b) on soils from the two sites described above confirmed the higher specific respiration rate of soil biomass in metal-contaminated soils. Although many of these soils had quite large metal concentrations, the investigations did demonstrate that proportionately more of added substrates (glucose, maize straw) were respired as CO_2 from contaminated soils than in comparable uncontaminated plots (Chander and Brookes, 1991b). Consequently, less new microbial biomass was formed from the added substrates, and they concluded that the lower efficiency of conversion of carbon into biomass is one of the explanations for the lower biomasses observed in metal-contaminated soils. The fact that specific respiration is greater in the experiments on metal-contaminated soils means that measurements of soil respiration *alone* cannot give a good indication of effects of metals on the size of the microbial biomass. In fact, increased respiration could be interpreted as an indication of an increased death rate of microbes.

Another contributing factor to the decrease in microbial biomass could be the decreased inputs of carbon from plant roots and root exudates (Chander and Brookes, 1991c). However, the same authors state that in their experiments with Woburn soil, the toxicity to the biomass, leading to decreased efficiency of formation of new biomass, was a more important factor. Also, the test plant used by Chander and Brookes (1991c) was sunflower (*Helianthus anuus* L.); it remains to be seen if other species (especially those normally grown in the Woburn experiment) also produce significantly less plant residues which enter the soil in the field.

A limitation of the Lee Valley and Luddington experiments was that they had only two rates of metal addition and therefore the soils did not contain steadily increasing amounts of Zn or Cu, so it is not possible to closely define the minimum concentrations at which decreases in biomass occurred. Nevertheless, Zn, Ni and Cd concentrations below current limits did not decrease the amounts of soil microbial biomass, but Cu decreased biomass when present at 10% more than the limit in the sandy loam soil. Nickel appeared to have no effect, even when present at up to about three times the limit.

Another experiment, at Gleadthorpe Experimental Husbandry Station, was also sampled (Chander, 1991). This started in 1982 on a sandy loam soil (9% clay), with one application of Zn- or Cu- or Ni-contaminated sludges, and some plots had a further application in 1985. These mono-metallic sludges were produced in a pilot plant by adding metal salts to separate batches of a relatively uncontaminated sludge. Zinc at 1.2 and 2.3 times the arable soil limit decreased biomass by 25 and 40%, respectively. Copper from 1.2 to 4.9 times the limit reduced biomass by 20–55%. Interestingly, Gleadthorpe has unique mixed metal treatments to examine the interactions of Zn + Cu or Zn + Ni. Adding Ni to Zn had no effect, but Zn + Cu at only 1.2 times and 1.4 times the respective limits reduced biomass by 34%. Another treatment with Zn at 1.4 times and Cu at 1.8 times the limits had a 57% smaller biomass compared with the soil treated with uncontaminated sludge. These effects were similar to the sum of those of either metal alone at comparable concentrations, suggesting that the effects of Zn and Cu are additive (Chander, 1991).

Measurements of microbial biomass in the clayey soils of an experiment started in 1957 at Ultuna in Sweden showed, like Woburn, decreased biomass in sludge-treated soil compared with an FYM-treated control (Witter et al., 1993). However, in this experiment the sludge-treated soil had a pH of 5.3 and the FYM-treated soil 6.6, making it difficult to attribute the effects solely to the heavy metals present in the soil. Approximate metal concentrations in the Swedish experiment (mg kg^{-1} dry soil) were: sludge-treated soil Cu 125, Zn 228, Ni 34, Cd 0.4; FYM-treated soil Cu 28, Zn 69, Ni 23, Cd < 0.3.

At Braunschweig in Germany, two field experiments were begun in 1980, with the following main treatments: inorganic fertiliser only; uncontaminated liquid sludge at 100 m^3 year^{-1} or 300 m^3 year^{-1}, or metal-contaminated sludge at the same two rates. The sludges used were naturally contaminated with the metals Cd, Zn, Cu, Ni, Pb and Cr in the first year of the experiment, but the contaminated sludge was later "spiked" to known composition with salts of these metals and then stored anaerobically for six weeks prior to application. The experiments were replicated on old arable (pH 6.1–6.8) or ex-woodland soils (pH 5.3–5.7) on the same field. Sludges were applied for 10 years, with the aim of reaching the German limits for metal concentrations in soils receiving sewage sludge (BMI, 1982), and was successful by 1989, except for Ni (see Table 6.4). On both, with either rate of uncontaminated sludge, increasing soil microbial biomass was observed with increased additions of sludge organic matter each year. This was less pronounced or even absent compared with the inorganic treatment, when the sludge was contaminated with

Table 6.4. Range of metal concentrations in soils in 1988 of two field experiments at Braunschweig that received 100 m^{-3} or 300 m^{-3} year^{-1} of metal-contaminated sewage sludge for 9 years, compared with the German limits for metals in sludge-treated soils

Element	German limit (BMI, 1982)	Arable field, pH 6.1–6.8	Ex-woodland, pH 5.3–5.7
Zinc	300	157–381	163–413
Copper	100	42–102	39–106
Nickel	50	12–25	14–28
Cadmium	3	0.7–2.5	0.8–2.9
Chromium	100	41–90	44–111
Lead	100	61–88	48–101

Source: Sauerbeck, D. and Rietz, E. (personal communication).

metals, especially at the highest rate. These effects were obvious after seven years of addition of contaminated sludge, even at the lower rate (Fliessbach *et al.*, 1989). The largest metal concentrations in the soil were (mg kg^{-1}): 2.7 Cd, 330 Zn, 90 Cu, 24 Ni, 100 Pb and 95 Cr. Comparison of Table 6.4 with Tables 6.1 and 6.2 shows that most of the levels in these soils were within CEC upper limits, but a few of the most contaminated plots were over the UK limits for Zn and Cu by small percentages, especially at the lower pH because the limits are smaller.

Measurements of microbial biomass do not show the activity of the biomass. By analogy, it is a measure of the "standing crop" of microscopic organisms in the soil, and can be a useful indicator of changes in soil management or other environmental factors. However, different measurements must be made to examine whether there are effects on specific microbially mediated processes in soil or on specific groups of organisms.

6.4   SOIL ENZYMES

The activity of various enzymes in soils has been suggested as an easy measure of microbial activity, giving an indication of change, but not an absolute measurement of any specific group of organisms. Near a brass smelter in Sweden, the activity of soil dehydrogenase decreased in direct relationship with soil metal concentrations (Tyler, 1974). Brookes *et al.* (1984) measured the activities of phosphatase and dehydrogenase in soils of the Woburn plots. Phosphatase activity was unaffected, but dehydrogenase activity decreased with increased metal loads. The difference between these two enzymes is that phosphatase is mainly extra-cellular in soils, but dehydrogenase only functions in living cells and therefore may be a proxy measure of the amount of biomass, or at least its state of activity. Reddy *et al.* (1987) also found that dehydrogenase activity was inhibited by increasing rates of sewage sludge addition, but there was no effect on phosphatase activity, and the response of urease depended on soil type.

In the experiment at Braunschweig discussed above, the activities of dehydrogenase, phosphatase, protease, glucosidase, catalase and urease were measured (Balzer and Ahrens, 1991). Most were stimulated by the addition of sludge organic matter, except urease at the highest rate of contaminated sludge and dehydrogenase at both rates of contaminated sludge, which decreased to below the activities in soil receiving only inorganic fertilisers. According to Fliessbach and Reber (1990), dehydrogenase activity was more sensitive to sludge metals than soil microbial biomass.

Recent work by Chander and Brookes (1991d) has, however, cast doubt on many of the dehydrogenase measurements published previously. They found that dehydrogenase decreased most sharply in soils contaminated with Cu-sludges—much more so than with other sludges contaminated with Zn, Cd or Ni alone. Investigation showed that the specific dehydrogenase activity of the biomass was two to three times smaller than in treatments with the other metals. This happened despite the fact that the specific respiration rate of the biomass actually increased in all treatments, including those with Cu. The cause of the apparently smaller amounts of triphenylformazan (TPF), formed from triphenyltetrazolium chloride (the basis of the dehydrogenase method), was shown to be a non-biological reaction between TPF and Cu. Decreased absorbance due to TPF occurred because of this, and would be falsely interpreted as a true decrease in dehydrogenase activity. Clearly, dehydrogenase measurements in Cu-contaminated soils which rely on the formation of TPF are in doubt. However, Chander and Brookes (1991d) noted that part of the apparent decrease in TPF formation in the Cu-contaminated soils could be due to real decreases in dehydrogenase activity.

6.5 NITROGEN AND CARBON MINERALISATION

Many organisms take part in processes releasing inorganic nitrogen as a result of the mineralisation of organic matter, leading initially to the formation of NH_4^+ ions. In contrast, relatively few genera of autotrophic bacteria, such as *Nitrosomonas* and *Nitrobacter* acting in sequence, take part in the transformation of NH_4^+ to NO_2^- and then to NO_3^-.

Summarising various studies of the effects of metals on mineralisation of nitrogen and nitrification, Doelman (1986) indicated that these processes start to become inhibited at around $100\ mg\,kg^{-1}$ of Zn, Cu and Ni, around $100–500\ mg\,kg^{-1}$ for Pb and Cr, and $10–100\ mg\,kg^{-1}$ for Cd. However, most of these experiments did not use sludge-treated soils and the measurements were made shortly after adding soluble salts, so these results should be viewed with caution. Bååth (1989) and Chander (1991) both noted the conflicting results on the sensitivity of nitrogen mineralisation to heavy metals that exist in the literature. This could be due to many factors, including the exact form of metals applied in the experiments and the lack of standardisation of methods used.

Brookes et al. (1984) found that nitrogen mineralisation was unaffected in the sludge-treated plots from the Woburn experiment. Nitrification of added NH_4^+ was less in sludge-treated soils than in FYM-treated soils, but nitrification of added NO_2^- was unaffected. They hypothesised that sudden very large inputs could not be dealt with by the reduced population of organisms present in the sludge-treated soil, but that mineralisation of native organic nitrogen was unaffected under normal conditions. Duxbury (1985) concluded that the information on the effects of heavy metals on these processes is quite conflicting and recommended caution in attempts at generalisation.

Mineralisation of the added sludge organic matter did not differ significantly between metal-contaminated and uncontaminated treatments of the Woburn Market Garden Experiment (Johnston et al., 1989). Straw added to these soils decayed at the same rate, regardless of metal concentration (Bowen, R. and McGrath, S. P., unpublished results). However, the addition of ^{14}C-labelled plant material to these soils showed that there is a smaller incorporation of added substrate into the microbial biomass, and this may partly explain the smaller biomass present (Chander and Brookes, 1991b).

In the other UK field experiments, described earlier, Chander (1991) and Chander and Brookes (1991a) found accumulations of organic matter, compared with control soil receiving uncontaminated sludge. At Luddington, there was no effect of Zn and Ni at the rates applied in sludges, but Cu at 3.7 times the limit for soils of pH 5.5−6 (UK, 1989) contained 32% more organic matter than soils given uncontaminated sludge (Chander and Brookes, 1991a). At Lee Valley, where the soils were again of pH 5.5−6 but with more clay and organic matter, those treated with Zn-contaminated sludge had 3.4 times the permitted concentration of Zn and contained 10% more organic matter than soils treated with uncontaminated sludge. Similar plots given Cu-sludge and having 3.8 times the limit had about 14% more organic matter. The results from these sites suggest that the metals at the concentrations tested (i.e. above the respective limits) decrease the rate of turnover of soil organic matter (Chander and Brookes, 1991a).

Chander (1991) also reported decreases in inorganic nitrogen formation in Woburn soil (pH 6.5) amended in the laboratory with single-metal sludges to give concentrations of one, two, three and four times the upper EC limits (CEC, 1986). The relative inhibition by individual metals of inorganic nitrogen formation, relative to that in soil amended with uncontaminated sludge, was in the order $Zn \gg Ni > Cu \gg Cd$ (Chander, 1991).

Balzer and Ahrens (1991) showed that mineralisation of both carbon and nitrogen in sewage sludge experiments at Giessen, Speyer and Braunschweig in Germany increased with increasing applications of sludge, regardless of whether it was contaminated. However, the metal concentrations in these soils from the first two sites were well below UK limits.

Decomposition of ^{14}C-labelled straw added to soil was less in the high rate treatment with contaminated sewage sludge at Braunschweig, and in the low rate of contaminated sludge, but only in the more acid ex-woodland soil (Fliessbach and Reber, 1991). The same authors also studied the breakdown of the following xenobiotic compounds added to soil from the Braunschweig experiment: dichlorprop, lindane, anthracene, parathion and pyrazophos. In the old arable soil only, at the highest rate of sludge addition (giving soil metal concentrations close to German limits; see Table 6.4), the contaminated soil showed decreased breakdown of parathion and pyrazophos (Fliessbach and Reber, 1991). This suggests that a loss of genetic diversity occurs along with the overall decrease in biomass, resulting in fewer or less efficient organisms that are able to break down some xenobiotic compounds.

6.5.1 NITROGEN FIXATION

Fixation of atmospheric dinitrogen by biological enzyme systems can be performed by three main types of organisms in soils. These vary in the potential amounts of nitrogen that can be fixed in temperate regions, as shown below:

	kg N ha^{-1} year^{-1}
Free-living heterotrophic bacteria	? 1–2
Phototrophic cyanobacteria	5–30
Symbiotic associations	100–200
(e.g. *Rhizobium*–clover)	

Representatives of each of these groups have been studied in sludge-treated soil.

6.5.2 FREE-LIVING HETEROTROPHS

Brookes *et al.* (1984) reported significantly decreased acetylene reduction by these organisms in soil of the Woburn Market Garden Experiment. When appropriate measurements of endogenous ethylene production by soil are made, the reduction of added acetylene to ethylene indicates activity of the nitrogenase enzyme, but cannot be used when the aim is to quantify the amounts of N_2 fixed (Giller and Day, 1985), for which measurement of the incorporation of the ^{15}N isotope is usually necessary.

Laboratory studies have confirmed the sensitivity of heterotrophic N_2-fixation at concentrations of 50 mg Cr kg^{-1} and between 50 and 200 mg Cu kg^{-1} soil (Skujins *et al.*, 1986) and between 2 and 4 mg Cd kg^{-1} (Coppola *et al.*, 1988). However, the former authors used metal salts and the latter Cd-spiked sludge.

Skujins *et al.* (1986) proposed that the sensitivity of heterotrophic N_2-fixation could be a simple biological indicator of metal-contamination of soils. A recent study by Lorenz *et al.* (1992) tested this possibility on four sludge-treated soils from the Woburn, Luddington, Lee Valley and Gleadthorpe experiments. Although the Woburn soils gave the same results as in Brookes *et al.* (1984), no activity could be found at Luddington and Lee Valley, and the sporadic activity at Gleadthorpe did not correlate with metal concentrations in soil. Apparently, heterotrophic bacteria are not ubiquitous or active enough in all soils to be suitable as indicator organisms (Lorenz *et al.*, 1992).

There was no difference in free-living N_2-fixation activity in soils taken directly from different treatments of the Ultuna experiment in Sweden. However, when glucose was added, the peat and sewage sludge-treated soils of similar pH differed greatly: acetylene reduction was more than 30 times less in the sludge-treated soil (Mårtensson and Witter, 1990). Balzer and Ahrens (1991) reported that numbers of *Azotobacter* in soils of treatments receiving $2.5\ t\ ha^{-1}$ of sewage sludge regularly at Giessen were tenfold greater than in soil receiving NPK; by contrast, they were only one-tenth of the numbers in the NPK control in a treatment receiving sludge at $5\ t\ ha^{-1}$. However, the pH of these soils differed, being 6.7, 5.4 and 5.8 for 2.5 t, 5 t sludge and NPK treatments, respectively.

6.5.3 CYANOBACTERIAL PHOTOTROPHS

Cyanobacteria or blue-green algae grow on the soil surface and use sunlight as an energy source. Growth, acetylene reduction and ^{15}N incorporation by these organisms was studied on soils from the Woburn Market Garden Experiment (Brookes *et al.*, 1986). Little growth or N_2-fixation was found in the metal-contaminated sludge-treated soil after 120 days. These authors also sampled soils along a field transect which gave a gradient of increasing metal concentrations in soil and found a smooth decrease in nitrogenase activity with increasing metal contamination. Nitrogenase activity decreased by 50% in soils containing the concentrations of metals shown in Table 6.5. As discussed later, it is not possible to say from these experiments which individual metal or metals are responsible for these effects.

Other studies in Sweden and Germany confirm the results obtained by Brookes *et al.* (1986). Thus, at Ultuna, blue-green algae were prevalent on FYM-treated soil but were virtually absent from sludge-treated soil (Mårtensson and Witter, 1990). However, as stated above, the sludge-treated soil was more acid than the FYM-treated soil. Fliessbach *et al.* (1989) reported large decreases in counts of cyanobacteria in two out of three samplings of the metal-contaminated sludge treatments at Braunschweig, in comparison with uncontaminated sludge at both rates of addition. It is also clear from their data that annual additions of uncontaminated sludge itself decreased

Table 6.5. Comparison of the metal concentrations in soils from the Woburn Market Garden Experiment giving 50% inhibition of nitrogenase activity by cyanobacteria with CEC and UK (SI 1263) limits for soils receiving sewage sludge ($mg\,kg^{-1}$ dry soil)

Element	50% inhibition	CEC (1986)	UK (1989)[a]
Zinc	114	150–300	300
Copper	33	50–140	135
Nickel	17	30–75	75
Cadmium	2.9	1–3	3
Chromium	80	100–150[b]	400[c]
Lead	40	50–300	300

[a] For soils of pH 6–7.
[b] Under discussion.
[c] Provisional.

abundance of cyanobacteria, compared with NPK alone. This would limit the usefulness of blue-green algae as indicator organisms on soils still receiving fresh additions of sewage sludge. It is not known whether this is a direct effect or due to the applications of sludges increasing the leaf growth of crops, leading to greater shading of the soil surface where the cyanobacteria grow.

6.5.4 SYMBIOTIC FIXATION

Symbiotic fixation is a type of N_2-fixation which takes place in bacteroids formed within root nodules of higher plants. Many species of plants in the Leguminosae family and bacteria (in the Rhizobiaceae) engage in this type of association. It should also be pointed out, however, that legume plants will also utilise inorganic nitrogen present in the soil instead of N_2-fixed. Soils receiving regular applications of sewage sludge or other organic manures, and in particular experimental plots, often contain large amounts of mineral nitrogen. If large enough, this alone will inhibit nodulation, or if nodulated, the plants will use the combined nitrogen, just like non-fixing species. Therefore, any negative effects on symbiotic fixation will not be detected in these soils until the readily decomposable material in sludge (and mineral nitrogen) has decreased.

6.5.4.1 Clover growth

The proportion of clover remaining in a two-year-old sown ryegrass–white clover sward at Lee Valley decreased with increased applications of Zn-contaminated sludge, decreased to a lesser extent with Cu, and not at all with Ni or Cr compared with uncontaminated sludge treatments (Vaidyanathan,

1975). The sludges had been applied seven years before these observations were made. Yields of white clover in monoculture on sludge-treated plots at Woburn decreased by up to 60% compared with FYM plots in 1984 although both treatments ended 20 years previously (McGrath *et al.*, 1987). In Figure 6.1, these results are shown in relation to the total Zn concentration in the soil, but it must be remembered that at least six heavy metals were present, derived from the sludges, and that the total concentrations of all these metals correlate highly with one another. The range of concentrations of total, soils solution and $[M^{2+}]$ in soil solutions typical of the sludge treatments of this experiment can be seen in Table 6.1.

Red clover failed almost completely at three harvests taken in 1985 from some of the most metal-contaminated plots at Luddington; specifically, those with 455 and 511 mg $Zn\,kg^{-1}$ soil (Jackson, 1985). In the "low" Zn treatments, with 238 mg Zn kg^{-1} soil, there was no decrease in yield. Other treatments with 118 or 92 mg Ni kg^{-1} soil also had poor yields, but only at the first harvest. However, these "nickel" plots also contained 128 and 104 mg $Zn\,kg^{-1}$ respectively, and it is likely that this combination of Zn and Ni caused the initial poor establishment. In 1988 at the Gleadthorpe experiment, yields of white clover were decreased at concentrations of Zn in soil of more than 170 $mg\,kg^{-1}$, and at more than 100 mg $Cu\,kg^{-1}$ (Table 6.6). no yield decreases resulted from the concentrations of Ni reached in this

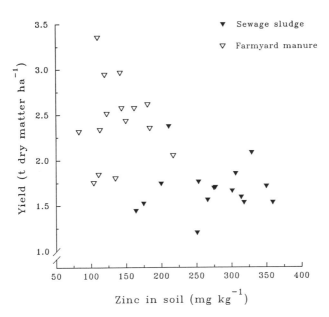

Figure 6.1. Yields of white clover on plots of the Woburn Market Garden Experiment (mean of two cuts)

Table 6.6. Mean metal concentrations in Gleadthorpe soil and average yields of white clover in 1988. Reproduced by permission of WRc plc from Royle *et al.* (1989)

Treatment	Soil (0–30 cm)—metals in strong acid extract (mg kg^{-1})			Yield (t ha^{-1} DM)
	Zn	Cu	Ni	
1 No sludge	34	7	7	3.09
2 Control^a	61	14	7	3.09
3 Control^a	43	8	5	2.71
4 Zn	173	15	7	2.83
5	325	14	7	2.61
6	214	13	6	nil
7	378	20	8	nil
8 Zn : Cu	84	38	6	3.20
9	172	47	7	2.38
10	209	94	7	0.99
11	173	107	7	2.04
12 Cu	71	117	7	2.27
13	43	94	5	3.63
14	60	145	7	1.97
15	42	187	6	0.96
16 Zn : Ni	61	9	9	3.38
17	101	10	13	3.60
18	128	10	16	3.54
19	142	11	26	2.95
20 Ni	37	9	12	3.18
21	47	9	22	3.91
22	37	10	31	3.44
23	48	9	31	3.17

^{a} Relatively uncontaminated sludge.

experiment (Royle *et al.*, 1989). There was some evidence of increased severity of yield reductions when both Zn and Cu were applied (Table 6.6). One Zn treatment (number 5 in Table 6.6) had fresh sludge in 1986 and a higher yield than expected for a soil containing 325 mg Zn kg^{-1}.

6.5.4.2 Nitrogen fixation by clovers

McGrath *et al.* (1988) showed that the above effects on clover growth are not due to direct phytotoxicity of the metals. Thus, adding nitrogen fertiliser restored the yield of clover on Woburn sludge-treated soil to that of the uncontaminated soil. A gradient of increasing metal concentrations revealed a drastic fall (50% or more) in N$_2$-fixation by clover plants growing in soils containing more than (mg kg^{-1}) 334 Zn, 99 Cu, 27 Ni and 10 Cd. Although, as with all sludges, it was not possible to discriminate which metal or metals were affecting fixation, current limits would not appear to give much protection as far as Zn, Cu and Ni are concerned. Concentrations of Cd were three times the present limits in this soil. However, there is no proof that Cd alone is responsible for the negative effects on N_2-fixation.

Smith et al. (1990) sampled soils from experimental pasture and arable sites that had received sewage sludge. Simple laboratory plant infection tests and acetylene reduction assays (ARA) were used to assess the presence or absence of effective nitrogen fixation in white clover. Whether or not effective ARA was found depended on the presence or absence of white clover at the site, and on the mixture and concentration of metals present. Thus, no effective ARA could be found in their laboratory tests with clover grown in sludge-amended soils from arable farms which had Zn, Cu, Ni, Cr and Pb concentrations below the current UK limits, but which had Cd concentrations 2–4 times the (3 mg kg^{-1}) limit. One sludged sample had all metal concentrations below the limit and positive ARA, but two "control" soils (presumably unsludged) showed no ARA. An experiment at Royston, Hertfordshire, UK, with free calcium carbonate in the soil (pH > 7.4) gave positive ARA on all treatments, although only one had a metal concentration, Cd, above the limits (3.4 mg Cd kg^{-1}). Two experiments at Cassington in Oxfordshire gave positive ARA in control plots, but none with sludge-treated soil, except for one treatment at the site with heavy clay soil and (mg kg^{-1}) 664 Zn, 218 Cu, 105 Ni, 392 Cr, 185 Pb and 4.7 Cd (Zn, Cu, Ni and Cd above limits). Smith et al. (1990) also sampled the Luddington experiment, but by then clover was present on all plots and all gave positive ARA. This is in direct contrast to the 1985 results from Luddington, referred to in the last section, and is thought to result from protection of Rhizobium when the bacteria are present inside host nodule tissue (Giller et al., 1989) and to the effect of clover on the size of the Rhizobium population in the soil (see below). Zinc and Cu decreased the rhizobial population in the soil at Luddington more than Ni, and Cr had no effect (Smith et al., 1990).

In two Water Research Centre (WRc) field trials, Stark and Lee (1988) found that N_2-fixation (ARA activity) was decreased by 50% at the following metal concentrations (mg kg^{-1}) in two soils: (a) sandy loam soil, pH 6.5, 300 Zn, 110 Cu, 60 Ni (no other analyses given); and (b) alluvial soil, possibly clay-rich, pH 6.7, 981 Zn, 158 Cu, 66 Ni, 4.6 Cd and 189 Pb. A 15% decrease was measured in soils from another treatment on the sandy loam with (mg kg^{-1}) 210 Zn, 70 Cu and 40 Ni. The concentrations for the sandy loam soil are much smaller than those misquoted from Stark and Lee (1988) in a later report by Smith (1989) and negative effects actually take place at or below the UK (1989) statutory limits for parable soils.

6.5.4.3 Effects on Rhizobium

McGrath et al. (1988) showed that on sludge-treated soils at Woburn nodulation of clover occurred, but the nodules were ineffective at fixing N_2. Isolation of bacteria from these ineffective nodules, and testing for N_2-fixation in the *absence* of toxic heavy metals, showed that these bacteria still did not fix nitrogen. It was, therefore, concluded that ineffectiveness on sludge-treated soil is not due to direct metal toxicity (Giller *et al.*, 1989). Rather, it is due to the survival of only one genotype of the *Rhizobium* population (McGrath and Hirsch, 1989). Giller *et al.* (1989) showed that effective rhizobia inoculated into the sludge-contaminated soil were killed in eight weeks. This happened when inocula of 10^7 or fewer cells were used. Larger doses than this did allow effective nodulation of clover, presumably because a sufficient number of cells survived to nodulate. Once nodules have formed, the host tissue protects the N_2-fixing bacteroids from metal toxicity.

This explains why effective bacteria are found in metal-contaminated soils that have clover growing on them for several years. For example, runners of clover growing on the small plots at Luddington may have carried effective rhizobia from control plots to neighbouring metal-contaminated plots. On unfertilised soils, plants that can fix N_2 are at a great competitive advantage over those that cannot. They are likely, therefore, to prevail over the original stunted plants, and to release more effective rhizobia into the soil. The lower counts of the free-living bacteria in the soil of metal-contaminated plots support the idea that these are continually being killed by the metals present, particularly by Zn and Cu (Smith *et al.*, 1990). Similarly, once clover/*Rhizobium* associations invade other metal-contaminated sites they are likely to flourish (Smith *et al.*, 1990), although attempts to re-vegetate old mine spoils with N_2-fixing legumes have shown that it is very difficult to establish clover in the first years.

At Ultuna in Sweden, compared with a soil receiving the same amount of FYM ($4\,t\,ha^{-1}\,year^{-1}$ since 1957), the sludge-treated soil contained similar numbers of *Rhizobium leguminosarum* biovar *trifolii* per gram of soil, but the time taken to reach maximum nodulation of red clover (*Trifolium pratense*) was delayed by about three weeks (Mårtensson and Witter, 1990), and this could lead to poor establishment of clover which depended on N_2-fixation for growth. But it must be remembered again that in this experiment the sludged plots have lower pH than the equivalent FYM ones (pH values of 5.3 and 6.6 respectively).

To try to answer the question "which metal or metals is most toxic to *Rhizobium*?" Chaudri *et al.* (1992) recently tested the tolerance to individual metals of the ineffective strain of *R. leguminosarum* bv. *trifolii* isolated from metal-contaminated plots at Woburn and exposed to metals under laboratory conditions, and compared the results with those for isolates of effective strains from the uncontaminated plots. The strains of *Rhizobium* from metal-contaminated soils showed increased tolerance to Zn, Cd, Cu and, albeit weakly, to Ni compared with effective strains. This would suggest that the strains that survive in the metal-contaminated soil have a non-specific tolerance mechanism, such as increased excretion of extracellular polysaccharides, rather than a special metal-specific mechanism (e.g. a metal efflux pump). This procedure indicates which metals the surviving bacteria have responded to, but cannot be related to the metal concentrations in soil that are inhibitory because of the many factors that affect the concentrations of metals present in soil solution. It cannot, therefore, be used directly for setting limits. Also, such tests are limited to the Woburn experiment, where ineffective rhizobia have been found. However, detailed examination of the data showed that the order of decreasing toxicity in solution for the effective (F strains) was Cu > Cd > Zn = Ni, but for the ineffective isolates it was Cu > Cd > Ni > Zn. From this it is concluded that the ineffective strains have particularly strong differential tolerance to Zn, compared to the other metals at the same concentrations. This fits with the type of exposure to metals that the organisms have had in these soils, which is dominated by Zn (see Table 6.1).

6.5.4.4   Other species of legumes/rhizobia

Little work has been done on associations other than with white clover, particularly with respect to metals from sewage sludge. Spring beans (*Vicia faba*) were grown on the Market Garden Experiment in 1968 and 1969 (more than 7 years after the last application of sewage sludge). The beans were not inoculated, but would have been infected with the indigenous population of *R. leguminosarum* bv. *viciae*. Yields in both years were almost exactly the same on plots that had received high or low rates of sludge or FYM, both of which were slightly larger than the inorganic (NPK) fertiliser treatment (Johnston and Wedderburn, 1974). This suggests that *R. leguminosarum* bv. *viciae* is not as sensitive to metals in the same experiment as bv. *trifolii* described above. Giller *et al.* (1993) tested the survival of inocula of *Rhizobium loti* and *R. meliloti*, using a gradient of soils with increasing metal concentrations from the Woburn experiment. They found that *R. loti* was as sensitive as *R. leguminosarum* bv. *trifolii* in the experiments reported above, but that *R. meliloti* survived well in the contaminated soil. This work was done with single introduced strains, but demonstrates that there may be great differences in sensitivity of different species of *Rhizobium*, leading to differences in N_2-fixation in metal-contaminated soils.

Reddy *et al.* (1983) observed a decrease in an introduced population of *Bradyrhizobium japonicum* nodulating soybean; however, the soils used were freshly amended with large amounts of sewage sludge in the laboratory. In a greenhouse experiment, Eivazi (1990) added the equivalent of 7 or 14 t dry matter ha^{-1} of contaminated or uncontaminated sewage sludges and found significantly decreased dry matter and N_2-fixation by both soybean and alfalfa with the contaminated sludge. Calculations show that, due to dilution with added sand, metal concentrations in the medium hardly increased at all, even with the largest application of the contaminated sludge. It is therefore difficult to relate this result to soil metal limits. Further investigations in the USA on yields and N_2-fixation by soybean have been conducted using two field experiments established in 1975 and 1976 respectively, at Fairland and Beltsville in Maryland (Heckman et al., 1986, 1987a,b; Kinkle et al., 1987). Heckman et al. (1986) used soils amended with heat-treated sludge at Beltsville in a pot experiment, and showed increasing yield and N_2-fixation by soybean with the following rates of sludge: 0, 56, 112 and 224 t ha^{-1}. Using the differences in nitrogen accumulation in non-nodulating and nodulating isolines of soybean, Heckman et al. (1987b) estimated the amounts of N_2 fixed in both experiments in the field in 1983 and 1984. No evidence of toxicity of metals was found at Beltsville, but % N from fixation decreased from 55 to 32% on the Fairland plots in 1983, suggesting toxicity due to metals. However, the availability of soil N was greater on the sludged soils in 1983, and, as the % N from fixation was calculated from the differences between the isolines, it is difficult to determine whether the apparent decrease in N_2-fixation with increasing sludge applications are real.

Numbers of B. japonicum in the soils at Fairland increased with increasing rates of sludge applied (Kinkle et al., 1987). They also showed that the numbers of serotypes and their metal-sensitivity did not change with the treatments. One of the difficulties in relating the results from Fairland and Beltsville with studies of other species on other soils is that the published papers do not give any measures of metal concentrations in the soils examined. It was possible to estimate the concentrations in the Fairland soil, assuming the following: (1) dry weight of plough layer soil of 3000 t ha^{-1} (23 cm deep, 1.3 g ml^{-1} bulk density; (2) the "background" (untreated) soil metal concentrations given by Sheaffer et al. (1979); (3) no losses of metals either below the plough layer or outside the original treated plots (McGrath and Lane, 1989), as the plots were surrounded by wooden frames; and (4) the largest rate of addition of sludges (112 t ha^{-1} at Fairland). It can be seen (Table 6.7) that in every case the concentrations were smaller than the CEC (1986) and UK (1989) limits for sludge-treated soils even in the most heavily contaminated plots at Fairland. Additional new data for the soils at Beltsville were obtained from Drs J. S. Angle and R. L. Chaney (personal communication). Again, these are less than the appropriate European limits, except for Cd which was 30–50% above the UK limit in a treatment with "Nu-Earth" sludge from Chicago (Table 6.7). However, it is worth noting that the only suggestion of toxicity noted from these experiments was at Fairland, which had a concentration of Cd below the lowest CEC limits (Table 6.5), but concentrations of Zn and Cu both above the lowest value in the CEC range. This may indicate that adverse effects could occur in soils below the limits set for Zn and Cu in the UK.

More work is needed in this area in the future, using both agronomically important crops (e.g. beans and peas) and species from grassland and semi-natural environments that have different species of Rhizobium present in the nodules.

Table 6.7. Concentrations of metals (mg kg^{-1}) in soils of soybean experiments in the USA: calculated for the plots at Fairland (see assumptions given in the text) and measured for Beltsville (UK limit for equivalent soil pHs shown in parentheses)

Site	pH	Zn	Cu	Ni	Cd
Fairland[a]	6.4–6.9	195	87	16	0.8
		(300)	(135)	(75)	(3)
Beltsville[b]	5.8–6.0	136	33	11	1.0
		(250)	(100)	(60)	(3)
Beltsville[b]	5.1–5.3	119	31	12	0.8
		(200)	(80)	(50)	(3)
Beltsville[c]	6.4–6.6	151	35	23	4.6
		(300)	(135)	(75)	(3)
Beltsville[c]	5.7–5.9	128	33	21	4.0
		(250)	(100)	(60)	(3)

[a] Sandy loam soil; anaerobically digested sludge at 112 t ha^{-1}; concentrations calculated using the assumptions described in the text and concentrations of metals in sludges and untreated soil given by Sheaffer *et al.* (1979).
[b] Fine sandy loam soil; heat-treated sludge from Annapolis at 224 t ha^{-1}.
[c] Fine sandy loam soil; Chicago Nu-Earth sludge at 100 t ha^{-1}.

6.6 MYCORRHIZAS

Few studies have been done on sludge-treated soils. Work using other metal-contaminated soils have shown that mycorrhizas can either increase or decrease the uptake of heavy metals by plants (Giller and McGrath, 1988). This appears to depend on the metal concentrations present in the soil (Heggo *et al.*, 1990) and the type of mycorrhizal infection. Thus, sheathing or "ecto" mycorrhizas envelop the roots and may give protection from toxic metals (Colpaert and van Assche, 1987); vesicular–arbuscular mycorrhiza (VAM) can either decrease metal concentrations in plants at high concentrations of metals, or increase them at low concentrations (Heggo *et al.*, 1990).

Koomen *et al.* (1990) studied VAM infection of white clover on sludge- or FYM-treated Woburn soil and found evidence of a delay in infection on the metal-contaminated soil which was not overcome by inoculation of the soil with VAM. This, and the lack of N$_2$-fixation, could result in poor establishment of clover on soils of low phosphate status. However, the soils of the Woburn experiment are relatively rich in phosphate.

On the sludge experiments at Braunschweig discussed earlier, there were indications of decreased VAM infection of maize with increased rates of metal-contaminated sludge. Linked to this is a relative shift of mycorrhizal infection to the uncontaminated subsoil (Kücke, 1989). However, these effects are confounded by two factors:

(i) the presence of large concentrations of phosphate which may inhibit infection at the largest rate of sludge addition, and

(ii) the roots showing least infection by mycorrhizas were from the plots with the lower pHs; therefore the confounding effect of pH on root and fungal growth cannot be ruled out.

More work is needed on mycorrhizal infection of crops grown on sludged soils, but it must be remembered that these soils are invariably loaded with phosphate from the sludge, and, unlike nitrogen, the phosphate residues do not decrease rapidly with time. Research on the effects of metals on mycorrhizas may be of greater significance in natural or semi-natural ecosystems (e.g. forests) if it is proposed to dispose of sludge in such places.

6.7 CONCLUSIONS: CONTROLLING METAL CONTAMINATION OF SOILS

Three types of information must be considered in a "whole system" approach to controlling metal contamination of soils, so as to minimise adverse effects on soil microorganisms:

1. ecotoxicological effects of metal contamination under conditions similar to those of interest—in this case sewage sludge-amended soils;
2. the concentrations of potentially toxic metals already in soils, in relation to proposed limiting concentrations in soil; and
3. to what extent soils are already polluted and the likely rate of increase due to diffuse pollution (i.e. atmospheric deposition).

The information given above shows that adverse biological effects of metals derived from sewage sludge can be observed at soil metal concentrations that are lower than previously envisaged. However, often we still do not have enough evidence to judge the long-term implications that these effects may have for various microbial parameters and hence agricultural productivity.

Biological processes in soil can be affected by metals at soil concentrations below current UK limits. It would therefore be prudent to lower the limits. But the information on which this statement is based often comes from a single treatment or plot at one site and type of soil, sometimes with other confounding factors. With such scant evidence it would not be sensible to set the limits at, for instance, the lowest concentration of each metal at which no effect was found. Indeed, these so-called NOAECs (no observed adverse effect concentrations) are sometimes less than those typical of uncontaminated soils (Meent et al., 1990; Witter, 1992).

It would be more sensible, given the present state of knowledge, to lower the limits to concentrations that are realistic and sustainable, given the current concentrations in soils; for example, the maximum permitted increases above background concentrations that are consistent with minimal biologically adverse effects. However, to establish these in a more rigorous fashion, much more evidence is needed for different groups of organisms, and from more experiments or sites with better ranges of metal concentrations and minimal confounding variables such as soil pH or organic matter status. Such evidence is needed from a wider range of soil types, and needs to be long-term in nature. However, it does appear from the results discussed above that light textured soils with small clay contents need greater protection. As suggested by Scheltinga and Cardinas (1986), a doubling of the background concentrations could be the basis of more protective limits. The implicit assumption here is that all organisms are adapted to, and their activity will not be impaired by, background concentrations of metals in soils and that they will also tolerate some increase above background.

Lighter textured soils do have smaller natural background concentrations of metals than most other soils (McGrath and Loveland, 1992). Thus, using the information available on metal concentrations in soils of different textures in England and Wales (McGrath and Loveland, 1992), it would, for example, be possible to set precautionary limits based on a doubling of the soil-specific background concentrations, whilst further ecotoxicological investigations proceed. However, if this is done, it must be recognised that concentrations of metals in soils of industrialised countries have already increased above their "natural" background. For example, Jones et al. (1987) measured an increase in the Cd concentration in an untreated silty clay loam soil at the semi-rural Rothamsted site from 0.5 $mg\,kg^{-1}$ in 1840 to 0.75 $mg\,kg^{-1}$ in 1980. So, it seems that, depending on the type of soil, we are probably half-way or more than half-way to a doubling of preindustrial concentrations of metals in soils already. This suggests an urgent need to limit the inputs of metals to soils from all sources. Finally, the discussion above relies on information about the total concentrations in soils, with some comments about the bioavailability changing because of variables such as pH or soil type. As pointed out in the Introduction, almost no studies measured the more biologically relevant speciation of metals in soil solution, possibly because of the great costs in time and instrumentation, or because of a lack of instrumentation. There are great methodological and analytical difficulties with this, but this must surely be the way forward in future. When enough information is available from this sort of approach, limits could be based on the resulting activities of metals in solution. The costs of putting this into practical application may, however, be high.

ACKNOWLEDGEMENTS

The author wishes to thank Drs Amar M. Chaudri and Ken E. Giller for their useful comments on a draft manuscript, and the Ministry of Agriculture, Fisheries and Food and the European Commission's Science and Technology for Environmental Protection programme for financial support.

REFERENCES

Anderson, T. H. and Domsch, K. H. (1989) Ratios of microbial biomass carbon to total organic carbon in parable soils. *Soil Biology and Biochemistry*, **21**, 471–479.

Bååth, E. (1989) Effects of heavy metals in soil on microbial processes and populations (a review). *Water, Air and Soil Pollution*, **47**, 335–379.

Balzer, W. and Ahrens, E. (1991) Mikrobiologische bewertung von böden aus dauerversuchen mit faulschlamm. In: Sauerbeck, D. and Lübben, S. (Eds) *Auswirkungen von Siedlungsabfällen auf Böden. Bodenorganismen und Pflanzen*, pp. 359–389. Berichte aus de Ökologische Forschung Band 6, Forschungszentrum, Jülich.

BMI: (Bundesministerium des Innern) (1982) Klärschlammverordnung—AbfKlärV vom 25.06.1982. *Bundesgesetzblatt* (1982), Teil I: 734–739.

Brookes, P. C. and McGrath, S. P. (1984) Effects of metal toxicity on the size of the soil microbial biomass. *Journal of Soil Science*, **35**, 341–346.

Brookes, P. C. and McGrath, S. P. (1987) Adenylate energy charge in metal contaminated soil. *Soil Biology and Biochemistry*, **19**, 219–220.

Brookes, P. C., McGrath, S. P., Klein, D. A. and Elliot, E. T. (1984) Effects of heavy metals on microbial activity and biomass in field soils treated with sewage sludge. In: *Environmental Contamination*, CEP, Edinburgh, pp. 574–585.

Brookes, P. C., McGrath, S. P. and Heijnen, C. (1986) Metal residues in soils previously treated with sewage sludge and their effects on growth and nitrogen fixation by blue-green algae. *Soil Biology and Biochemistry*, **18**, 345–353.

CEC (Commission of the European Communities) (1986) Council Directive of 12 June 1986 on the protection of the environment, and in particular of the soil, when sewage sludge is used in agriculture. *Official Journal of the European Communities*, No. L181 (86/278/EEC), pp. 6–12.

Chander, K. (1991) The effects of heavy metals from past applications of sewage sludge on soil microbial biomass and microbial activity. PhD thesis, University of Reading.

Chander, K. and Brookes, P. C. (1991*a*) Effects of heavy metals from past applications of sewage sludge on microbial biomass and organic matter accumulation in a sandy loam and a silty loam UK soil. *Soil Biology and Biochemistry* **23**, 927–932.

Chander, K. and Brookes, P. C. (1991*b*) Microbial biomass dynamics during decomposition of glucose and maize in metal-contaminated and non-contaminated soils. *Soil Biology and Biochemistry*, **23**, 917–925.

Chander, K. and Brookes, P. C. (1991*c*) Plant inputs of carbon to metal-contaminated soil and effects on the soil microbial biomass. *Soil Biology and Biochemistry*, **24**, 1169–1177.

Chander, K. and Brookes, P. C. (1991*d*) Is the dehydrogenase assay invalid as a method to estimate microbial activity in copper-contaminated soils? *Soil Biology and Biochemistry*, **23**, 909–915.

Chaudri, A. M., McGrath, S. P. and Giller, K. E. (1992) Metal tolerance of isolates of *Rhizobium leguminosarum* biovar *trifolii* from soil contaminated by past applications of sewage sludge. *Soil Biology and Biochemistry*, **24**, 83–88.

Colpaert, J. V. and van Assche, J. A. (1987) Heavy metal tolerance in some ectomycorrhizal fungi. *Functional Ecology*, **1**, 415–421.

Coppola, S., Dumontet, S., Portonio, M., Basile, G. and Marino, P. (1988) Effect of cadmium-bearing sewage sludge on crop plants and microorganisms in two different soils. *Agriculture, Ecosystems and Environment*, **20**, 181–194.

Cornfield, A. H., Beckett, P. H. T. and Davis, R. D. (1976) Effect of sewage sludge on mineralisation of carbon in soil. *Nature*, **260**, 518–520.

Doelman, P. (1986) Resistance of soil microbial communities to heavy metals. In: Jensen, V., Kjoller, A. and Sørensen, L. H. (Eds) *FEMS Symposium No. 33, Microbial Communities in Soil*, Copenhagen, 4–8 August 1985, pp. 369–398. Elsevier Applied Science, London.

D.o.E. (1989) *Code of Practice for the Agricultural Use of Sewage Sludge*, HMSO, London.

Duxbury, T. (1985) Ecological aspects of heavy metal responses in micro-organisms. In: Marshall, K. C. (Ed.) *Advances in Microbial Ecology*, Vol. 8, pp. 185–235. Plenum Press, New York.

Eivazi, F. (1990) Nitrogen fixation of soybean and alfalfa on sewage sludge-amended soils. *Agriculture, Ecosystems and Environment*, **30**, 129–136.

Fliessbach, A. and Reber H. (1990) Effects of long term sewage sludge applications on soil microbial parameters. Joint Seminar of the Commission of the European Community and German Federal Res. Centre of Agriculture (FAL), 6–8 June 1990, Braunschweig.

Fliessbach, A., and Reber, H. (1991) Auswirkungen einer langjährigen Zufuhr von Klärschlamm auf Bodenmikroorganismen und ihre Leistungen. In: Sauerbeck, D. and Lübben, S. (Eds) *Auswirkungen von Siedlungsabfällen auf Böden. Bodenorganismen und Pflanzen*, pp. 327–358. Berichte aus de Ökologische Forschung Band 6, Forschungszentrum, Jülich.

Fliessbach, A., Reber, H. and Martens, R. (1989) Auswirkungen einer langjährigen Zufuhr von Klärschlamm auf Bodenmikroorganismen und ihre Leistungen. Report on project 0339059 to BMFT, Germany.

Giller, K. E. and Day, J. M. (1985) Nitrogen fixation in the rhizosphere; significance in natural and agricultural systems. In: *Ecological Interactions in Soil*, Fitter, A. H. (Ed.), pp. 127–147. Special Publication No. 4 of the British Ecological Society. Blackwell Scientific, Oxford.

Giller, K. E. and McGrath, S. P. (1988) Pollution by toxic metals on agricultural soils. *Nature*, **335**, 676.

Giller, K. E., McGrath, S. P. and Hirsch, P. R. (1989) Absence of nitrogen fixation in clover grown on soil subject to long-term contamination with heavy metals is due to survival of only ineffective *Rhizobium*. *Soil Biology and Biochemistry*, **21**, 841–848.

Giller, K. E., Nussbaum, R., Chaudri, A. M. and McGrath, S. P. (1993) *Rhizobium meliloti* is less sensitive to heavy-metal contamination in soil than *R. leguminosarum* bv. *trifolii* or *R. loti. Soil Biology and Biochemistry*, **25**, 273–278.

Heckman, J. R., Angle, J. S. and Chaney, R. L. (1986) Soybean nodulation and nitrogen fixation on soil previously amended with sewage sludge. *Biology and Fertility of Soils*, **2**, 181–185.

Heckman, J. R., Angle, J. S. and Chaney, R. L. (1987*a*) Residual effects of sewage sludge on soybean: II. Accumulation of heavy metals. *Journal of Environmental Quality*, **16**, 113–117.

Heckman, J. R., Angle, J. S. and Chaney, R. L. (1987*b*) Residual effects of sewage sludge on soybean: II. Accumulation of soil and symbiotically fixed nitrogen. *Journal of Environmental Quality*, **16**, 118–124.

Heggo, A., Angle, J. S. and Chaney, R. L. (1990) Effects of vesicular–arbuscular mycorrhizal fungi on heavy metal uptake by soybeans. *Soil Biology and Biochemistry*, **22**, 865–869.

Hopkin, S. P. (1989) *Ecophysiology of Metals in Terrestrial Invertebrates*. Elsevier Applied Science, London.

Jackson, D. R. (1985) Tolerance of red clover to toxic metals in sewage treated soil. In: *Research and Development in the Midlands and Western Region. 1985,* pp. 173–176. Great Britain Ministry of Agriculture, Fisheries and Food/Agricultural Development and Advisory Service, Wolverhampton.

Jenkinson, D. S. and Ladd J. N. (1981) Microbial biomass in soil: Measurement and turnover. In: Paul, E. A. and Ladd, J. N. (Eds) *Soil Biology and Biochemistry*, Vol. 5, pp. 415–471. Marcel Dekker, New York.

Johnston, A. E. and Wedderburn, R. W. M. (1974) The Woburn Market Garden Experiment, 1942–1969. I. History of the experiment, details of the treatments and yields of the crops. *Rothamsted Report for 1974*, Part 2, 79–101.

Johnston, A. E., McGrath, S. P., Poulton, P. R. and Lane, P. W. (1989) Accumulation and loss of nitrogen from manure, sludge and compost: long term experiments at Rothamsted and Woburn. In: Hansen, J. A. and Henriksen, K. (Eds) *Nitrogen in Organic Wastes applied to Soils*, pp. 126–128. Academic Press, London.

Jones, K. C., Symon, C. J. and Johnston, A. E. (1987) Retrospective analysis of an archived soil collection. II. Cadmium. *The Science of the Total Environment*, **67**, 75–89.

Kinkle, B. K., Angle, J. S. and Keyser, H. H. (1987) Long-term effects of metal-rich sewage sludge application on soil populations of *Bradyrhizobium japonicum*. *Applied and Environmental Microbiology*, **53**, 315–319.

Koomen, I., McGrath, S. P. and Giller, K. E. (1990) Mycorrhizal infection of clover is delayed in soils contaminated with heavy metals from past sewage sludge applications. *Soil Biology and Biochemistry*, **22**, 871–873.

Kücke, M. (1989) Effect of long-term sewage sludge application and heavy metal enrichment in soils on growth and VA mycorrhiza infection of maize roots. FAL Braunschweig Jahresbericht, G22.

Lorenz, S. E., McGrath, S.P. and Giller K.E. (1992) Assessment of free-living nitrogen fixation activity as a biological indicator of heavy metal toxicity in soil. *Soil Biology and Biochemistry*, **24**, 601–606.

Ma, W. (1988) Toxicity of copper to lumbricid earthworms in sandy agricultural soils amended with Cu enriched organic waste materials. *Ecological Bulletins*, **39**, 53–56, Copenhagen.

Mårtensson, A. M. and Witter, E. (1990) The influence of various soil amendments on nitrogen-fixing soil microorganisms in a long-term field experiment, with special reference to sewage sludge. *Soil Biology and Biochemistry*, **22**, 977–982.

McGrath, S. P. (1984) Metal concentrations in sludges and soil from a long-term field trial. *Journal of Agricultural Science. Cambridge*, **103**, 25–35.

McGrath, S. P. (1987) Long term studies of metal transfers following application of sewage sludge. In: Coughtrey, P. J., Martin, M. H. and Unsworth, M. H. (Eds) *Pollutant Transport and Fate in Ecosystems*, pp. 301–318. Special Publication No. 6 of the British Ecological Society. Blackwell Scientific, Oxford.

McGrath, S. P. and Hirsch, P. R. (1989) Effects of pollutants on diversity of soil microbes. *Institute of Arable Crops Research Report for 1989*, pp. 77–78.

McGrath, S. P. and Lane, P. W. (1989) An explanation for the apparent losses of metals in a long-term field experiment with sewage sludge. *Environmental Pollution*, **60**, 235–256.

McGrath, S. P. and Loveland, P. J. (1992) *Soil Geochemical Atlas of England and Wales*. Blackie, Glasgow.

McGrath, S. P., Sanders, J. R., Laurie, S. H. and Tancock, N. P. (1986) Experimental determinations and computer predictions of trace metal ion concentrations in dilute complex solutions. *Analyst*, **111**, 459–465.

McGrath, S. P., Giller, K. E. and Brookes, P. C. (1987) *Rothamsted Experimental Station Report for 1986*, p. 154.

McGrath, S. P., Brookes, P. C. and Giller, K. E. (1988) Effects of potentially toxic metals in soil derived from past applications of sewage sludge on nitrogen fixation by *Trifolium repens* L. *Soil Biology and Biochemistry*, **20**, 415–424.

Meent, D. van de, Aldenberg, T., Canton, J. H., Gestel, C. A. M. van, and Sloof, W. (1990) *Desire for Levels*. National Institute of Public Health and Environmental Protection Report no. 670101.002, Bilthoven.

Powlson, D. S. and Jenkinson D. S. (1981) A comparison of organic matter, biomass, adenosine triphosphate and mineralisable nitrogen contents of ploughed and direct drilled soils. *Journal of Agricultural Science*, **97**, 713–721.

Powlson, D. S., Brookes P. C. and Christensen B. T. (1987) Measurement of soil microbial biomass provides an early indication of changes in total soil organic matter due to straw incorporation. *Soil Biology and Biochemistry*, **19**, 159–164.

Reddy, G. B., Cheng, C. N. and Dunn, S. J. (1983) Survival of *Rhizobium japonicum* in soil–sludge environment. *Soil Biology and Biochemistry*, **15**, 343–345.

Reddy, G. B., Faza, A. and Bennett, R. (1987) Activity of enzymes in rhizosphere and non-rhizosphere soil amended with sludge. *Soil Biology and Biochemistry*, **19**, 203–205.

Royle, S. M., Chandrasekhar, N. C. and Unwin, R. J. (1989) The effect of zinc, copper and nickel applied to soil in sewage sludge on the growth of white clover. WRc York Symposium, 5–7 September, 1989, WRc, Medmenham.

Sanders, J. R. (1984) Simultaneous determination of ions of several metals in soil solutions. In: Perry, R. (Ed.) *Environmental Contamination*, pp. 589–594. CEP Consultants, Edinburgh.

Sanders, J. R., McGrath, S. P. and Adams, T. M. (1987) Zinc, copper and nickel concentrations in soil extracts and crops grown on four soils treated with metal-loaded sewage sludges. *Environmental Pollution*, **44**, 193–210.

Sauerbeck, D. and Styperek, P. (1989) Heavy metals in soils and plants of 25 long term field experiments treated with sewage sludge. In: Welte, E. and Szabolcs, I. (Eds) *Proceedings of the 4th International CIEC Symposium on Agricultural Waste Management and Environmental Protection*, Braunschweig, pp. 439–451. International Scientific Centre of Fertilizers (CIEC), Belgrade, Goettingen, Vienna.

Scheltinga, H. M. J. and Cardinas, T. (1986) Political and administrative consider-ations in the formulation of guidance for sludge utilization. In: Davis, R. D., Haeni, H. and L'Hermite, P. (Eds) *Factors Influencing Sludge Utilization Practices in Europe*, pp. 90–102. Elsevier Applied Science, London.

Sheaffer, C. C., Decker, A. M., Chaney, R. L. and Douglas, L. W. (1979) Soil temperature and sewage sludge effects on corn yield and macronutrient content. *Journal of Environmental Quality*, **8**, 450–459.

Skujins, J., Nohrstedt, H. O. and Odén, S. (1986) Development of a sensitive biological method for the determination of a low level toxic contamination in soils. *Swedish Journal of Agricultural Science*, **16**, 113–118.

Smith, S. (1989) Effects of sewage sludge applications on soil microbial processes and soil fertility. Report FR0034 to the Foundation for Water Research.

Smith, S. R., Obbard, J. P., Kwan, K. H. M. and Jones, K. C. (1990) Symbiotic N_2-fixation and microbial activity in soils contaminated with heavy metals resulting from long-term sewage sludge applications. *Foundation for Water Research, Report FR 0128*. FWR, Marlow, Bucks.

Sommers, L., Volk, V. van, Giordano, P. M., Sopper, W. E. and Bastian, R. (1987) Effects of soil properties on accumulation of trace elements by crops. In: Page, A. L., Logan, T. G. and Ryan, J. A. (Eds) *Land Application of Sludge*, pp. 5–24. Lewis Publishers, Chelsea, Michigan.

Stark, J. H. and Lee, D. H. (1988) Sites with a history of sludge deposition. Final report on rehabilitation field trials and studies relating to soil microbial biomass (LDS 9166 SLD) Final Report to the Department of the Environment. WRc Report DoE 1768-M, WRc, Medmenham.

Tyler, G. (1974) Heavy metal pollution and soil enzymatic activity. *Plant and Soil*, **41**, 303–311.

Tyler, G. (1981) Heavy metals in soil biology and biochemistry. In: Paul, E. A. and Ladd, J. N. (Eds) *Soil Biochemistry*, Vol. 5, pp. 371–414. Marcel Dekker, New York.

UK (United Kingdom Statutory Instrument) (1989) *The Sludge (Use in Agriculture) Regulations, 1989*. SI No. 1263, HMSO, London.

Vaidyanathan, L. V. (1975) Residual effects of metal contaminated sewage sludge additions to soil on grass-clover ley. In: *Experiments and Development in the Eastern Region, 1975*, pp. 53–55. Great Britain Agricultural Development and Advisory Service, Cambridge.

Witter, E. (1989) Agricultural use of sewage sludge—controlling metal contamination of soils. *Staten Naturvårdverket Rapport, 3620*, Sweden.

Witter, E. (1992) Heavy metal concentrations in agricultural soils critical to microorganisms. *Staten Naturvårdverket Rapport, 4079*, Sweden.

Witter, E., Mårtensson, A. M. and Garcia, F. V. (1993) Size of the soil microbial biomass in a long-term field experiment as affected by different N-fertilizers and organic manures. *Soil Biology and Biochemistry*, **25**, 659–669.

7 Mechanisms of Ecosystem Recovery Following 11 Years of Nutrient Enrichment in an Old-field Community

SUSAN R. BREWER AND GARY W. BARRETT
Miami University, Oxford, Ohio, USA

MARY BENNINGER-TRUAX
Hiram College, Hiram, Ohio, USA

ABSTRACT

Subplots in each of three former long-term (11-year) treatments (sludge, fertiliser, and control) were tilled and/or limed in 1989 to address biological (e.g. seed bank disturbance) and chemical (e.g. soil pH) mechanisms of recovery. Differences in plant community structure after three years reflected long-term treatments rather than short-term manipulations. During nutrient enrichment, fertiliser and sludge-treated plots arrested in an early stage of succession. Control plots remained dominated by perennials and biennials such as *Solidago canadensis* and *Daucus carota* throughout the study. Although fertiliser plots remained dominated by the summer annual *Ambrosia trifida* throughout much of the enrichment and recovery stages, perennials such as *Solidago canadensis* increased in 1991, indicative of the onset of secondary succession. Sludge plots, previously dominated by *Ambrosia artemisiifolia*, were dominated by summer annuals *Ambrosia trifida*, *Chenopodium alba* and *Polygonum pennsylvanicum* in 1991. Plant species richness was greater in control than nutrient-enriched plots, whereas annual above-ground NPP was greater in nutrient-enriched than control plots throughout 14 years.

Active soil bacterial biomass was greater in control and fertiliser plots than in sludge plots. It appears that soil pH and nutrient availability affected bacterial activity and nutrient cycling which, in turn, controlled plant community structure and metal bioaccumulation in plants. Plants collected from sludge plots tended to accumulate higher levels of Cd, Cu, Pb and Zn than plants collected from fertiliser or control plots. However, bioaccumulation of heavy metals varied with each metal and each plant species. Subplot manipulations, however, had little effect on heavy metal bioaccumulation. Soil nitrate levels increased significantly in sludge plots, but remained constant in control and fertiliser plots during recovery. It is likely that this accounted for the increased abundance of *Ambrosia trifida* in sludge plots. These findings support Tilman's resource-ratio hypothesis.

Toxic Metals in Soil–Plant Systems. Edited by S. M. Ross
© 1994 John Wiley & Sons Ltd

7.1 INTRODUCTION

Ecological toxicology is a rapidly emerging field of study in applied ecology that combines practical application with ecological theory (Barrett, 1987). The application of sewage sludge addresses problems associated both with waste disposal and the successional changes related to nutrient enrichment of plant communities. Sewage sludge contains both phosphorous and nitrogen, elements essential for plant community growth and development. Sewage sludges also represent an inexpensive fertiliser for agricultural fields; this application process represents a potentially inexpensive means of sludge disposal. However, sludges contain significant levels of heavy metals and trace elements (e.g. Cd, Cu, Pb and Zn) which may pose health risks at higher trophic levels.

The accumulation of metals through food webs poses potential health risks to consumers and may alter ecosystem-level functions. Culliney and Pimentel (1986) found that the application of sewage sludge to agricultural land was not a suitable means of disposal due to the potential for metal bioaccumulation. Likewise, Burton and Hook (1979) found that significant levels of NO_3-N were released into the groundwater supplies following application of sewage sludge to forest ecosystems. However, old-field communities may provide ecologically safe and economically viable sites for sludge disposal and recycling (Carson and Barrett, 1988).

Although numerous investigators have examined rates of metal uptake in plant communities (e.g. Chang et al., 1987), there is need for research that: (a) monitors metal uptake on a greater temporal scale; (b) evaluates the response of plant communities to long-term nutrient enrichment; and (c) continues to monitor ecosystem structural and functional changes following the cessation of chronic sludge application. Of particular interest is the relationship between soil and plants, and the resulting effects that soil–plant interactions have on plant community structure and function. Little is known concerning the mechanisms of recovery following long-term nutrient enrichment. The purpose of this study was to examine the effects of long-term (11-year) nutrient enrichment on old-field structure and function, and to examine mechanisms of ecosystem recovery following a long-term perturbation.

7.2 STUDY SITE AND RESEARCH DESIGN

A long-term (14-year) study evaluating the effects of nutrient enrichment (municipal sludge or fertiliser) on old-field communities was established in 1977 at the Miami University Ecology Research Center located near Oxford, Ohio, USA. An eight-plot design was established to compare the effects of municipal sludge and commercial fertiliser application on old-field communities (Figure 7.1). This study provided the opportunity to examine both the

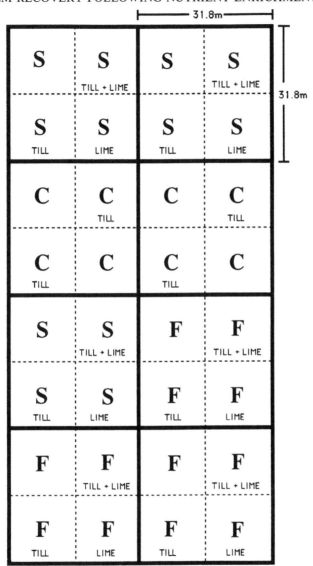

Figure 7.1. Diagram depicting the original long-term (11-year) research design including fertiliser (F), sludge (S), and control (C) treatments. Recent (1989) manipulations included tilling and/or liming

effects of differing types of nutrient enrichment on old-field communities, as well as the effects of heavy metal uptake on ecosystem dynamics. Winter wheat (*Triticum aestivum* var. Ranger) was planted in each 0.1-ha plot in October 1977. The wheat was not harvested and plots were permitted to proceed into secondary succession in 1978.

The soil at the study site is a Xenia silt-loam (Aquic Hapludalfs) over limestone bedrock, with a mean soil pH of 5.1 ± 1.0 SE in 1978 (Ohio Department of Natural Resources, 1978; Carson and Barrett, 1988). No initial heavy metal analyses were conducted on soil prior to sludge and fertiliser application. A commercial sludge, Milorganite (6:2:0, N:P:K), was applied to three plots at a rate of 1792 mg kg^{-1} every four weeks for a total annual application of 8960 kg ha^{-1} (Levine *et al.*, 1989). A commercial urea–phosphate fertiliser (34:11:0, N:P:K) was also applied to three separate plots at a rate of 314 kg ha^{-1} to supply an equivalent nutrient subsidy. Two remaining plots were left untreated to serve as controls. Sludge and fertiliser were applied annually (May–September) for 11 years (1978–1988).

In 1989, each plot was equally divided into four subplots and manipulated by tilling and/or liming (Figure 7.1). Subplots were tilled once on 17–18 May 1989, whereas subplots were limed annually (1989–1991) to restore pH values to control levels. Lime was applied 7–12 April 1989, 12–13 February 1990, and 14–15 February 1991 at rates recommended by the Ohio Agricultural Research and Development Center, Wooster, Ohio, based upon measurements of soil pH during the previous growing season. Lime was not applied in 1992 because soil pH values in limed subplots had reached control values.

Subplot manipulations were conducted in an attempt to restore biologically (i.e. by disturbing the seed bank) or chemically (i.e. by raising the soil pH) ecosystem structure and function to control levels. This design permitted the authors to evaluate mechanisms of old-field recovery. It was hypothesised that disturbing the seed bank by tilling would increase the rate of secondary succession by accelerating the establishment of old-field plant species. Long-term nutrient enrichment, in the form of both sludge and fertiliser applications, was found to acidify the soil (Levine *et al.*, 1989). Lime was applied to restore soil pH to control levels and, theoretically, to permit the establishment of plant and microbial species that could not exist under more acidic conditions. Liming may also play a critical role concerning the bioavailability of metals for uptake into higher trophic levels, as the solubility of heavy metals increases with decreasing soil pH (Kiekens, 1984).

Plant community response to nutrient enrichment over 11 years was determined by measuring plant species richness and diversity, above-ground standing crop biomass, and annual above-ground net primary production (ANPP). In addition, soil nutrient and pH levels were measured, as well as the levels of heavy metals in soils, plants, primary and secondary consumers, and detritivores. This chapter will first summarise changes in the structure and function of an old-field community following long-term nutrient enrichment, including metal accumulation. The focus of this chapter, however, will be primarily to describe the three years of community recovery following biological and chemical manipulations.

7.3 MATERIALS AND METHODS

7.3.1 SAMPLING METHODS

Nine markers were placed in each subplot in a 3×3 design, arranged 4.1 m apart. Samples were collected along diagonal transects from each marker. Each sample was collected from a randomly chosen position 0.5 m, 1.0 m or 1.5 m from a marker in 1989, 1990 and 1991, respectively. This sampling regime avoided sampling any site more than once. All samples were collected at least 1.0 m from the subplot edge.

7.3.2 SOIL AND MICROBIAL ANALYSES

Soil samples were collected in May and October each year (1989–1991). Three 10-cm cores were collected from random sites in each subplot on each sampling date. Each sample was air-dried for one week; one-half of each sample was shipped to the Ohio Agricultural Research and Development Center for analysis of soil pH, nitrates, phosphorous, potassium, magnesium, cation exchange capacity, and per cent organic matter. The remainder of each sample was frozen at $-20\,^\circ$C for heavy metal analysis at a later date.

At the time of metal analysis, samples were oven-dried at $80\,^\circ$C for 48 h and then weighed to 1.0 g samples. Each sample was digested repeatedly in concentrated HNO_3 over heat to remove organic matter. Following digestion, samples were diluted in 10% HNO_3, filtered, and the supernatant analysed for Cd, Cu, Pb and Zn concentrations using atomic absorption spectrophotometry (IL157 flame spectrophotometer). Methods were similar to those reported in Levine *et al.* (1989).

Soil samples were collected on 21 August 1991 and shipped to the Microbial Biomass Service, Oregon State University, Corvallis, Oregon, USA. These soil samples were analysed for total and active bacterial and fungal biomass.

7.3.3 VEGETATION

Vegetation was sampled at six-week intervals from April to September (1978–1988), June to October (1989), and May to September (1990 and 1991). Above-ground plant biomass was harvested from four randomly selected 0.25 m^2 sites in each subplot on each sample date. Vegetation was sorted into litter, standing dead, and live biomass; the last of these was further separated by species. All vegetation was oven-dried at $80\,^\circ$C for 72 h, then weighed to the nearest 0.01 g.

Plant species diversity was determined using indices of species richness (mean number species per 0.25 m^2 per subplot per date), species apportionment (Shannon-Wiener Index; Shannon and Weaver, 1964), and absolute diversity (total number of species per subplot per year). Above-ground annual net primary productivity (ANPP, $g m^{-2} year^{-1}$) was calculated by summing the peak biomass of each species per subplot.

Three specimens each of *Setaria faberii*, *Ambrosia artemisiifolia*, *Ambrosia trifida* and *Solidago canadensis*, were harvested during the reproductive stage of growth for heavy metal analysis. Each plant was collected from random sites from the interior of each subplot. Plants were frozen at $-20\,^{\circ}C$ until metal analysis at a later date.

As with soils, plants were oven-dried at $80\,^{\circ}C$ for 48 h, then weighed to 1.5 g samples. Each sample consisted of 0.5 g each of stems, leaves and flowers. Plant samples were repeatedly digested in concentrated HNO_3 and then diluted in 10% HNO_3 and filtered. The supernatant was analysed for Cd, Cu, Pb and Zn concentrations using atomic absorption spectrophotometry, as described above.

7.3.4   DATA ANALYSIS

Analysis of variance (ANOVA) and Duncan's New Multiple Range Test were used to determine significant differences ($p < 0.05$) between long-term treatments and between subplot manipulations regarding soil pH, nitrates and organic matter; plant and soil heavy metal concentrations; and plant species richness, apportionment and productivity.

7.4   LONG-TERM NUTRIENT ENRICHMENT

7.4.1   PLANT SPECIES RICHNESS AND DIVERSITY

During the first year of nutrient enrichment (1978), no significant differences were observed in the mean number of plant species per treatment (Carson and Barrett, 1988). Although not significantly different, fertiliser-treated plots consistently maintained a greater number of plant species than control or sludge-treated plots; species richness peaked in all plots in June and declined thereafter.

Species richness increased in control plots during 1979, and was significantly greater in control plots than in nutrient-enriched plots. Neither plant species richness nor community composition changed significantly in nutrient-enriched plots between 1978 and 1979. In 1980, however, plant species richness increased in control plots to approximately 20 species per m^2, whereas sludge- and fertiliser-treated plots maintained approximately 10–13 species per m^2 throughout the growing season (Carson and Barrett, 1988).

Differences in species richness between control and nutrient-enriched plots remained relatively consistent for the remainder of the 11 years (1978–1988) of annual (May–September) nutrient enrichment (Maly and Barrett, 1984; Hyder and Barrett, 1986; Bollinger *et al.*, 1991). Species richness (i.e. the mean total number of species per m^2) in control treatments rose steadily, peaking at 20 species per plot in 1981 (Figure 7.2). Thereafter, the mean number of plant species in control plots remained consistent until subplot manipulations were introduced in 1989. The absolute mean diversity of sludge and fertiliser-treated plots, however, remained consistently low (approximately 10 species per plot) until subplot manipulations were introduced. Total absolute species richness (number of species plot^{-1} year^{-1}) was approximately 40 in control plots and 20–25 in nutrient-enriched plots. Although subplot manipulations increased plant species diversity of all treatments, these changes will be discussed when we consider mechanisms of the old-field recovery.

7.4.2 PLANT PRODUCTIVITY

Mean annual above-ground net primary productivity (ANPP) and above-ground standing crop biomass were significantly greater in fertiliser plots than in control or sludge plots in 1978. ANPP in sludge plots was also significantly greater than in control plots in 1978. Fertiliser plots were dominated by the fast-growing summer annual *Ambrosia artemisiifolia*. Both nutrient-enriched and control plots were dominated by *Ambrosia artemisiifolia* and *Triticum aestivum*. Species were considered dominant if they comprised greater than 10% of the total ANPP of a treatment.

Species composition began to diverge between control and enriched plots in 1979 and 1980. Annual ANPP for *Ambrosia artemisiifolia* and *Setaria faberii* was significantly greater in enriched plots than control plots for 1978 through

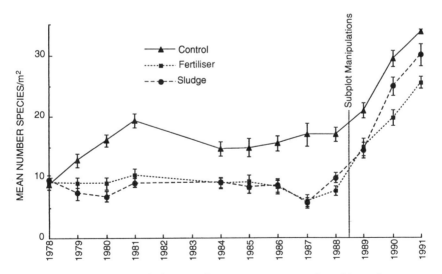

Figure 7.2. Mean number of plant species per square metre found in each treatment during the 14 year study. Data are expressed as mean ± SE per treatment

1980. ANPP for perennials such as *Aster pilosis* and *Trifolium pratense*, as well as the winter annual *Erigeron annuus*, was greatest in control plots, reflecting the shift to a dominance by perennials in control plots during the second and third years of secondary succession (Carson and Barrett, 1988). This trend continued throughout the 11-year nutrient enrichment study.

Conversely, the summer annual *Ambrosia artemisiifolia* had a greater ANPP in enriched plots than control plots in 1979, and both *Polygonum persicaria* and *Setaria faberii* had a greater ANPP in enriched plots than control plots for both 1979 and 1980. *Ambrosia trifida*, giant ragweed, had a higher ANPP in fertiliser plots than control or sludge plots in 1979 and 1980, indicating a shift in community composition from *Ambrosia artemisiifolia* to *Ambrosia trifida* (Carson and Barrett, 1988). Total ANPP in 1980, however, was significantly greater in control and fertiliser treatments than in sludge treatments. The high ANPP in control plots was attributed to two peak biomass productions in one growing season. Biennials in control plots peaked early in the growing season, while perennials peaked late in the season. This contrasts one peak biomass production of annuals in nutrient-enriched plots.

In 1981, Maly and Barrett (1984) examined changes in primary productivity during the fourth year of nutrient enrichment. Interestingly, they found that ANPP was highest in fertiliser plots (2005 $g\,m^{-2}$ $year^{-1}$), followed by sludge plots (1638 $g\,m^{-2}$ $year^{-1}$), and control plots (742 $g\,m^{-2}$ $year^{-1}$).

Fertiliser-treated plots maintained the greatest ANPP and above-ground biomass (Figure 7.3) between 1984 and 1987. A peak biomass production was reached in all treatments in 1986, followed by a decline until 1989. All three

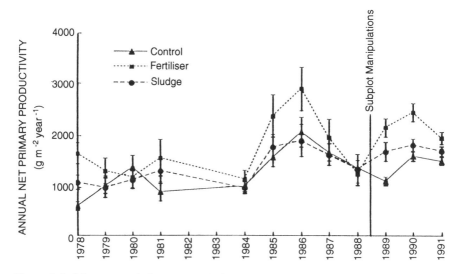

Figure 7.3. Mean annual above-ground biomass per square metre during 14 years. The three treatments are represented. Data are expressed as mean ± SE per treatment

treatments contained similar amounts of above-ground biomass in 1988 due to drought conditions (Bollinger *et al.*, 1991).

7.4.3   PLANT COMMUNITY COMPOSITION

Nutrient-enriched plots were dominated by summer annuals (90% of total ANPP) (e.g. *Ambrosia artemisiifolia*) during the first year of enrichment in 1978, whereas control plots were dominated by perennials (30%) (e.g. *Solidago canadensis* and *Trifolium pratense*) and summer annuals (70%) (e.g. *Ambrosia artemisiifolia*) (Carson and Barrett, 1988). A change in community composition occurred in control plots during 1979, when summer annuals were replaced by herbaceous perennials and winter annuals, such as *Aster pilosus* and *Erigeron annuus*, respectively. It is likely that this shift in community composition resulted from plant species that are better competitors in preempting resources, particularly limited nutrients (Tilman, 1985).

Nutrient-enriched plots remained dominated by the summer annuals *Ambrosia artemisiifolia* and *Setaria faberii*, with the addition of *Ambrosia trifida* in fertiliser plots during 1979 to 1987 (Maly and Barrett, 1984; Hyder and Barrett, 1986; Carson and Barrett, 1988; Bollinger *et al.*, 1991). The continued dominance by summer annuals suggests that nutrient preemption did not occur, and that nitrogen and phosphorus resources were maintained at constant levels during this time. This supports Tilman's resource-ratio hypothesis of plant competition in which old-field plants compete for limiting resources along gradients (Tilman, 1985). This hypothesis states that each species is a superior competitor at one point along the nutrient gradient.

Trends observed from 1979 to 1981 continued between 1984 and 1987. Control plots remained dominated by perennials such as *Solidago canadensis* and *Rubus frondosus*. Fertiliser plots remained dominated by the summer annual *Ambrosia trifida*, whereas sludge-treated plots were dominated by the summer annuals *Ambrosia artemisiifolia* and *Setaria faberii*.

In summary, nutrient-enriched plots seemed to be arrested in an early seral stage of old-field succession, as represented by the large number of summer annuals present (Maly and Barrett, 1984). Bakelaar and Odum (1978) suggested that short-term disturbances would not alter community development and that a disturbance of at least four years would be necessary to alter the trajectory of ecosystem development. This hypothesis was supported by the results of Maly and Barrett (1984), following four years of nutrient enrichment. Because nitrogen and phosphorus levels were maintained by enrichment, several opportunistic species (e.g. *Ambrosia trifida*, *Ambrosia artemisiifolia*, and *Setaria faberii*) were able to compete successfully with perennials for nutrients.

Overall, nutrient enrichment altered species richness and apportionment, shifting dominance towards summer annuals, such as *Ambrosia artemisiifolia* and *Ambrosia trifida*. Although summer annuals are common in recently disturbed old-fields (Hartnett *et al.*, 1987), winter annuals, biennials and perennials normally dominate old-fields during later stages of secondary succession (Hartnett *et al.*, 1987). The abundance of *Ambrosia* spp. in nutrient-enriched plots supports the hypothesis (Tilman, 1985) that enrichment maintained plant composition representative of an early seral stage of community development.

7.4.4 METAL BIOACCUMULATION AND TOXICITY

Metals were found to accumulate in the plants growing in this old-field community (Maly and Barrett, 1984; Levine *et al.*, 1989). However, heavy metal toxicity was not responsible for plant composition changes in the old-field community. Metal toxicity was minimised due to the following: (1) nutrients and elements in sludge suspected as potentially toxic under field conditions are released over long time periods (Ryan *et al.*, 1973); (2) heavy metals in sludge rarely reach toxic levels in plants (Kelling *et al.*, 1977; Soon *et al.*, 1980) and may instead accumulate in higher trophic levels (Anderson and Barrett, 1982; Levine *et al.*, 1989; Brueske and Barrett, 1991); and (3) the toxic effects attributed to metals in sewage sludge would not account for similar responses in plant community structure observed in both sludge and fertiliser treatments.

During each year of nutrient enrichment (1978–1988), concentrations of all four metals (Cd, Cu, Pb and Zn) were higher in Milorganite than fertiliser additions (Levine *et al.*, 1989). Mean cadmium concentrations in Milorganite ranged from 19.8 $mg\,kg^{-1}$ to 59.0 $mg\,kg^{-1}$, whereas Cd concentrations in fertiliser ranged between 1.2 $mg\,kg^{-1}$ and 3.9 $mg\,kg^{-1}$. Mean copper concentrations in Milorganite ranged from 320.0 $mg\,kg^{-1}$ to 380.6 $mg\,kg^{-1}$. Fertiliser Cu concentrations remained low, ranging between 1.0 $mg\,kg^{-1}$ and 1.8 $mg\,kg^{-1}$. Mean lead concentrations in Milorganite ranged from 243.0 $mg\,kg^{-1}$ to 473.0 $mg\,kg^{-1}$, whereas fertiliser concentrations were between 5.4 $mg\,kg^{-1}$ and 12.5 $mg\,kg^{-1}$. Mean zinc concentrations in Milorganite ranged from 865.8 $mg\,kg^{-1}$ to as high as 1281.0 $mg\,kg^{-1}$. However, fertiliser zinc concentrations remained low, ranging from 16.4 $mg\,kg^{-1}$ to 33.3 $mg\,kg^{-1}$.

Metal concentrations of Cd, Cu, Pb and Zn were determined in 1981 for the summer annual *Setaria faberii*, the biennial *Barbaria vulgaris*, and the perennial *Cirsium arvense* (Maly and Barrett, 1984). Cadmium and zinc levels were significantly higher in plants collected from sludge-treated plots than those collected from fertiliser or control plots for all three species. Zinc uptake in *Barbaria vulgaris* was also significantly greater in fertiliser than control plots. No differences were found regarding levels of copper for *Cirsium arvense* collected from control and sludge plots. Copper accumulation in *Setaria faberii* and *Barbaria vulgaris*, however, was greater in sludge plots than control plots and greater than fertiliser plots for *Setaria faberii*. Interestingly,

Maly and Barrett (1984) found that copper levels in *Cirsium arvense* were significantly lower in fertiliser plots than in control or sludge plots. No significant differences between treatments were found in the levels of lead in plants.

It appears that each plant species has a different affinity for uptake and accumulation of different metals. No trends were observed between plants of differing life histories. The biennial species *Daucus carota* accumulated the highest levels of Cd, Pb and Zn, whereas the perennial species *Cirsium arvense* accumulated the highest level of Cu. The annual *Setaria faberii* accumulated higher levels of Zn than the perennial *Cirsium arvense*. The lack of any trend or consistency in metal accumulation may be a result of factors which alter metal activity and availability, such as organic matter content in the soil, soil pH, and ionic or microbial interactions associated with specific plant species.

Levine *et al.* (1989) summarised metal analyses for 1984–1987. Metal analyses were conducted on the sludge (Milorganite), soil, and the following plant species: *Rubus frondosus*, *Setaria faberii*, *Bromus japonicum* and *Poa* spp. (*S. faberii* was not collected from control plots). Cadmium levels in sludge decreased between 1977 and 1987, whereas soil levels of Cd rose significantly during the 11 years. Copper levels within Milorganite were consistent throughout the 11 years (320–330 mg kg^{-1}) with the exception of a peak of approximately 380 mg kg^{-1} observed in 1984. Soil Cu levels rose gradually to a peak of 36 mg kg^{-1} in 1986 and decreased thereafter.

Lead levels in sludge decreased from 1978 through 1987, whereas soil Pb levels increased consistently to a peak of 48 mg kg^{-1} in 1986. Soil zinc levels also rose gradually over 11 years. However, zinc levels within the sludge rose to a peak of approximately 1270 mg kg^{-1} in 1984 and then decreased significantly afterwards.

Levine *et al.* (1989) found that Cd concentrations in 1987 were significantly higher in vegetation collected from sludge plots compared to control or fertiliser plots. Copper concentrations were greater in *Setaria faberii* collected from sludge plots than fertiliser plots. *Setaria faberii* was not available in control plots in 1987. Copper concentrations in brome and bluegrass, however, were significantly greater in sludge plots than fertiliser plots, with control plots intermediate and not significantly different from sludge. *Setaria faberii* was the only species found to accumulate higher levels of Pb from sludge plots. The low accumulation of Cu and Pb in some species supports the idea that there exists a soil–plant barrier to these metals (Chaney, 1980). Copper and Pb also occur in the soil primarily in unextractable forms, particularly at soil pH levels above 3.0 (Kiekens, 1984). Zinc concentrations were found to be significantly greater in *Bromus japonica* and *Setaria faberii* collected from sludge plots compared to control or fertiliser plots, and in *Poa* spp. collected from sludge and fertiliser plots compared to control plots.

Although all heavy metals accumulated to some degree in the plant species analysed, plants collected from sludge plots did not concentrate Cd above levels in the soil. However, plants collected from control and fertiliser plots concentrated Cd from 11 to 32 times the levels found in the soil.

The increasing levels of metals within soils and low concentration in plants from sludge plots may indicate a lag period between sludge application and metal uptake and concentration. This suggests that a major portion of the organic matter in sludge was resisting decomposition (Levine et al., 1989). The slow decomposition of sludge and slow release of metals and nutrients may have important consequences on the trajectory and/or recovery of an old-field community.

7.4.5 DROUGHT CONDITIONS IN 1988

Drought conditions prevailed during the early part of the growing season in 1988 (Bollinger et al., 1991). Drought conditions were most intense in June and early July, yet from mid-July until 15 August 1988, rainfall increased to 146% of the normal rainfall for previous years. Nutrient-enriched plots, however, remained dominated by summer annuals such as Ambrosia trifida, Ambrosia artemisiifolia and Setaria faberii, and control plots remained dominated by the perennials Solidago canadensis, Poa compressa and Poa pratensis.

As found during previous years, species diversity was significantly greater in control plots than in nutrient-enriched plots. Due to drought, species diversity was significantly lower in control plots in 1988 than 1985–1987, whereas fertiliser and sludge plots did not change significantly from previous years. The reason for this response was that control plots were typically more diverse than nutrient-enriched plots, allowing greater opportunity for drought to have an effect. However, in July there was no significant difference in species diversity of any treatments compared to previous years.

Bollinger et al. (1991) found that the major effect of the drought was exerted on community function (e.g. net primary productivity) of the three treatments rather than on community structure (e.g. species diversity). During July, for example, at the height of the drought stress, the average daily NPP for all treatments was only 16% of 1985–1987 averages. Yet the alleviation of drought effects observed following 15 July 1988 resulted in an increase in NPP to levels similar to those observed in 1985–1987. Interestingly, the more diversified control plots differed least during drought stress compared to the less diversified nutrient-enriched plots. Daily NPP in sludge and fertiliser plots during August 1988 were 224% and 140%, respectively, of the 1985–1987 average for August, compared to 100% in control plots. Results suggest that communities dominated by annuals manifest greater resilience, but less resistance, to stress than communities dominated by perennials. Although no plant species showed a significant decline in ANPP, Chenopodium album and Poa compressa increased in ANPP, in spite of drought conditions (Bollinger et al., 1991).

7.5 MECHANISMS OF RECOVERY FOLLOWING LONG-TERM NUTRIENT ENRICHMENT

Recent research has focused on mechanisms to increase the rate of recovery of this old-field community following cessation of long-term nutrient enrichment. After 11 years of nutrient enrichment, will this old-field community recover to control levels both structurally and functionally? What will be the fate of the heavy metals deposited in the soil? Experimental manipulations have been employed to explore theoretically the recovery process. This section will focus on rates of recovery following these manipulation strategies.

Nutrient enrichment ceased in 1989. Each long-term plot (C, F and S; Figure 7.1) was subdivided into four equal subplots. During May 1989, one subplot within each former long-term plot was tilled to disturb the seed bank and to reinitiate secondary succession. During April, two subplots in each former nutrient-enriched plot were treated with lime to restore soil pH to control levels and to chemically accelerate the recovery process. One subplot was tilled and limed to examine interaction effects (Figure 7.1).

Long-term nutrient enrichment had decreased soil pH from a mean of 5.5 in control plots to a mean of 4.4 in nutrient-enriched plots. It is likely that this decreased pH also provided for increased availability of metals in the soil. The more acidic the soil, the more metals are free in solution in ionic form and, consequently, more available for uptake into producer and consumer trophic levels (Kabata-Pendias and Pendias, 1984). Also, lime subsidises Ca^{2+} content in the soil. Calcium ions are readily available to complex with metals, thus making them unavailable for uptake by plants. It has also been suggested (Kabata-Pendias and Pendias, 1984) that raising soil pH may indirectly alter metal availability by affecting soil microbial populations.

7.5.1 SOIL MICROBIAL POPULATIONS

The role of bacteria and fungi in decomposing organic matter in old-field communities remains poorly understood. Recent studies, however, indicate that long-term nutrient enrichment has affected microbial activity and rates of decomposition in this old-field community (Sutton *et al.*, 1991). Bacteria produce humic acid as a by-product of decomposition when actively decomposing organic matter. Humic acid, like Ca^{2+}, has the affinity to complex with metal ions and remove the metals from solution. Therefore, increasing soil pH values to control levels should cause soil bacteria to increase rates of plant decomposition in these old-field communities.

In 1989, Sutton *et al.* (1991) determined the rate of soil respiration in each of these long-term treatment and short-term manipulation subplots. They found increased rates of soil respiration in control plots as compared to fertiliser- or sludge-treated plots. They also found that liming, as expected, decreased soil acidity, and increased rates of soil respiration, supporting the hypothesis that increasing soil pH should increase soil microbial activity and litter decomposition. The increased soil respiration in limed subplots suggested an increase in soil microbial activity and organic decomposition (Sutton *et al.* 1991). However, the composition of the microbial community was not determined at that time.

In 1991, soil samples collected from each experimental subplot were analysed for bacterial and fungal biomass and density by the Microbial Biomass Service, Oregon State University, Corvallis, Oregon. Five soil cores per subplot were collected and analysed for microbial content. No differences were found in total fungi biomass and number, as well as active fungi biomass, between long-term treatments or short-term manipulations. Differences were found, however, in soil bacterial communities between treatments. Control and fertiliser plots had the significantly greater total bacterial biomass (mg g^{-1} dry wt soil) than sludge plots (Figure 7.4). Similar results were determined for total active bacterial biomass and count (number active bacteria per gramme dry weight soil).

Tilling appeared to have a negative effect on active bacterial density and biomass because tilled and interaction subplots contained the lowest mean bacterial biomass. Unmanipulated ("control") fertiliser and control subplots not tilled contained the highest bacterial biomass. Liming did not affect bacterial biomass within subplot manipulations. This contradicts Sutton *et al.* (1991), who found a greater soil respiration rate in limed subplots. Although liming did not alter total bacterial biomass within treatments, activity levels of bacteria may have been affected by lime addition. However, three years of

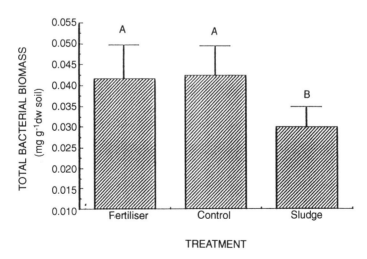

Figure 7.4. Total biomass of bacteria within the various treatments in 1991. Data are expressed as mean ± SE per treatment. Bars with letters in common are not significantly different ($p > 0.05$)

"recovery" following nutrient enrichment has not been sufficient time to restore the bacterial community structure or function to control levels. Thus, there is a need to monitor these communities in future years, as a mean control pH value was not reached in nutrient-enriched treatments until 1991 (i.e. allowing only one year for direct recovery comparisons). There is also a need to examine levels of bacterial activity and the composition of bacterial populations within treatments.

The low bacterial biomass in sludge-treated plots (Figure 7.4) indicates that sludge application is negatively affecting bacterial populations. Whereas low soil pH values, present in both sludge and fertiliser plots, did not directly decrease soil bacterial populations, interactions between low soil pH and heavy metal availability may affect bacterial populations and activity. This further emphasises the need for research concerning the effects of sludge and heavy metal toxicity on microbial populations and microbial activity.

7.5.2   SOIL CHEMISTRY

7.5.2.1   Soil pH

Soil pH values were found to be significantly higher in control plots than in nutrient-enriched plots in 1989. Within former nutrient-enriched treatments, limed subplots had significantly greater mean pH values than those only tilled or left unmanipulated. In former sludge-treated plots, those subplots both tilled and limed had significantly higher pH values than those only limed (Figure 7.5(a)). However, no nutrient-enriched values reached pH values determined for control plots.

In 1990, control subplots maintained significantly greater soil pH values than former nutrient-enriched treatments, with the exception of one sludge subplot both tilled and limed. Subplots limed, or limed and tilled, had soil pH values significantly greater than subplots not limed (Figure 7.5(b)).

Control subplots in 1991 continued to have significantly greater pH values than tilled fertiliser or sludge subplots. Subplots limed, or tilled and limed, also had significantly greater pH values than tilled subplot manipulations. Subplots both tilled and limed even exceeded long-term control subplot values (Figure 7.5(c)). It appears that tilling soil, in addition to liming, enhances the effect of lime on soil pH values.

7.5.2.2   Soil nitrates and organic matter

Levels of soil nitrates ($kg\,NO_3\text{-}N\,ha^{-1}$) were determined for each former long-term treatment and for each short-term manipulation at the Ohio Agricultural Research and Development Center by diluting soil samples both in distilled water and Orion nitrate interference suppressor. Each sample was then analysed on a pH/ion meter and compared to a standard curve containing

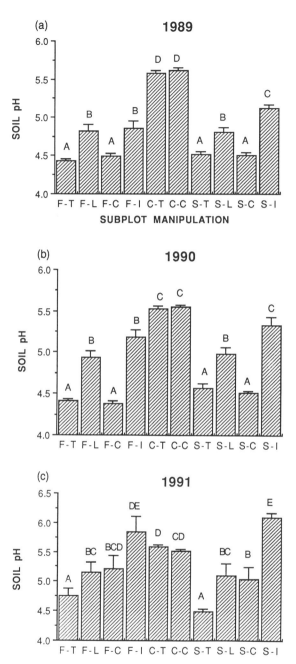

Figure 7.5. Mean soil pH values in the various subplot manipulations for (a) 1989, (b) 1990, and (c) 1991. Standard error bars with letters in common are not significantly different ($p > 0.05$). F represents long-term fertiliser-treated plots, S represents long-term sludge-treated plots, and C represents long-term control plots. Subplot manipulations are designated by T (tilled subplots), L (limed), I (tilled and limed), and C (unmanipulated)

known concentrations of $NaNO_3$. Nitrate levels in 1989, 1990 and 1991 were significantly greater in sludge plots than in control or fertiliser plots. Soil from fertiliser plots contained significantly higher nitrate levels than levels determined for control soil (Figure 7.6(a)). Subplot manipulations, however, had no effect on nitrate levels in soil. No differences were found between subplot manipulations within treatments regarding nitrate levels present.

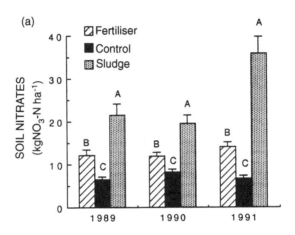

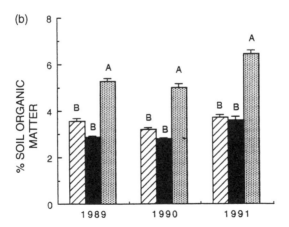

Figure 7.6. Mean (a) soil nitrate and (b) organic matter content for former long-term treatments during 1989, 1990 and 1991. Bars ($\pm$ SE) within a year with letters in common are not significantly different ($p > 0.05$)

Nitrate levels in control and fertiliser plots have changed little during three years of recovery, whereas nitrate levels in sludge-treated plots increased between 1989 and 1991 (Figure 7.6(a)). Because organic matter (i.e. plant biomass) was not harvested or removed from any treatment, this suggests that organic sludge is still being decomposed and releasing nitrates into the soil. This assumption is also supported by the high levels of organic matter found in sludge-treated soils compared to fertiliser-treated or control soils (Figure 7.6(b)). In 1989, 1990 and 1991, former sludge-treated soils contained significantly more organic matter than former fertiliser-treated soils; fertiliser plots contained more organics than control plots. Organic matter, however, increased in sludge plots during 1991 compared to 1989 and 1990 values. It is likely that this increase is related to differences in plant productivity, rates of litter decomposition, and amounts of organic matter applied in sludge-treated soils. Low bacterial densities in sludge plots may have contributed to the high organic matter levels found in sludge-treated soils in 1991. Low soil pH values and metal toxicity in sludge plots may decrease microbial activity, in turn, lowering rates of organic matter decomposition.

7.5.3 PLANT COMMUNITY RESPONSE

7.5.3.1 Community richness

Levels of soil nutrients, pH and organic matter not only affected bacteria and rates of decomposition, but also appear to have affected the plant community structure (richness and composition) between treatments. For example, control plots consistently had significantly greater plant species richness than former sludge or fertiliser plots throughout the growing season. Also, plant species richness gradually increased in each former long-term treatment during the three years of recovery. The interaction of both disturbing the seed bank, as well as increasing soil pH, accelerated the emergence of additional plant species, thus possibly accelerating succession. The sudden increase in species richness of control subplots after 1988 is a result of tilling, thereby increasing species richness in control tilled subplots (Figure 7.2).

7.5.3.2 Community productivity and composition

Annual above-ground production (ANPP; g $0.25\ m^{-2}\ year^{-1}$) was greatest in fertiliser plots, followed by sludge plots and then control plots in 1989, 1990 and 1991 (Figure 7.3). No consistent differences were observed between subplot manipulations in fertiliser or sludge treatments during 1989 or 1990. In 1991, however, when pretreatment (i.e. control) soil pH values had been achieved in former nutrient-enriched plots, limed and interaction subplots had the greatest ANPP. Liming and both tilling and liming may have increased ANPP because soil pH reached levels either optimum or competitive for additional plants species (e.g. *Erigeron canadensis* and *Solidago canadensis*) not previously found in these treatments.

During 1989 and 1990, control plots were dominated primarily by the perennial *Solidago canadensis*, the biennial *Daucus carota*, and the annual *Setaria faberii* (Table 7.1). *Setaria faberii*, however, was replaced by dominant perennial species during 1991. The presence of summer annuals is characteristic of disturbed communities and may persist for numerous years before being replaced by perennials (Hartnett *et al.*, 1987). However, *Setaria faberii* was replaced by dominant perennials such as *Solidago canadensis* in tilled control subplots three years following disturbance (Table 7.1).

Setaria faberii remained a dominant species in former nutrient-enriched treatments. The dominant plant species in former fertiliser plots during 1989, 1990 and 1991 was clearly *Ambrosia trifida*, especially in limed subplots. *Ambrosia artemisiifolia* was the dominant species in former sludge-treated plots during 1990, with *Chenopodium alba*, *Polygonum pennsylvanicum* and *Polygonum persicaria* also abundant in 1990 and 1991. *Chenopodium alba* and *Polygonum pennsylvanicum* in tilled and limed subplots each had a significantly greater ANPP than other subplot manipulations in 1990 (Table 7.1).

Control plots remained dominated by the perennials *Solidago canadensis* and *Rubus frondosus*, as well as the biennial *Daucus carota*, in 1990 and 1991. Although an attempt was made each winter to hand remove *Rubus frondosus* from all experimental treatments, it remained a dominant species in long-term control plots through 1991. It is believed that *Rubus frondosus* was introduced by birds perching on plot markers and walls. This rapid emergence of *R. frondosus* would probably not occur under natural conditions. In 1991, the perennial *R. frondosus* also became abundant in fertiliser and sludge plots. Fertiliser plots, however, remained dominated by the annuals *Ambrosia trifida* and *Setaria faberii*, and the perennial *Solidago canadensis* (Table 7.1).

No differences in ANPP were found among subplot manipulations within control or fertiliser plots in 1991. However, limed subplots within former sludge-treated plots were dominated largely by *Ambrosia trifida*. Although this species was present in all sludge-treated subplots, it accounted for significantly greater ANPP in limed subplots than other subplot manipulations. Sludge-treated subplots were dominated by the annuals *Chenopodium alba*, *Setaria faberii* and *Polygonum pennsylvanicum*, the perennials *Solidago canadensis*, *Poa compressa* and *Festuca elatior*, and the biennial *Daucus carota*.

Although biennials and perennials became important in former nutrient-enriched plots during 1991, and *Ambrosia artemisiifolia* all but disappeared, these plots remained dominated by annuals (i.e. plant species characteristic of recently disturbed or early successional old-fields). Interestingly, *Ambrosia trifida* was found to be increasing in sludge plots, rather than decreasing as earlier successional models predicted (Odum, 1969). This increased ANPP was attributed to increased nitrogen levels in sludge-treated soils, supporting Tilman's resource-ratio hypothesis (Tilman, 1985).

Table 7.1. Annual above-ground net primary productivity (g m^{-2} year^{-1}) for dominant plant species found in former fertiliser, sludge and control treatments during 1989, 1990 and 1991. Letters in common within a year indicate no significant difference ($p > 0.05$) between treatments

Plant species	1989			1990			1991		
	Fertiliser	Control	Sludge	Fertiliser	Control	Sludge	Fertiliser	Control	Sludge
Ambrosia artemisiifolia	48[a]	23[a]	193[b]	4[a]	1[a]	364[b]	5[a]	7[a]	5[a]
Ambrosia trifida	490[b]	0[a]	3[a]	1017[c]	0[a]	76[b]	753[c]	0[a]	62[b]
Chenopodium alba	4[a]	0[a]	1[a]	2[a]	1[a]	65[b]	2[a]	0[a]	80[b]
Daucus carota	0[a]	0[a]	0[a]	0[a]	86[c]	18[b]	0[a]	159[c]	13[b]
Festuca elatior	0[a]	5[b]	1[a]	0[a]	20[b]	3[a]	0[a]	8[b]	44[c]
Poa compressa	0[a]	23[b]	0[a]	12[a]	41[c]	17[b]	4[a]	42[b]	30[b]
Polygonum spp.	0[a]	48[b]	50[b]	8[a]	37[b]	140[c]	20[b]	0[a]	192[c]
Rubus frondosus	0[a]	4[a]	0[a]	8[a]	49[b]	9[a]	30[a]	74[b]	23[a]
Setaria faberii	188[b]	98[a]	66[b]	113[a]	142[b]	297[c]	37[a]	17[a]	269[b]
Solidago canadensis	0[a]	60[b]	4[a]	8[a]	159[c]	27[b]	38[a]	211[b]	53[a]

Therefore, the r-selected, opportunistic species, such as the annuals *Ambrosia trifida* and *Setaria faberii*, remained dominant in these high-nutrient environments. Clearly, three years of recovery was not sufficient for plant composition or community development to reach control values (i.e. these plots remain representative of an early seral stage of secondary succession).

7.5.3.3 Metal uptake during old-field recovery

Between 1989 and 1991, soil and plant samples were collected for analysis of heavy metal (Cd, Cu, Pb and Zn) concentrations. Metals were found in the highest concentration in soil collected from former sludge-treated plots, compared to control or fertiliser plots, during all three years. Subplot manipulations did not affect metal concentrations in the soil.

Three plant species of differing life histories were chosen in 1989 for examining heavy metal uptake. A summer annual (monocot), *Setaria faberii*, a summer annual (dicot), *Ambrosia artemisiifolia*, and a perennial (dicot), *Solidago canadensis*, were chosen for detailed analysis. *Ambrosia artemisiifolia*, during 1991, could not be found in sufficient numbers in any treatment and was, therefore, replaced by the annual dicot, *Ambrosia trifida*.

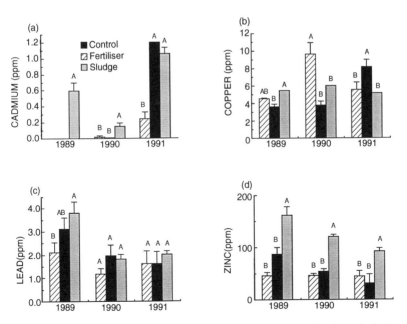

Figure 7.7. Mean concentrations of (a) cadmium, (b) copper, (c) lead, and (d) zinc in *Setaria faberii* collected from former sludge, fertiliser and control treatments. Bars (± SE) within a year with letters in common are not significantly different ($p > 0.05$)

All four metals were found in the highest concentrations in *S. faberii* collected from former sludge plots in 1989. Interestingly, Pb and Cu concentrations in control and fertiliser samples, respectively, increased above sludge concentrations in 1990. *Setaria* collected from sludge plots in 1991 contained the highest levels of lead and zinc (Figure 7.7). Levels of heavy metals in *Setaria* collected from sludge plots tended to decrease with time, with the exception of cadmium, which rose dramatically in 1991.

Solidago canadensis, goldenrod, collected from sludge plots in 1989 contained the highest concentrations of all four metals. However, levels of all four metals decreased in *Solidago* collected from sludge plots in 1990 and 1991 (Figure 7.8). In 1991, goldenrod collected from sludge plots were found to contain lower levels of each metal, whereas levels from control and fertiliser plots varied (Figure 7.8).

Ambrosia artemisiifolia, common ragweed, collected from sludge plots also contained the greatest concentrations of Cd and Zn, whereas Cu and Pb concentrations were greatest in plants collected from control and fertiliser plots, respectively. Levels of each metal in plants from sludge plots decreased continuously over the two years of recovery (Figure 7.9). However, levels of Cu and Zn in common ragweed collected from fertiliser plots increased over time. Levels of Pb and Cd in common ragweed collected from control plots

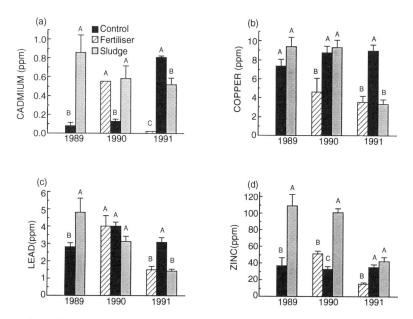

Figure 7.8. Mean concentrations of (a), cadmium, (b) copper, (c) lead, and (d) zinc in *Solidago canadensis* collected from former sludge, fertiliser and control treatments. Bars (± SE) within a year with letters in common are not significantly different ($p > 0.05$)

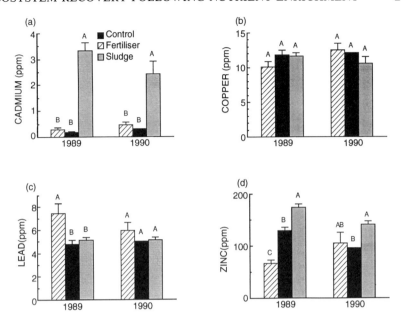

Figure 7.9. Mean concentrations of (a) cadmium, (b) copper, (c) lead, and (d) zinc in *Ambrosia artemisiifolia* collected from former sludge, fertiliser, and control treatments. This species, not present in sufficient numbers in 1991, was replaced with *Ambrosia trifida*. Bars ($\pm$ SE) within a year with letters in common are not significantly different ($p > 0.05$)

remained steady over time, whereas zinc and copper concentrations decreased (Figure 7.9).

Ambrosia trifida, giant ragweed, collected from sludge plots, contained the highest levels of Cd, Cu and Zn. This species, however, had the highest concentrations of lead in plants collected from control plots (Figure 7.10). Levels of lead in *A. trifida* collected from all treatments, however, were low due to its unextractable nature at low pH values (Kiekens, 1984). Giant ragweed collected from fertiliser plots accumulated the lowest levels of Cd, Cu and Zn.

No significant trends in heavy metal uptake in plants were attributed to subplot manipulations during 1989 or 1990. Plants collected from sludge-treated tilled and unmanipulated subplots, however, accumulated significantly higher levels of Cd, compared to limed and sludge-treated interaction subplots (Figure 7.10). This supports the hypothesis that liming reduced the availability of metals for uptake.

Based on plant life histories examined, the fast-growing annual dicots, *Ambrosia trifida* and *A. artemisiifolia*, tended to accumulate higher levels of all four metals compared to the annual monocot, *Setaria faberii*, or the

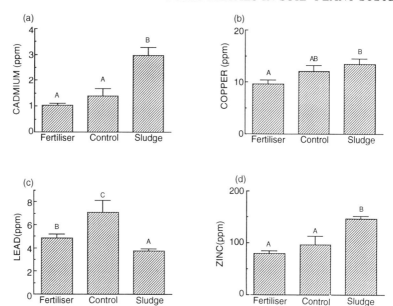

Figure 7.10. Mean concentrations of (a) cadmium, (b) copper, (c) lead, and (d) zinc in *Ambrosia trifida* collected from former sludge, fertiliser, and control treatments. Data expressed as mean ± SE per treatment

perennial dicot, *Solidago canadensis*. A soil–plant barrier to heavy metals, however, may be associated with *S. faberii* and *S. canadensis*. In addition, only above-ground plant structures were analysed for heavy metals. Future studies will focus on the various plant components, including root systems, involved in metal storage.

7.6   CONCLUSIONS

Old-field communities appear to provide a safer means of disposal for sewage sludge than agricultural fields or forest systems. It is obvious, however, that even this mode of disposal has profound effects on growth, species composition, and community development of this old-field system. Rates of secondary succession in fields treated with fertiliser or sludge were affected by long-term nutrient enrichment. Nutrient-enriched fields remained in early seral stages of succession for several years following enrichment. This was attributed to the increased abundance of essential nutrients (especially nitrogen) in the soil. It appears that increased competition for these resources favours species that are superior competitors at that location along a resource-ratio nutrient gradient (Tilman, 1985). This hypothesis best describes trends observed at this study site. High levels of nitrogen in the soil were found in plots dominated by fast-growing annuals. These plots remained dominated by annuals, and only after additional tilling and the application of lime, did species richness begin to increase. Although perennials and biennials have been found to invade these former nutrient-enriched plots, it is important to remember that these plots were still largely dominated by the annuals which have been present for 14 years.

Although heavy metal uptake into the primary producer trophic level has decreased since sludge application ceased, metals are still present at significant levels. Cadmium appears to be the most important heavy metal in this community. Copper and lead are present at low concentrations and are not accumulated to high concentrations in plants. Although zinc accumulated to high levels in plants, being an essential element, zinc is not considered to be as toxic to producers and consumers as cadmium.

Raising the soil pH may have contributed to the decrease in metals found in plants collected from these sites, yet many complicating factors (e.g. soil organic matter, soil–plant barriers and cation exchange capacity) need to be addressed. More information is also necessary concerning the microbial population present in the soil and their response between treatments. The relationship of bacteria and fungi to rates of metal uptake could be profound, and may be altered by changes in soil pH resulting from long-term nutrient enrichment or short-term manipulations. While applying calcium in the form of lime may precipitate metals out of solution and make them unavailable for plant uptake, it is not known how long the effects of liming will persist, or how lime application will alter ecosystem processes on a long-term basis.

Clearly, more information is needed to understand long-term effects of sewage sludge application on the structure, function and recovery of old-field communities. Implications of such studies for resource management policy may be critical. The need for long-term research in the field of ecology, particularly applied ecology and ecological toxicology, is vital to understand fully how a system functions following chronic long-term perturbations.

ACKNOWLEDGEMENTS

The authors thank the staff of the Miami University Ecology Research Center for their assistance in the field and laboratory. This study was supported by EPA Grant R-8105033-01-0 and NSF Grant BSR-8818086, awarded to G. W. Barrett and D. H. Taylor.

REFERENCES

Anderson, T. J. and Barrett, G. W. (1982) Effects of dried sewage sludge on meadow vole (*Microtus pennsylvanicus*) populations in two grassland communities. *Journal of Applied Ecology*, **19**, 759–772.

Bakelaar, R. G. and Odum, E. P. (1978) Community and population level responses to fertilization in an old-field ecosystem. *Ecology*, **59**, 660–665.

Barrett, G. W. (1987) Applied ecology at Miami University: an integrative approach. *Bulletin of the Ecological Society of America*, **68**, 154–155.

Bollinger, E. K., Harper, S. J. and Barrett, G. W. (1991) Effects of seasonal drought on old-field plant communities. *The American Midland Naturalist*, **125**, 114–125.

Brueske, C. C. and Barrett, G. W. (1991) Dietary heavy metal uptake by the Least shrew, *Cryptotis parva*. *Bulletin of Environmental Contamination and Toxicology*, **47**, 845–849.

Burton, T. M. and Hook, J. E. (1979) A mass balance study of the application of municipal waste water to forests in Michigan. *Journal of Environmental Quality*, **4**, 267–273.

Carson, W. P. and Barrett, G. W. (1988) Succession in old-field plant communities: Effects of contrasting types of nutrient enrichment. *Ecology*, **69**, 984–994.

Chaney, R. L. (1980) Health risks associated with toxic metals in municipal sludge. In: Britton, G. (Ed.) *Sludge: Health Risks of Land Application*, pp. 58–83. Ann Arbor Science Publications, Ann Arbor, Michigan, USA.

Chang, A. C., Hinesly, T. D., Bates, T. E., Doner, H. E., Dowdy, R. H and Ryan J. A. (1987) Effects of long-term sludge application on accumulation of trace elements in crops. In: Page, A. L., Logan, T. J. and Ryan, J. A. (Eds) *Land Application of Sludge: Food Chain Implications*, pp. 53–66. Lewis Publishers, Chelsea, Michigan.

Culliney, T. W. and Pimentel, D. (1986) Effects of chemically contaminated sewage sludge on an aphid population. *Ecology*, **67**, 1665–1669.

Hartnett, D. C., Hartnett, B. B. and Bazzaz, F. A. (1987) Persistence of *Ambrosia trifida* populations in old field and responses to successional changes. *American Journal of Botany*, **74**, 1239–1248.

Hyder, M. B. and Barrett, G. W. (1986) Effects of nutrient enrichment on the producer trophic level of a six-year old-field community. *Ohio Journal of Science*, **86**, 10–14.

Kabata-Pendias, A. and Pendias, H. (1984) *Trace Elements in Soil and Plants*. CRC Press, Boca Raton, Florida.

Kelling, K. A., Keeney, D. R., Walsh, L. M. and Ryan, J. A. (1977) A field study of the agricultural use of sewage sludge. III. Effects on uptake and extractability of sludge-borne metals. *Journal of Environmental Quality*, **6**, 352–358.

Kiekens, L. (1984) Behaviour of heavy metals in soils. In: Berlund, S., Davis, R. D. and L'Hermite, P. (Eds) *Utilization of Sewage Sludge on Land: Rates of Application and Long-term Effects of Metals*, pp. 126–134. D. Reidel Publishing, Dordrecht.

Levine, M. B., Hall, A. T., Barrett, G. W. and Taylor, D. H. (1989) Heavy metal concentrations during ten years of sludge treatment to an old-field community. *Journal of Environmental Quality*, **18**, 411–418.

Maly, M. S. and Barrett, G. W. (1984) Effects of two types of nutrient enrichment on the structure and function of contrasting old-field communities. *American Midland Naturalist*, **111**, 342–357.

Odum, E. P. (1969) The strategy of ecosystem development. *Science*, **164**, 262–270.

Ohio Department of Natural Resources (1978) *An Inventory of Ohio Soils: Butler County*. Progress Report 53. Ohio Department of Natural Resources, Division of Lands and Soils, Columbus, Ohio.

Ryan, J. A., Keeney, D. R. and Walsh, L. M. (1973) Nitrogen transformations and availability of an anaerobically digested sewage sludge in soil. *Journal of Environmental Quality*, **2**, 489–492.

Shannon, C. E. and Weaver, W. (1964) *The Mathematical Theory of Communication*. University of Illinois Press, Urbana, Illinois.

Soon, Y. K., Bates, T. E. and Moyer, J. R. (1980) Land application of chemically treated sewage sludge. III. Effects on soil and plant heavy metal content. *Journal of Environmental Quality*, **9**, 497–504.

Sutton, S. D., Barrett, G. W. and Taylor, D. H. (1991) Microbial metabolic activities in soils of old-field communities following eleven years of nutrient enrichment. *Environmental Pollution*, **72**, 1–10.

Tilman, D. (1985) The resource-ratio hypothesis of plant succession. *American Midland Naturalist*, **125**, 827–852.

8 Effects of Metal Pollutants on Decomposition Processes in Terrestrial Ecosystems with Special Reference to Fungivorous Soil Arthropods

STEPHEN P. HOPKIN
University of Reading, UK

ABSTRACT

Environmental quality standards for soils are often derived from laboratory-based experiments that disregard interactions between species. These interactions make effects of pollutants in the field difficult to predict from single species tests. A good example is the complexity of invertebrate–microbial relationships. One of the best-studied interactions is that between fungal hyphae and fungivorous springtails (Insecta: Collembola). Collembola are "primitive" wingless micro-arthropods that are important components of most terrestrial decomposer communities. This review concentrates on the relationships between fungal hyphae and Collembola. It aims to show how knowledge of the interactions between species helps us to understand the effects of metal pollutants on individuals, populations, and ecosystem processes such as decomposition of plant remains. The review concludes with a discussion of the ecological implications for invertebrate–microbial relationships of setting "95% protection levels" for metals in soils.

8.1 INTRODUCTION

The routes of exposure of soil animals to metal pollutants in contaminated terrestrial ecosystems are extremely complex. On an ecosystem level, pollution of soils can arise from aerial deposition from sources such as smelting works or car exhausts, or from persistence of contamination following closure of metalliferous mines. However, the factor which in many cases determines the severity of the disruption of normal ecological processes in soil and leaf litter ecosystems, is the extent to which metals are accumulated and transferred between soil organisms (Hopkin, 1989). It is important to understand these

---

Toxic Metals in Soil–Plant Systems. Edited by S. M. Ross
© 1994 John Wiley & Sons Ltd

processes at every level of organisation from the ecosystem down to individual organelles (Figures 8.1 and 8.2). Only then will it be possible to set critical concentrations for metals that we can be confident will protect the majority of species (Van Straalen and Ernst, 1991; Van Straalen, 1993).

Understanding all aspects of metal dynamics in soil and leaf litter may seem like an impossible task. Nevertheless, much progress has been made in recent years and unifying concepts are beginning to emerge. For example, numerous studies have shown that copper and lead have an extremely high affinity for organic matter and are much less mobile than cadmium and zinc at normal to slightly acid pHs (Scokart *et al.*, 1983; Bergkvist *et al.*, 1989; see Chapter 9, this volume).

Several ecotoxicologists have described processes of metal detoxification that appear to be common across a wide range of species (Depledge and Rainbow, 1990; Rainbow *et al.*, 1990; Beeby, 1991; Dallinger, 1993). Among these is the propensity of particular groups of terrestrial invertebrates to retain metals in organs associated with the digestive system (Hopkin, 1989), and to store these contaminants as intracellular granules of three main types (Hopkin *et al.*, 1989; Hopkin, 1990*a*). However, it is more difficult to predict the effects of metal contamination on populations due to the complexity of soil

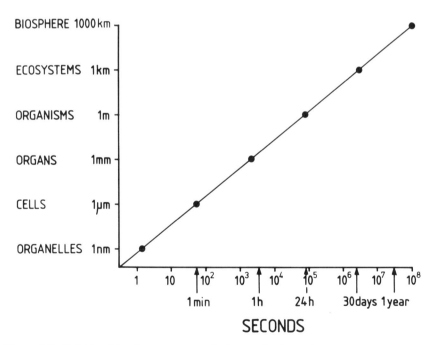

Figure 8.1. Relationship between complexity and size of natural systems and "compartments" (= "black boxes") and typical response times to metal "insults"

Zinc (μg)

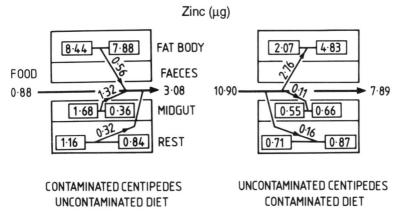

CONTAMINATED CENTIPEDES
UNCONTAMINATED DIET

UNCONTAMINATED CENTIPEDES
CONTAMINATED DIET

Figure 8.2. Net fluxes in amounts of zinc (μg) through *Lithobius variegatus*, and between the midgut, fat body (sub-cuticular tissues) and "rest" of the tissues of centipedes from a contaminated and an uncontaminated site. Centipedes were fed over a 28 day period on the hepatopancreas of six uncontaminated or six contaminated specimens of the woodlouse *Oniscus asellus* collected from the same sites (mean of seven centipedes in both cases). The value given in the left-hand box for each tissue fraction represents the amount of zinc estimated to have been present at the start of the experiment. The values in the right-hand boxes represent the amounts present at the end. Most of the zinc excreted in the faeces of contaminated centipedes was lost from the midgut (1.32 μg). Most of the net zinc assimilation by uncontaminated centipedes was accounted for by an increase in the amount in the fat body (2.76 μg). After Hopkin and Martin (1984)

and leaf litter communities. It is important to quantify these effects since absence of particular species may disrupt decomposition or other important ecological processes.

In this chapter these topics will be discussed using, as the main example, the importance of fungal hyphae as a critical pathway of exposure to metals for fungivorous soil and leaf litter arthropods. After a general discussion on invertebrate–microbial interactions during decomposition (Section 8.2), the effects of metals on microbial processes (Section 8.3) and Collembola will be described (Section 8.4). Collembola are wingless "primitive" insects that are extremely common in soil and leaf litter (Figures 8.3, 8.4). Transfer of metals to the predators of Collembola will be covered in Section 8.5, and the chapter concludes (Section 8.6) with a discussion of the ecological implications of setting critical concentrations for individual metal pollutants in soils.

The literature on metals and microorganisms is huge. Several excellent and comprehensive reviews on the subject have been published in recent years (e.g. Duxbury, 1985; Gadd, 1988, 1990; Hughes and Poole, 1989; Wilkins, 1991) but interactions between microorganisms, invertebrates and metal pollution have been less well covered. Consequently, this chapter will concentrate on

Figure 8.3. A euedaphic (soil-dwelling) collembolan (*Onychiurus armatus*) of 2 mm in length

Figure 8.4. An edaphic (litter-dwelling) collembolan (*Tomocerus longicornis*) of 5 mm in length

these interactions, especially food chain transfer and effects of metals in relation to the feeding biology of fungivorous soil arthropods.

8.2 INVERTEBRATE–MICROBIAL INTERACTIONS DURING DECOMPOSITION

Many soil and leaf litter invertebrates including isopods, millipedes, termites and mites, as well as Collembola, have evolved in intimate association with fungi and bacteria. One role of microorganisms in the association is to make nutrients in dead plant material available for assimilation by supplying enzymes such as cellulases which the animals are unable to produce themselves (Gunnarsson and Tunlid, 1986; Gunnarsson, 1987; Hopkin and Read, 1992; Hopkin, 1993*a*). A well-known example is that of leaf-cutting ants which culture fungus in "gardens" which they feed to their larvae (Holldobler and Wilson, 1990). Most early colonisers of the land in the Devonian and Carboniferous periods probably grazed fungal hyphae from the surfaces of leaf litter. The hyphae acted as an "external rumen" by degrading plant material which the digestive systems of the invertebrates were unable to break down (Price, 1988; Piearce, 1989; Little, 1990).

Fungus is a rich source of readily-available nutrients (Cromack *et al.*, 1977) and there are numerous species of Collembola which feed extensively on hyphae in soil and leaf litter (Van Straalen, 1989; Faber, 1991*a,b*; Hedlund *et al.*, 1991). *Sinella curviseta* grazes hyphae from roots and has been put forward as a potential biological control agent of the fungus that causes wilt disease of cucumber seedlings (Nakamura *et al.*, 1992). However, grazing of the vesicular–arbuscular mycorrhiza (VAM) *Glomus fasciculatus* by *Folsomia candida* leads to a reduction in growth rate of leek (*Allium porrum*) (Warnock *et al.*, 1982). The volume edited by Fitter (1985) includes several other examples of interactions between mycorrhizae and Collembola.

There is no evidence that Collembola possess a permanent symbiotic microflora like that of termites (Ponge, 1991*a*). The bacteria present in the lumen of the digestive system of Collembola are derived from their food although these microbes may proliferate in the gut and contribute enzymes to digestive processes before being lost in the faeces. Bacteria associated with food material may also be digested during gut passage.

Thus, if fungal hyphae were to accumulate metals to higher concentrations than the material on which they were growing, the potential exists for much greater transfer of pollutants to fungivores than might be predicted from analysis of pooled samples of leaf litter.

Soil invertebrates are responsible directly for only about 5–10% of the chemical decomposition of leaf litter (Petersen and Luxton, 1982). However, they act as "catalysts" by stimulating the activities of bacteria and fungi which conduct the majority of chemical decomposition (Anderson and Ineson, 1984;

Anderson 1988; Shaw *et al.*, 1991). They do this by fragmenting leaf litter into small particles which are voided as faecal pellets. These pellets provide a more favourable substrate for microbial breakdown (Eisenbeis and Wichard, 1987). Decomposition is promoted further if the faeces are deposited in deeper, moister litter layers where decomposition is faster and bacterial and fungal spores can germinate (Hassall *et al.*, 1986*a*; Hames and Hopkin, 1989; Van Wensem, 1989).

The presence of invertebrates has been shown to have a profound effect on mineralisation and subsequent fluxes of major nutrients (Griffiths *et al.*, 1989; Morgan *et al.*, 1989; Verhoef and Brussard, 1990; Faber and Verhoef, 1991). Clearly, they must also influence fluxes of metal pollutants, although as far as the author is aware, this has not been quantified in detail. Van Straalen *et al.* (1985) demonstrated that food chain transfer of lead by the collembolan *Orchesella cincta* was very small (about 11 times the population standing pool per year) in comparison to the flux of lead through consumption and defecation (about 10 000 times the standing pool per year).

One of the characteristic features of metal-contaminated ecosystems is a reduction in the rate of decomposition of dead plant material. Accumulation of leaf litter occurs in deciduous woodlands (Coughtrey *et al.*, 1979) and coniferous forests (Bengtsson *et al.*, 1988*a*) and is due primarily to a reduction in the activities of microorganisms and litter-consuming invertebrates (Strojan, 1978), not to an increase in the amount of litter falling onto the forest floor (Jackson and Watson, 1977). Bengtsson *et al.* (1988*a*) showed that the rate of leaf litter decomposition in metal-contaminated sites near to a brass mill in Sweden was only about one-tenth of the rate in uncontaminated areas. This difference was attributed to decreases in the activities of microorganisms, Collembola, enchytraeids and mites as a result of metal pollution.

8.3 METALS AND MICROORGANISMS

The effects of metal pollution on microbial communities are extremely difficult to quantify (Hughes and Poole, 1989; Tyler *et al.*, 1989). Many studies have examined species assemblages (treating the microbial community as a "black box"), studying macro effects such as reduction or stimulation of respiration, or rates of decomposition of simple substances such as sugars (Ruhling and Tyler, 1973; Tyler, 1974, 1975; Bond *et al.*, 1976; Strojan, 1978; Hattori, 1989, 1991; Wilke, 1991). Babich *et al.* (1983) proposed that the inhibition of microbe-mediated ecological processes (mineralisation of carbon, respiration and nitrification) could be used to quantify the sensitivity of natural ecosystems to pollutants. Such studies indicate the effects of metal pollutants on *processes*. However, the black box approach disguises changes in the composition of species. For example, Nordgren *et al.* (1983) showed that the numbers of colony-forming units of microfungi did not change along a gradient of metal pollution but that the composition of species changed drastically.

Some researchers have isolated species for study. Clint *et al.* (1991) measured a wide range of species-specific differences in influx rates of ^{137}Cs into fungal hyphae. However, results derived from this approach must be treated with care if they are extrapolated to the complex situation that occurs in the field. For example, Doelman and Haanstra (1979*a,b*) showed that there was an order of magnitude difference between sandy and peaty soils in the concentration of lead that significantly reduced respiration and dehydrogenase activities of microbes.

Total microbial biomass in soil may be reduced by metal pollution (Bisessar, 1982). For example, soil biomass carbon as a proportion of total carbon was reduced from its normal level of around 2%, to 1% in smelter-polluted soil in Sweden (Baath *et al.*, 1991). Species that are tolerant to metal pollution may proliferate under such circumstances (Duxbury and Bicknell, 1983). Indeed, there are several species of fungi that are tolerant to copper (Baath, 1991; Table 8.1). Metal pollution may lead to selection for resistant (i.e. genetically distinct) strains (Jordan and Lechevalier, 1975; Doelman and Haanstra, 1979*c*). Selection may be rapid and can occur by transfer of metal resistance genes between species (Top *et al.*, 1990).

The basis of resistance to cadmium, copper or zinc may be duplication of the gene that codes for the detoxifying protein metallothionein (Gadd and

Table 8.1. Concentration of copper (mg Cu l^{-1}) in media giving 50% reduction of radial growth rate (LD$_{50}$), and highest Cu concentration where growth was present after 1 month incubation. Highest concentration tested was 400 mg Cu l^{-1}. Swedish isolates from soil were used unless otherwise stated (reproduced by permission of the British Mycological Society from Baath 1991)

Fungal species	LD$_{50}$	Growth
Cordyceps militaris (L.:Fr.) Link CBS 110.70	>400	400
Paecilomyces farinosus (Holm:Fr.) A. H. Brown & G. Smith	>400	400
Verticillium lecanii (Zimm.) Viégas CBS 546.81	263	200
Beauveria bassiana (Balsamo) Vuill.	204	400
Metarhizium anisopliae (Metschn.) Sorok. CBS 459.75	195	200
Verticillium chlamydosporium Goddard	174	200
Aureobasidium pullulans (de Bary) Arnaud	160	200
Verticillium suchlasporium Gams & Dackman	105	200
Microdochium bolleyi (Sprague) de Hoog & Hermanides-Nijhof	74	100
Fusarium oxysporum Schlecht. CBS 267.50	72	100
Alternaria alternata (Fr.:Fr.) Keissler	71	200
Mucor hiemalis Wehmer CBS 201.65	<72	100
Rhizopus oryzae Went & Prinsen Geerligs	<59	100
Paecilomyces lilacinus (Thom) Samson	54	100
Paecilomyces variotii Bain.	<50	25
Trichoderma polysporum (Link:Fr.) Rifai	<41	100

White, 1989). The consequence of the duplication is that metallothioneins may be produced in larger amounts in response to a metal "insult". There may also be increased secretion of metal-binding ligands which precipitate metals extracellularly (Wood et al., 1984; Mullen et al., 1992), or a decrease in the rate of assimilation across the cell wall (Budd, 1991).

Fungal hyphae have a remarkable ability to accumulate metals to exceptionally high concentrations (Gadd, 1988, 1990). For copper, much of the accumulated metal is adsorbed onto the external surfaces of the hyphae (Gadd and White, 1989). Concentrations of more than $3 \, mg \, Cu \, g^{-1}$ have been measured in *Verticillium bulbillosum* grown on a medium containing only $150 \, \mu g \, Cu \, ml^{-1}$ (Bengtsson et al., 1983). Mitani and Misic (1991) measured levels of $16 \, mg \, Cu \, g^{-1}$ of mycelium (=1.6% of the dry weight) in *Penicillium* sp. grown on a medium containing $1 \, mg \, Cu \, ml^{-1}$; at least 65% of this copper was present on the surface. Hopkin (1993a) measured $6720 \, \mu g \, Cu \, g^{-1}$ dry weight in fungal hyphae scraped from a leaf to which copper nitrate had been applied, to give a total leaf concentration of only $107 \, \mu g \, g^{-1}$. There is strong evidence for the presence of energy-dependent transport mechanisms for other essential elements such as zinc (Starling and Ross, 1991).

Hopkin (1993a) has suggested that basidiomycete fungal hyphae may need to accumulate high concentrations of essential metals since they undergo massive growth dilution of nutrients when sporophores are produced. Non-essential elements such as cadmium, which are also accumulated to very high concentrations by hyphae (Gadd and White, 1989), may follow similar biochemical pathways to essential metals such as copper and zinc. For example, the transport system for zinc in *Penicillium notatum* is inhibited competitively by cadmium, but not by copper or other cations (Starling and Ross, 1991).

Many fungi in soil and leaf litter form mycorrhizal associations with the roots of higher plants (Newman, 1988; Brundrett, 1991). The hyphae are extremely important in enhancing nutrient uptake and, in exchange, may obtain up to 40% of the photosynthate produced by the host plants (Gehring and Whitham, 1991). In white clover (*Trifolium repens*), more than 50% of the copper assimilated by the plants was via its mycorrhizal fungus (Li et al., 1991). In two species of grass, VAM infection "protected" the plants against zinc poisoning (Dueck et al., 1986). However, in pigeon pea, VAM infection increased assimilation of zinc by the plants (Wellings et al., 1991).

In natural situations, it is difficult to quantify the potential effects of metal pollution on these associations (Wilkins, 1991). For example, in experiments on the effects of lead on six-month old loblolly pine (*Pinus taeda*) and its ectomycorrhizum *Cenococcum geophilum*, growth was greatest at the lowest and highest concentrations of lead in the soil with least growth at intermediate concentrations (Chappelka et al., 1991). Mycorrhizal fungi of pine are involved also in the decomposition of old pine needles when the fine roots penetrate among them. Indeed, there is evidence that their presence may inhibit bacterial colonisation (Ponge, 1991*b*).

Tyler (1991), in a simple but elegant experiment, examined the effects on fungi of removal of the annual litter fall in a Swedish beech forest. Removal of litter for two consecutive years increased the sporophore production of mycorrhizal *Russula* species in both years whereas sporophore production by most decomposer agarics was greatly reduced. It is possible that the lower availability of nutrients stimulated the development of ectomycorrhizae. Thus in metal-polluted sites where the rate of decomposition of leaf litter is reduced, growth of metal-tolerant fungal hyphae may be stimulated, with increased potential for food chain transfer of pollutants. However, at present, it is difficult to come to any firm conclusions as to the effects of metal pollution on fungal mycorrhizae.

In contrast, Stahl and Christensen (1992) showed that for non-mycorrhizal species, competition between different soil fungi was greatest when they were colonising resource-rich habitats in comparison to nutrient-poor media.

8.4   EXPOSURE OF FUNGAL-FEEDING COLLEMBOLA TO METALS

Collembola may reach densities of several thousand per square metre in temperate woodlands. Many species consume a diet that is mainly, if not exclusively, composed of fungal hyphae. Collembola graze fungus from the surfaces of plant roots, soil particles and leaf litter (Saur and Ponge, 1988; Faber, 1991*a*; Ponge, 1991*a*). The grazing can be quite selective in terms of species preferred (Hassall *et al.*, 1986*b*; Schultz, 1991). It can also alter the extent to which leaf litter is colonised by different species of fungi (Parkinson *et al.*, 1979; Seastedt, 1984), may inhibit or stimulate hyphal growth (Hanlon and Anderson, 1979; Hanlon, 1981; Faber, 1991*b*; Leonard and Anderson, 1991*a,b*; Faber *et al.*, 1992) and may have important implications for the success or failure of populations of particular species of Collembola in metal-contaminated sites.

The study of food preferences is important in understanding the population dynamics of Collembola. Usher *et al.* (1982) reported that the growth rate of *Folsomia candida* depended on the species of fungus on which it was fed. However, more research is needed on food preferences since it is not clear to what extent collembolan diets are dictated by availability rather than choice in soil and litter microhabitats (Gilmore and Raffensperger, 1970; Anderson and Healey, 1972; Vegter, 1983).

Isotoma olivacea was more abundant in a lead-polluted site in Norway than in adjacent uncontaminated areas because it was able to take advantage of the lack of competition from other species of Collembola which were more sensitive to the metal (Hagvar and Abrahamsen, 1990). Similar community effects were also observed near to the Gusum brass mill in Sweden (Bengtsson and Rundgren, 1988).

These phenomena may be due to species-specific differences in sensitivity to the same levels of pollution, or to a choice of diets containing different concentrations of metals. For example, *Folsomia fimetaroides* dominates *Isotomiella minor* in metal-polluted soils due to a combination of being able to avoid consuming metal-rich fungi, the prevalence of its preferred fungal diet in the polluted sites and, possibly, more efficient detoxification mechanisms (Bengtsson and Rundgren, 1988; Tranvik and Eijsackers, 1989). Indeed, total densities of Collembola may be higher in metal-polluted woodlands if the thick accumulation of leaf litter provides a refuge from predation (Hopkin *et al.*, 1985).

Van Straalen (1989) calculated that Collembola in a Dutch pine forest assimilated the equivalent in energy terms of 6% of the annual litterfall (although much of this must have been via consumption of fungal hyphae). The grazing activities of Collembola increase nitrogen mobilisation from pine litter, but this effect is often overlooked as most of the nitrogen is assimilated rapidly by ectomycorrhizal fungi on tree roots (Faber and Verhoef, 1991). Thus by implication, Collembola may be of great importance in releasing metals bound in metal-rich fungi and making these available to trees.

Two species of Collembola, *Onychiurus armatus* and *Orchesella cincta*, have been examined in the greatest detail with regard to their metal dynamics. The research has been conducted primarily by groups in Sweden and The Netherlands respectively. Earlier research by these workers was reviewed by Hopkin (1989) and Joosse and Verhoef (1987) but the summary below includes more recent publications which have shed new light on the effects of metal pollution on Collembola.

8.4.1 ONYCHIURUS ARMATUS

Onychiurus armatus (Figure 8.3) is edaphic and lives permanently in the humus layer or mineral soil where it ingests mycorrhizal and/or saprophytic fungal hyphae or spores (Faber, 1991*a*). This species has been studied extensively in relation to the effects of copper, lead and zinc pollution from a brass mill at Gusum in Sweden in a series of papers by Bengtsson, Rundgren and co-workers at the University of Lund. *Onychiurus armatus* locates fungal hyphae in soil by following concentration gradients of volatile compounds released from the mycelium (Bengtsson *et al.*, 1991). These compounds are in the range C_5 to C_{18}, and in olfactometer experiments are released at a rate of about 250 pg h^{-1} from a patch of fungus 175 mm^2 representing about 400 μg of mycelium (Bengtsson *et al.*, 1988*b*). In these experiments, the collembolan showed a distinct preference for particular fungal species, although the order of preference changed depending on whether the fungus was cultivated on agar or soil.

Verticillium bulbillosum was most attractive when fungi were cultured on agar (Bengtsson *et al.*, 1988*b*). This species of fungus contained concentrations of copper and lead of at least an order of magnitude greater than the substrate on which they were growing. Experiments on the growth of *Onychiurus armatus* fed on this diet showed that reductions could be detected at concentrations of copper and lead in the collembolans that were found in field populations near to the brass mill. However, despite a prediction of extinction at the site (Bengtsson *et al.*, 1985*a*), Collembola are able to survive at the site because the situation in the field is more complex than in the laboratory. For example, *Onychiurus armatus* can tolerate higher levels of metals if plenty of protein-rich food is supplied (Bengtsson *et al.*, 1985*b*). A further interesting observation was that the collembolans performed best when their food was slightly contaminated with metals (known as "hormesis"; Stebbing, 1982). This effect may have been due to the elimination of a metal-sensitive parasite, or in response to a copper deficiency, or due to the stimulation of fungal growth following an increase in available nutrients after lysis of metal-sensitive microorganisms.

The Swedish work has shown that fungal hyphae accumulate metals to much higher concentrations than those in the substrate. Collembola feed on the hyphae and accumulate metals. The metals inhibit growth and reproduction at tissue concentrations that can be found in the field, but these effects may be mitigated by the presence of protein-rich food and/or avoidance by the Collembola of the most contaminated fungus. The pollution can also alter the normal species composition of an area, favouring those able to tolerate the metals.

8.4.2 ORCHESELLA CINCTA

Orchesella cincta is edaphic and lives among recently fallen leaf litter where it feeds on saprophytic fungal hyphae and spores (Faber, 1991*a*). The species has been studied in the field and laboratory by Van Straalen and co-workers at the Vrije University, Amsterdam. The ease with which *Orchesella cincta* can be cultured in the laboratory has led to its nickname of the *"Drosophila"* of soil invertebrate zoologists (although *Folsomia candida* is another likely candidate when pollutants other than metals are considered). Verhoef *et al.* (1988) have shown that the preferred diet of *Orchesella cincta* is fungal mycelium, although it can be reared successfully on algae, the diet chosen by the Dutch group on which to feed their experimental animals.

Orchesella cincta assimilates 8.3% of the cadmium and 0.4% of the lead from a contaminated diet of algae, but is able to lose 30% of the assimilated cadmium and 48% of the assimilated lead at the following moult when the lining of the gut epithelium is shed (Joosse and Buker, 1979; Van Straalen and Van Meerendonk, 1987; Van Straalen *et al.*, 1987). This tissue contains metals in the form of intracellular granules (Humbert, 1978). Rates of assimilation and excretion are both affected by temperature changes (Janssen and Bergema, 1991).

Physiological tolerance to lead and cadmium in *Orchesella cincta* has been shown by breeding experiments to have a genetic basis. Offspring of adults from clean areas grow less well on a metal-contaminated diet than offspring from adults from polluted areas (Posthuma, 1990). The physiological basis of the tolerance appears to be an increased excretion efficiency (Posthuma *et al.*, 1992).

Similar problems to the Swedish group have been experienced in trying to relate the findings of laboratory experiments to the situation in the field. Populations of *Orchesella cincta* exist in forest soils which are contaminated with cadmium to levels far above the "no effect level" for individual growth (Van Straalen *et al.*, 1989). Van Straalen and De Goede (1987) have suggested that the high natural mortality from predation obscures the sublethal effects on growth. It is clear that factors such as availability and choice of food, climate, and competition from other species may have a profound effect on the success or failure of an organism to resist pollution.

8.5 TRANSFER OF METALS TO PREDATORS OF COLLEMBOLA

For Collembola, as with many groups of soil (and other) animals, there is information on the identity of predators, but little quantitative data on the biomass consumed by particular species. Laboratory experiments using "model" food chains have been adopted to analyse specific pathways (e.g. yeast to the collembolan *Folsomia candida* to the carabid beetle *Nebria brevicollis*; Gruttke *et al.*, 1988) but relating this to the field is difficult.

Van Straalen (1987) has made the important point that because some species accumulate pollutants, more metal can be transferred from a prey population to predators than would be deduced from the transfer of biomass only. Thus in *Orchesella cincta*, because there is a tenfold increase in the concentration of lead during their lifetime, overall turnover of lead is about 1.5 times the turnover of biomass. In addition, because many predators are able to regulate their internal concentrations of metals (Hopkin, 1989), bioconcentration does not necessarily take place at higher trophic levels and predators are not subject to chronic poisoning due to long-term accumulation (Van Straalen and Ernst, 1991). However, this does not mean that predators are not affected by acute poisoning if they consume a single toxic dose from a highly contaminated individual prey item (Depledge, 1990; Hopkin, 1993*a*,*b*).

A wide range of arthropods prey on Collembola in soil and leaf litter. The major ones are centipedes, pseudoscorpions, spiders, carabid beetles, and harvestmen spiders. Birds may also consume Collembola. Thus, it is clear that Collembola provide a route of exposure to metals for their predators.

However, at present we are unable to set critical concentrations in Collembola that will protect their predators from poisoning because we lack the detailed ecological information on their biology in the field.

8.6 CONCLUSIONS

It may seem like an impossible task to set critical levels of metals in soils in the light of the complexities described in the previous sections of this review. However enormous progress has been made in understanding the dynamics of metals in soils in the past ten years. The new science of "ecotoxicology" has emerged which aims to describe the effects of environmental pollutants on natural populations (Calow, 1989). It is certain that many problems that seem insoluble today will be solved by the turn of the century.

One of the most interesting developments has been the recognition that the toxicity of environmental pollutants may differ widely between species. Even within closely-related taxonomic groups such as terrestrial isopods, there are large differences in toxicity and rates of assimilation of metals between species (Van Straalen and Van Wensem, 1986; Hopkin, 1990b; Hames and Hopkin, 1991a; Van Straalen and Ernst, 1991). This may be due to selective feeding or differential rates of excretion resulting from metals being stored in different internal compartments that have different rates of turnover (Hames and Hopkin, 1991b; Figure 8.2). Furthermore some species apparently benefit from mild pollution (Bengtsson et al., 1985b).

Thus, it is an impossible task to set a critical concentration for a pollutant in soil that will protect all species to the same degree. Indeed, the author has previously argued (Hopkin, 1990a) that the value for the critical concentration is largely a political decision as to what is an "acceptable level" of environmental damage. This theme was developed originally by Van Straalen and others who have promoted the idea of setting critical concentrations of metals in soils that will protect 95% of species (Van Straalen and Denneman, 1989). This 95% target has found favour with politicians, particularly in The Netherlands.

However, the present 95% protection values have been calculated on the assumption that the relationship between the number of species (y axis) and $\log_{10}$ no observed effect concentrations (NOECs, x axis) is a normal curve (Figure 8.5). Since there have been very few experiments on NOECs, the left-hand region of the curve approaches zero. Consequently, the 95% protection levels suggested for cadmium, copper and lead by Van Straalen (1993) are extremely low (much lower than those of Bengtsson and Tranvik, 1989).

The "hazardous concentration for 5% of the species" (HC5) in soil for cadmium is only 0.2 μg Cd g^{-1} and for lead is 77 μg Pb g^{-1}. One would be hard-pressed to find soil anywhere in an industrialised country which contained levels of cadmium and lead that were below these values. Similarly,

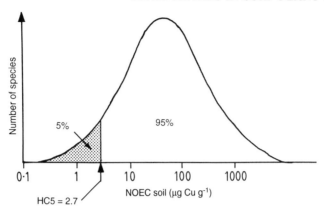

Figure 8.5. Schematic diagram of the proposed relationship between the number of species and the No Observed Effect Concentration (NOEC) of copper for soil invertebrates. Van Straalen (1993) has suggested that the hazardous concentration below which 5% of species are affected by copper poisoning (HC5) may be only 2.7 $\mu g\,g^{-1}$

the HC5 for copper is only 2.7 μg Cu g^{-1} (Figure 8.5), about half the value that defines a copper-deficient soil for agricultural purposes ($<5\ \mu g$ Cu g^{-1}).

Van Straalan's figures are sure to be revised upwards as more experiments on NOECs are conducted (indeed the values contain an uncertainty margin due to the fact that only a small number of test species were used to estimate the distribution of sensitivities). However, if they are accurate, it is interesting to make two speculations as to the implications of these extremely low HC5 values. First, at least 5% of species are being affected by low level pollution of non-essential elements such as cadmium and lead in most soil habitats in areas which have hitherto been regarded as uncontaminated. Second, where essential elements such as copper are concerned, the NOEC level in soil for *at least* 5% of species may be below that which is needed to sustain the dietary requirements of the other 95% (Van Straalen, 1993).

The idea that some species in a habitat may be dying through copper poisoning while others are suffering from copper deficiency may seem rather far-fetched (Figure 8.6). However, many researchers measure total acid-soluble copper in leaf litter and soil without taking biological availability into account. Metals in soils exist in a wide range of abiotic compartments including pore water, organic acids (Kuiters and Mulder, 1992) and adsorbed onto humus and soil particles, the extent of which depends critically on pH (see Chapter 9, this volume). Biotic compartments include fungal hyphae and other living and dead microorganisms, Protozoa, plant roots, and invertebrates and their faecal material.

As has been shown in previous sections of this review, the soil and leaf litter "black box" contains fungal hyphae in which concentrations of copper may

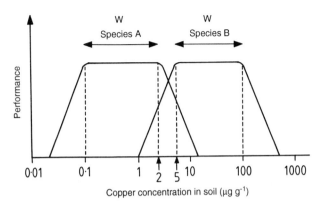

Figure 8.6. Schematic diagram of the relationships between performance (fecundity, growth, survival) and concentration of copper in soil for two species of soil invertebrate. Species A falls within the HC5 category (Figure 8.5); its performance is reduced by concentrations of copper in soil of only $2\ \mu g\, g^{-1}$. This level is insufficient to supply the minimum dietary requirement of species B ($5\ \mu g\, g^{-1}$). Note that the "windows of essentiality" (W) for the two species do not overlap

be more than an order of magnitude higher than the substrate on which they are growing. It is not inconceivable that hyphae of some fungi contain very low concentrations of copper (possibly as "nutritional defence" as has been suggested for birch-feeding insects by Haukioja *et al.* (1991)). Fungivores that have evolved to eat these hyphae may not be able to tolerate mild enrichment of their diet with copper and may be killed by concentrations that would be insufficient to supply the dietary requirements of the majority of soil invertebrates (Hopkin, 1993*c*). In terms of other trace elements, competition between fungal species for iron is well established. One species may outcompete the other because it produces siderophores that bind all freely-available iron (Weinberg,1984; Misaghi *et al.*, 1988).

Complications also arise when deciding which components of the ecosystem are most suitable in which to set critical concentrations. To date, most legislators have recommended critical levels of pollutants for abiotic components such as air, water, soil or sediment. However, setting a critical concentration for a 1 g sample of soil assumes that there is a close relationship between the level of the pollutant in soil and effects on the biota. This level is only directly relevant to an animal that consumes soil at a rate of about 1 g per day. For other species, the relationship may not be close at all (see, e.g. Jones and Hopkin, 1991; Hopkin, 1993*b*). Thus, to protect a particular species effectively, critical concentrations must be determined in the diet of that species. For fungivorous Collembola this may be the concentration in leaf litter above which the hyphae growing on its surface contain harmful levels of a pollutant.

Setting NOECs for soil animals is very difficult if other factors are to be taken into account. These include seasonal changes (Janssen and Bedaux, 1989; Janssen *et al.*, 1990, 1991; Janssen and Bergema, 1991) and the complexity of the interactions that occur between organisms in soil and leaf litter (Read *et al.*, 1987; Bengtsson *et al.*, 1988a; Verhoef and Brussard, 1990; Ponge, 1991*a*,*b*; Wilke, 1991). It is important to remember that most harmful effects may occur when the additional stress due to pollution combines with other stresses (e.g. cold or drought) that the animal would normally tolerate (i.e. the "straw that breaks the camel's back" theory of stress; see Welden and Slauson (1986), Sibly and Calow (1989) and Hopkin (1990*a*) for further definitions and discussions of stress). Furthermore the common toxicological practice of dividing the NOEC by 10 to obtain an acceptable safety margin for dietary exposure is not possible with essential elements such as copper and selenium because their "window of essentiality" is relatively small (Williams, 1981).

An alternative approach to setting critical concentrations for metals in soil would be to determine critical levels in a suite of biological indicator organisms (Samiullah, 1990). One could then establish relationships between concentrations in organisms that accumulate metals, and effects on others that do not (Hopkin, 1993*b*). It has been shown in an earlier publication that the concentrations of cadmium in the snail *Helix aspersa* can be predicted more accurately from the concentrations in the isopods *Oniscus asellus* or *Porcellio scaber* at metal-contaminated sites than from soil (Jones and Hopkin, 1991). In the future it may be possible, following cross-species laboratory experiments, to set critical concentrations for metals in *Porcellio scaber* that experiments have shown will protect 95% of other soil organisms (including Collembola) from poisoning (Hopkin *et al.*, 1993). Since isopods have a fairly cosmopolitan diet of fungal hyphae, leaf litter and other plant material (Hopkin, 1991), this approach would be more relevant than setting a critical concentration for soil that does not take into account biological availability. It would also go some way towards satisfying the desire to improve the links between laboratory and field studies (Anderson *et al.*, 1991), the "Holy Grail" of soil (and other) ecologists.

ACKNOWLEDGEMENT

The author is grateful to the Natural Environment Research Council who provided financial support for this study.

REFERENCES

Anderson, J. M. (1988) Invertebrate-mediated transport processes in soils. *Agriculture Ecosystems and Environment*, **24**, 5–19.

Anderson, J. M. and Healey, I. N. (1972) Seasonal and interspecific variations in major components of the gut contents of some woodland Collembola. *Journal of Animal Ecology*, 41, 359–368.

Anderson, J. M. and Ineson, P. (1984) Interactions between microorganisms and soil invertebrates in nutrient flux pathways of forest ecosystems. In: Anderson, J. M., Rayner, A. D. M. and Walton, D. W. H. (Eds) *Invertebrate–Microbial Interactions*, pp. 59–88. British Mycological Society Symposium Number 6, Cambridge University Press, Cambridge.

Anderson, J., Knight, D. and Elliot, P. (1991) Effects of invertebrates on soil properties and processes. In: Veeresh, G. K., Rajagopal, D. and Viraktamath, C. A. (Eds) *Advances in Management and Conservation of Soil Fauna*, pp. 473–484. Oxford & IBH, New Delhi.

Baath, E. (1991) Tolerance of copper by entomogenous fungi and the use of copper-amended media for isolation of entomogenous fungi from soil. *Mycological Research*, 95, 1140–1142.

Baath, E., Arnebrant, K. and Nordgren, A. (1991) Microbial biomass and ATP in smelter-polluted forest humus. *Bulletin of Environmental Contamination and Toxicology*, 47, 278–282.

Babich, H., Bewley, R. J. F. and Stotzky, G. (1983) Application of the "ecological dose" concept to the impact of heavy metals on some microbe-mediated ecologic processes in soil. *Archives of Environmental Contamination and Toxicology*, 12, 421–426.

Beeby, A. (1991) Toxic metal uptake and essential metal regulation in terrestrial invertebrates: a review. In: Newman, M. C. and McIntosh, A. W. (Eds) *Metal Ecotoxicology: Concepts and Applications*, pp. 65–89. Lewis, Michigan.

Bengtsson, G. and Rundgren, S. (1988) The Gusum case: a brass mill and the distribution of soil Collembola. *Canadian Journal of Zoology*, 66, 1518–1526.

Bengtsson, G. and Tranvik, L. (1989) Critical metal concentrations for forest soil invertebrates. *Water, Air, and Soil Pollution*, 47, 381–417.

Bengtsson, G., Gunnarsson, T. and Rundgren, S. (1983) Growth changes caused by metal uptake in a population of *Onychiurus armatus* (Collembola) feeding on metal polluted fungi. *Oikos*, 4, 216–225.

Bengtsson, G., Gunnarsson, T. and Rundgren, S. (1985*a*) Influence of metals on reproduction, mortality and population growth in *Onychiurus armatus* (Collembola) *Journal of Applied Ecology*, 22, 967–978.

Bengtsson, G., Ohlsson, L. and Rundgren, S. (1985*b*) Influence of fungi on growth and survival of *Onychiurus armatus* (Collembola) in a metal polluted soil. *Oecologia*, 68, 63–68.

Bengtsson, G., Berden, M. and Rundgren, S. (1988*a*) Influence of soil animals and metals on decomposition processes: a microcosm experiment. *Journal of Environmental Quality*, 17, 113–119.

Bengtsson, G., Erlandsson, A. and Rundgren, S. (1988*b*) Fungal odour attracts soil Collembola. *Soil Biology and Biochemistry*, 20, 25–30.

Bengtsson, G., Hedlund, K. and Rundgren, S. (1991) Selective odour perception in the soil collembola *Onychiurus armatus*. *Journal of Chemical Ecology*, 17, 2113–2125.

Bergvist, B., Folkeson, L. and Berggren, D. (1989) Fluxes of Cu, Zn, Pb, Cd, Cr and Ni in temperate forest ecosystems. *Water, Air, and Soil Pollution*, 47, 217–286.

Bisessar, S. (1982) Effects of heavy metals on microorganisms in soils near a secondary lead smelter. *Water, Air, and Soil Pollution*, 17, 305–308.

Bond, H., Lighthart, B., Shimabuku, R. and Russel, L. (1976) Some effects of cadmium on coniferous soil and litter microcosms. *Soil Science*, 121, 278–287.

Brundrett, M. (1991) Mycorrhizas in natural ecosystems. *Advances in Ecological Research*, **21**, 171–313.

Budd, K. (1991) A cadmium-tolerant strain of *Neocosmospora vasinfecta* shows reduced cadmium influx. *Canadian Journal of Botany*, **69**, 1296–1301.

Calow, P. (1989) Ecotoxicology? *Journal of Zoology*, **218**, 701–704.

Chappelka, A. H., Kush, J. S., Runion, G. B., Meier, S. and Kelly, W. D. (1991) Effects of soil-applied lead on seedling growth and ectomycorrhizal colonization of loblolly pine. *Environmental Pollution*, **72**, 307–316.

Clint, G. M., Dighton, J. and Rees, S. (1991) Influx of ^{137}Cs into hyphae of basidiomycete fungi. *Mycological Research*, **95**, 1047–1051.

Coughtrey, P. J., Jones, C. H., Martin, M. H. and Shales, S. W. (1979) Litter accumulation in woodlands contaminated by Pb, Zn, Cd and Cu. *Oecologia*, **39**, 51–60.

Cromack, K., Sollins, P., Todd, R. L., Crossley, D. A., Fender, W. M., Fogel, R. and Todd, A. W. (1977) Soil microorganism-arthropod interactions: fungi as major calcium and sodium sources. In: Mattson, W. J. (Ed.) *The Role of Arthropods In Forest Ecosystems*, pp. 78–84. Springer Verlag, New York.

Dallinger, R. (1993) Strategies of metal detoxification in terrestrial invertebrates. In: Dallinger, R. and Rainbow, P. S. (Eds.) *Ecotoxicology of Metals in Invertebrates*, pp. 245–289. Lewis Publishers, Chelsea, MA.

Depledge, M. H. (1990) New approaches in ecotoxicology: can inter-individual physiological variability be used as a tool to investigate pollution effects? *Ambio*, **19**, 251–252.

Depledge, M. H. and Rainbow, P. S. (1990) Models of regulation and accumulation of trace metals in marine invertebrates. *Comparative Biochemistry and Physiology*, **97C**, 1–7.

Doelman, P. and Haanstra, L. (1979*a*) Effect of lead on soil respiration and dehydrogenase activity. *Soil Biology and Biochemistry*, **11**, 475–479.

Doelman, P. and Haanstra, L. (1979*b*) Effects of lead on the decomposition of organic matter. *Soil Biology and Biochemistry*, **11**, 481–485.

Doelman, P. and Haanstra, L. (1979*c*) Effects of lead on the soil bacterial microflora. *Soil Biology and Biochemistry*, **11**, 487–491.

Dueck, T. A., Visser, P., Ernst, W. H. O. and Schat, H. (1986) Vesicular–arbuscular mycorrhizae decrease zinc toxicity in grasses growing in zinc-polluted soil. *Soil Biology and Biochemistry*, **18**, 331–333.

Duxbury, T. (1985) Ecological aspects of heavy metal responses in microorganisms. *Advances in Microbial Ecology*, **8**, 185–235.

Duxbury, T. and Bicknell, B. (1983) Metal tolerant bacterial populations from natural and metal polluted soils. *Soil Biology and Biochemistry*, **15**, 243–250.

Eisenbeis, G. and Wichard, W. (1987) *Atlas on the Biology of Soil Arthropods*. Springer Verlag, Berlin, Heidelberg.

Faber, J. H. (1991*a*) Functional classification of soil fauna: a new approach. *Oikos*, **62**, 110–117.

Faber, J. H. (1991*b*) The interaction of Collembola and mycorrhizal roots in nitrogen mobilization in a Scots pine forest soil. In: Veeresh, G. K., Rajagopal, D. and Viraktamath, C. A. (Eds) *Advances in Management and Conservation of Soil Fauna*, pp. 507–515. Oxford & IBH, New Delhi.

Faber, J. H. and Verhoef, H. A. (1991) Functional differences between closely-related soil arthropods with respect to decomposition processes in the presence or absence of pine tree roots. *Soil Biology and Biochemistry*, **23**, 15–23.

Faber, J. H., Teuben, A., Berg, M. P. and Doelman, P. (1992) Microbial biomass and activity in pine litter in the presence of *Tomocerus minor* (Insecta, Collembola). *Biology and Fertility of Soils*, **12**, 233–240.

Fitter, A. H. (1985) (Ed.) *Ecological Interactions in Soil*. Blackwell Scientific, Oxford.

Gadd, G. M. (1988) Accumulation of metals by microorganisms and algae. In: Rehm, H. J. and Reed, G. (Eds) *Biotechnology*, Vol. 6b, pp. 401–433. VCH, Weinheim.

Gadd, G. M. (1990) Heavy metal accumulation by bacteria and other microorganisms. *Experentia*, **46**, 834–840.

Gadd, G. M. and White, C. (1989) Heavy metal and radionuclide accumulation and toxicity in fungi and yeasts. In: Poole, R. K. and Gadd, G. M. (Eds) *Metal–Microbe Interactions*, pp. 19–38. IRL Press, Oxford.

Gehring, C. A. and Whitham, T. G. (1991) Herbivore-driven mycorrhizal mutualism in insect-susceptible pinyon pine. *Nature*, **353**, 556–557.

Gilmore, S. K. and Raffensperger, E. M. (1970) Foods ingested by *Tomocerus* spp. (Collembola, Entomobryidae), in relation to habitat. *Pedobiologia*, **10**, 135–140.

Griffiths, B. S., Wood, S. and Cheshire, M. V. (1989) Mineralisation of ^{14}C-labelled plant material by *Porcellio scaber* (Crustacea, Isopoda). *Pedobiologia*, **33**, 355–360.

Gruttke, H., Kratz, W., Weigmann, G. and Haque, A. (1988) Terrestrial model food chain and environmental chemicals. I. Transfer of sodium [^{14}C] pentachlorophenate between springtails and carabids. *Ecotoxicology and Environmental Safety*, **15**, 253–259.

Gunnarsson, T. (1987) Selective feeding on a maple leaf by *Oniscus asellus* (Isopoda). *Pedobiologia*, **30**, 161–165.

Gunnarsson, T. and Tunlid, A. (1986) Recycling of fecal pellets in isopods: microorganisms and nitrogen compounds as potential food for *Oniscus asellus* L. *Soil Biology and Biochemistry*, **18**, 595–600.

Hagvar, S. and Abrahamsen, G. (1990) Microarthropoda and Enchytraeidae (Oligochaeta) in naturally lead-contaminated soil: a gradient study. *Environmental Entomology*, **19**, 1263–1277.

Hames, C. A. C. and Hopkin, S. P. (1989) The structure and function of the digestive system of terrestrial isopods. *Journal of Zoology*, **217**, 599–627.

Hames, C. A. C. and Hopkin, S. P. (1991*a*) Assimilation and loss of [109]Cd and [65]Zn by the terrestrial isopods *Oniscus asellus* and *Porcellio scaber*. *Bulletin of Environmental Contamination and Toxicology*, **47**, 440–447.

Hames, C. A. C. and Hopkin, S. P. (1991*b*) A daily cycle of apocrine secretion by the B cells in the hepatopancreas of terrestrial isopods. *Canadian Journal of Zoology*, **69**, 1931–1937.

Hanlon, R. D. G. (1981) Influence of grazing by Collembola on the activity of senescent fungal colonies grown on media of different nutrient concentration. *Oikos*, **36**, 362–367.

Hanlon, R. D. G. and Anderson, J. M. (1979) The effects of Collembola grazing on microbial activity in decomposing leaf litter. *Oecologia*, **38**, 93–99.

Hassall, M., Parkinson, D. and Visser, S. (1986*a*) Effects of the collembolan *Onychiurus subtenuis* on decomposition of *Populus tremuloides* leaf litter. *Pedobiologia*, **29**, 219–225.

Hassall, M., Visser, S. and Parkinson, D. (1986*b*) Vertical migration of *Onychiurus subtenuis* (Collembola) in relation to rainfall and microbial activity. *Pedobiologia*, **29**, 175–182.

Hattori, H. (1989) Influence of cadmium on decomposition of sewage sludge and microbial activities in soils. *Soil Science and Plant Nutrition*, **35**, 289–299.

Hattori, H. (1991) Influence of cadmium on decomposition of glucose and cellulose in soil. *Soil Science and Plant Nutrition*, **37**, 39–45.

Haukioja, E., Ruohomaki, K., Suomela, J. and Vuorisalo, T. (1991) Nutritional quality as a defense against herbivores. *Forestry Ecology and Management*, **39**, 237–245.

Hedlund, K., Boddy, L. and Preston, C. M. (1991) Mycelial responses of the soil fungus, *Mortierella isabellina*, to grazing by *Onychiurus armatus* (Collembola). *Soil Biology and Biochemistry*, **23**, 361–366.

Holldobler, B. and Wilson, E. O. (1990) *The Ants*. Springer Verlag, Berlin, Heidelberg.

Hopkin, S. P. (1989) *Ecophysiology of Metals in Terrestrial Invertebrates*. Elsevier Applied Science, London and New York.

Hopkin, S. P. (1990*a*) Critical concentrations, pathways of detoxification and cellular ecotoxicology of metals in terrestrial arthropods. *Functional Ecology*, **4**, 321–327.

Hopkin, S. P. (1990*b*) Species-specific differences in the net assimilation of zinc, cadmium, lead, copper and iron by the terrestrial isopods *Oniscus asellus* and *Porcellio scaber*. *Journal of Applied Ecology*, **27**, 460–474.

Hopkin, S. P. (1991) A key to the woodlice of Britain and Ireland. *Field Studies*, **7**, 599–650.

Hopkin, S. P. (1993*a*) Deficiency and excess of copper in terrestrial isopods. In: Dallinger, R. and Rainbow, P. S. (Eds) *Ecotoxicology of Metals in Invertebrates*, pp. 359–382. Lewis Publishers, Chelsea, MA.

Hopkin, S. P. (1993*b*) *In situ* biological monitoring of pollution in terrestrial and aquatic ecosystems. In: Calow, P. (Ed.) *Handbook of Ecotoxicology*, Vol. 1, pp. 397–427. Blackwell, Oxford.

Hopkin, S. P. (1993*c*) Ecological implications of "95% protection levels" for metals in soil. *Oikos*, Vol. 66, pp. 137–141.

Hopkin, S. P. and Martin, M. H. (1984) The assimilation of zinc, cadmium, lead and copper by the centipede *Lithobius variegatus* (Chilopoda), *Journal of Applied Ecology*, **21**, 535–546.

Hopkin, S. P. and Read, H. J. (1992) *Biology of Millipedes*. Oxford University Press, Oxford.

Hopkin, S. P., Matson, K., Martin, M. H. and Mould, M. L. (1985) The assimilation of heavy metals by *Lithobius variegatus* and *Glomeris marginata* (Chilopoda: Diplopoda). *Bijdragen tot de Dierkunde*, **55**, 88–94.

Hopkin, S. P., Hames, C. A. C. and Dray, A. (1989) X-ray microanalytical mapping of the intracellular distribution of pollutant metals. *Microscopy and Analysis*, **14**, 23–27.

Hopkin, S. P., Jones, D. T. and Dietrich, D. (1993) The terrestrial isopod *Porcellio scaber* as a monitor of the bioavailability of metals: towards a global "woodlouse watch" scheme. *Science of the Total Environment* Supplement 1993, 357–365.

Hughes, M. N. and Poole, R. K. (1989) *Metals and Micro-organisms*. Chapman and Hall, London and New York.

Humbert, W. (1978) Cytochemistry and X-ray microprobe analysis of the midgut of *Tomocerus minor* Lubbock (Insecta, Collembola) with special reference to the physiological significance of the mineral concretions. *Cell and Tissue Research*, **187**, 397–416.

Jackson, D. R. and Watson, A. P. (1977) Disruption of nutrient pools and transport of heavy metals in a forested watershed near a lead smelter. *Journal of Environmental Quality*, **6**, 331–338.

Janssen, M. P. M. and Bedaux, J. J. M. (1989) Importance of body size for cadmium accumulation by forest litter arthropods. *Netherlands Journal of Zoology*, **39**, 194–207.

Janssen, M. P. M. and Bergema, W. F. (1991) The effect of temperature on cadmium kinetics and oxygen consumption in soil arthropods. *Environmental Toxicology and Chemistry*, **10**, 1493–1501.

Janssen, M. P. M., Joosse, E. N. G. and Van Straalen, N. M. (1990) Seasonal variation in concentration of cadmium in litter arthropods from a metal contaminated site. *Pedobiologia*, **34**, 257–267.

Janssen, M. P. M., Bruins, A., De Vries, T. H. and Van Straalen, N. M. (1991) Comparison of cadmium kinetics in four soil arthropod species. *Archives of Environmental Contamination and Toxicology*, **20**, 305–312.

Jones, D. T. and Hopkin, S. P. (1991) Biological monitoring of metal pollution in terrestrial ecosystems. In: Ravera, O. (Ed.) *Terrestrial and Aquatic Ecosystems: Perturbation and Recovery*, pp. 148–152. Ellis Horwood, London.

Joosse, E. N. G. and Buker, J. B. (1979) Uptake and excretion of lead by litter-dwelling Collembola. *Environmental Pollution*, **18**, 235–240.

Joosse, E. N. G. and Verhoef, H. A. (1987) Developments in ecophysiological research on soil invertebrates. *Advances in Ecological Research*, **16**, 175–248.

Jordan, M. J. and Lechevalier, M. P. (1975) Effects of zinc smelter emissions on forest floor microflora. *Canadian Journal of Microbiology*, **21**, 1855–1865.

Kuiters, A. T. and Mulder, W. (1992) Gel permeation chromatography and Cu-binding of water soluble organic substances from litter and humus layers of forest soils. *Geoderma*, **52**, 1–15.

Leonard, M. A. and Anderson, J. M. (1991*a*) Growth dynamics of Collembola (*Folsomia candida*) and a fungus (*Mucor plumbeus*) in relation to nitrogen availability in spatially simple and complex laboratory systems. *Pedobiologia*, **35**, 163–173.

Leonard, M. A. and Anderson, J. M. (1991*b*) Grazing interactions between a collembolan and fungi in a leaf litter matrix. *Pedobiologia*, **35**, 239–246.

Li, X. L., Marschener, H. and George, E. (1991) Acquisition of phosphorus and copper by VA-mycorrhizal hyphae and root-to-shoot transport in white clover. *Plant and Soil*, **136**, 49–57.

Little, C. (1990) *The Terrestrial Invasion: An Ecophysiological Approach to the Origins of the Land Animals*. Cambridge University Press, Cambridge.

Misaghi, I. J., Olsen, M. W., Cotty, P. J. and Donndelinger, C. R. (1988) Fluorescent siderophore-mediated iron deprivation—a contingent biological control mechanism. *Soil Biology and Biochemistry*, **20**, 573–574.

Mitani, T. and Misic, D. M. (1991) Copper accumulation by *Penicillium* sp. isolated from soil. *Soil Science and Plant Nutrition*, **37**, 347–349.

Morgan, C. R., Schindler, S. C. and Mitchell, M. J. (1989) The effects of feeding by *Oniscus asellus* (Isopoda) on nutrient cycling in an incubated hardwood forest soil. *Biology and Fertility of Soils*, **7**, 239–246.

Mullen, M. D., Wolf, D. C., Beveridge, T. J. and Bailey, G. W. (1992) Sorption of heavy metals by the soil fungi *Aspergillus niger* and *Mucor rouxii*. *Soil Biology and Biochemistry*, **24**, 129–135.

Nakamura, Y., Matsuzaki, I. and Itakura, J. (1992) Effect of grazing by *Sinella curviseta* (Collembola) on *Fusarium oxysporum* f.sp. *cucumerinum* causing cucumber disease. *Pedobiologia*, **36**, 168–171.

Newman, E. I. (1988) Mycorrhizal links between plants: their functioning and ecological significance. *Advances in Ecological Research*, **18**, 243–270.

Nordgren, A., Baath, E. and Soderstrom, B. (1983) Microfungi and microbial activity along a heavy metal gradient. *Applied Environmental Microbiology*, **45**, 1829–1837.

Parkinson, D., Visser, S. and Whittaker, J. B. (1979) Effects of collembolan grazing on fungal colonization of leaf litter. *Soil Biology and Biochemistry*, **11**, 529–535.

Petersen, H. and Luxton, M. (1982) A comparative analysis of soil fauna populations and their role in decomposition processes. *Oikos*, **39**, 287–388.

Piearce, T. G. (1989) Acceptability of pteridophyte litters to *Lumbricus terrestris* and *Oniscus asellus* and implications for the nature of ancient soils. *Pedobiologia*, **33**, 91–100.

Ponge, J. F. (1991*a*) Food resources and diets of soil animals in a small area of Scots pine litter. *Geoderma*, **49**, 33–62.

Ponge, J. F. (1991*b*) Succession of fungi and fauna during decomposition of needles in a small area of Scots pine litter. *Plant and Soil*, **138**, 99–113.

Posthuma, L. (1990) Genetic differentiation between populations of *Orchesella cincta* (Collembola) from heavy metal contaminated sites. *Journal of Applied Ecology*, **27**, 609–622.

Posthuma, L., Hogervorst, R. F. and Van Straalen, N. M. (1992) Adaptation to soil pollution by cadmium excretion in natural populations of *Orchesella cincta* (L.) (Collembola) *Archives of Environmental Contamination and Toxicology*, **22**, 146–156.

Price, P. W. (1988) An overview of organismal interactions in ecosystems in evolutionary and ecological time. *Agriculture, Ecosystems and Environment*, **24**, 369–377.

Rainbow, P. S., Phillips, D. J. H. and Depledge, M. H. (1990) The significance of trace metal concentrations in marine invertebrates. A need for laboratory investigation of accumulation strategies. *Marine Pollution Bulletin*, **21**, 321–324.

Read, H. J., Wheater, C. P. and Martin, M. H. (1987) Aspects of the ecology of Carabidae (Coleoptera) from woodlands polluted by heavy metals. *Environmental Pollution*, **48**, 61–76.

Ruhling, A. and Tyler, G. (1973) Heavy metal pollution and decomposition of spruce needle litter. *Oikos*, **24**, 402–417.

Samiullah, Y. (1990) *Biological Monitoring of Environmental Contaminants: Animals*. Monitoring and Assessment Research Centre, Kings College, University of London.

Saur, E. and Ponge, J. F. (1988) Alimentary studies on the Collembolan *Paratullbergia callipygos* using transmission electron microscopy. *Pedobiologia*, **31**, 355–379.

Schultz, P. A. (1991) Grazing preference of two collembolan species, *Folsomia candida* and *Proisotoma minuta*, for ectomycorrhizal fungi. *Pedobiologia*, **35**, 313–325.

Scokart, P. O., Meeus-Verdinne, K. and De Borger, R. (1983) Mobility of heavy metals in polluted soils near zinc smelters. *Water, Air, and Soil Pollution*, **20**, 451–463.

Seastedt, T. R. (1984) The role of microarthropods in decomposition and mineralization processes. *Annual Reviews of Entomology*, **29**, 25–46.

Shaw, C. H., Lundkvist, H., Moldenke, A. and Boyle, J. R. (1991) The relationships of soil fauna to long-term forest productivity in temperate and boreal ecosystems: processes and research strategies. In: Dyke, W. J. and Mees, C. A. (Eds) *Long-Term Field Trials to Assess Environmental Impacts of Harvesting*, pp. 39–77. Proceedings of the IEA/BE T6/A6 Workshop, Florida, USA, February 1990. Report Number 5. Forest Research Institute Bulletin 161, Rotorua, New Zealand.

Sibly, R. M. and Calow, P. (1989) A life-cycle theory of responses to stress. *Biological Journal of the Linnean Society*, **37**, 101–116.

Stahl, P. D. Christensen, M. (1992) *In vitro* mycelial interactions among members of a soil microfungal community. *Soil Biology and Biochemistry*, **24**, 309–316.

Starling, A. P. and Ross, I. S. (1991) Uptake of zinc by *Penicillium notatum*. *Mycological Research*, **95**, 712–714.

Stebbing, A. R. D. (1982) Hormesis—the stimulation of growth by low levels of inhibitors. *Science of the Total Environment*, **22**, 213–234.

Strojan, C. L. (1978) The impact of zinc smelter emissions on forest litter arthropods. *Oikos*, **31**, 41–46.

Top, E., Mergeay, M., Springael, D. and Verstraete, W. (1990) Gene escape model: transfer of heavy metal resistance genes from *Escherichia coli* to *Alcaligenes eutrophus* on agar plates and in soil samples. *Applied and Environmental Microbiology*, **56**, 2471–2479.

Tranvik, L. and Eijsackers, H. (1989) On the advantage of *Folsomia fimetarioides* over *Isotomiella minor* (Collembola) in a metal polluted soil. *Oecologia*, **80**, 195–200.

Tyler, G. (1974) Heavy metal pollution and soil enzymatic activity. *Plant and Soil*, **41**, 303–311.

Tyler, G. (1975) Heavy metal pollution and mineralisation of nitrogen in forest soils. *Nature*, **255**, 701–702.

Tyler, G. (1991) Effects of litter treatments on the sporophore production of beech forest macrofungi. *Mycological Research*, **95**, 1137–1139.

Tyler, G., Balsberg Pahlsson, A. M., Bengtsson, G., Baath, E. and Tranvik, L. (1989) Heavy-metal ecology of terrestrial plants, microorganisms and invertebrates. *Water Air and Soil Pollution*, **47**, 189–215.

Usher, M. B., Booth, R. G. and Sparkes, K. E. (1982) A review of progress in understanding the organization of communities of soil arthropods. *Pedobiologia*, **23**, 126–144.

Van Straalen, N. M. (1987) Turnover of accumulating substances in populations with weight structure. *Ecological Modelling*, **36**, 195–209.

Van Straalen, N. M. (1989) Production and biomass turnover in two populations of forest floor Collembola. *Netherlands Journal of Zoology*, **39**, 156–168.

Van Straalen, N. M. (1993) Soil and sediment quality criteria derived from invertebrate toxicity data. In: Dallinger, R. and Rainbow, P. S. (Eds) *Ecotoxicology of Metals in Invertebrates*, pp. 427–441. Lewis Publishers, Chelsea, MA.

Van Straalen, N. M. and De Goede, R. G. M. (1987) Productivity as a population performance index in life-cycle toxicity tests. *Water Science and Technology*, **19**, 13–20.

Van Straalen, N. M. and Denneman, C. A. J. (1989) Ecotoxicological evaluation of soil quality criteria. *Ecotoxicology and Environmental Safety*, **18**, 241–251.

Van Straalen, N. M. and Ernst, W. H. O. (1991) Metal biomagnification may endanger species in critical pathways. *Oikos*, **62**, 255–256.

Van Straalen, N. M. and Van Meerendonk, J. H. (1987) Biological half-lives of lead in *Orchesella cincta* (L.) (Collembola). *Bulletin of Environmental Contamination and Toxicology*, **38**, 213–219.

Van Straalen, N. M. and Van Wensem, J. (1986) Heavy metal content of forest litter arthropods as related to body-size and trophic level. *Environmental Pollution*, **42A**, 209–221.

Van Straalen, N. M., Burghouts, T. B. A. and Doornof, M. J. (1985) Dynamics of heavy metals in populations of Collembola in a contaminated pine forest soil. *Proceedings of the International Conference on Heavy Metals in the Environment*, Athens 1985, Vol. 1, pp. 613–615. CEP Consultants, Edinburgh.

Van Straalen, N. M., Burghouts, T. B. A., Doornhof, M. J., Groot, G. M., Janssen, M. P. M., Joosse, E. N. G., Van Meerendonk, J. H., Theeuwen, J. P. J. J., Verhoef, H. A. and Zoomer, H. R. (1987) Efficiency of lead and cadmium excretion in populations of *Orchesella cincta* (Collembola) from various contaminated forest soils. *Journal of Applied Ecology*, **24**, 953–968.

Van Straalen, N. M., Schobben, J. H. M. and De Goede, R. G. M. (1989) Population consequences of cadmium toxicity in soil microarthropods. *Ecotoxicology and Environmental Safety*, **17**, 190–204.

Van Wensem, J. (1989) A terrestrial micro-ecosystem for measuring effects of pollutants on isopod-mediated litter decomposition. *Hydrobiologia*, **188/189**, 507–516.

Vegter, J. J. (1983) Food and habitat specialization in coexisting springtails (Collembola, Entomobryidae). *Pedobiologia*, **25**, 253–262.

Verhoef, H. A. and Brussard, L. (1990) Decomposition and nitrogen mineralization in natural and agroecosystems: the contribution of soil animals. *Biogeochemistry*, **11**, 175–211.

Verhoef, H. A., Prast, J. E. and Verweij, R. A. (1988) Relative importance of fungi and algae in the diet and nitrogen nutrition of *Orchesella cincta* (L.) and *Tomocerus minor* (Lubbock) (Collembola). *Functional Ecology*, **2**, 195–201.

Warnock, A. J., Fitter, A. H. and Usher, M. B. (1982) The influence of a springtail *Folsomia candida* (Insecta, Collembola), on the mycorrhizal association of leek *Allium porrum* and the vesicular-arbuscular mycorrhizal endophyte *Glomus fasciculatus*. *New Phytologist*, **90**, 285–292.

Weinberg, E. (1984) Iron witholding: a defense against infection and neoplasia. *Physiological Reviews*, **64**, 65–102.

Welden, C. W. and Slauson, W. L. (1986) The intensity of competition versus its importance: an overlooked distinction and some implications. *Quarterly Reviews of Biology*, **61**, 23–44.

Wellings, N. P., Wearing, A. H. and Thompson, J. P. (1991) Vesicular–arbuscular mycorrhizae (VAM) improve phosphorus and zinc nutrition and growth of pigeonpea in a vertisol. *Australian Journal of Agricultural Research*, **42**, 835–845.

Wilke, B. M. (1991) Effect of single and successive additions of cadmium, nickel and zinc on carbon dioxide evolution and dehydrogenase activity in a sandy luvisol. *Biology and Fertility of Soils*, **11**, 34–37.

Wilkins, D. A. (1991) The influence of sheathing (ecto-) mycorrhizas of trees on the uptake and toxicity of metals. *Agriculture, Ecosystems and Environment*, **35**, 245–260.

Williams, R. J. P. (1981) Natural selection of the chemical elements. *Proceedings of the Royal Society of London*, **213B**, 361–397.

Wood, J. M. (1984) Microbial strategies in resistance to metal ion toxicity. In: Sigel, H. (Ed.) *Metal Ions in Biological Systems*, Vol. 18, pp. 333–351. Marcel Dekker, New York.

9 The Impact and Fate of Heavy Metals in an Oak Woodland Ecosystem

MICHAEL H. MARTIN AND RICHARD J. BULLOCK
University of Bristol, UK

ABSTRACT

The distribution of the heavy metals Pb, Zn, Cd and Cu within a deciduous woodland ecosystem contaminated by aerial deposition of these metals is described. The effects of this pollution on plants, invertebrates and decomposition are considered. Certain plant species have been shown to have evolved tolerance to heavy metals, and some invertebrate species, notably snails (Mollusca) and woodlice (Isopoda), show very high accumulation of these metals. The concentrations of heavy metals in the woodland are sufficient to cause alteration in the organic matter decomposition process which is linked to adverse affects on the detrivore invertebrate populations. Studies of the distribution of metals in soil profiles with depth and time have revealed relatively rapid mobilisation and redistribution of particularly Zn and Cd, with little effect on Cu and Pb. The extent of the mobilisation of Zn and Cd appears to be related to changes in soil pH and progressive soil acidification.

9.1 INTRODUCTION

Environmental contamination of terrestrial ecosystems by heavy metals is not a recent phenomenon (Freedman and Hutchinson, 1981); however, emissions of certain heavy metals to the atmosphere as a result of human activity now rival or exceed the scale of natural emissions (Hughes, 1981). Interference factors (IF) show that anthropogenic sources exceed the flux from natural sources (IF > 1.0) of metal emission into the atmosphere for Zn, Cd and Pb (Nriagu, 1989). Indeed, the ratio of anthropogenic to natural Pb flux is so large (IF = 18) that no place on earth is now free from lead pollution: IF values are as follows (after Nriagu, 1989):

Pb = 18	Cd = 4.8	Zn = 2.3	V = 1.7
As = 1.6	Hg = 1.4	Ni = 1.4	Sb = 1.0
Cu = 0.85	Cr = 0.71	Mo = 0.66	Se = 0.63
Mn = 0.08			

Toxic Metals in Soil–Plant Systems. Edited by S. M. Ross
© 1994 John Wiley & Sons Ltd

Table 9.1. TP values calculated as tonnes mined/mg kg^{-1} in the earth's crust (multiplied by 5×10^7 to normalise data relative to Mn)

Metal	TP value	Metal	TP value
Cd	140	Zn	10
Au	60	Cr	4
Pb	30	Ni	2
Hg	30	Fe	1
Ag	20	Mn	1
Cu	20		

Source: Forstner (1987).

Likewise the ratio of the annual metal mining activity to the mean metal concentration in the earths crust (the Technophility Index or TP; see Table 9.1), suggests that the highest degree of changes in the geochemical budget caused by man's activities has occurred for metals such as Cd and Pb (Forstner, 1987).

The atmosphere is a key medium in the dispersal of metal contaminants, and as such plays an important role in the global contamination by heavy metals (Nriagu, 1989). Smelting of base metals accounts for over 50% of the Cd, Pb, Cu and Zn released from anthropogenic sources, excluding car exhaust emissions of Pb (Nriagu and Pacyna, 1988). Thus anthropogenic metal inputs due to industrial activities are a major source of metal pollution in terrestrial ecosystems and soils (Hughes et al., 1980; Hughes 1981; Mattigod and Page, 1983). Primary production of Cu, Pb and Zn accounts for 5.52×10^3 t year^{-1} Cd, 22.75×10^3 t year^{-1} Cu, 45.5×10^3 t year^{-1} Pb and 70.73×10^3 t year^{-1} Zn in world-wide emissions (Nriagu and Pacyna, 1988). These pyrometal-lurgical processes for the production of Pb, Cu and Zn produce the largest emissions of atmospheric Cd, Cu and Zn (Pacyna, 1989) and the second largest emissions of atmospheric Pb.

9.2 THE AVONMOUTH INDUSTRIAL AREA

This chapter focuses attention on the extent and consequences of heavy metal contamination of deciduous woodland ecosystems (see Figure 9.1) in the Avonmouth area, near Bristol, UK. During the First World War, Britain found itself cut off from its primary source of zinc bullion in Belgium. After the war, strategic policy decided that a primary zinc smelter was needed in Britain. As a result the first of the Avonmouth smelters was commissioned in 1929 (Cocks and Walters, 1968). The site smelted zinc and lead, but in 1939 the increasing demand for high purity zinc resulted in a process allowing the

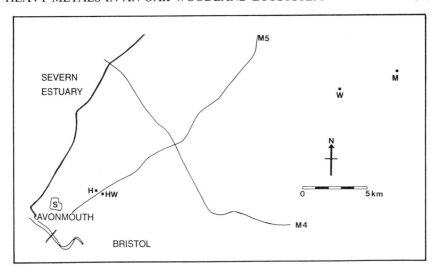

Figure 9.1. A map of the region north of Bristol and Avonmouth indicating the relative position of the smelting works and the four woodland sites mentioned in the text. S = Smelting works, H = Hallen Wood, HW = Haw Wood, W = Wetmoor Wood, M = Midger Wood, M4 and M5 are motorways

recovery of cadmium. In 1969 a new and much larger smelter design, the Imperial Smelting Furnace No. 4, began operation. This revolutionary smelting process, developed at Avonmouth and called the Imperial Smelting Process, smelted lead and zinc ores simultaneously, combining the exothermic reaction of smelting lead oxide with the endothermic reaction of smelting zinc oxide with coke as the fuel.

The production capacities of the current Avonmouth smelter are:

Cd	0.6×10^3 t year^{-1}
Pb	40×10^3 t year^{-1}
Zn	90×10^3 t year^{-1}
Sulphuric acid	200×10^3 t year^{-1}
Phosphoric acid	73×10^3 t year^{-1}

Despite efficiencies in excess of 99.9% for the containment and arrestment of particulate matter, emissions to the atmosphere amount to:

Cd	0.0035	$\times 10^3$ t year^{-1}
Pb	0.035	$\times 10^3$ t year^{-1}
Zn	0.052	$\times 10^3$ t year^{-1}
SO$_3$	3.5	$\times 10^3$ t year^{-1}

(calculated from data in Coy, 1984).

If emission : production ratios are calculated as percentages,

Cd   0.58%
Pb   0.09%
Zn   0.06%

Cd stands out amongst the three metals as potentially the most serious contaminant of the surrounding environment, bearing in mind the relative toxicities of the three metals. These emission : production data are related to the differing boiling points of the three metals (Cd, 765 °C; Pb, 910 °C; Zn, 1751 °C). According to Hutton and Symon (1986, 1987), the Avonmouth smelter accounts for an estimated 25% of UK emissions of Cd (from all sources), with the comparable figure for Pb being 0.36%; however, using data in Pacyna et al. (1991), the values would be 11.4% Cd, 2.26% Zn and 0.40% Pb.

Other industries within the Avonmouth area include fertiliser plants, nitric and sulphuric acid plants, brick works and carbon black works. In addition there are refuse disposal tips, a sewage works and Bristol's municipal incinerator. The incinerator also produces significant particulate emissions of the heavy metals Pb, Cu, Cd, Ni and Cr (Scott, 1987) from a 76 m high stack (data for Zn were not given in Scott, 1987).

9.3 THE WOODLAND STUDY SITES

Avonmouth is located on the east bank of the Severn Estuary, immediately to the north of the mouth of the River Avon. The study sites are two woodlands, called Hallen Wood (OS grid reference ST 554 802) and Haw Wood (OS grid reference ST 558 800), which lie 4.5 km NE from the centre of Avonmouth at approximately 60 m above sea level (see Figure 9.1). The two woodlands are situated 2.9 km and 3.1 km ENE of the smelting complex at Avonmouth. They are old oak–hazel (Quercus robur, Corylus avellana) woodlands with field layer species including Hyacinthoides non scripta, Milium effusum, Holcus lanatus, Mercurialis perennis and Rubus fruticosus agg. Both woodlands have been greatly disturbed but the National Vegetation Classification (Rodwell, 1991) appears to be Type W8 Fraxinus excelsior, Acer campestre, Mercurialis perennis woodland.

The two woodlands are situated on heavy clay soils derived principally from Rhaetic and Lias clays. Soil Survey maps of the area indicate that the chief soil series belong to the Denchworth and Worcester Series. The Denchworth Series clay soils contain in excess of 50% clay by weight (Findlay, 1976).

9.4 GEOGRAPHICAL SPREAD AND RELATIVE INTENSITIES OF FALLOUT

Previous work on the impact of heavy metals in the area has included detailed mapping of the geographical extent and relative intensities of heavy metal particulate deposition. Such studies have utilised biological monitoring techniques including tree leaves (Little and Martin, 1972; Bewley and Campbell, 1980), grass (Burkitt *et al.*, 1972; Bewley, 1979), *Sphagnum* moss bags (Little and Martin, 1974; Gill *et al.*, 1975; Martin and Coughtrey, 1981, 1982; Cameron and Nickless, 1977) and woodlice (Coughtrey and Martin, 1977*c*; Hopkin *et al.*, 1986). The studies have shown highest deposition of heavy metals close to the smelting works and an exponential decline in relative deposition with increasing distance from the smelter. The patterns of distribution show a close correspondence to wind direction and wind speeds integrated over the sample exposure period. In general, the annual wind direction patterns show a dominance of winds from the WSW; however, there is marked variation between prevailing wind directions when sampling periods are reduced to monthly intervals (Little and Martin, 1974). Correlation of total deposition to moss bags with wind direction during the exposure period have provided directional data pointing to the smelting complex (and perhaps the domestic incinerator) as the main source of Pb, Zn and Cd deposition at the two woodland sites (Coughtrey, 1978; Martin and Coughtrey, 1982).

9.5 EFFECTS ON THE WOODLAND

The contamination of vegetation in the area around Avonmouth is primarily through the processes of atmospheric deposition so that leaf surfaces become coated with heavy metal particles, many of which may be sub-micron in size (Beckett, 1989) and which are difficult to wash off except by the use of detergents or dilute acids (Little 1973, 1974). According to Coy (1984), the mean aerodynamic diameter of the high level stack emissions was $<2.5\ \mu$m; other data on particle size and chemical composition are given by Allen *et al.* (1974), Bewley and Campbell (1978), Harrison and Williams (1983) and Roberts (1972).

At the two woodland sites concentrations of heavy metals vary between different species and between different parts of individuals of the same species. Thus some of the highest concentrations have been observed in bryophytes, ferns and on the small woody twigs and bark of shrubs and trees (see Figure 9.2 and Martin and Coughtrey (1981, 1982) for details of Hallen Wood, and Martin *et al.* (1982) for details of Haw Wood). Similar data are available for other forest ecosystems, e.g. Tyler (1972), Denaeyer-de Smet and Duvigneaud (1974), Fangmeier and Steubing (1986), Heinrichs and Mayer (1977, 1980), Mayer (1981), Parker *et al.* (1978) and Van Hook *et al.* (1977).

332

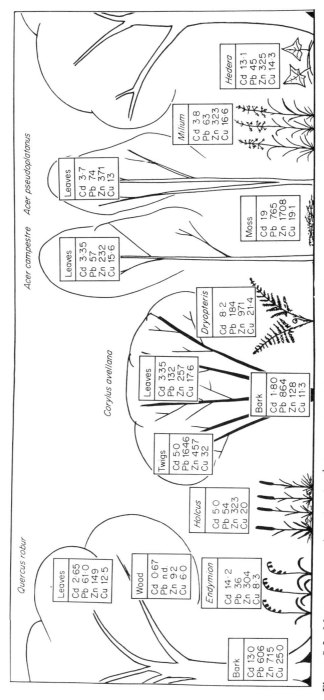

Figure 9.2. Mean concentrations (mg kg^{-1} dry weight) of heavy metals in major components of Hallen Wood. The data are means calculated from a monthly sampling programme conducted over the period 1976–1979. Reproduced by permission of Elsevier Applied Science Publishers Ltd from Martin and Coughtrey (1981)

Quercus robur

Leaves
Cd 2·65
Pb 61·0
Zn 149
Cu 12·5

Wood
Cd 0·67
Pb n.d.
Zn 92
Cu 6·0

Bark
Cd 130
Pb 606
Zn 715
Cu 250

Endymion
Cd 14·2
Pb 36
Zn 304
Cu 8·3

Holcus
Cd 5·0
Pb 54
Zn 323
Cu 20

Corylus avellana

Twigs
Cd 5·0
Pb 1646
Zn 457
Cu 32

Leaves
Cd 3·35
Pb 132
Zn 257
Cu 17·6

Bark
Cd 1·80
Pb 864
Zn 128
Cu 11·3

Dryopteris
Cd 8·2
Pb 184
Zn 971
Cu 21·4

Acer campestre

Leaves
Cd 3·35
Pb 57
Zn 232
Cu 15·6

Acer pseudoplatanus

Leaves
Cd 3·7
Pb 74
Zn 371
Cu 13

Moss
Cd 19
Pb 765
Zn 1708
Cu 19·1

Milium
Cd 3·8
Pb 63
Zn 323
Cu 16·6

Hedera
Cd 13·1
Pb 45
Zn 325
Cu 14·3

Concentration data of the type shown in Figure 9.2 present a relatively static picture of what in reality is a dynamic system. Thus, for example, concentrations of heavy metals in leaves of deciduous trees show progressive increases with age of the leaf (Martin and Coughtrey, 1982). In addition, such data give no indication of the total deposition nor of the partitioning of such deposition between the different components of the woodland. Annual inputs to Hallen Wood over the period 1976–1979 were estimated to be:

Cd 9.2 mg m^{-2} year^{-1}

Pb 285.3 mg m^{-2} year^{-1}

Zn 600 mg m^{-2} year^{-1}

Cu 26.3 mg m^{-2} year^{-1}

and Martin and Coughtrey (1981, 1982; see also Hutton, 1984) presented data showing inputs, transfers and accumulations of heavy metals. These inputs were much greater than those of an oak–hornbeam forest in southern Poland studied by Weiner and Grodzinski (1984) where values of 123.1 mg m^{-2} year^{-1} Zn, 31.5 mg m^{-2} year^{-1} Pb and 1.5 mg m^{-2} year^{-1} Cd were recorded. Similarly they were also greater than those recorded by Van Hook et al. (1977) for the relatively unpolluted forest ecosystem at Walker Branch Watershed in Tennessee (values in mg m^{-2} year^{-1}) of 53.8 Zn, 28.6 Pb and 2.1 Cd). Table 9.2 shows data for the total amounts of heavy metals in the major

Table 9.2. The amounts of the heavy metals Cd, Pb, Zn and Cu found in various components of Hallen Wood. Data are mg m^{-2} and are not corrected for background concentrations present in each component

Component	Cd	Pb	Zn	Cu
Trees				
leaves	0.456	9.04	35.449	1.94
branches	31.920	1 415.48	1 618.59	74.22
bark	21.19	987.78	1 165.45	40.75
wood	8.01	2.39	110.03	83.13
total	61.576	2 414.69	2 929.519	201.13
Shrubs				
leaves	0.186	7.32	14.246	0.98
branches	0.610	200.81	55.75	3.90
bark	0.320	154.66	22.91	2.02
wood	0.920	442.37	65.54	5.79
total	2.036	804.66	158.446	12.69
Ground flora	1.611	18.131	91.312	2.20
Organic litter	531.608	30 598	27 107	1 590
Mineral soil[a]	4 348	48 327	255 500	9 265
Total	4 944.831	82 162.481	285 786.277	11 062.02

Source: Martin and Coughtrey (1981).
[a]Calculated to a depth of 31 cm.

Table 9.3. Percentage partition of data in Table 9.2

	Cd	Pb	Zn	Cu
Trees	1.25	2.94	1.03	1.82
Shrubs	0.04	0.98	0.06	0.12
Ground flora	0.03	0.02	0.03	0.02
Litter	10.75	37.24	9.49	14.37
Soil	87.93	58.82	89.40	83.76

components of Hallen Wood, and Table 9.3 shows the percentage distribution of the total metal burden in the woodland system. Comparing data in Table 9.2 with those of Weiner and Grodzinski (1984) for an oak–hornbeam forest in Poland reveals that Hallen Wood was $17\times$ more contaminated in the above-ground biomass with Pb and Zn, $4.2\times$ with Cd and $3.3\times$ with Cu. The data in Tables 9.2 and 9.3 highlight two main features:

1. By far the largest burden of all four metals appears in the soil (if the organic litter layer and the mineral soil combined are considered, this is 98.7% Cd, 96.1% Pb, 98.9% Zn and 98.1% Cu). It should be noted that these calculations have not been corrected for the uncontaminated concentrations of the four metals.
2. For the organic components, the litter biomass contains significant amounts, whilst the living biomass holds only small amounts of the total metal burden.

9.6 EFFECTS ON SOIL LITTER LAYER

A notable feature associated with heavy metal contamination of woodlands is the accumulation of abnormally large amounts of organic matter (litter) on the soil surface (Watson, 1975; Strojan, 1978; Coughtrey et al., 1979; Wiener and Grodzinski, 1984). Coughtrey et al. (1979) showed that the standing crop of litter at Hallen Wood was nine times greater than at a similar type of woodland 23 km away from the smelter (see also Table 9.4). The high accumulations of litter were found not to be due to higher annual inputs of organic matter but to interferences in the processes of decomposition (see also Tyler, 1975a,b,c, 1984). By studying a number of woodlands at differing distances from the smelter and showing varying degrees of heavy metal contamination (see Table 9.5), evidence was obtained that litter accumulation was closely related to both cadmium and zinc concentrations in the litter. Attempts to link the lack of decomposition with effects on microbial populations were unsuccessful under experimental conditions using field-collected litter. Soil and leaf surface microbial populations showed greater tolerance to heavy metals when Pb, Zn

Table 9.4. Mean quantities of litter $(kg\,m^{-2})$ collected at three woodlands in the vicinity of the Avonmouth smelter. For relative locations of the three woodlands see Figure 9.1

Year	Wetmoor	Haw	Hallen
1977	—	—	8.34
1978	—	—	9.00
1979	0.91	7.91	11.96
1982	1.35	—	14.28
1984	—	7.90	14.15
1987	—	8.21	12.35

Table 9.5. Mean concentrations of heavy metals in different size fractions of litter $(\mu g\,g^{-1}$ dry weight)

	Size fraction (mm)	Cd	Pb	Zn
Hallen	0.5–1.0	60 ± 7.9	2908 ± 274	3320 ± 306
	1.0–2.0	60 ± 7.7	3004 ± 252	3094 ± 336
	2.0–4.0	53 ± 7.7	3052 ± 407	2555 ± 270
	4.0–8.0	50 ± 9.4	2748 ± 248	2392 ± 261
	>8.0	39 ± 6.2	2334 ± 210	2048 ± 191
Haw	0.5–1.0	112 ± 15	1547 ± 149	3511 ± 740
	1.0–2.0	104 ± 7.5	1636 ± 246	3490 ± 385
	2.0–4.0	86 ± 6.5	1615 ± 77	3384 ± 358
	4.0–8.0	59 ± 5.3	1201 ± 174	2565 ± 249
	>8.0	50 ± 7.3	1016 ± 142	1884 ± 194
Wetmoor	0.5–1.0	0.93 ± 0.22	107 ± 9.1	204 ± 5.9
	1.0–2.0	1.00 ± 0.10	84 ± 6.6	196 ± 37
	2.0–4.0	0.96 ± 0.11	64 ± 5.0	118 ± 9.4
	4.0–8.0	0.89 ± 0.06	52 ± 3.5	99 ± 5.8
	>8.0	0.77 ± 0.11	45 ± 5.5	86 ± 6.3

Source: Reproduced from Coughtry *et al.* (1979) by permission of Springer-Verlag.

and Cd were incorporated into agar culture plates (Gingell *et al.*, 1976; Martin *et al.*, 1980), but there was little evidence of any changes in abundance or diversity of microorganisms including fungi.

9.7 EFFECTS ON INVERTEBRATE FAUNA

Studies of the soil/litter fauna (Hopkin *et al.*, 1985) showed (see Table 9.6) that there were some marked differences in the taxa of invertebrates present in the contaminated compared to uncontaminated woodlands. In particular, groups

Table 9.6. The numbers of litter-inhabiting invertebrates in litter collected from Hallen Wood compared to a relatively uncontaminated woodland (Wetmoor Wood). The data are presented as numbers per m^2 and as numbers per kg of litter

	Hallen Wood		Wetmoor Wood	
	No. m^{-2}	No. kg^{-1}	No. m^{-2}	No. kg^{-1}
Isopoda				
Oniscus asellus	56	3.9	20	14.8
Trichoniscus pusillus	0	0	151	112
Diplopoda				
Polydesmidae	8	0.6	79	58.5
Julidae	0	0	11	8
Glomeridae	0	0	21	15.6
Chilopoda				
Lithobiidae	112	7.8	116	86
Geophilomorpha	328	23	263	195
Arachnida				
Acari	129 000	9 034	194 00	14 370
Aranae	248	17.4	81	60
Pseudoscorpionidae	200	14	67	50
Insecta				
Collembola	20 800	1 457	8 688	6 436
Coleoptera	902	63	48	36
Coleoptera (larvae)	120	8.4	4	3
Diptera (larvae)	4 590	321.4	291	215.6
Annelida[a]				
Lumbricus rubellus	17	1.2	3	2.2
Lumbricus terrestris	0	0	29	21.5
Aporrectodea longa	0	0	4	3
Aporrectodea caliginosa	0	0	30	22
Octoclasium cyaneum	0	0	9	6.7

Source: Data taken from, and in part recalculated from, Hopkin *et al.* (1985).
[a]Annelids were sampled using the application of dilute formalin solution to the soil surface.

such as millipedes and earthworms were greatly reduced in numbers per unit area in the polluted conditions. The comparisons are even more striking when recalculated on numbers per unit weight of litter basis. Thus woodlice (Isopoda), millipedes (Diplopoda) and earthworms (Annelida) were all less abundant in the contaminated litter/soil. These groups were also amongst the most important invertebrates responsible for the comminution and subsequent decomposition of organic matter.

Work with woodlice (Isopoda) and with snails (Mollusca) has demonstrated that very high concentrations of heavy metals are accumulated in the hepato-pancreas (Coughtrey and Martin, 1976, 1977a; Hopkin and Martin, 1982). In woodlice the hepatopancreas contains 76.3% of the zinc, 95.8% of the cadmium, 83.4% of the lead and 85.6% of the copper in the whole body, whilst the dry weight of this organ is just 7% of the total dry weight of the animal (Hopkin and Martin, 1982). The concentrations of the four metals in the hepatopancreas may reach 1% zinc, 0.5% cadmium, 2.5% lead and 3% copper on a dry weight basis without apparent adverse affects. However, if zinc concentrations in the hepatopancreas reach about 2.5% (about 2500 μg g^{-1} in the whole animal) the animals become moribund and the hepatopancreas becomes enlarged and discoloured (Hopkin and Martin, 1982, 1984a; Hopkin, 1990, 1991). Effects on other invertebrates in these situations have also been studied see (Hopkin and Martin, 1983, 1984b, 1985; Hopkin et al., 1988).

An interesting further development results from a study of the species of Carabidae in a range of woodlands with increasing distance (and decreasing degree of contamination) from Avonmouth. Read et al. (1986, 1987) showed that whilst the numbers of species and total number of individuals were not affected by change in degree of contamination, there were more subtle changes in the life cycle of at least one of the species. The effect on the life cycle appeared to be a delaying of the breeding cycle in the more contaminated sites.

In the case of millipedes, *Tachypodoiulus niger* was absent from Haw Wood but occurred in high numbers in cleaner woodlands, whereas *Glomeris marginata* was caught in largest numbers from the contaminated Haw Wood but was absent from the even more contaminated Hallen Wood. It was shown that in juvenile *T. niger* and *G. marginata* fed under laboratory conditions on leaves from Wetmoor (clean) and from Haw Wood (contaminated), there were lower rates of survival, slower growth rates and faster accumulation of metals when fed contaminated litter (Read and Martin, 1990).

Figure 9.3 shows a simplified food web for Hallen Wood. Of particular note are the high concentrations of Cd and Zn in earthworm (Annelida), woodlice (Isopoda), and snail and slug (Mollusca) species. In general the data show that, apart from the vegetation : animal and the soil : invertebrate steps in the food web, there is no other evidence of metal concentrations increasing along the food chain. Data of a similar nature, for Zn, Cd and Pb in a forested ecosystem in Poland were published by Dmowski and Karolewski (1979), and for Cd and Pb in a German forest by Fangmeier and Steubing (1986).

9.8 EFFECTS ON PLANTS

Plants that are present in the ground flora of the woodlands often have rather shallow root systems compared to the more woody species of the shrub and tree layer. Species such as *Holcus lanatus* (Yorkshire Fog) are often found rooting entirely in the deep organic litter layer present in these contaminated woodlands. Experiments in which populations of species from Hallen Wood and from an uncontaminated area have been grown in solution culture containing cadmium have shown that the Hallen population grows appreciably

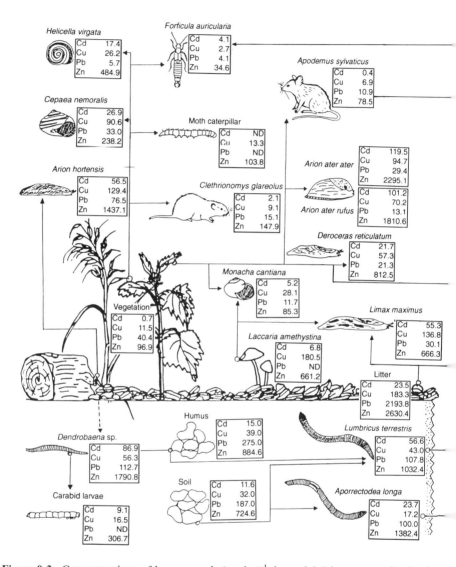

Figure 9.3. Concentrations of heavy metals (mg kg^{-1} dry weight) in a range of animal species from Hallen Wood. The data represent a collection period of October–December 1980 and each value represents the mean of a variable number of replicates; in some cases the replicates were for bulked individuals where individual animal weights were low. ND = not detected

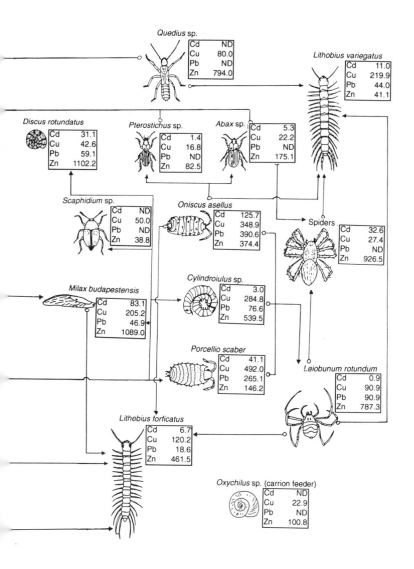

Quedius sp.

Cd	ND
Cu	80.0
Pb	ND
Zn	794.0

Lithobius variegatus

Cd	11.0
Cu	219.9
Pb	44.0
Zn	41.1

Discus rotundatus

Cd	31.1
Cu	42.6
Pb	59.1
Zn	1102.2

Pterostichus sp.

Cd	1.4
Cu	16.8
Pb	ND
Zn	82.5

Abax sp.

Cd	5.3
Cu	22.2
Pb	ND
Zn	175.1

Scaphidium sp.

Cd	ND
Cu	50.0
Pb	ND
Zn	38.8

Oniscus asellus

Cd	125.7
Cu	348.9
Pb	390.6
Zn	374.4

Spiders

Cd	32.6
Cu	27.4
Pb	ND
Zn	926.5

Cylindroiulus sp.

Cd	3.0
Cu	284.8
Pb	76.6
Zn	539.5

Milax budapestensis

Cd	83.1
Cu	205.2
Pb	46.9
Zn	1089.0

Porcellio scaber

Cd	41.1
Cu	492.0
Pb	265.1
Zn	146.2

Leiobunum rotundum

Cd	0.9
Cu	90.9
Pb	90.9
Zn	787.3

Lithobius forficatus

Cd	6.7
Cu	120.2
Pb	18.6
Zn	461.5

Oxychilus sp. (carrion feeder)

Cd	ND
Cu	22.9
Pb	ND
Zn	100.8

better (see Table 9.7). This adaptation or tolerance of the Hallen population to cadmium and to other metals has been documented, for *H. lanatus*, by Coughtrey and Martin (1977*b*, 1978*a*, 1979). Further work on this species suggests that the tolerance shown by the Hallen population to cadmium may be induced by the presence of the metal (Brown and Martin, 1981); but also subsequently lost if the same individuals are grown on in the absence of cadmium (Baker *et al.*, 1985, 1986).

One of the attributes of the tolerant plants of *H. lanatus* is that when grown on the same concentrations of cadmium in the culture solution (Coughtrey and Martin, 1978*b*), or in soil (see Table 9.8), the concentrations of cadmium in the plant shoots are lower than in non-tolerant plants grown under the same conditions. There was no evidence that this restriction of concentration in shoots of tolerant individuals was linked to decreased total uptake, but restricted translocation from roots to shoots was suggested (Coughtrey and Martin, 1978*b*).

Table 9.7. Percentage tolerance of three species of grass collected from Hallen Wood and Midger Wood, an uncontaminated woodland (see Figure 9.1 for relative locations). The tolerance was calculated from the mean length of roots in 2 ppm Cd/mean length of roots with no Cd. The measurements were made after 14 days in full strength Hoaglands culture solution

	Hallen plants	Midger plants
Dactylis glomerata	105.9	57.4
Holcus lanatus	113.8	28.1
Deschampsia cespitosa	82.2	37.5

Table 9.8. Dry weight and cadmium concentrations of shoots of *Holcus lanatus* grown in soil litter collected from Hallen Wood with additions of cadmium sulphate. Plants were grown from small tillers. Midger Wood is an uncontaminated site

Population	Treatment	Cd in soil $(\mu g\,g^{-1})$	Dry weight (g)	Cd concentration $(\mu g\,g^{-1})$
Hallen	No addition	55	0.6947	5.21
	1	108	0.7523	10.47
	2	263	0.5480	29.32
	3	390	0.6833	46.25
Midger	No addition	55	0.6753[a]	14.08[*]
	1	108	0.5170[a]	35.34[**]
	2	263	0.5087[a]	80.19[**]
	3	390	0.4470[*]	147.14[**]

Significant differences between corresponding treatments for the two populations: [a]non-significant, [*]$p < 0.05$, [**]$p < 0.01$.

9.9 EFFECTS ON SOILS

Soils and sediments on the earth's surface can act as vast reservoirs for the containment of heavy metals. Soils are particularly important in this respect as they include surface-active mineral and humic constituents which are involved in reactions that affect metal retention (Evans, 1989). It is clear from Tables 9.2 and 9.3 that the soil is the major site of heavy metal accumulation in these contaminated woodlands. Processes such as wet and dry deposition, canopy leaching, throughfall, stem flow and litterfall all result in the addition of heavy metals to the soil surface. Thus, at sites subjected to heavy metal contamination through aerial deposition and with otherwise uncontaminated soil parent material the fate of metals entering the woodland ecosystem is ultimately the soil. The extent of contamination, distribution, bioavailability, chemical speciation and mobility of the heavy metals are then legitimate areas of study.

9.9.1 SOIL PROFILES

Early studies in the 1970s at Hallen Wood recorded heavy metal concentration profiles in the soil which showed very high concentrations in the organic layers (L,F,H) and also in the top few centimetres of the mineral soil. However, the concentration of Pb, Zn and Cd fell rapidly with increasing soil depth so that sampling beyond 30 cm depth was considered unnecessary because background concentrations had been reached (see Table 9.9 for general data on background concentrations of heavy metals in soils). Later studies showed that the metal profiles were progressively changing such that a wave-like movement of Zn and Cd concentrations could be detected over successive sampling periods (see Figures 9.4–9.7). In addition, whilst the apparent movement of

Table 9.9. Concentrations (in $mg\,kg^{-1}$) of heavy metals in uncontaminated soils. The Clarke value is the average concentration of the element in the earth's crust

	Mean soil[a]	Mean soil[b]	Median soil[c]	Range soil[a]	Range soil[d]	Clarke value[e]
Cd	0.06	0.62	0.35	0.01 –0.7	0.06 –1.1	0.2
Cu	20	25.8	30	2 –100	6 –80	55
Pb	10	29.2	35	2 –200	10.4 –84	12.5
Zn	50	59.8	90	10 –300	17 –125	70

Sources:
[a]Bowen (1966)
[b]Ure and Berrow (1982)
[c]Bowen (1979)
[d]Kabata-Pendias and Pendias (1984)
[e]Plant and Raiswell (1983).

342

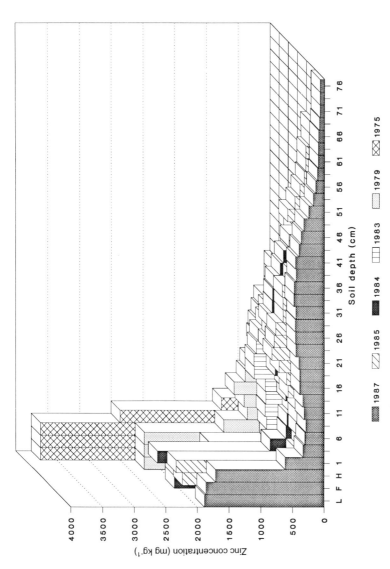

Figure 9.4. Concentrations of zinc (mg kg^{-1} dry weight) in Hallen Wood soil profiles over the period 1975–1987. Each value for the mineral soil represents analysis (with a minimum of three replicates) of a block of soil collected at depths of 0–1 cm, and then at 2.5 cm intervals to the final depth. The profiles show two main features: a reduction over time in concentration in the L, F and H organic layers, and a progressive wave of zinc moving down the profile. Data in Figures 9.4–9.6 are for "total" metal contents as extracted with boiling concentrated nitric acid; all determinations were made using atomic absorption spectroscopy in flame or flameless mode, as appropriate

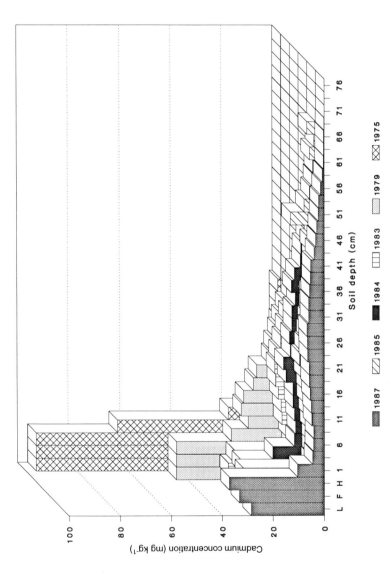

Figure 9.5. Concentrations of cadmium (mg kg^{-1} dry weight) in Hallen Wood soil profiles over the period 1975–1987. Each value for the mineral soil represents analysis of a block of soil collected at depths of 0–1 cm, and then at 2.5 cm intervals to the final depth. The profiles show two main features: a reduction over time in concentration in the L, F and H organic layers and a progressive wave of cadmium moving down the profile

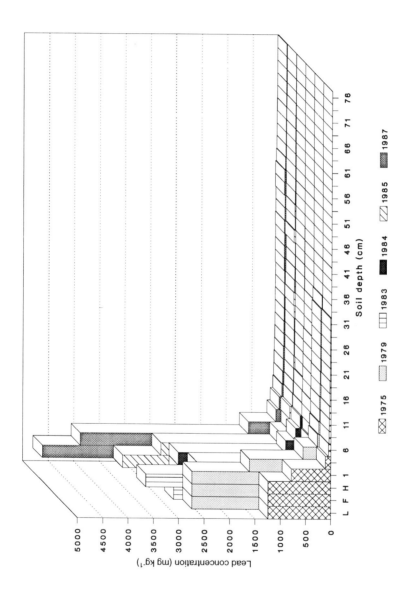

Figure 9.6. Concentrations of lead (mg kg^{-1} dry weight) in Hallen Wood soil profiles over the period 1975–1987. Note the date order is reversed compared to Figures 9.4 and 9.5 The profiles show increasing concentrations of lead in the L, F and H organic layers with time, and little or no evidence of mobility of lead in the mineral soil

345

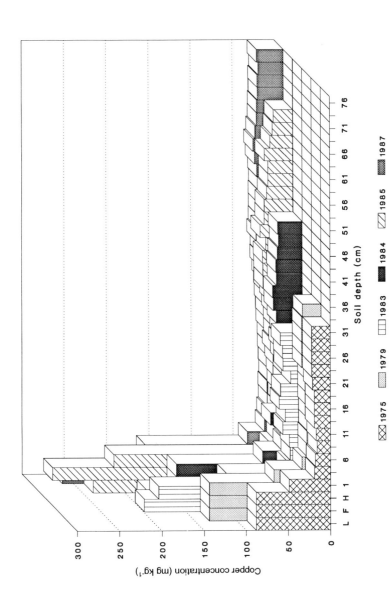

Figure 9.7. Concentrations of copper (mg kg^{-1} dry weight) in Hallen Wood soil profiles over the period 1975–1987. Note the date order is reversed compared to Figures 9.4 and 9.5. The profiles show increasing concentrations of copper in the L, F and H organic layers with time and little or no evidence of movement of copper in the mineral soil although concentrations generally show some increase in the more compact and less weathered subsoil

metals down the profile occurred, concentrations in the uppermost layers, including the organic layers, were decreasing. In contrast to Zn and Cd, little change was observed for Pb; although little vertical movement could be detected, there was an increase in the Pb concentrations in the surface layers.

Comparing the percentage of metals in each of the profile layers (0–1 cm, then 2.5 cm depths to 46 cm), relative to the total metal load in the mineral soil (0–46 cm but ignoring the organic L, F and H layers), highlights the movement of Zn and Cd and the relative immobility of Pb (see Figures 9.8–9.11). The progressive increase of Cu with increasing depth, where concentrations are slightly higher in the subsoil compared to the surface mineral horizons and an increase in soil density with increasing depth, is shown in Figure 9.11.

In the case of Cu profiles, the concentrations in the mineral soil remain remarkably constant with depth, although concentrations in the organic layers are much higher. Copper is an essential element and is a natural component of organic matter; however, the degree of Cu contamination at HallenWood is very low compared to Zn, Pb and Cd, and unlike these metals, Cu is present at concentrations in the mineral soil close to natural background levels (see Table 9.9).

The patterns of soil metal concentrations in Haw Wood, up to and including 1987, are interesting in that they compare very closely with profiles from Hallen Wood collected in the early 1970s. This suggests that the slightly less exposed position of Haw Wood has meant less rapid acidification and less severe deposition of heavy metals. The evidence collected so far (Bullock, 1992) indicates a delay in the movement of metals in the soil profiles of this wood compared to Hallen Wood.

9.9.2 THE CHEMICAL SPECIATION OF THE MOBILE PHASES

For ecological considerations, the concentration and chemical form of heavy metal species in the soil solution is of immense importance, because metal mobility and bioavailability are closely linked to the composition of the liquid phase (Brummer, 1986), usually referred to as the soil solution.

Chemical speciation of a heavy metal may be defined as the partition of the total concentration into various chemical forms; the most important factors controlling speciation of metals in the soil solution are the concentration of various ligands (inorganic and organic) and the stability of the resulting complexes (Bergkvist and Folkeson, 1988; Bergkvist et al., 1989).

Metals within the soil solution exist chemically as either free (uncomplexed) soil solution ions or as a range of complexes with both inorganic and organic ligands. In the soil as a whole the metals are present in a variety of phases and associated with a range of solid constituents:

1. Free metal cations, e.g. Pb^{2+}
2. Inorganic complexes, e.g. $CdCl^+$

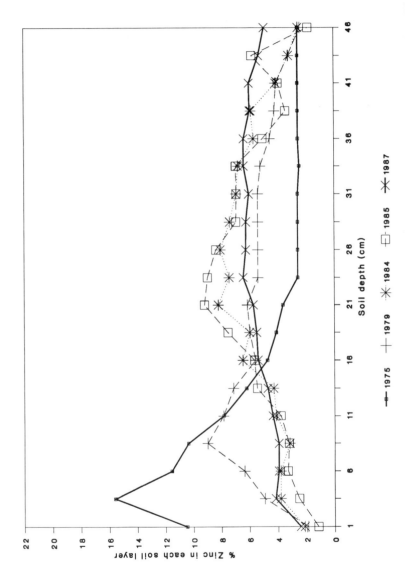

Figure 9.8. Percentage distribution of total zinc in the mineral soil of Hallen Wood in profiles from 1975 to 1987. The data were calculated from total amount of zinc in the whole profile sampled

348

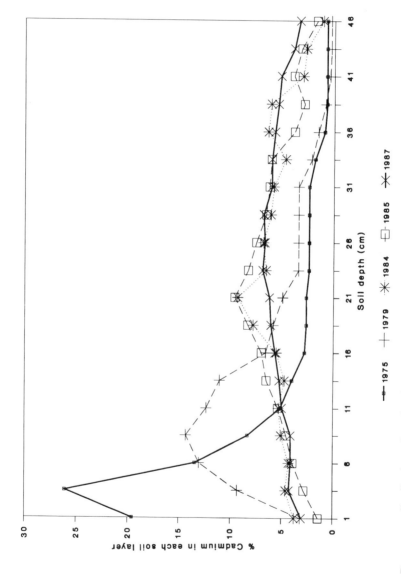

Figure 9.9. Percentage distribution of total cadmium in the mineral soil of Hallen Wood in profiles from 1975 to 1987. The data were calculated from the total amount of cadmium in the whole profile sampled

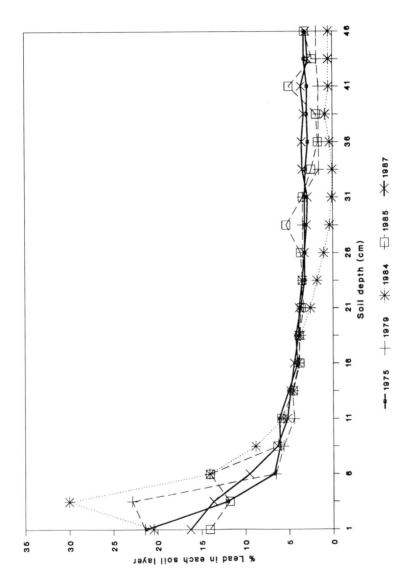

Figure 9.10. Percentage distribution of total lead in the mineral soil of Hallen Wood in profiles from 1975 to 1987. The data were calculated from the total amount of lead in the whole profile sampled

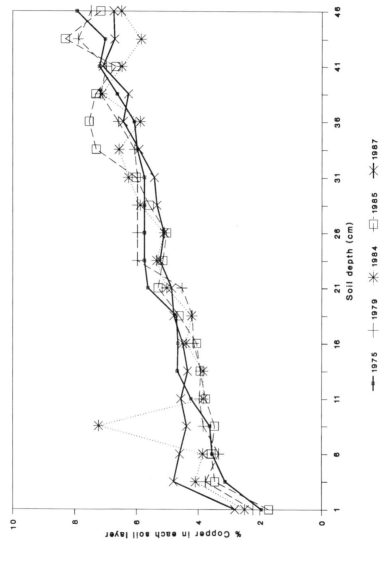

Figure 9.11. Percentage distribution of total copper in the mineral soil of Hallen Wood in profiles from 1975 to 1987. The data were calculated from the total amount of copper in the whole profile sampled

3. Organo-metal complexes, e.g. $(CH_3)4Pb$
4. Organic complexes, chelates
5. Metal species bound on high molecular weight organic materials, e.g. metal lipids
6. Metal species in the form of dispersed colloids
7. Metal adsorbed on colloids
8. Large particles of insoluble materials, e.g. silicates

In the soil solution the first four phases are normally expected to be present. These metal forms in solution will not be equally available to plants or for mobilisation within the soil profile; it is therefore important to determine the concentrations of heavy metals in the soil solution and to partition them into charged and neutral chemical species. In assessing the mobile phase of heavy metals in soils, it is important that the methodology for obtaining representative solutions from the soil are as close to the field situation as possible. Thus the use of aggressive chemical extractants will remove more than the soluble and easily exchangeable metals from the soil (see above). In this study (Bullock, 1992), it was considered that water extracts were the most appropriate, although ammonium extracts were also studied.

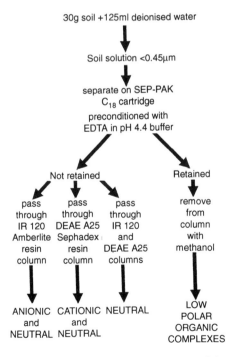

Figure 9.12. Schematic representation of the procedure used for speciation of metals in soil samples. Redrawn and modified from Tills and Alloway (1983)

Water and ammonium extracts from Hallen Wood and Haw Wood soil profiles were subjected to chemical speciation using the methodology described by Tills and Alloway (1983), with slight modifications (see Figure 9.12). By using suitable ion-exchange systems in this way, it was possible to separate the metals present in the extracts into cationic forms, anionic forms,

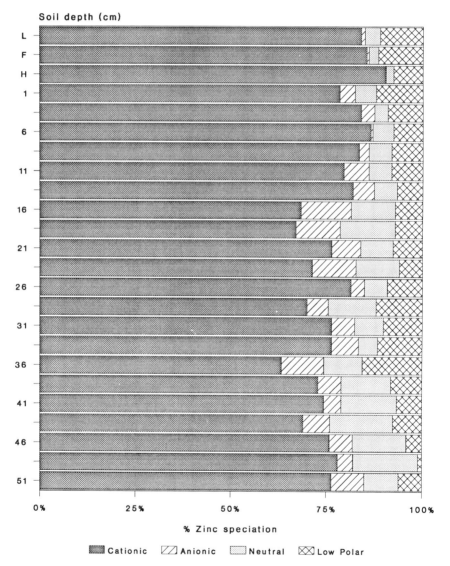

Figure 9.13. Percentage zinc speciation into cationic, anionic, neutral and low polar zinc in the 1987 Hallen Wood soil profile. The data are based on water extracts with three replicates per soil depth

neutral forms and organic forms (as low polar organics) (see also Table 3.7 on p. 91).

The results of the speciation are shown in Figures 9.13–9.16. Generally they show that there was a strong predominance of cationic forms of particularly Cd and Zn, whilst in the case of Cu and Pb there was an equal or higher

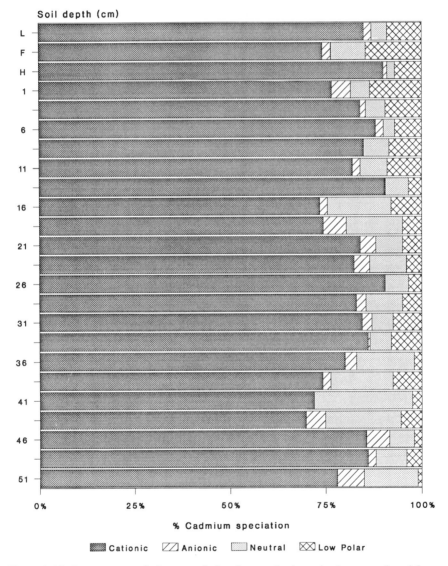

Figure 9.14. Percentage cadmium speciation into cationic, anionic, neutral and low polar cadmium in the 1987 Hallen Wood soil profile. The data are based on water extracts with three replicates per soil depth

proportion of organic complexes (anionic and low polar organics) in the extracts. Referring to Table 3.7 on page 91 it is seen that the observed results fit very closely those proposed by Sposito (1983) and Sposito and Page (1984). There are thus clearly two groupings of metals in the extracts; Cu and Pb

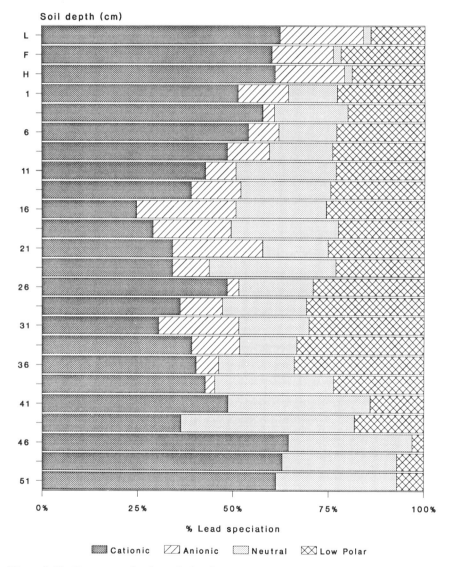

Figure 9.15. Percentage lead speciation into cationic, anionic, neutral and low polar lead in the 1987 Hallen Wood soil profile. The data are based on water extracts with three replicates per soil depth

which are associated to an appreciable degree with organic complexes, and Zn and Cd which are primarily present in cationic form. These two groupings also coincide with observations on the mobility of the four metals in the soil profile, Pb and Cu being relatively immobile and Cd and Zn highly mobile.

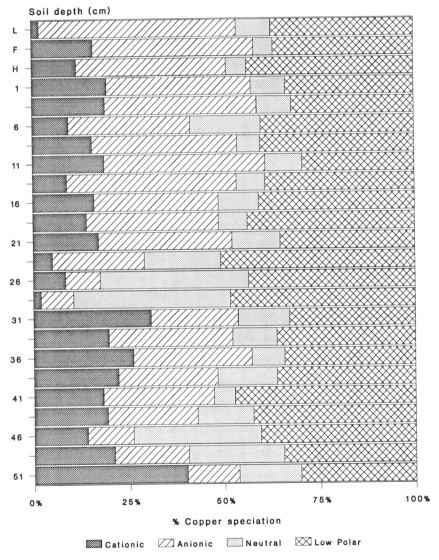

Figure 9.16. Percentage copper speciation into cationic, anionic, neutral and low polar copper in the 1987 Hallen Wood soil profile. The data are based on water extracts with three replicates per soil depth

9.9.3 EFFECT OF SOIL PH ON HEAVY METAL SOLUBILITY

In general, the solubility and availability of these four heavy metals all increase with increasing acidity and decrease with increasing alkalinity. This is a well-known effect (e.g. McKenzie, 1980; Lobersli *et al.*, 1991) and such results have been obtained by Adams and Sanders (1984), Tyler *et al.* (1987) and Brummer and Herms (1983), using a variety of soil types.

In order to demonstrate this effect in the context of Hallen Wood soil, additions of metal oxides were added to samples from a 68.5–71 cm soil depth to give total soil concentrations of Cd 78.8, Cu 128.1, Pb 102.8 and Zn 143.0 μg g^{-1}. The soils were adjusted to pH 8.0 with calcium hydroxide and left to equilibrate. A slurry of 1:40 (soil : water) was prepared and the pH progressively lowered by the addition of hydrochloric acid. Samples of the liquid phase were removed for analysis at 0.5 pH unit intervals. The results of this experiment are shown in Figure 9.17 and show large changes in the

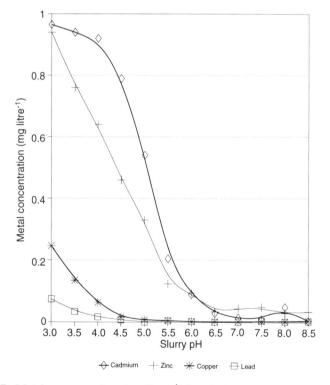

Figure 9.17. Metal concentrations (mg litre^{-1}) in solution, at different pH values, in a slurry of Hallen Wood soil amended with additions of Cd, Pb, Zn and Cu. The slurry was initially adjusted to pH 8.5 and then progressively acidified by additions of hydrochloric acid

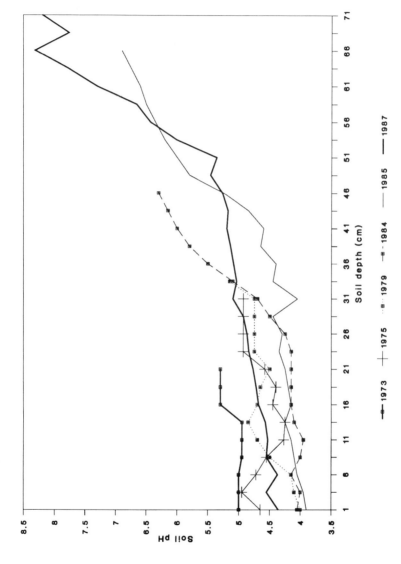

Figure 9.18. Soil pH profiles for Hallen Wood over the period 1973–1987. Each measurement used a 1 : 2.5 soil:deionised water mixture. The sample collection for each profile was not standardised with regard to time of year and total sampling depth

solubility of Cd and Zn to occur in the pH range 5.5–6.0; and for Cu and Pb in the 4.0–4.5 pH range. There are strong reasons, therefore, to associate increasing mobility of heavy metals in Avonmouth soils with increasing soil acidification. The Avonmouth area includes major sources of acidifying substances including nitric acid producers, fertiliser manufacturers and a large sulphuric acid plant, associated with the heavy metal smelter, utilising waste gases from the roasting of high sulphide metal ores.

Martin and Coughtrey (1987) investigated a range of possible causes for the redistribution of heavy metals in soils but decided that the only plausible explanation in the case of the Avonmouth woodland soils was increasing acidification. Data collected for soil pH profiles over the period of study are illustrated in Figure 9.18. This clearly shows that there has been a progressive decrease in the pH of the surface horizons of the Hallen Wood profile and that the acidification has moved progressively to greater soil depths which seems to parallel the movement of Cd and Zn.

9.10 DISCUSSION

Metal budget studies (principally for temperate forest ecosystems) have been reviewed by several authors (e.g. Hughes, 1981; Hughes et al., 1980; Kabata-Pendias and Pendias, 1984; Bergkvist et al., 1989). Bergkvist et al. (1989) have indicated that metals such as Cu and Pb are associated with positive ecosystem budgets, whereas Cd and Zn, in some examples, were found to have negative ecosystem budgets. In the latter case, negative ecosystem budgets were associated with larger outputs of Cd and Zn (as $g\,ha^{-1}\,year^{-1}$) from the deepest baseline from which the soil was sampled, relative to inputs from atmospheric deposition. Smith (1981) regarded forest soils as important sinks for a variety of air contaminants, with the retention of Pb by organic material on the forest floor as a most dramatic example. He suggested that forest soils could act as less efficient sinks for other metals such as Cu, Cd and Zn when compared to Pb retention. However, he maintained that forest soil sinks were still important for such metals, particularly where the forest systems were close to primary source emissions. The case studies of the Hallen and Haw woodlands detailed in the present chapter lend support to and appear to largely agree with the views of Bergkvist et al. (1989) and Smith (1981) outlined above.

Studies of the various ecosystem components and processes have shown that the impact of heavy metals on these woodlands is considerable. The impact varies from mere accumulation of high concentrations in certain components with potential implications to consumers, to changes in the invertebrate fauna, to alteration of the decomposition process, to changes in the ability of organisms to tolerate the presence of metals and to redistribution of metals in the soil profile.

A feature of the Hallen and Haw wood studies at Avonmouth is that records of soil profile metal concentrations are available for at least a 12 year period. It is also fortuitous that during the study period soil processes showed a dramatic change. In particular, the acidification of surface mineral horizons, resulting from long-term acid deposition, appears to have exhausted the buffering capacity of the original soil such that the switch from base saturated (or near base saturated) to base deficient conditions occurred to increasing soil depths over a relatively short period. It must be stressed that the relatively rapid movements of Cd and Zn seen in these woodland soils are occurring in heavy clay soils in strict contrast to other studies where mobilisation has occurred on acid sandy soils (Dowdy and Volk, 1983; Miller *et al.*, 1983; Scokart *et al.*, 1983) or acid organic soils (Tyler, 1978).

Changes in the distribution of heavy metals will affect the vegetation by causing differences in the availability of those metals at different depths in the soil profile. There are therefore likely to be changes in the uptake patterns in plant species of varying rooting depths. The long-term impact of these metals will largely depend on rates of acidification, which in time may drive the mobility of Cd and Zn to such depths as to contaminate groundwaters.

ACKNOWLEDGEMENTS

The authors wish to thank the Natural Environment Research Council for funding much of the case-study work described, and past undergraduates and postgraduates who have contributed to the sum of knowledge of the Avonmouth sites.

REFERENCES

Adams, T. McM. and Sanders, J. R. (1984) The effect of pH on the release to solution of zinc, copper and nickel from a metal-loaded sewage sludge. *Environmental Pollution (Series B)*, **8**, 85–99.

Allen, G., Nickless, G. and Pickard, J. (1974) Heavy metal particle characterisation. *Nature (London)*, **252**, 571–572.

Baker, A. J. M., Grant, C. J., Martin, M. H., Shaw, S. C. and Whitebrook, J. (1985) Induction and loss of cadmium tolerance in the grass *Holcus lanatus* L. In: *Proceedings International Conference Heavy Metals in the Environment*, Athens, Vol. 2, pp. 229–231. CEP Consultants, Edinburgh.

Baker, A. J. M., Grant, C. J., Martin, M. H., Shaw, S. C. and Whitebrook, J. (1986) Induction and loss of cadmium tolerance in *Holcus lanatus* L. and other grasses. *New Phytologist*, **102**, 575–587.

Beckett, A. (1989) Backscattered electron imaging and low temperature SEM of plant surfaces. *Microscopy and Analysis*, January 1989, 27–29.

Bergkvist, B. and Folkeson, L. (1988) *Fluxes of Cu, Zn, Pb, Cd, Cr and Ni in Forest Ecosystems*. Dept. of Plant Ecology, University of Lund, Sweden.

Bergkvist, B., Folkeson, L. and Berggren, D. (1989) Fluxes of Cu, Zn, Pb, Cd, Cr, and Ni in temperate forest ecosystems. *Water, Air, and Soil Pollution*, **47**, 217–286.

Bewley, R. J. F. (1979) The effects of zinc, lead and cadmium pollution on the leaf surface microflora of *Lolium perenne* L. *Journal of General Microbiology*, **110**, 247–254.

Bewley, R. J. F. and Campbell, R. (1978) Scanning electron microscopy of oak leaves contaminated with heavy metals. *Transactions British Mycological Society*, **71**, 508–511.

Bewley, R. J. F. and Campbell, R. (1980) Influence of zinc, lead, and cadmium pollutants on the microflora of hawthorn leaves. *Microbial Ecology*, **6**, 227–240.

Bowen, H. J. M. (1966) *Trace Elements in Biochemistry*. Academic Press, London.

Bowen, H. J. M. (1979) *Environment Chemistry of the Elements*. Academic Press, London.

Brown, H. and Martin, M. H. (1981) Pretreatment effects of cadmium on the root growth of *Holcus lanatus* L. *New Phytologist*, **89**, 621–629.

Brummer, G. W. (1986) Heavy metal species, mobility and availability in soils. In: Bernhard, M., Brinckman, F. E. and Sadler, P. J. (Eds) *The Importance of Chemical "Speciation" in Environmental Processes*. pp. 169–192. Springer-Verlag, Berlin.

Brummer, G. W. and Herms, U. (1983) Influence of soil reaction and organic matter on the solubility of heavy metals in soils. In: Ulrich, B. and Pankrath, J. (Eds) *Effects of Air Accumulation of Pollutants in Forest Ecosystems*, pp. 233–243. Reidel Publishing, Dordrecht.

Bullock, R. J. (1992) Mobility, Chemical Form and Bioavailability of Cd, Zn, Pb and Cu in Woodland soils Contaminated by Aerial Fallout. PhD Thesis, University of Bristol.

Burkitt, A., Lester, P. and Nickless, G. (1972) Distribution of heavy metals in the vicinity of an industrial complex. *Nature (London)*, **238**, 327–328.

Cameron, A. J. and Nickless, G. (1977) Use of mosses as collectors of airborne heavy metals near a smelting complex. *Water, Air, and Soil Pollution*, **7**, 117–125.

Cocks, E. J. and Walters, B. (1968) *History of the Zinc Smelting Industry in Britain*. Harrap, London.

Coughtrey, P. J. (1978) Cadmium in Terrestrial Ecosystems: A Study at Avonmouth. Bristol, England. PhD Thesis, University of Bristol.

Coughtrey, P. J. and Martin, M. H. (1976) The distribution of Pb, Zn, Cd and Cu within the pulmonate mollusc *Helix aspersa* Muller. *Oecologia (Berl.)*, **23**, 315–322.

Coughtrey, P. J. and Martin, M. H. (1977*a*) The uptake of lead, zinc, cadmium and copper by the mollusc *Helix aspersa* Muller, and its relevance to the monitoring of heavy metal contamination of the environment. *Oecologia (Berl.)*, **27**, 65–74.

Coughtrey, P. J. and Martin, M. H. (1977*b*) Cadmium tolerance of *Holcus lanatus* from a site contaminated by aerial fallout. *New Phytologist*, **79**, 273–280.

Coughtrey, P. J. and Martin, M. H. (1977*c*) The woodlouse, *Oniscus asellus*, as a monitor of environmental cadmium levels. *Chemosphere*, **6**, 827–832.

Coughtrey, P. J. and Martin, M. H. (1978*a*) Tolerance of *Holcus lanatus* to lead, zinc and cadmium in factorial combination. *New Phytologist*, **81**, 147–154.

Coughtrey, P. J. and Martin, M. H. (1978*b*) Cadmium uptake and distribution in tolerant and non-tolerant populations of *Holcus lanatus* grown in solution culture. *Oikos*, **30**, 555–560.

Coughtrey, P. J. and Martin, M. H. (1979) Cadmium, lead and zinc interactions and tolerance in two populations of *Holcus lanatus* grown in solution culture. *Environmental and Experimental Botany*, **19**, 285–290.

Coughtrey, P. J., Jones, C. H., Martin, M. H. and Shales, S. W. (1979) Litter accumulation in woodlands contaminated by Pb, Zn, Cd and Cu. *Oecologia (Berl.)*, **39**, 51–60.

Coy, C. M. (1984) Control of dust and fume at a primary zinc and lead smelter. *Chemistry in Britain*, May 1984, 418–420.

Denaeyer-de Smet, S. and Duvigneaud, P. (1974) Accumulation de metaux lourds dans divers ecosystemes terrestres pollue par des retombees d'origine industrielle. *Bulletin Societe Royale de Botanique de Belgique*, **107**, 147–156.

Dmowski, K. and Karolewski, M. A. (1979) Cumulation of zinc, cadmium and lead in invertebrates and in some vertebrates according to the degree of an area contamination. *Ekologia Polska*, **27**, 333–349.

Dowdy, R. H. and Volk, V. V. (1983) Movement of heavy metals in soils. In: *Chemical Mobility and Reactivity in Soil Systems*. SSSA Special Publication No. 11. pp. 229–240. Soil Society of America and American Society of Agronomy, Madison, Wisconsin.

Evans, L. J. (1989) Chemistry of metal retention by soils. *Environmental Science and Technology*, **23**, 1046–1056.

Fangmeier, A. and Steubing, L. (1986) Cadmium and lead in the food web of a forest ecosystem. In: Georgii, H. W. (Ed.) *Atmospheric Pollutants in Forest Areas: Their Deposition and Interception*, pp. 223–234. D. Reidel, Dordrecht.

Finlay, D. C. (1976) *The Soils of the Southern Cotswolds and Surrounding Country*. Memoir Soil Survey Great Britain, Soil Survey, Harpenden.

Forstner, U. (1987) Changes in metal mobilities in aquatic and terrestrial cycles. In: Patterson, J. W. and Passino, R. (Eds) *Metals Speciation, Separation, and Recovery*, pp. 3–25. Lewis Publishers, Chelsea, Michigan.

Freedman, B. and Hutchinson, T. C. (1981) Sources of metal and elemental contamination of terrestrial environments. In: Lepp, N. W. (Ed.) *The Effects of Heavy Metal Pollution on Plants*, Vol. 2, pp. 35–94. Applied Science Publishers, London and New York.

Gill, R., Martin, M. H., Nickless, G. and Shaw, T. L. (1975) Regional monitoring of heavy metal pollution. *Chemosphere*, **4**, 113–118.

Gingell, S. M., Campbell, R. and Martin, M. H. (1976) The effect of zinc, lead and cadmium pollution on the leaf surface microflora. *Environmental Pollution*, **11**, 25–37.

Harrison, R. M. and Williams, C. R. (1983) Physico-chemical characterization of atmospheric trace metal emissions from a primary zinc-lead smelter. *The Science of the Total Environment*, **31**, 129–140.

Heinrichs, H. and Mayer, R. (1977) Distribution and cycling of major and trace elements in two central European forest ecosystems. *Journal Environmental Quality*, **6**, 402–407.

Heinrichs, H. and Mayer, R. (1980) The role of forest vegetation in the biogeochemical cycle of heavy metals. *Journal Environmental Quality*, **9**, 111–118.

Hopkin, S. P. (1990) Species-specific differences in the net assimilation of zinc, cadmium, lead, copper and iron by the terrestrial isopods *Oniscus asellus* and *Porcellio scaber*. *Journal Applied Ecology*, **27**, 460–474.

Hopkin, S. P. (1991) Critical concentrations, pathways of detoxification and cellular ecotoxicology of metals in terrestrial arthropods. *Functional Ecology*, **4**, 321–327.

Hopkin, S. P. and Martin, M. H. (1982) The distribution of zinc, cadmium, lead and copper within the woodlouse *Oniscus asellus* (Crustacea, Isopoda). *Oecologia (Berl.)*, **54**, 224–232.

Hopkin, S. P. and Martin, M. H. (1983) Heavy metals in the centipede *Lithobius variegatus* (Chilopoda). *Environmental Pollution (Series B)*, **6**, 309–318.

Hopkin, S. P. and Martin, M. H. (1984*a*) Heavy metals in woodlice. In: Sutton, S. L. and Holdich, D. M. (Eds) *The Biology of Terrestrial Isopods*, pp. 143–166. Symposium of the Zoological Society of London, No. 53, Oxford University Press, Oxford.

Hopkin, S. P. and Martin, M. H. (1984b) Assimilation of zinc, cadmium, lead and copper by the centipede *Lithobius variegatus* (Chilopoda). *Journal of Applied Ecology*, **21**, 535–546.

Hopkin, S. P. and Martin, M. H. (1985) Assimilation of zinc, cadmium, lead, copper, and iron by the spider *Dysdera crocata*, a predator of woodlice. *Bulletin Environmental Contamination Toxicology*, **34**, 183–187.

Hopkin, S. P., Watson, K., Martin, M. H. and Mould, M. L. (1985) The assimilation of heavy metals by *Lithobius variegatus* and *Glomeris marginata* (Chilopoda; Diplopoda). *Bijdragen tot de Dierkunde*, **55**, 88–94.

Hopkin, S. P., Hardisty, G. N. and Martin, M. H. (1986) The woodlouse *Porcellio scaber* as a "Biological Indicator" of zinc, cadmium, lead and copper pollution. *Environmental Pollution (Series B)*, **11**, 271–290.

Hughes, M. K. (1981) Cycling of trace metals in ecosystems. In: Lepp, N. W. (Ed.) *The Effects of Heavy Metal Pollution on Plants*, Vol. 2, pp. 95–118. Applied Science Publishers, London and New York.

Hughes, M. K., Lepp, N. W. and Phipps, D. A. (1980) Aerial heavy metal pollution and terrestrial ecosystems. *Advances in Ecological Research*, **11**, 217–327.

Hutton, M. (1984) Impact of airborne metal contamination on a deciduous woodland system. In: Sheehan, P. J., Miller, D. R., Butler, G. C. and Bourdeau, Ph. (Eds) *Effects of Pollutants at the Ecosystem Level*. SCOPE 22, pp. 365–375. Wiley and Sons, New York.

Hutton, M. and Symon, C. (1986) The quantities of cadmium, lead, mercury and arsenic entering the UK environment from human activities. *The Science of the Total Environment*, **57**, 129–150.

Hutton, M. and Symon, C. (1987) Sources of cadmium discharge to the UK environment. In: Coughtrey, P. J., Martin, M. H. and Unsworth, M. H. (Eds) *Pollutant Transport and Fate in Ecosystems*, pp. 223–237. Special Publication No. 6 of the British Ecological Society, Blackwell Scientific Publications, Oxford.

Kabata-Pendias, A. and Pendias, H. (1984) *Trace Elements in Soils and Plants*. CRC Press, Florida.

Little, P. (1973) A study of heavy metal contamination of leaf surfaces. *Environmental Pollution*, **5**, 159–172.

Little, P. (1974) Airborne zinc, lead and cadmium pollution and its effects on soils and vegetation. PhD Thesis, University of Bristol.

Little, P. and Martin, M. H. (1972) A survey of zinc, lead and cadmium in soil and natural vegetation around a smelting complex. *Environmental Pollution*, **3**, 241–254.

Little, P. and Martin, M. H. (1974) Biological monitoring of heavy metal pollution. *Environmental Pollution*, **6**, 1–19.

Lobersli, E., Gjengedal, E. and Steinnes, E. (1991) Impact of soil acidification on the mobility of metals in the soil–plant system. In: Vernet, J.-P. (Ed.) *Heavy Metals in the Environment*, pp. 37–53. Elsevier, Amsterdam.

McKenzie, R. M. (1980) The adsorption of lead and other heavy metals on oxides of manganese and iron. *Australian Journal Soil Research*, **18**, 61–73.

Martin, M. H. and Coughtrey, P. J. (1981) Impact of heavy metals on ecosystem function and productivity. In: Lepp, N. W. (Ed.) *The Effects of Heavy Metal Pollution on Plants*, Vol. 2, pp. 119–158. Applied Science Publishers, London and New York.

Martin, M. H. and Coughtrey, P. J. (1982) *Biological Monitoring of Heavy Metal Pollution: Land and Air*. Applied Science Publishers, London and New York.

Martin, M. H. and Coughtrey, P. J. (1987) Cycling and fate of heavy metals in a contaminated woodland ecosystem. In: Coughtrey, P. J., Martin, M. H. and Unsworth, M. H. (Eds) *Pollutant Transport and Fate in Ecosystems*, pp. 319–336. Special Publication No. 6 of the British Ecological Society, Blackwell Scientific, Oxford.

Martin, M. H., Coughtrey, P. J., Shales, S. W. and Little, P. (1980) Aspects of airborne cadmium contamination of soils and natural vegetation. In: *Inorganic Pollution and Agriculture*, pp. 56–69. MAFF Reference Book 326, HMSO, London.

Martin, M. H., Duncan, E. M. and Coughtrey, P. J. (1982) The distribution of heavy metals in a contaminated woodland ecosystem. *Environmental Pollution (Series B)*, 3, 147–157.

Mattigod, S. V. and Page, A. L. (1983) Assessment of metal pollution in soils. In: Thornton, I. (Ed.) *Applied Environmental Geochemistry*, pp. 355–394. Academic Press Geology Series, Academic Press, London.

Mayer, R. (1981) Naturliche und anthropogene komponenten des schwermetall-haushalts von waldokosytemen. *Gottinger Bodenkundliche Berichte*, 70, 1–152.

Mayer, R. (1983), Interaction of forest canopies with atmospheric constituents aluminium and heavy metals. In: Ulrich, B. and Pankrath, J. (Eds) *Effects of Accumulation of Air Pollutants in Forest Ecosystems*, pp. 47–55. Reidel Publishing, Dordrecht.

Miller, W. P., McFee, W. W. and Kelly, J. M. (1983) Mobility and retention of heavy metals in sandy soils. *Journal Environmental Quality*, 12, 579–584.

Nriagu, J. O. (1989) Natural versus anthropogenic emissions of trace metals to the atmosphere. In: Pacyna, J. M. and Ottar, B. (Eds) *Control and Fate of Atmospheric Trace Metals*, pp. 3–13. Kluwer Academic, Dordrecht.

Nriagu, J. O. and Pacyna, J. M. (1988) Quantitative assessment of worldwide contamination of air, water and soils with trace metals. *Nature (London)*, 333, 134–139.

Pacyna, J. M. (1989) Technological parameters affecting atmospheric emissions of trace elements from major anthropogenic sources. In: Pacyna, J. M. and Attar, B. (Eds) *Control and Fate of Atmospheric Trace Metals*, pp. 15–31. Kluwer Academic, Dordrecht.

Pacyna, J. M., Munch, J. and Axenfeld, E. (1991) European inventory of trace metal emissions to the atmosphere. In: Vernet, J.-P. (Ed.) *Heavy Metals in the Environment*, pp. 1–20. Elsevier, Amsterdam.

Parker, G. R., McFee, W. W. and Kelly, J. M. (1978) Metal distributions in forested ecosystems in urban and rural Northwestern Indiana. *Journal of Environmental Quality*, 7, 337–342.

Plant, J. A. and Raiswell, R. (1983) Principles of environmental chemistry. In: Thornton, I. (Ed.) *Applied Environmental Geochemistry*, pp. 1–39. Academic Press, London.

Read, H. J. and Martin, M. H. (1990) A study of myriapod communities in woodlands contaminated with heavy metals. In: Minelli, A. (Ed.) *Proceedings of the 7th International Congress of Myriapodology*, pp. 289–298. E.J. Brill, Leiden.

Read, H. J., Wheater, C. P. and Martin, M. H. (1986) The effects of heavy metal pollution on woodland communities of surface active Carabidae (Coleoptera). *Proceedings of the 3rd European Congress of Entomology, Amsterdam*, pp. 295–298. Nederlandsc Entomologische, Vereniging. Amsterdam.

Read, H. J., Wheater, C. P. and Martin, M. H. (1987) Aspects of the ecology of Carabidae (Coleoptera) from woodlands polluted by heavy metals. *Environmental Pollution*, 48, 61–76.

Roberts, T. M. (1972) The spread and accumulation in the environment of toxic non-ferrous metals from urban and industrial sources. PhD Thesis, University of Wales.

Rodwell, J. S. (Ed.) (1991) *British Plant Communities: Vol. 1, Woodlands and Scrub*. Cambridge University Press, Cambridge.

Scokart, P. O., Meeus-Verdinne, K. and De Borger, R. (1983) Mobility of heavy metals in polluted soils near zinc smelters. *Water, Air, and Soil Pollution*, **20**, 451–463.

Scott, D. W. (1987) The measurement of suspended particular material, heavy metal and selected organic emissions at the Avonmouth municipal refuse incinerator. Report No. LR 614 (PA), Warren Spring Laboratory, Stevenage.

Smith, W. H. (1981) *Air Pollution and Forests: Interactions between Air Contaminants and Forest Ecosystems*. Springer-Verlag, Berlin and New York.

Sposito, G. (1983) The chemical forms of trace metals in soils. In: Thornton, I. (Ed.) *Applied Environmental Geochemistry*, pp. 123–170. Academic Press, London.

Sposito, G. and Page, A. L. (1984) Cycling of metal ions in the environment: metal ions in biological systems. In: Siegel, H. (Ed.) *Circulation of Metals in the Environment*, Vol. 18, pp. 287–332. Dekker, New York.

Strojan, C. L. (1978) Forest litter decomposition in the vicinity of a zinc smelter. *Oecologia (Berl.)*, **32**, 203–212.

Tills, A. R. and Alloway, B. J. (1983) The speciation of cadmium and lead in soil solutions from polluted soils. *Proceedings International Conference Heavy Metals in the Environment*, Vol. 2, pp. 1211–1214. CEP, Edinburgh.

Tyler, G. (1972) Heavy metals pollute nature, may reduce productivity. *Ambio*, **1**, 52–59.

Tyler, G. (1975*a*) Effect of heavy metal pollution on decomposition and mineralization rates in forest soils. In: Hutchinson, T. C. (Ed.) *International Conference on Heavy Metals in the Environment, Symposium Proceedings*, Vol. 2, part 1, pp. 217–226. Toronto.

Tyler, G. (1975*b*) *Effects of Heavy Metal Pollution on Decomposition in Forest Soils. I.1. Introductory Investigations*. National Swedish Environment Protection Board, Report SNV PM 443E, Lund.

Tyler, G. (1975*c*) *Effects of Heavy Metal Pollution on Decomposition in Forest Soils. II. Decomposition Rate, Mineralization of Nitrogen and Phosphorus, Soil Enzyme Activity*. National Swedish Environment Protection Board, Report SNV PM 542E, Lund.

Tyler, G. (1978) Leaching rates of heavy metal ions in forest soil. *Water, Air, and Soil Pollution*, **9**, 137–148.

Tyler, G. (1984) The impact of heavy metal pollution on forests: a case study of Gusum, Sweden. *Ambio*, **13**, 18–24.

Tyler, G., Berggren, D., Bergkvist, B., Falkengren-Grerup, U., Folkeson, L. and Ruhling, A. (1987) Soil acidification and metal mobility in forests of southern Sweden. In: Hutchinson, T. C. and Meema, K. M. (Eds) *Effects of Atmospheric Pollutant on Forests, Wetlands and Agricultural Ecosystems*. NATO ASI Series, Series G: Ecological Series, Vol. 16, pp. 347–359. Springer-Verlag, Berlin.

Ure, A. M. and Berrow, M. L. (1982) The elemental constituents of soil. In: Bowen, H. J. M. (Ed.) *Environmental Chemistry*, Vol. 2, Royal Society of Chemistry, London.

Van Hook, R. I., Harris, W. F. and Henderson, G. S. (1977) Cadmium, lead, and zinc distributions and cycling in a mixed deciduous forest. *Ambio*, **6**, 281–286.

Watson, A. P. (1975) Trace element impact on forest floor litter in the new lead belt region of South eastern Missouri. *Trace Elements in Environmental Health*, **9**, 227–236.

Weiner, J. and Grodzinski, W. (1984) Energy, nutrients, and pollutant budgets of a forest ecosystem. In: Grodzinski, W., Weiner, J. and Maycock, P. F. (Eds) *Forest Ecosystems in Industrial Regions: Studies on the Cycling of Energy Nutrients and Pollutants in the Niepolomice Forest Southern Poland*, pp. 203–229. Springer-Verlag, Berlin.

10 Fungicide-derived Copper in Tropical Plantation Crops

NICHOLAS W. LEPP AND NICHOLAS M. DICKINSON
Liverpool John Moores University, UK

ABSTRACT

This review considers the relationship between various tropical plantation crops and elevated copper levels resulting from the long-term use of copper-based fungicides. The history of copper fungicide development and use is considered, together with the crop/disease combinations for which copper fungicides are currently used. The use and impact of copper in tea, cocoa, coffee and banana cultivation are reviewed, together with other plantation crops. Copper fungicide usage in beverage crops increases Cu levels in harvested tea and cacao, but there is no accumulation in coffee beans. The long-term use of copper fungicides in coffee cultivation and its influence on the coffee bushes and their associated soils is fully discussed. Effects of increased rates of fungicide application on yield, growth and interactions with other pests are reviewed, together with the influence of copper on reproductive structures in coffee. Alternatives to copper are discussed, and the problems of copper tolerance in pathogenic bacteria, as a consequence of long-term fungicide use, are also considered. Future land use and the influence of elevated soil copper levels on this are given full consideration.

10.1 INTRODUCTION

10.1.1 HISTORY OF COPPER FUNGICIDES

The world's first effective fungicidal chemicals were based on copper. The efficiency of this element was discovered by chance during a major outbreak of mildew (*Plasmopora viticola*) in the vineyards of the Bordeaux region in France. This disease was new to Europe, and quite devastating in its effect on the vines. At the time, no treatment would alleviate the disease, and the Bordeaux wine industry was on the verge of collapse. In 1882, Millardet, Professor of Botany at the Faculté des Sciences at Bordeaux, noted an unusual response of vines in the vineyard of Chateau Beaucaillou in the St. Julien district of Bordeaux. The vineyards were badly affected by mildew, and the majority of infected vines had lost their leaves. However, one or two rows adjacent to the roadside still possessed leaves. On closer examination,

Toxic Metals in Soil–Plant Systems. Edited by S. M. Ross
© 1994 John Wiley & Sons Ltd

Millardet observed that the leaves were covered with a tacky blue substance. The bailiff at Chateau Beaucaillou told Millardet that the blue substance was verdigris, a copper compound applied to easily accessible vines as a deterrent to local grape thieves. This observation convinced Millardet of the potential role of copper as a fungicidal agent. He had discovered earlier, again quite by accident, that copper affected fungal spores of vine mildew, finding that spores did not germinate in water from one of his wells; a subsequent analysis showed that the old copper pump through which the water was raised gave the well water a copper content of 5 mg litre^{-1} which was actually about 20 times greater than the lethal dose for mildew spores.

Throughout the next two years, Millardet worked on developing copper as a practical fungicidal treatment; this work was given added impetus by further observations on the relationship between copper and vine mildew. Scientists in other vine-growing regions of France noted that vinestocks supported by wooden stakes treated with a copper-based preservative remained immune from mildew, and a powder based on copper sulphate was claimed to give good control. Millardet's trials were not very conclusive. Many of his test formulations, including the copper sulphate solution, were ineffective, and only the oily "Medoc" mixture, used on the vines at Chateau Beaucaillou, showed any promise. He had also advised the bailiff at Chateau Beaucaillou, M. David, who was conducting his own trials. David established that the "Medoc" mixture alone gave control of mildew, and not a combination of this material and "roadside dust". In a letter to Millardet, he indicated the most widely used formulation of this mixture: "into 100 litres of water (well, river or rain), 8 kilos of commercial copper sulphate is dissolved. Make up separately from 30 litres of water and 15 kg of granulated lime a milky lime solution, which is then mixed into the copper sulphate solution. A bluish deposit then forms." The method of application was also described: "The workman pours part of this mixture into a bucket, which he holds in his left hand, whilst with his right, using a small broom, he sprinkles the leaves with this preserving wash, being careful not to sprinkle the grape bunches." The broom referred to was a bunch of heather (*Calluna vulgaris*) twigs, tied at the base, and this represents the first recorded example of a fungicide applicator (Anon., 1985a).

10.1.2 APPLICATION AND FUNGICIDAL PROPERTIES

Bordeaux mixture, as the preparation was subsequently named, has been in use ever since this time as a commercial fungicide, utilising a copper sulphate content of 2% (500 g Cu ha^{-1}) which gives good disease control. A freshly prepared Bordeaux mixture contains complex forms of copper, gypsum ($CaSO_4$) and excess lime. Lime neutralises the copper sulphate, giving rise to a fine blue precipitate which forms small blisters around the lime particles. The mixture dries on the leaf to which it has been applied as a deposit of insoluble cupric forms of copper. Carbon dioxide present in precipitation slowly dissolves these forms, releasing soluble forms of copper into the water film on the leaf surface. In this way copper reaches fungal spores and inhibits either germination, or the growth of mycelium from germinated spores.

Since the discovery of its anti-fungal properties, copper has been widely used as a plant protection agent. At present, copper formulations can be placed into one of four groups:

1. *Soluble copper salts*: sulphate or acetate are commonly used. Such treatments show limited adhesion to plant surfaces and may also be phytotoxic.
2. *Copper mixtures*: prepared by neutralising copper sulphate solutions with ground lime (Bordeaux mixture) or sodium carbonate (Burgundy mixture). The latter produces a fine and homogeneous mixture which is easier to apply than Bordeaux mixture, but which must be precisely prepared, as either ingredient present in excess may cause a phytotoxic reaction. It also weathers more easily than Bordeaux mixture, and is much less widely used.
3. *Copper suspensions*: prepared by adding sparingly soluble copper salts (oxides, oxychlorides, carbonates) to water. Recent advances in the preparation of finely divided and micronised powders of these materials have greatly improved their efficacy as fungicides.
4. *Copper powders*: these are finely ground powders of copper salts, bulked out with an inert carrier, and sometimes containing sulphur, for application as fungicidal dusts. Persistence and efficacy depend on: (a) the plant surfaces these are applied to, and (b) the solubility of the copper compound used.

Copper acts as a fungicide by binding to various organic constituents of the cell cytoplasm (phosphate, sulphydryl and amine groups, for example) and by interfering with enzyme activity (pyruvate dehydrogenase, α-ketoglutarate dehydrogenase) intimately associated with energy production. At the present time, there are no systemic copper-based formulations (Anon., 1975).

10.1.3 DISEASE CONTROL

Since the introduction of Bordeaux mixture, copper-based fungicides have been used to control fungal and bacterial diseases which affect a wide range of temperate and tropical crops (Smith, 1985). Copper is widely used on vines for the control of Downy mildew (*Plasmopara viticola*), anthracnose (*Elsinoë ampelina*), black rot (*Guignardia bidwellii*) and bacterial necrosis *(Xanthomonas anpelina)*, which became a serious problem in vines where copper-based treatments were replaced with organic formulations. In top fruits, such as apples, copper formulations control apple scab (*Venturia inaequalis*) and canker (*Nectria galligena*) together with bacterial canker of store fruits (*Pseudomonas syringea* pv. *morsprunorum*) and bark canker of apricots (*Fusicoccum amygdali*) (Anon., 1975). Among herbaceous crops,

Table 10.1. Fungal diseases of tropical plantation crops, and forms of copper usage in these crops

Crop	Disease and agent	Cu treatment
Coffee	Coffee rust (*Hemileia vastatrix*)	Bordeaux mixture or Cu oxychloride
	Seedling droop (*Rhizoctonia solani*)	Bordeaux mixture or Cu oxychloride
	Coffee berry disease[a] (*Colletotrichum coffeanum*)	Cu oxychloride (only partly effective)
	Bacterial blight/Elgon Die back[a] (*Pseudomonas syringiae* pv. *Garcae*)	Cu oxychloride
Cacao	Black pod (*Phytophthora palmivora*)	Bordeaux mixture CuO, Cu oxychloride
	Pink disease (*Corticum salmonicolor*)	Bordeaux mixture CuO, Cu oxychloride
	Die back (*Botryodiplodia theobromae*)	Bordeaux mixture CuO, Cu oxychloride
Tea	Blister blight[b] (*Exobasidium vexans*)	CuO suspension
	Black rot (*Corticium invisum*)	Cu oxychloride dust
Banana	Sigatoka disease (*Mycosphaerella musicola*)	Mixed Cu-organic formulation
	Black rot/Die back (*Botryodiplodia theobromae*)	Bordeaux mixture Cu oxychloride
Pineapple	Heart and stem rot (*Phytophthora* spp.)	Bordeaux mixture
Rubber	Leaffall (*Phytophthora palmivora*)	Bordeaux mixture "Copper dust"
	Increase in latex flow	CuSO₄ injection into trunk
Citrus	Mal Secco (*Deutorophoma tracheiphila*)	Bordeaux mixture Cu oxychloride
	Brown rot (*Phytophthora citrophthora*)	Bordeaux mixture Cu oxychloride
Stone fruits[c] (Peach, apricot, etc.)	Bacterial canker (*Pseudomonas syringae* pv. *morsprunorum*)	Bordeaux mixture Bordeaux mixture
	Bark canker (apricot) (*Fusicoccum amygdali*)	
Olive	Leaf spot (*Spilocaea oleaginum*)	Copper products
Vines	Downy mildew (*Plasmopara viticola*)	Bordeaux mixture Cu oxychloride (in balance with organic agents)

Reproduced from Anon. (1975).
[a] In East Africa only.
[b] Not in East Africa.
[c] Most species sensitive to copper sprays, resulting in foliar damage.

copper is used on beets (sugar and fodder), celery, potatoes, tomatoes and rice.

Many important diseases of tropical crops are controlled using copper formulations (Table 10.1). The plantation crops (coffee, cacao, tea, banana and rubber) all receive routine copper applications, based on various formulations, and applied with varying frequency to effect full or partial control. Due to climatic conditions, plant diseases may flourish year round in these crops, and the low cost of copper products, as opposed to more complex organic compounds, combined with a need for frequent applications to these perennial crops, is a major factor in their continued widespread use.

10.1.4 WORLD USAGE

An estimate of world use of copper fungicides is shown in Figure 10.1, based on data for 1984 (this is incomplete due to lack of information from some African, Asian and South American countries, and estimates from others (Anon., 1985b). Some 40% of fungicides are used in Europe, 28% in the Americas and 13% in both Asia/Australasia and Africa respectively. Individual countries consume more than others (Figure 10.2). European usage includes areas of extensive vine, olive and, in the case of Spain, citrus culture. The Asian examples grow tea and rice, the African states are coffee or cocoa

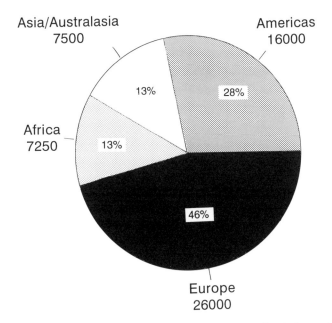

Figure 10.1. World use of copper fungicides in tonnes (1984 values). Reproduced from Anon. (1985b)

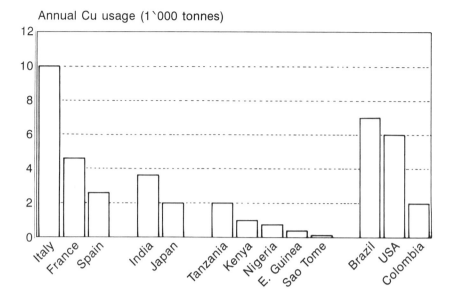

Figure 10.2. Annual use of selected fungicides in selected countries (1984 values). Reproduced from Anon. (1985b)

producers and the American examples are for two coffee growing regions (Brazil, Colombia) and the USA, where a wide variety of temperate and subtropical crops, notably citrus, are grown. In comparison with the listed data, the UK used 100 tones of Cu fungicide and the Netherlands 10 tones at this time, both less than Equatorial Guinea.

Copper fungicides have been applied to many of these crops for an extended period of time. In the case of vines, routine use extends to almost 100 years, and for many of the tropical crops, copper has been applied for at least half a century. Rates and frequency of application have changed with time, reflecting changes in agronomic practice, differing patterns of disease development or the advent of new pathogens, and the development of newer application techniques. At the same time, many developing countries have come to depend on crops such as coffee and cocoa as major sources of foreign currency and the maintenance of crop yields has become an integral part of their economic development. Commodity crops such as coffee, cocoa and rubber have been subject to international quota and pricing systems, some of which are now abolished. This has either led to expansion in cultivation of hard currency crops with increased fungicide use, or to removal of such crops and replacement with high value seasonal crops of soft fruit and vegetables for export to markets in the developed nations. In both cases, the past use of fungicides can be considered to present a problem to future development.

10.1.5 PROPERTIES OF COPPER IN SOILS

Copper fungicides, unlike many of their organic counterparts, do not degrade with time. Copper has a strong affinity for organic matter, and tends to bind preferentially to a wide range of organic materials. In consequence, its migration through soil profiles tends to be slow, but the strength of binding makes for a depleted pool of plant-available copper in most soils in comparison to the initial input. On light soils, with a low organic content, copper phytotoxicity may arise, and there are well-documented examples of this where some crop rotation has occurred on soils of this type with a history of copper fungicide application (Reuther and Smith, 1952; Reuther et al., 1953; Corder and Ramirez, 1979; Tiller and Merry, 1981).

10.2 COPPER AND TROPICAL CROPS

In considering the potential impact of copper on tropical plantation crops, it should be remembered that each of the main crops to which copper is applied have very different growth patterns, agronomic characteristics, diseases and end uses. As a result, it is impossible to generalise with respect to copper impacts, and each crop must be considered separately.

10.2.1 TEA

Tea is a perennial shrub, cultivated in acidic soils under cooler conditions than cocoa or coffee. Only the young leaves of the plant are harvested and eventually used, the fermented and dried leaves being used to prepare an infusion. In many parts of its extensive cultivated range, tea is affected by blister blight (*Exobasidium vexans*), a basidiomycete fungus which affects the stems of the shrub. Control is often achieved by spraying with CuO fungicides, with applications made at monthly intervals. As the end product of this crop is leaves, which become contaminated with fungicide, it may be expected that the incidence of blister blight may be reflected in the copper content of the end product. Analysis of the copper content of tea samples from nine tea-growing regions reveals significant differences (Figure 10.3). Normal plant Cu concentrations range between $5-20$ $\mu g\,g^{-1}$ dry wt (Lepp and Dickinson, 1985). All tea samples exceeded this range, with the exception of the Kenyan leaves. Blister blight has been excluded from Kenya due to strict phytosanitary methods and, as a result, copper fungicides are not applied to tea. Little is known of the copper status of tea soils, or of the circulation of copper within the tea plantation ecosystem.

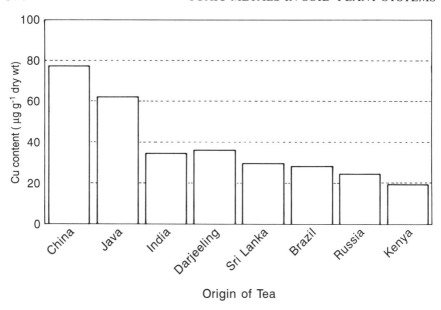

Origin of Tea

Figure 10.3. Copper content of tea from different regions, collected as processed and dried tea from a packaging merchant in Liverpool

10.2.2 COCOA

Cocoa is a tree crop of the humid tropics, generally cultured at low altitudes. The beans (seeds) are the end product and the crop is cultivated in West Africa, South and Central Central America and South-East Asia. Cocoa is affected by a range of pathogens, the most serious of which is Black Pod, caused by *Phytophthora palmivora*. This disease is prevalent in West Africa and Brazil, where most major growing regions are affected (Wood, 1975). Other important diseases are Pink disease (*Corticum salmonicolor*) and Die-back (*Botryopdiplodia theobromae*), which are found in most production regions. All diseases are controlled by copper fungicides, but rates and methods of application vary. In Nigeria, 1% Bordeaux mixture is applied at 200–500 litres ha^{-1} on a routine basis. Other susceptible regions have different systems; on Fernando Po (Equatorial Guinea), a complex system of fixed sprays was installed in the plantations, and spraying with between 1000 and 2000 litres ha^{-1} of 2% copper formulation was carried out 3–4 times per annum. Spraying at this rate was carried out for approximately 40 years, initiated during the Spanish colonial period, and the accumulation of copper in the plantation soils is probably reflected in the copper content of the testa of the cocoa beans (Figure 10.4). Levels in Fernando Po beans are threefold higher than from other cocoa-growing regions in West Africa, the Caribbean and South-East Asia (Wood, 1975).

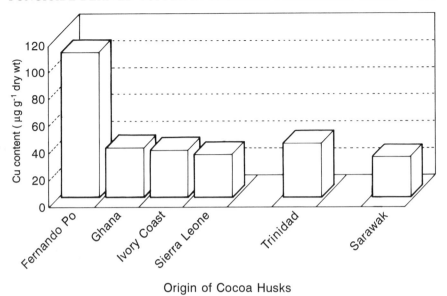

Origin of Cocoa Husks

Figure 10.4. Copper content of cocoa husks (μg g^{-1} dry wt) from different countries. Data from Wood (1975)

Applications of copper also vary with altitude. In West Africa, high altitude cocoa requires more frequent copper application at greater strength, due to higher rainfall. There are no reports on the influence of altitude on soil copper levels in plantations. It is usual to spray only the pods, but in some regions (Latin America) the canopy is also routinely sprayed. Targeting may not always be effective. Where large fungicide volumes are applied, as in Fernando Po, it has been noted that "much falls on the ground" (Wood, 1975).

There have been very few detailed studies on the fate of copper in cocoa plantations. Ayanlaja (1983), working in Nigeria, demonstrated a threefold increase in HCl-extractable copper in plantation soils in comparison with levels in adjacent forest soils. These increased levels were implicated as part of the complex of factors responsible for replanting disease, where new cocoa plants fail to thrive when planted in soils at the sites of former plantations. Lima (1993) has investigated copper budgets in three different-aged cocoa stands in Bahia state, Brazil. A threefold increase in the reservoir of copper (from Cu fungicides) was demonstrated in a stand treated with 224 kg Cu ha^{-1} over a 16 year period, and a 2.5-fold increase in a stand receiving 48 kg Cu ha^{-1} over a 5 year period. Trees stored 20% of the Cu reservoir in the 5 year stand, and 30% of the Cu reservoir in the 16 year stands, as compared with <5% in an unsprayed stand. What effects this has on productivity or tree vitality is not known.

10.2.3 COFFEE

Coffee (*Coffea arabica*, *C. robusta*, *C. canephora*) can be cultivated in tropical highlands or lowlands. It was first developed as a crop in the mountains of southern Arabia and Ethiopia (*C. arabica*) and has since been introduced to many countries in the tropical and subtropical belt on both sides of the Equator. Lowland cultivation is mainly via the inferior *C. robusta* and *C. canephora*; the production of all these types of coffee makes the crop the most important commodity crop in world trade, and the most important commodity after oil. Coffee represents a significant source of currency-earning exports for many countries in the developing world (e.g. 25% in Kenya). Such a valuable investment repays careful protection against potentially injurious diseases (Cannell, 1983).

Historically, the most important coffee pathogen is rust (*Hemileia vastatrix*). This was responsible for the complete demise of coffee cultivation in Sri Lanka in 1860 and its subsequent replacement with tea. Similar problems occurred in Natal and in the north-east of South Africa, where rust is still a potentially important yield-reducing pathogen, found wherever coffee is cultivated.

The disease is controlled with copper, either in the past as Bordeaux mixture or now, more commonly, as Cu-oxychloride (Purseglove, 1974). In East Africa, two other serious pathogens exist. Coffee Berry disease (*Colletotrichum coffeanum*) first developed in the 1920s, following widespread use of copper sprays (Purseglove, 1974; Acland, 1975). The causal agent is a common saprophyte, frequently present on coffee throughout the tropics, but the pathogenic strain, which attacks the developing coffee fruit, is confined to East Africa. The reasons for the development of pathogenesis are not known, but speculations link this to the increased incidence of fungicidal and "tonic" spraying of copper. Copper fungicides are only partially effective against *C. coffeanum*, and are used in conjunction with organics (Purseglove, 1974). The third major disease is Elgon Die-back or Bacterial Blight of Coffee (BBC). This appeared in the 1950s in Kenya, and has spread throughout the coffee-gowing regions in that part of Africa. The causal agent is *Pseudomonas syringea* pv. *garcae*, and the disease causes the wilting and death of young shoots. Frequent spraying with copper is necessary to control this problem; currently Cu-oxychloride applied between 8 and 12 times per annum is considered effective. In Kenya, copper fungicide use varies between different types of producers; in 1982–83, inputs to small holder coffee plantations were 7.8 kg ha^{-1} but in estates inputs were 25.4 kg ha^{-1} (Whittaker, 1984). Almost 60% of Kenyan coffee is produced by the former group; estates have a 70% higher yield per hectare, due, in some part, to use of effective fungicides.

All coffee bushes follow a well-defined routine of pruning, known as a multiple stem system. Old shoots are removed every 4–7 years, to be replaced by suckers on the main trunk (Cannell, 1971). The prunings are used as mulch material, together with grasses cut from poor soil adjacent to the coffee stands.

10.2.4 BANANAS

The most important pathogen of bananas is the fungus *Mycosphaerella musicola* (*Cercospora musae*), the causal agent of Sigatoka disease. In the past, Bordeaux mixture, copper oxides and oxychlorides were the main fungicides used against this disease, but these have now been replaced by Cuivral, a mixed copper–organic formulation. Significant agricultural engineering has been used against Sigatoka disease in some production areas. Up to 390 km^2 of plantations in Honduras and neighbouring Central American states have been fitted with pipe-line spray systems connected to central fungicide reservoirs. These are capable of delivering regular spraying with fungicide at 2–5 week intervals (10–25 sprays year^{-1}) mainly with Bordeaux mixture. The intensity of spraying can be gauged by the need to remove fungicide residues from fruit prior to shipping from the plantation (Wardlaw, 1972).

The consequences of such routine copper applications on the chemical properties of banana plantation soils have been described by Cordero and Ramirez (1979). Applications of 100 kg ha^{-1} Cu as Bordeaux mixture over a 20 year period (1930–1950) produced acid-extractable Cu levels of over 1000 ppm in the upper 20 cm of the soil. Rice subsequently grown on these soils showed significant copper toxicity. These authors estimate that some 50 000 ha of agricultural soils in Costa Rica are affected to a similar extent as a result of long-term Cu fungicide applications.

10.2.5 OTHER PLANTATION CROPS

Copper is used to protect various citrus crops against fungal diseases, such as Mal Secco (*Deutorophoma tracheiphila*) and Brown Rot (*Phytophthora citriphthora*), as well as bacterial infections such as Black pit (*Pseudomonas syringae*) (Anon., 1975). Excessive copper use in Florida citrus orchards, established on acid, sandy soils, resulted in copper toxicity problems in various annual vegetable crops grown on former orchard sites (Reuther and Smith, 1952).

Rubber trees are frequently treated with Bordeaux mixture to combat "abnormal leaf fall" caused by *Phytophthora palmivora*. In addition, periodic copper sulphate injections into the tree trunk (3–6 g CuSO$_4$ twice a year) cause a significant increase in latex flow. Consequences of routine copper input to rubber plantations have received no investigation to date (Anon., 1975).

10.3 IMPACT OF COPPER ON INTENSIVE COFFEE PLANTATIONS

In coffee cultivation the fate of copper, applied for fungicidal purposes, has been extensively studied. In Kenya, coffee was introduced from Mauritius in the early years of this century, and has been in continuous cultivation as a plantation crop since 1915. Normally coffee bushes have a productive life span of 30–40 years, punctuated by routine pruning to maintain the population of young, productive shoots on a 4–7 year rotation. On large-scale coffee estates, bushes are planted in small (usually 1 ha) plots, evenly spaced and situated to take advantage of the productive murram soils (eutric Nitisols) on volcanic footridges. The pruning, yield and treatment of each coffee plot is coordinated by the estate manager, and their records have provided the basis for the accurate modelling of the behaviour of copper in these plots with respect to inputs and losses over an estimated time course. It is usual to find that an estate has been progressively planted over a number of years, and thus consists of numerous plots of different-aged bushes (Lepp and Dickinson, 1987).

Studies in Kenya have examined many aspects of copper cycling. Initial studies focused on the loci of copper concentration within a treated coffee stand. Samples were collected from one of the oldest known stands of coffee, which was still productive after nearly 70 years and had received continuous copper inputs since the 1950s. Various components of the bushes were analysed, together with the associated soil and litter fractions (Dickinson *et al.*, 1984). It is quite evident that some tissues accumulate significant amounts of copper, whereas concentrations in others do not appear to be raised above normal background levels (Figure 10.5). The root system provides a good example: coffee possesses three main types of root; the primary tap root which may grow down 3–6 m to reach the watertable, secondary roots which act as anchors, and an extensive network of branched

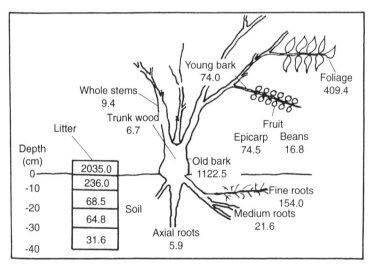

Figure 10.5. Distribution of copper (μg g^{-1} dry wt.) in plant components and soils in a 68-year old coffee (*Coffea arabica*) plantation in Kenya. Reproduced by permission of CEP Consultants Ltd from Dickinson and Lepp (1983)

feeding roots, which form a dense mat in the upper soil horizons. These latter fine roots show significant copper accumulation as opposed to the larger roots from lower down the soil profile. Copper also accumulates in the bark, in unwashed foliage, and in young shoots (primarily in the bark). In the associated soils, copper levels are very high in the litter, and significantly elevated in soils to a depth of 40 cm.

These initial findings led to a comparative study of copper accumulation in three different-aged coffee stands (24, 14 and 4 years old), located on the same plantation and subject to a common management regime (Lepp *et al.*, 1984). A range of plant tissues, soils and litter were analysed from each stand, which clearly indicated accumulation of copper in litter, soil horizons down to 80 cm, fine roots and bark (Table 10.2). The lack of mobility of copper from soil to the plant, and from bark to the remainder of the plant via lateral exchange with the functional system is clearly illustrated by the lack of increase in trunk

Table 10.2. HNO_3-extractable copper concentrations ($\mu g\,g^{-1}$ dry wt) in plant tissues, soil profiles and surface litter in different-aged stands of coffee (*Coffea arabica*) in Kenya (Values are means $\pm$ SD)

Tissue	Stand age (years)			Unsprayed control site
	24	14	4	
Fine root	332.82 ± 15.13	118.73 ± 5.51	46.44 ± 1.08	
Foliage				
Unopened leaves	36.26 ± 0.34	38.46 ± 8.69	54.81 ± 1.75	
Young leaves	71.63 ± 3.06	41.01 ± 2.68	51.56 ± 3.47	
Mature leaves	103.92 ± 3.24	121.5 ± 2.58	129.08 ± 3.34	
Lateral branches	300.78 ± 15.84	151.57 ± 13.52	33.12 ± 1.14	
Main trunk[a]				
Bark	415.4 ± 51.76	283.83 ± 14.97	96.97 ± 3.48	
(unwashed)	690.71 ± 64.00	517.63 ± 51.84	249.51 ± 1.94	
Wood	9.27 ± 0.80	24.38 ± 1.12	8.25 ± 0.36	
	12.02 ± 2.09	10.26 ± 2.86	8.38 ± 0.95	
Soil depth				
0–20 cm	136.06 ± 4.73	47.98 ± 4.81	24.5 ± 0.67	11.09 ± 0.53
20–40 cm	56.6 ± 1.57	25.25 ± 1.34	15.95 ± 0.31	10.11 ± 0.95
40–60 cm	22.79 ± 1.15	25.79 ± 0.57	19.38 ± 0.34	9.78 ± 0.46
60–80 cm	30.2 ± 0.33	27.16 ± 0.89	13.54 ± 0.39	9.66 ± 0.08
Litter fraction				
Leaf	424.13 ± 38.3	211.87 ± 2.53	170.29 ± 5.61	
Twig	220.44 ± 14.5	100.82 ± 5.33	158.38 ± 14.46	
Total litter[a]	883.69 ± 397.41	320.41 ± 150.16	—[b]	

Reproduced by permission of Kluwer Academic Publishers from Lepp *et al.* (1984).
[a] Total litter including amorphous fraction.
[b] Insufficient litter had accumulated to obtain samples for analysis.

wood copper content with time, and the lack of difference between foliar copper contents in the bushes from different stands. These results clearly indicated that much of the copper was bound in surface tissues, suggesting that any effects of contamination may be superficial.

In further studies, the copper content of the reproductive structures and fruit from several stands of differing ages, showed accumulations of copper in the flowers and the outer pulp of the coffee fruit (cherry), but no significant increase in copper content of the fresh or roasted seeds (coffee beans) (Figure 10.6). The copper content of beans was very similar to values obtained in earlier studies in 1949 before copper fungicides were used (Taylor, 1949), again indicative of a lack of mobility of copper within the plant. This superficial contamination of the coffee fruits posed some interesting questions relating to the treatment of fruit to remove coffee beans. Harvested fruit are soaked for 24–48 h to allow fermentation to start, following which the beans are easily separated from the pulp. The beans are then treated to remove the testa epicarp, and finally sun-dried prior to shipment (Lepp and Dickinson, 1987). The residual pulp is collected in large mounds and eventually disposed of, some of it as mulch, applied between the rows of bushes. At present, there is

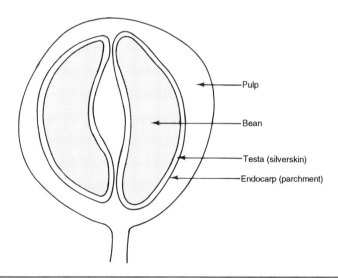

	Flowers	Cherry pulp	Hull	Silverskin	Coffee beans			Roasted coffee beans
					First	Seconds	Lights	
Copper conc.	172.5 ± 5.3	74.5 ± 3.9	48.6 ± 0.1	68.7 ± 3.7	17.6 ± 0.5	16.1 ± 0.4	16.8 ± 1.0	16.8 ± 0.5

Figure 10.6. Transverse section of coffee berry, showing the components of the fruit and copper concentrations (μg g^{-1} dry wt). Adapted from Dickinson *et al.* (1984) and Lepp and Dickinson (1987)

no good use for this residue (Bressiani, 1979) which acts as a significant focus for copper accumulation.

In addition to copper contaminating the fruit pulp, the water from washing and fermentation is also removed from the site; in the case of the plantation under study, this involved removal via a series of drainage channels and lagoons, with eventual discharge into a local river. The sediments from the lagoons and channels are routinely dredged, and spread on to grassland areas on poorer soils in the valleys between the volcanic footridges. The sediments are applied for their fertiliser value, and the grasses are subsequently harvested and used as mulching material, applied between the rows of coffee bushes. Dredged sediments from channels and lagoons that are often applied to mulch grass areas contain high levels of copper (Table 10.3).

A model was developed of copper circulation in the coffee plantation, based on application, removal from the system, and fluxes due to the prevailing agronomic practices associated with the cultivation. The routine application of the organic mulch, in order to conserve water in the dry season, can be seen as a significant factor in the copper cycle. The provision of an additional layer of organic matter to bind excess copper ensures that decomposition is slowed and that copper accumulates in this layer with time (Table 10.4). The slow release of copper from this litter means that potential contamination of the soil at depth is less probable, but the incorporation of the contaminated mulch into the upper soil horizons may be a problem in the future (Lepp and Dickinson, 1987).

Table 10.3. Copper content (μg g^{-1} dry wt) of sediments from watercourses receiving waste water from "wet" coffee processing

Location	Copper content
Drainage channel (adjacent to factory)	605.2 ± 2.5
Settlement lagoon	435.5 ± 12.2

Reproduced from Lepp and Dickinson (1987).

Table 10.4. Copper content (μg g^{-1} dry wt) of mulch applied to coffee soils, as related to age of the sample

Mulch age (time after application)	Copper content
Fresh grass	13.6 ± 0.5
8.5 weeks	421.0 ± 25.7
15 months	943.9 ± 18.1

Reproduced by permission of Blackwell Scientific Publications from Lepp and Dickinson (1987).

In conjunction with information on rates of fungicide application from 1955 to 1983 (Figure 10.7), the plantation records for frequency of pruning, and yields of coffee beans from each plot, analyses of soils and vegetation enabled models to be produced for the input, retention and loss of copper from different-aged stands (Dickinson and Lepp, 1985; Lepp and Dickinson, 1987). It is clear from these data (Figure 10.8), that patterns of retention and loss of copper differ with the age of the stand to which copper is applied. In the oldest stand (24 years), which has a long history of treatment, the bulk of the applied copper is located in the soil, and losses from the system are negligible. The younger stands show a significant loss of copper from the system, with a lower proportional retention by soil. In each case, less than 1% of the applied copper is removed by harvesting and pruning. The reasons for this difference are twofold. First, the rate of fungicide input showed a significant increase in the late 1960s and 1970s; the younger stands have received greater doses in shorter time intervals. Secondly, the litter layer in the older stand is better established and contains significantly more copper than that in the two younger stands. This lack of an organic layer, coupled with the increased rate of application, may account for more rapid penetration into the soil horizons, and subsequent downward migration or losses in runoff.

Further studies have been carried out on the temporal accumulation of soil copper as a result of routine fungicide application. A study of 14 different-aged coffee stands in the same area of Kenya examined the accumulation of copper in the upper soil horizons (0–5 cm) (Figure 10.9) and at 10 cm increments from 0 to 40 cm (Figure 10.10). In all cases, copper content increased with stand age; however, the duration of copper application is not the only factor responsible for this accumulation. Doubling times for copper content in these soils have been calculated (Table 10.5) ranging between 9 and 16 years, assuming an increase of fungicide inputs by 1% year^{-1} (Dickinson et al., 1988).

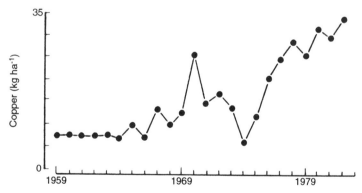

Figure 10.7. Annual rates of copper fungicide application (kg ha^{-1}) into coffee plantation stands, 1959–1983. Reproduced by permission of CEP Consultants Ltd from Dickinson and Lepp (1983)

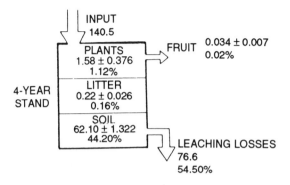

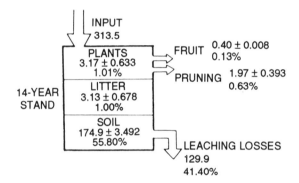

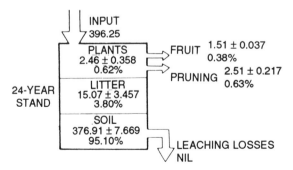

Figure 10.8. Copper budgets for three different-aged stands of coffee, aged 4, 14 and 24 years (Values are kg Cu ha^{-1} ± SE). Reproduced by permission of Elsevier Science Publishers BV from Dickinson and Lepp (1985)

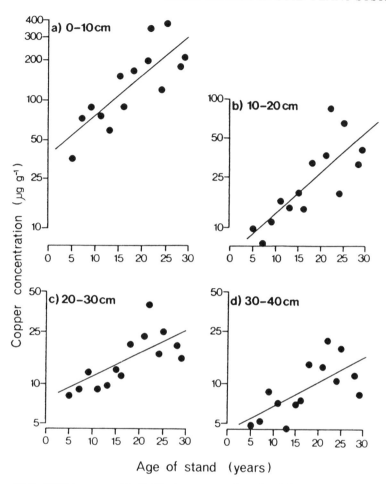

Age of stand (years)

Figure 10.9. EDTA-extractable (0.05 M) copper at four depth increments in soils in 14 different-aged stands of coffee (log scales). Reproduced by permission of Elsevier Science Publishers BV from Dickinson *et al.* (1988)

Other studies have shown there may be further consequences of increased rates of fungicide application. A small-scale field trial, again carried out in Kenya, investigated the response of coffee bushes to a doubling of normal rates of fungicide input (eight applications of 2.5 kg ha^{-1} Cu over a 16 week period). During this period, treated and control bushes (which received normal rates of Cu input) were scored for a variety of parameters and, at the conclusion of the treatment, fruit production and yield, together with dry matter production was assessed (Lepp and Dickinson, 1986*a*). Results showed that extensions of the first internodes on treated shoots were significantly reduced by excess copper, but other internodes remained unaffected. Leaf dry matter content was reduced in treated bushes, but the number of leaves retained was

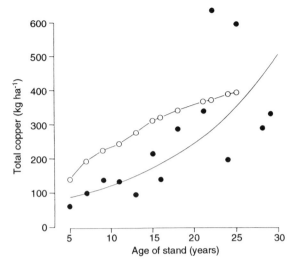

Figure 10.10. Total application since planting of fungicide copper to different-aged stands of coffee (open circles) and estimated total copper in soil (0–40 cm depth) (closed circles) minus background levels. Reproduced by permission of Elsevier Science Publishers BV from Dickinson *et al.* (1988)

Table 10.5. Predicted time required for a doubling of copper levels in coffee soils, from a model developed by Dickinson *et al.* (1988)

Parameter	Soil depth (cm)	Predicted doubling time (years)
EDTA-extractable Cu	0–5	12.6
concentrations	0–10	9.6
(μg g^{-1} dry wt)	10–20	9.4
	20–30	16.5
	30–40	16.5
Total Cu in soil (kg ha^{-1})	0–40	9.6

increased and effects on fruit production were also evident (Table 10.6). Foliar abnormalities (corrugation, tip curl and necrotic margins) were more apparent in the treated bushes, and the incidence of leaf miner damage (caused by larvae of the Lepidopteran *Leucoptera meyricki*) was increased.

The treated bushes showed a reduced rate of fruit maturation together with shrivelling and premature abscission. There was no difference in the dry matter content of mature fruit from each treatment; this is not what might be expected given (a) the reduced numbers of fruit on treated bushes and (b) the increased leaf retention, giving the potential for more photosynthetic production. Increased leaf retention, coupled with reduced dry matter content

Table 10.6. Effect of excessively high rates of application of copper fungicide on leaf weight and number, number of fruits per branch and mean fruit dry weight. Bushes received eight applications of 2.5 kg Cu ha^{-1} at 14 day intervals over a 16 week period

Time after treatment commenced (weeks)	Leaf dry matter content (g^{-1})		Leaf number	
	Treated	Control	Treated	Control
10	0.23 ± 0.002	0.28 ± 0.005	657	517
13	0.26 ± 0.007	0.28 ± 0.004	329	259
16	0.23 ± 0.003	0.26 ± 0.002	395	261

Weeks after treatment commenced	Mean fruit number $\pm$ SD	Dry weight (g^{-1})
10		
Treated	28.3 ± 6.5	—
Control	54.3 ± 10.3	—
13		
Treated	28.3 ± 7.6	—
Control	59.3 ± 12.2	—
16		
Treated	23.3 ± 6.7	0.8 ± 0.003
Control	69.3 ± 14.0	0.85 ± 0.18

Reproduced by permission of Selper Ltd from Lepp and Dickinson (1986a).

is probably a response to the persistent coating of fungicide which developed on treated leaves, reducing light penetration and inducing the development of "shade" leaves, giving a greater canopy to harvest light energy. The increase in canopy could develop into a reservoir for leaf-mining insect larvae. Copper penetrates the leaf cuticle poorly, hence internal tissue feeders will be little affected; herbivorous insects seem capable of coping with increases in dietary copper with few adverse effects. Perhaps the most serious aspect of this trial was the effect of the increased copper treatment on fruit production.

It is well known that shading depresses floral initiation in coffee, although the above trial was not conducted at a major harvest time. Reduced photosynthetic efficiency, due to fungicide deposits, may have affected fruit yield, but other experiments showed that copper may interfere with fruit production in a different manner. Previous studies on copper distribution in mature coffee bushes (Dickinson *et al.*, 1984) showed an accumulation of copper in the flowers, and levels of copper were of an order of magnitude that could potentially affect floral biology, especially the behaviour of pollen following anther dehiscence. Tests on the effects of a range of copper concentrations on pollen germination (Lepp and Dickinson, 1986*b*) revealed that increasing the concentration of copper caused a reduction in pollen germination with an LC_{50} between 50 and 75 ppm applied copper. Different cultivars of coffee are planted on the same estate, and pollen from three cultivars, SC28, SC34 and K7, showed a remarkable uniformity in pollen response to copper (Table 10.7). However, results from the original bushes tested (cv. SC28, planted in 1934) differed from the same cultivar planted in 1963. Coffee cultivars show a high degree of genetic homogeneity; coffee is the only major plantation crop routinely multiplied by seed—a reflection of the genetic stability of named cultivars—so the possibility that the history of copper exposure that bushes had received being a contributory factor to copper tolerance of their pollen deserved investigation. The effect of copper on pollen germination in cv. SC28 (planted 1934) and Catimore (planted 1983, never sprayed) was investigated. There were considerable differences in sensitivity to copper between the two cultivars with different histories of copper treatment; the unsprayed variety is an order of magnitude more sensitive to copper effects on pollen germination than the variety with a lengthy history of fungicide application. Pollen is known to express the same tolerance to metals as the parent plant from which it was produced (Searcy and Mulcahy, 1985); coffee cultivars are genetically

Table 10.7. Effect of copper on *Coffea arabica* pollen germination (a) of three different cultivars, and (b) on cultivar SL28 from two different-aged stands. Values represent % germination ± SD

(a)

Cultivar	ppm Cu					
	0	5	10	20	50	100
SL28	72.6 ± 0.3	63.3 ± 1.2	57.0 ± 0.8	50.3 ± 0.3	34.0 ± 1.2	8.3 ± 0.3
SL34	73.0 ± 0.4	63.7 ± 0.3	55.3 ± 1.1	51.3 ± 2.1	33.0 ± 0.6	8.0 ± 0.6
K7	71.6 ± 0.3	65.3 ± 0.3	54.3 ± 2.0	50.6 ± 2.8	34.3 ± 2.8	9.3 ± 0.7

Two-way AOV indicates SD between Cu concentrations ($p < 0.001$) but not between cultivars.

(b)

Date of planting	ppm Cu					
	0	5	10	20	50	100
1934	75.6 ± 0.3	69.6 ± 0.3	63.9 ± 0.3	59.5 ± 0.8	40.3 ± 0.3	15.3 ± 0.3
1963	70.6 ± 0.3	63.6 ± 0.3	57.0 ± 0.3	50.3 ± 0.3	34.0 ± 1.2	8.3 ± 0.3

Two-way AOV indicates SD between Cu concentration ($p < 0.001$) and date of planting ($p < 0.001$).
Reproduced by permission of CEP Consultants Ltd from Lepp and Dickinson (1986*b*).

Increased copper inputs also affect other aspects of the plantation ecosystem. There is evidence for development of copper tolerance in a common weed found in the coffee stands. Seed from populations of an annual composite, *Bidens pilosa* (Black Jack), established in different-aged coffee stands, germinated and grew best in the soils of the parent stand of each population (Dickinson and Lepp, 1988). Control seed (from unsprayed soil) grew poorly in each of the plantation soils. As *B. pilosa* has been shown to evolve herbicide tolerance in other parts of the world, and given that local conditions favour four or five generations per annum, evolution of copper tolerance as a result of repeated fungicide treatment may be taking place (Dickinson and Lepp, 1988).

10.3.1 CONSEQUENCES OF COPPER ACCUMULATION IN PLANTATION SOILS

Routine use of copper fungicides in coffee causes an accumulation of copper in soils, in excess of background levels, to depths of up to 80 cm. Substantially elevated levels are found in the upper 10 cm and the litter layer. Copper phytotoxicity in most plants occurs at critical EDTA-extractable soil copper levels between 50 and 100 $\mu g\,g^{-1}$ dry wt of soil. Data for copper in the soils of different-aged coffee stands (Figures 10.9 and 10.10) indicate that this threshold is exceeded in all stands aged between 4 and 30 years at the 0–5 cm depth, and in all stands aged between 10 and 30 years if the 0–10 cm values are considered (Dickinson *et al.*, 1988).

Coffee bushes are long-lived, and replacement is only considered after 25–30 years of fruit production, given that stands less than four years old do not produce significant yields of beans (Cannell, 1971). Stands that are removed will be those with a lengthy history of fungicide treatments, where copper has accumulated to potentially phytotoxic levels in the upper soil horizons. Coffee has always been planted on good quality agricultural land, and the re-utilisation of this for crop production is intended; the type of crop replanted will depend on political agricultural priorities and policies. The majority of alternative crops have to be established as seeds or transplanted seedlings (coffee). All will have to grow in soil horizons containing phytotoxic copper concentrations. Yields of maize are depressed at soil levels in excess of 100 $\mu g\,g^{-1}$ dry wt EDTA-extractable copper (Lexmond, 1980), and seedling growth is depressed at such levels. Poor root growth may cause problems with water and nutrient stress. Phytotoxicity following copper fungicide use is well known, occurring following citrus and tomato culture in Florida, orchards in Australia, in old banana plantations in Costa Rica, and probably in the "replanting sickness" of cocoa in West Africa.

Reclamation and rehabilitation of these soils would not be easy. Liming would reduce the plant-available soil copper pool, although lime is already routinely applied to the coffee stands. New plantings of coffee bushes are usually established in holes filled with less contaminated soil, but outgrowth of roots could be affected later. Deep ploughing to mix the soil layers is also possible, but this could merely even out copper levels to unacceptably high levels further down the soil profile. The economics of any form of soil amelioration must also be weighed against potential income. What is clear is that the legacy of copper fungicide use may be very persistent. Studies in Australia estimate that a period of 80–120 years would be necessary to achieve natural reductions of copper content in soil containing up to 1280 $\mu g\,g^{-1}$ Cu, derived from fungicide application (Tiller and Merry, 1981). It is certain that copper fungicides will continue to be used; indeed their use could well expand as alternative organic formulations are little tested in coffee and are much more expensive. Future use of copper should, however, be tempered by a knowledge of the consequences for long-term agricultural use of the soils supporting the treated crops, and strategies should be developed that reduce the serious detrimental effects.

10.4   FUTURE USE OF COPPER FUNGICIDES

Current use of copper fungicides, and future changes depend on several factors:

1. Development of alternative products

Developing alternative products is an increasingly costly process, with the need to demonstrate a significant return on investment. Target crops need to be those of high value located in regions where increased fungicide cost can be met by the agricultural producer and, eventually, the consumer. The length of time for product development, the lengthy process of certification, and the possibility of evolution of tolerant strains of pathogenic organisms must also be seriously considered.

2. Changes in agronomic practice

New techniques in crop treatment, spacing and fertilisation may reduce areas under cultivation whilst producing the same or increased yield in comparison to present-day methods. In the case of coffee, it is estimated that Kenya could meet its quota by closer planting and more intensive cultivation, leaving a surplus of land in the region of 50 000 ha (50% of that currently under coffee). Whilst rates of copper fungicide application may alter, the loci of application would become more concentrated.

3. Changing strategies of disease control

Rates of copper fungicide application have increased steadily in Kenya (Figure 10.7). It is evident that this trend will continue. Field trials have demonstrated better control of Coffee Berry Disease and Bacterial Blight with the equivalent of 52 kg ha^{-1} year^{-1} of copper hydroxide-based formulation (Kairu *et al.*, 1985); at this rate, some toxicity of symptoms were observed in treated bushes, but yields were not affected. By comparison, untreated bushes showed an 80% reduction in fruit yield. There is clearly a very fine economic balance between rates of copper application, costs and yield, which may vary with changes in quotas or with demand for the end product.

4. Changes in resistance of pathogens

Evolution of metal tolerance in bacteria and fungi is a widely recognised phenomenon. Rates and patterns of the evolution of such properties are very dependent on the manner of input of the metals to the microorganisms. Tolerance development is favoured by a gradual input over lengthy time periods, or by selection of tolerant individuals over many generations following colonisation of metal-polluted substrates. Copper fungicide inputs are concentrated and may not follow a regular pattern. Timing is coincident with the germination, rather than reproduction of the fungal propagules so, in many cases, the microorganism cannot complete its life cycle. Only highly tolerant individuals survive, and that incidence within a population may be very small. In addition, increasing rates of input may place further strains on tolerant individuals, breaching limits of tolerance and reducing the rate of development of more tolerant individuals. Bacteria have a more rapid generation time than fungi and have the potential to evolve tolerant populations more rapidly.

Sudin *et al.* (1989) have demonstrated the presence of copper-resistant *Pseudomonas syringae* pv. *syringae* in cherry orchards previously sprayed with copper sulphate to control cherry leaf spot (*Coccomyces hyemalis* Higgins) and bacterial canker (*P. syringae* pv. *syringae* van Hall and *P. syringae* pv. *morsprunorum*). Seventeen copper-resistant isolates were identified; the persistence of copper resistance was demonstrated by reinoculating Cu-sensitive and Cu-resistant isolates onto bean leaves previously sprayed with cupric hydroxide. Only copper-resistant strains showed little or no reduction in the presence of cupric hydroxide residues. Similar studies on *P. syringae* isolates from citrus orchards in California (Anderson *et al.*, 1991) revealed the presence of copper-tolerant isolates. Prior exposure of tolerant strains to sublethal copper concentrations caused a thousand-fold increase in the fraction of cells that could survive a subsequent exposure to elevated media copper levels. Non-tolerant strains did not respond in a similar manner. LC$_{50}$ of tolerant strains (23 ppb Cu^{2+}) was increased to

160 ppb Cu^{2+} following exposure to sublethal copper levels. Copper-tolerant strains also grew rapidly on $Cu(OH)_2$-treated leaves, in contrast to non-tolerant strains which failed to flourish in these conditions. Copper resistance in *P. syringae* pv. Tomato is plasmid-mediated (Cooksey, 1987). Resistant bacteria accumulate copper as part of the resistance mechanism encoded on the *cop* operon located on the plasmid (Cha and Cooksey, 1991). Studies investigating accumulation of other metals by a range of copper-resistant *Pseudomonas* spp. show several metals may be accumulated by the bacteria, but the presence of copper in the medium, which activates the *cop* operon, induced no further increase in metal uptake. This suggests that copper resistance and uptake of other metals are not linked, and that there is a common mechanism for copper resistance in *Pseudomonas* spp. found on plants which have received anti-microbial copper sprays (Cooksey and Azad, 1992).

There are now several instances where phytopathogenic bacteria show resistance to copper levels which kill sensitive strains (Marco and Stall, 1983; Adaskaveg and Hine, 1985; Bender and Cooksey, 1986). *P. syringae* pv. *Garcae*, the causal agent of bacterial blight of coffee has yet to be investigated; however, the increased rates of copper application needed to maintain control (Kairu *et al.*, 1985) would indicate that some resistance to copper may be developing.

REFERENCES

Acland, J. C. (1975) *East African Crops*. Longman, London.

Adaskaveg, H. G. and Hine, R. B. (1985) Copper tolerance and zinc sensitivity of Mexican strains of *Xanthomonas campestris* pv. *vesicatoria*, causal agent of black spot of pepper. *Plant Diseases*, **69**, 993–996.

Anderson, G. L., Menkissoglou, O. and Lindow, S. E. (1991) Occurrence and properties of copper-tolerant strains of *Pseudomonas syringae* isolated from fruit trees in California. *Phytopathology*, **81**, 448–656.

Anon. (1975) *Copper in Agriculture*. Conseil International pour le Développment du Cuivre, Geneve.

Anon. (1985*a*) La Médoc et la découverte de la bouille bordelaise. *Medoc (Bulletin d'Information du GIE des Vins du Médoc)* No. 31, July 1985, pp. 5–32.

Anon. (1985*b*) Promising outlook for copper sulphate in fungicides. *CRU Copper Studies* **13**, No. 5 p. 4.

Ayanlaja, S. A. (1983) Rehabilitation of cocoa (*Theobroma cacao*) in Nigeria: major problems and possible solutions. 1) Causes of difficulty of seedling establishment. *Plant and Soil*, **73**, 403–409.

Bender, C. L. and Cooksey, D. A. (1986) Indigenous plasmids in *Pseudomonas syringae* pv. Tomato: conjugative transfer and role in copper resistance. *Journal of Bacteriology*, **165**, 534–541.

Bressiani, R. (1979) The by-products of coffee berries. In: Braham, J. E. and Bressiani, R. (Eds) *Coffee Pulp. Composition, Technology and Utilisation*, pp. 5–10. IDRC, Ottawa.

Cannell, M. G. R. (1971) Production and distribution of dry matter in trees *Coffea arabica* L. in Kenya as affected by seasonal climatic differences and the presence of fruits. *Annals of Applied Biology*, **67**, 99–120.

Cannell, M. G. R. (1983) Exploited crops—coffee. *Biologist*, **30**, 257–263.

Cha, J-S. and Cooksey, D. A. (1991) Copper resistance in *Pseudomonas syringae* mediated by periplasmic and outer membrane proteins. *Proceedings of the National Academy of Sciences, USA*, **88**, 8915–8919.

Cooksey, D. A. (1987) Characterisation of a copper resistance plasmid conserved in copper resistant strains of *Pseudomonas syringae* pv. Tomato. *Applied Environmental Microbiology*, **53**, 454–456.

Cooksey, D. A. and Azad, H. R. (1992) Accumulation of copper and other metals by copper-resistant plant pathogenic and saprophytic pseudomonads. *Applied Environmental Microbiology*, **58**, 274–278.

Cordero, A. and Ramirez, G. F. (1979) Acumulamiento de cobre en los suelos del Pacifico sur de Costa Rica y sus efectos detrimentales en la agricultura. *Agronomica Costarricense*, **3**, 63–78.

Dickinson, N. M. and Lepp, N. W. (1983) Copper contamination of plants and soils associated with the cultivation of coffee (*Coffea arabica*) in Kenya. In: *Proc. Int. Conf. on Heavy Metals in the Environment* (Heidelberg), pp. 797–800. CEP, Edinburgh.

Dickinson, N. M. and Lepp, N. W. (1985) A model of retention and loss of fungicide-derived copper in different-aged stands of coffee in Kenya. *Agriculture, Ecosystems and Environment*, **14**, 15–23.

Dickinson, N. M. and Lepp, N. W. (1988) Accumulation, mobility and plant availability of copper in Kenyan soils. In: *Proc. 3rd Int. Symposium on Micronutrients in Agriculture*, Brussels, 1988, pp. 291–296.

Dickinson, N. M., Lepp, N. W. and Ormand, K. C. (1984) Copper contamination of a 68-year old coffee (*Coffea arabica* L.) plantation. *Environmental Pollution (Series B)*, **7**, 223–231.

Dickinson, N. M., Lepp, N. W. and Surtan, G. K. (1988) Further studies on copper accumulation in Kenyan *Coffee arabica* plantations. *Agriculture, Ecosystems and Environment*, **21**, 181–190.

Kairu, G. M., Nyangena, C. M. S. and Crosse, T. C. (1985) The effect of copper sprays on bacterial blight and coffee berry disease in Kenya. *Plant Pathology*, **34**, 207–213.

Lepp, N. W. and Dickinson, N. M. (1985) The consequences of routine long-term copper fungicide usage in tropical beverage crops: Current status and future trends. In: Lekkas, T. D. (Ed.) *Proc. Int. Conf. on Heavy Metals in the Environment*, Athens, pp. 274–276. CEP, Edinburgh.

Lepp, N. W. and Dickinson, N. M. (1986*a*) Some aspects of excess copper fungicide application on growth and yield of coffee (*Coffea arabica* L.) bushes on a Kenyan plantation. In: Lester, J. N., Perry, R. and Sterrit, R. M. (Eds) *Proc. Int. Conf. on Chemicals in the Environment*, Lisbon, pp. 384–390. Selper Publications, London.

Lepp, N. W. and Dickinson, N. M. (1986*b*) Effect of copper on germination of *Coffea* pollen: Possible induction of copper tolerance. In: *Proc. 2nd Int. Conference on Environmental Contamination*, Amsterdam, pp. 33–35. CEP, Edinburgh.

Lepp, N. W. and Dickinson, N. M. (1987) Partitioning and transport of copper in various components of Kenyan *Coffea arabica* stands. In: Coughtrey, P. G., Martin, M. H. and Unsworth, M. H. (Eds) *Pollution Transport and Fate in Ecosystems*, pp. 289–299. Blackwell, Oxford.

Lepp, N. W., Dickinson, N. M. and Ormand, K. C. (1984) Distribution of fungicide-derived copper in soils, litter and vegetation of different-aged stands of coffee (*Coffea arabica* L.) in Kenya. *Plant and Soil*, **77**, 263–270.

Lexmond, Th. M. (1980) The effect of soil pH on copper toxicity in forage maize grown under field conditions. *Netherlands Journal of Agricultural Sciences*, **28**, 164–183.

Lima, J. S. (1993) A balance of copper in cocoa agrarian ecosystems under cupric fungicide treatments. *Agriculture, Ecosystems, Environment* (in press).

Marco, G. M. and Stall, R. E. (1983) Control of bacterial spot of pepper initiated by strains of *Xanthomonas campestris* pv. *vesicatoria* that differ in sensitivity to copper. *Plant Diseases*, **67**, 779–781.

Purseglove, J. W. (1974) *Tropical Crops: Dicotyledons*. Longmans, London.

Reuther, W. and Smith, P. F. (1952) Iron chlorosis in Florida citrus groves in relation to certain soil constituents. *Proceedings of the Florida State Horticultural Society*, **65**, 62–69.

Reuther, W., Smith, P. F. and Scudder, G. K. Jr (1953) Relation of pH and soil type to toxicity of copper to citrus seedlings. *Proceedings of the Florida State Horticultural Society*, **66**, 73–80.

Searcy, K. B. and Mulcahy, D. L. (1985) The parallel expression of metal tolerance in pollen and sporophytes of *Silene dioica* (L) Clairv., *S. alba* (Mill) Krause and *Mimulus guttatus* Dc. *Theoretical and Applied Genetics*, **69**, 597–602.

Smith, I. M. (Ed.) (1985) *Fungicides for Crop Protection: 100 Years of Progress*, Vol. 1. Monograph No. 31, British Crop Protection Council, Croydon.

Sudin, G. W., Jones, A. L. and Fulbright, D. W. (1989) Copper resistance in *Pseudomonas syringae* pv. *syringae* from cherry orchards and its associated transport *in vitro* and *in planta* with a plasmid. *Phytopathology*, **79**, 861–865.

Taylor, A. (1949) The copper content of coffee and coffee products. *Chemical Industry*, **43**, 737–738.

Tiller, K. G. and Merry, R. H. (1981) Copper pollution of agricultural soils. In: Loneragan, J. F., Robson, A. D. and Graham, R. D. (Eds) *Copper in Soils and Plants*, pp. 119–137. Academic Press, London.

Wardlaw, C. W. (1972) *Banana Diseases*. Longman, London.

Whittaker, M. J. (1984) Studies of inputs to smallholder coffee farmers through the co-operative distribution system in Kenya from 1981 to 1983. *Kenya Coffee*, **49**, 227–238.

Wood, G. A. R. (1975) *Cocoa*. Longman, London.

11 Caesium Cycling in Heather Moorland Ecosystems

A. DAVID HORRILL AND GILL CLINT

Merlewood Research Station, Grange-over-Sands, Cumbria, UK

ABSTRACT

Radiocaesium, deposited by rainfall following the Chernobyl accident, represents a single pulse injection of a radio tracer into semi-natural ecosystems. The mobility of radiocaesium in the semi-natural systems of the western and north-western regions of the UK is much greater than predicted by models derived from agricultural systems. Investigations have been carried out in both the field and the laboratory. Field studies on heather-dominated upland systems have shown that nearly 50% of the Chernobyl-derived radiocaesium is still in the vegetation and upper 5 cm of the rooting zone after four years, and the bryophyte and litter layers contain appreciable amounts. Laboratory studies indicate that soil mobility of radiocaesium is affected by the soil fungi both binding and enhancing movement depending on circumstances. Mycorrhizal fungi can affect the uptake and mobility of radiocaesium in plants and, although reducing the overall plant uptake, have been shown to enhance translocation from root to shoot.

11.1 INTRODUCTION

Radiocaesium is present in many natural and semi-natural ecosystems. Its origins include nuclear weapons testing, reprocessing of reactor fuel by the nuclear industry and reactor accidents. The Chernobyl accident injected a single pulse of material with a unique signature in terms of its radioisotope composition, into many ecosystems. This accident has therefore provided scientists with a tracer that can be used to study the pathways of materials and the processes controlling those pathways.

A plume forming part of the Chernobyl release passed over the UK in May 1986 and this led to the deposition of a range of radionuclides by both wet and dry processes. Radionuclides reaching the UK were primarily those volatilised during the fire after the initial explosion and included ^{103}Ru, ^{106}Ru, ^{131}I, ^{134}Cs, ^{137}Cs and ^{140}Ba. After the radioactive decay of the short-lived radionuclides, those causing most concern were radioisotopes of caesium, ^{134}Cs and ^{137}Cs, with respective half-lives of 2 and 30 years. Whilst several releases of

Toxic Metals in Soil–Plant Systems. Edited by S. M. Ross
© 1994 John Wiley & Sons Ltd

material occurred from the Chernobyl plant, that which dominated the UK deposition was released on 27 April and initially forced southwards by a weather system situated over Scandinavia. It then passed over the UK in a south-east to north-westerly direction on 2–4 May. The release passed over the sea and returned a few days later on May 6, 7 and 8 (Smith and Clarke, 1989) (Figure 11.1). Caesium deposition was closely correlated with rainfall (Pierson *et al.*, 1960) and the highest areas of deposition were in north Wales, Cumbria and western Scotland (Horrill and Lindley, 1990) (Figure 11.2). These regions contain some of the largest areas of semi-natural vegetation in the UK, often with soils of low pH and high organic content. The radiocaesium recycling in vegetation resulted in concentrations above statutory levels in the tissues of grazing animals used for food.

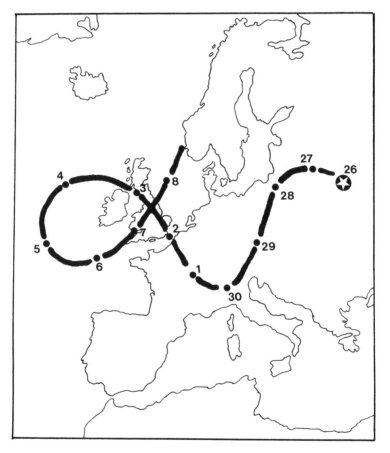

Figure 11.1. Approximate track of the Chernobyl material that affected the United Kingdom. From 26 April 1986 to 8 May 1986. Reproduced with the permission of the Controller of HMSO from Smith and Clarke (1989)

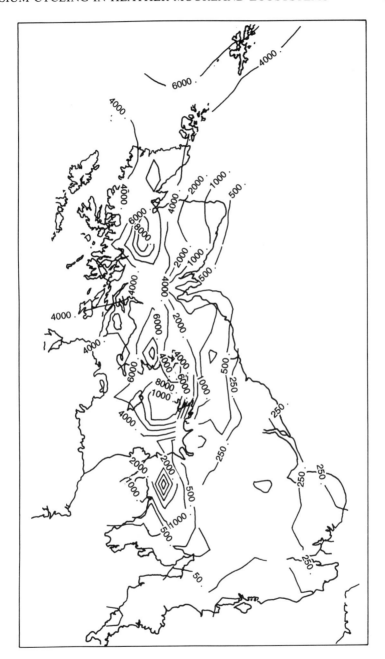

Figure 11.2. Caesium-137 concentrations (Bg kg^{-1} dry wt) measured on graminoid vegetation in UK, May 1986. Reproduced by permission of the International Atomic Energy Agency from Horrill and Lindley (1990).

11.2 ECOSYSTEMS AFFECTED

The high correlation between rainfall, landform and altitude means that many of the ecosystems contaminated are characteristic of the Atlantic and Boreal regions of Western Europe. The ecosystems are varied and the soils derived from a range of rock types. Common amongst these systems are the heaths and uplands of the UK. Much of this land is in areas where agriculture is marginal and semi-natural plant communities dominate. There are large areas dominated by heather (*Calluna vulgaris*) and these areas are grazed not only by domestic sheep but by wild populations of red deer (*Cervus elaphus*) and grouse (*Lagopus lagopus*); the red deer and grouse are both shot for sport and can find their way into the human foodchain (Lowe and Horrill, 1991).

11.3 THE PROBLEM

Shortly after the Chernobyl deposition it became apparent that predictions, largely based on models from agricultural systems in lowland habitats, were inaccurate. These predictions were that within a short time the radiocaesium would become tightly bound in the soil and its availability for plant uptake greatly reduced. This proved not to be the case for the soils of many upland systems—mostly those with highly organic soils although this was not universally the case. In many of these upland systems the radiocaesium was more mobile than had been anticipated, and was rapidly taken up into the vegetation and eventually into animal tissues. Sheep farming was the most affected activity, with restrictions on the sale and slaughter affecting areas of Wales, Cumbria and Scotland. Even more serious was the situation in Scandinavia, in which an entire economy depends on Reindeer. Contamination levels were many times those found in the UK.

The situation is further complicated by the fact that different species of plants take up radiocaesium at different rates even within the same habitat (Horrill *et al.*, 1990) (Table 11.1). Grazing animals are very selective, only certain plant species within a sward being favoured. The grazing choice of both sheep and cattle has been shown to vary seasonally, and choice will also vary according to spatial and temporal plant distribution. Some animals such as the red grouse may feed almost exclusively on one plant, in this case heather. Work prior to the Chernobyl accident (Bunzle and Kracke, 1984) had established that the heather family, the Ericaceae, could accumulate high concentrations of radiocaesium. Surveys carried out in the Lake District have demonstrated that of the ericaceous species found on a series of sites, *Calluna vulgaris* usually accumulates the highest radiocaesium burden (Table 11.2).

Once the survey phases of work on the Chernobyl deposition were complete, the emphasis changed to try and discover the mechanisms controlling the movement and retention of radiocaesium in ecosystems and the possibility of

Table 11.1. Range of radiocaesium activities for species of frequent occurrence in upland habitats in Bq kg^{-1} dry weight

| Species | N | Activity ^{137}Cs | | | |
		Max	Min	Mean	SE
Juncus effusus	14	1341	14	378	116
Nardus stricta	13	868	10	237	86
Calluna vulgaris	7	4464	39	915	598
Pteridium aquilinum	7	700	11	236	96
Juncus squarrosus	6	1989	101	832	301
Erica cinerea	5	93	28	52	12
Erica tetralix	3	2324	179	1212	620
Vaccinium myrtillus	4	220	78	141	33
Polytrichum commune	10	4722	339	1589	410
Sphagnum spp.	10	5918	411	1992	551

Source: After Horrill *et al.* (1990).

Table 11.2. Comparative performance of ericoid species on dry heathland sites in Cumbria in Bq kg^{-1} dry weight

Site	Species	^{137}Cs
Grasmoor (360 m)[a]	*Calluna vulgaris*	469
	Vaccinium myrtillus	220
	Erica cinerea	93
Hindscarth (490 m)	*Calluna vulgaris*	366
	Empetrum nigrum	107
	Vaccinium myrtillus	78
	Erica cinerea	33
Yewbarrow (85 m)	*Calluna vulgaris*	227
	Erica cinerea	40
Robinson (335 m)	*Calluna vulgaris*	126
	Vaccinium myrtillus	96
	Erica cinerea	65
Gawthwaite (200 m)	*Calluna vulgaris*	39
	Vaccinium myrtillus	170
	Erica cinerea	28

Source: After Horrill *et al.* (1990).
[a]Altitude of site in metres.

any remedial measures that might reduce levels in animals, either native or domestic used by man as food. There was a suspicion that the soil microflora might play a part in the cycling of radiocaesium as well as chemical and physical processes. Fungal fruiting bodies proved to be high in radiocaesium and have been shown to be important contributors to the body burden of wild animals in Scandinavia (Hove *et al.*, 1990). Investigations were therefore started on upland heather-dominated systems to study the mobility of radiocaesium in the field situation. At the same time, laboratory studies were initiated to investigate the role that fungal mycelia might play in controlling the movement of radiocaesium in the soil.

11.4   FIELD RESEARCH

Field research has been carried out on a number of heather systems contaminated by the Chernobyl deposit both in northern England and Scotland. Two radioisotopes of caesium were present in the fallout. Although a little ^{134}Cs is released in low level liquid waste by the nuclear industry, this is highly unlikely to penetrate to inland sites. It is thus possible to use the ^{134}Cs as a marker of the movement of radiocaesium in the environment. The ratio ^{134}Cs: ^{137}Cs can be used to separate ^{137}Cs originating from Chernobyl from that coming from other sources. This ratio is not constant over the UK, possibly due to inhomogeneities in the depositing cloud, and a range of 1:1.4 to 1:2.5 has been reported (Cambray *et al.*, 1987). However, early determinations at the experimental sites ranged from 1:1.9 to 1:2 and the actual values determined at each site have been used in any calculations.

Table 11.3. Percentage of radiocaesium found in ecosystem components from a deep peat site in Scotland

(a)	Whole system: total radiocaesium burden 14 kBq m^{-2} ^{137}Cs	
	Soil	52%
	Vegetation	48%
(b)	Vegetation component	
	Heather (current year's growth)	14.5%
	Heather (past growth)	11.5%
	Heather (old stems)	11.0%
	Dead plant material	11.5%
	Cryptogams	49.0%
	Other plant species	2.5%
(c)	Soil component (by depth)	
	Litter layer	45.5%
	0.0–2.5 cm	17.2%
	2.5–5.0 cm	16.2%
	5.0–7.5 cm	7.1%
	7.5–10.0 cm	3.0%
	10.0–12.5 cm	4.0%
	12.5–15.0 cm	3.0%
	15.0–20.0 cm	4.0%

Soils investigated under many of the heather systems are either nearly pure peats or podsolised soils with a highly organic surface layer. Many of the soils investigated are subject to seasonal or permanent waterlogging. The systems studied represent extremes with respect to the high content of organic matter and the acidity of the soils. The first stage of the investigation has been to determine the radiocaesium distribution in the systems as a whole. Over a series of heather systems investigated, on highly organic soils, the vegetation contained between 30 and 50% of total radiocaesium inventory. At one study site in Scotland on deep peat (Table 11.3), nearly half the radiocaesium inventory is contained in the vegetation layer. Breaking this down further for this site (Table 11.3(b)), it can be seen that the layer of cryptogams (mosses, liverworts and lichens) represent an important reservoir of material. Heather contains over 33% of the material with a similar concentration in both the past year's growth (PYG) and current year's growth (CYG). The rapid translocation into the new tissues is important when grazing is considered as it is the new growth that forms an important component of the diet of grouse and to a lesser extent deer and sheep.

If the soil profile is considered, the litter layer is a major contributor to the soil inventory, containing nearly half the material (Table 11.3(c)). Below this the upper 5 cm still contains the bulk of the radiocaesium, with smaller amounts as one passes down the profile. The samples for these figures were taken in June 1989 and it was found that the Chernobyl material, as identified by the presence of ^{134}Cs, had penetrated to at least 20 cm in the peat profiles. In podsolic soils, penetration was markedly slowed by the mineral layers and little Chernobyl material was detected below the organic/mineral boundary. Radiocaesium levels in the soil pore waters were very low, almost at the limits of detection. However, in the sites studied, concentrations between 1 and 10 mBq litre^{-1} have been recorded and these represent a constant supply of soluble material to the rooting systems.

The distribution differences between pre- and post-Chernobyl material are illustrated in Table 11.4 for the whole ecosystem. For the whole Scottish site, deposition calculated by summing soil and vegetation content gave a mean of ~19 kBq m^{-2} with a range of between 16 and 21 kBq m^{-2} between samples. This variability may be accounted for by local variations in soil drainage, the rainfall often "puddling" in localised areas. It is estimated for the deep peat site used in this example that 17% of the ^{137}Cs is of pre-Chernobyl origin. In this area it can be seen that three years after deposition half the Chernobyl material is still contained either in the vegetation or the root mat (top 5 cm of litter and soil). In contrast, a very small proportion of the pre-Chernobyl material (~2%) is in this layer and the majority lower in the soil profile.

An important management tool on heather moors is burning. Experiments are in progress to investigate this as a mechanism for remobilising radiocaesium in heather systems. Early results show that if the temperature of a fire under dry conditions reaches 650–700 °C, then about 20% of the

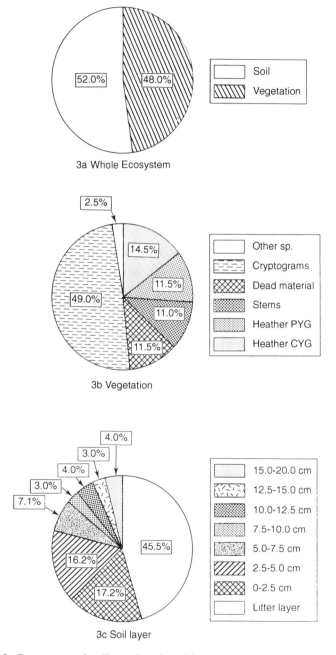

Figure 11.3. Percentage of radiocaesium found in ecosystem components from a deep peat site in Scotland. (a) The whole system, (b) the vegetation layer, (c) the soil layer

Table 11.4. Chernobyl and pre-Chernobyl ^{137}Cs in peatland ecosystem

Component	% pre-Chernobyl	% Chernobyl
Soil	97.8	50.0
Root-mat	2.0	36.3
Vegetation	0.2	13.7

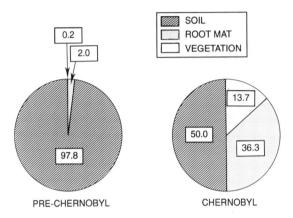

PRE-CHERNOBYL   CHERNOBYL

Figure 11.4. Percentage of pre-Chernobyl and Chernobyl caesium-137 in ecosystem components

radiocaesium is lost in the smoke. Under cooler conditions when temperatures in the range 550–600 °C were attained, the loss was only of the order of 10%. The resulting ash will, of course, contain easily soluble material and this will be immediately available for plant uptake on leaching by rainfall. Burning can take place either in the spring or autumn with different consequences. In the spring there will be a greater potential for plant uptake, particularly if the burning stimulates new growth. With an autumn burn situation there will be greater potential for the soluble material to be washed down the profile and lost to the plants. Studies still in progress investigating the movement of ash leachate down soil cores indicate that the vertical movement is much reduced in mineral-rich podsolic profiles.

11.5   THE RELEASE OF RADIOCAESIUM FROM PLANT LITTERS

In upland environments the plant litter lying on the soil surface is a major source of nutrients for plant uptake and growth. It is therefore important to investigate the fate of radiocaesium released from decaying plant litter of heather and other plant species in heather-dominated ecosystems.

The main processes by which nutrients and radiocaesium are removed from plant litters are by leaching (due to water flow through the litter) and by decomposition of the organic matter by microorganisms (Witkamp and Frank, 1970). Release of minerals including ^{137}Cs may be retarded by retention of the elements by the microorganisms themselves in a process referred to as microbial immobilisation (Witkamp and Barzansky, 1968). Litter decomposition and ^{137}Cs release is likely to be influenced, therefore, by the nature and level of microbial activity in the soil and by the biochemical composition of the litter, some plant litters being more "recalcitrant" than others. In addition physical factors such as degree of waterlogging of the site would be expected to affect the rate of release of nutrients and of radiocaesium.

Studying the release of radiocaesium from litter under field conditions is difficult (though not impossible) because of the problems of spatial variation of soils and because, in general, the levels of radiocaesium in field sites are not high enough to provide a sensitive assessment of the factors affecting release rate in short-term experiments. The authors therefore chose to study the release of ^{137}Cs from artificially labelled litters under controlled laboratory conditions using small microcosms, following the design of Anderson and Ineson (1982). The plant species used were heather (*Calluna vulgaris*), white clover (*Trifolium repens*) and common bent grass (*Agrostis capillaris*). Measurements were made of ^{137}Cs activity and potassium content of leachates from the litter samples taken at fortnightly intervals over a three-month period. Microbial activity was assessed by measuring the respiration rate of each litter sample by infrared gas analysis. (Clint *et al.*, 1992).

Differences were observed in the rates of loss of ^{137}Cs from litter samples of the different species. The rate of release from heather was slower than that from clover and common bent grass, as shown in Table 11.5, which summarises the time-course data. The rate of release of potassium was also significantly lower from heather than from the other species and, as with the release of ^{137}Cs, stabilised at 60–70% remaining within the litter after 3 months, as illustrated in Table 11.6. In continually moist litters, common bent grass had the highest respiration rate throughout the experiment, with heather litter generally exhibiting the lowest respiration rates of the three plant species.

Thus rates and patterns of ^{137}Cs release from plant litters varied with plant species and with physical treatment of the litter. Of the species studied, heather had the lowest rate of release of both ^{137}Cs and potassium, and the lowest levels of microbial activity throughout the experimental period. This suggests that litter from this species will tend to be broken down at a relatively slow rate under field conditions, and that, as a consequence, a large pool of ^{137}Cs may be trapped in this component of the ecosystem. There are clear differences in the behaviour of caesium and potassium. In all litters a larger

Table 11.5. Rate of leaching of radiocaesium from plant litters

	Mean % of original activity remaining	
Plant species	After 14 days	After 84 days
Heather	87	72
White clover	79	40
Common bent grass	76	30

Data shown are for litter samples kept in continually moist conditions and leached with distilled water every 14 days. For other treatments and full statistical analysis see Clint et al., (1992)

Table 11.6. Rate of leaching of potassium from plant litters

	Mean % of original content remaining	
Plant species	After 14 days	After 84 days
Heather	86	63
White clover	30	10
Common bent grass	35	10

Data shown are for litter samples kept in continually moist conditions and leached with distilled water every 14 days. For other treatments and full statistical analysis see Clint et al., (1992).

percentage of the ^{137}Cs was retained than potassium, so that in field conditions a slower leaching of ^{137}Cs than of potassium is likely to occur.

Faster leaching of potassium than ^{137}Cs from litter was indicated by the observation of an increasing ^{137}Cs:K ratio in the sequence: live shoots < dead shoots still attached to plant < litter on the soil surface, for some British heather-dominated ecosystems (Harrison et al., 1990). There is a clear process of discrimination between the two ions in their release from litter, so that the rate of recycling of potassium from litter cannot be used to predict the behaviour of radiocaesium in its release from litters.

To sum up the field work side of the investigation, it has been found that in heather-dominated upland systems. and almost certainly under other vegetation types, the litter and cryptogam layers contain considerable reserves of radiocaesium. These accumulations do not appear to be passing rapidly down the soil profile. Slow release into the rooting zones immediately underneath could provide a supply of material for root uptake at such a rate that material is recycled into the vegetation layer rather than leached from the soil profile.

The soil is not just a black box in which physical and chemical reactions take place. There is a strong biological element in all soils. It is therefore possible that the soil microflora might have a profound effect on the fate of radio-caesium in the soil. The microflora in upland organic soils is dominated by soil fungi (Dighton, J., personal communication), and to demonstrate the part they could play in radiocaesium cycling, a number of laboratory experiments have been carried out.

11.6 LABORATORY EXPERIMENTS

As mentioned previously, deposition patterns of radiocaesium fallout in the UK following the Chernobyl accident in 1986 appear to be strongly correlated with areas of high rainfall. Areas with the highest deposition include therefore the uplands of Wales, Cumbria and Scotland which support, in the main, nutrient-deficient organic soils. The availability of fallout radiocaesium for plant uptake in these regions is potentially greater than in the more fertile lowland soils by virtue of upland soils having a lower clay content, a lower pH and lower potassium concentrations.

It is in such nutrient-deficient soils that mycorrhizas play a significant role in the mineral nutrition of plants. A mycorrhiza is an association between a soil fungus and a plant root which, in normal circumstances, is beneficial to both partners. The plant (often inaccurately termed the "host") receives extra mineral nutrients from the fungus, whilst the fungus receives carbohydrate, usually in the form of sucrose, from the plant. It is generally thought that the fungal hyphae are able to exploit a greater soil volume than the root system alone, and with a lower energetic cost to the host plant (Harley, 1969; Harley and Smith, 1983). The examination of most field-grown plant material will reveal evidence of mycorrhizal infection, although the mycorrhizal type and the degree of infection vary. The distribution of different mycorrhizal types appears to be related to climate and to altitude, with both these factors in turn strongly influencing the rate of turnover of nutrients in soil (Dighton, 1986; Read, 1986). Ectomycorrhizas, in which the fungal material is present largely as a sheath surrounding the outside of the plant root, are dominant in temperate and upland soils. Endomycorrhizas, in which the fungal hyphae penetrate the cortical layers of the plant root and invade the root cells themselves, occur mainly in the tropics and in lowland fertile soils. The comparative efficiencies of these types of mycorrhizas in obtaining nutrients from different soil types with differing rates of inorganic nutrient supply is, therefore, of importance when considering the availability of pollutant radionuclides to plants.

Indications that ectomycorrhizal fungi were efficient accumulators of radiocaesium came from the analysis of fungal fruiting bodies which first appeared some months after the deposition from Chernobyl. Dighton and Horrill (1988) detected 105 pCi g^{-1} dry weight of ^{137}Cs in a sample of *Lactarius rufus* fruit bodies. Interestingly, they calculated that about 90% of this activity originated from radiocaesium fallout from weapons testing in the 1950s and 1960s, and not from the more recent Chernobyl fallout. This suggested that fungal tissues could be very "conservative" with respect to caesium, i.e. show appreciable accumulation but slow turnover. Other reports of radiocaesium accumulation by basidiomycete fungi after the Chernobyl accident have appeared (Elstner *et al.*, 1987; Haselwandter, 1987; Byrne, 1988; Haselwandter *et al.*, 1988; Oolbekkink and Kuyper, 1989). These reports give data on levels of activity of ^{137}Cs and ^{134}Cs in fruit bodies collected from a variety of habitats and include both saprotrophic and mycorrhizal fungi. All showed evidence of long-term storage of radiocaesium, but the tendency to conserve caesium was particularly marked in the mycorrhizal species.

In addition to their role in aiding nutrient acquisition, mycorrhizal fungi may sometimes be said to exert a protective role by accumulating phytotoxic chemicals in their hyphal tissues. For instance, accumulation of copper and zinc by mycorrhizal root systems of ericacious species allowed the continued growth of plants in concentrations of up to 75 mg litres^{-1} Cu and 150 mg litres^{-1} Zn, whereas non-mycorrhizal plants did not grow (Bradley *et al.*, 1982). Similarly, accumulations of zinc in the extramatrical hyphae and sheath of the ectomycorrhizal fungus *Paxillus involutus* in association with birch has been shown by Denny (1986).

Observations made in the immediate aftermath of the Chemobyl accident thus established that mycorrhizal fungi were likely to be extremely important components in the transfer pathway between the contaminated soil and the plant shoot. Several key questions clearly warranted particular attention:

1. Do high levels of radiocaesium in fungal fruiting bodies result from
    (a) an unusually high transport rate of caesium from the soil solution into the fungal hypha (i.e. a high influx across the fungal plasmalemma), or
    (b) a moderate rate of uptake but a high accumulation capacity, or
    (c) a combination of (a) and (b)?
2. How conservative are fungi for caesium? Is accumulated caesium easily re-translocated?
3. Does the presence of a mycorrhizal infection enhance plant uptake and accumulation of caesium?
4. What is the fate of the radiocaesium accumulated in plant tissues when the plant dies and the litter is decomposed?

Until recently, work on the uptake of environmental radionuclides by plants has relied heavily on the use of transfer factors or concentration ratios, which are derived from some form of the following expression:

$$\text{Concentration ratio} = \frac{\text{concentration in plant}}{\text{concentration in soil}}$$

The quantity is easy and quick to measure and has been used extensively by regulatory and advisory bodies to assess, for example, the likely risk to livestock from grazing on contaminated pasture. The measurement of transfer factors is, however, of little value as an experimental approach in a scientific study of the sort required here. The main problem with the use of transfer factors is that they do not address the compartmentation of the radionuclide within the system. For instance, activity in the bulk soil is not synonymous with activity available for plant uptake, since the physical and chemical composition of the soil determines the mobility of elements within it. (A more reliable parameter is the soil *solution* activity, which is rarely determined.) In addition, the transfer factor does not take into account the length of time from the initial deposition during which accumulation into the plant might occur, or the effects of plant growth during the uptake period.

In order to address the key questions listed above the authors therefore considered it necessary to adopt a physiological approach and to investigate the complex soil-to-plant pathway by splitting it up into discrete steps as follows.

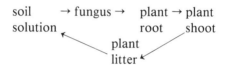

The aim was to measure the unidirectional fluxes of caesium at each of the steps which could be isolated experimentally. Additional data were to be acquired concerning the accumulation capacity of each of the compartments within the system.

A series of laboratory experiments was therefore carried out to investigate the operation of mycorrhizas in terrestrial ecosystems, and the importance of the fungal component in the accumulation and transport of radiocaesium in soil.

11.6.1 UPTAKE OF RADIOCAESIUM BY FUNGAL HYPHAE IN A RANGE OF BASIDIOMYCETES

Short-term influx of caesium into fungal hyphae was measured for 18 fungal species by experiments in which hyphal mats were incubated in ^{137}Cs-labelled medium (Clint et al., 1991). The species investigated included saprotrophic and mycorrhizal basidiomycete fungi together with examples of non-basidiomycete species. Uptake was measured over a 15 min period using buffered Pachlewska's medium containing 5 μM CsCl labelled with ^{137}Cs at 1000 Bq ml^{-1} (200 Bq nmol^{-1} Cs). Caesium influx on a dry weight basis was calculated for each species from the radioactivity in the sample, the specific activity of the labelled solution and the weight of fungal hyphae present. The data were also transformed to give caesium influx on a surface area basis by estimating hyphal length per unit dry weight of each species, and determining hyphal surface area per unit length microscopically.

Species were ranked according to the magnitude of the measured influx. The range of mean influx values observed for all the species was large, from 85 to 276 nmol g^{-1} dry wt h^{-1}. However, the values of the standard errors associated with each mean value were small, confirming that true species differences in caesium influx occur. With data expressed on a dry weight basis, the tendency was for saprotrophic species such as *Mycena polygramma* and *Mycena sanguinolenta* to exhibit the highest values of influx, whilst the non-basidiomycete mycorrhizal fungi such as *Cenococcum graniforme* and *Hymenoscyphus ericae* (the fungal species which forms ericaceous mycorrhizas with heather) had the lowest values. The basidiomycete mycorrhizal fungi had intermediate values.

In short-term influx measurements, where the plasmalemma is the organelle of prime interest, it is perhaps more relevant to consider uptake expressed on a surface area basis. Examples can be found in the literature of influxes in biological tissue expressed on a fresh weight basis, a dry weight basis, or on a surface area basis, depending on the experimental method used. In many cases there is a constant and direct relationship between surface area and weight of the tissue, and the choice of unit is not critical. When the present fungal influxes were expressed on a surface area basis, the range of mean values was 0.01–2.81 nmol Cs m^{-2} h^{-1}, and the ranking of species was markedly changed. The relationship between influx and fungus type was less clear. The change in ranking results from the different surface area to dry weight ratios exhibited by the different fungal species, i.e. the growth form of the different fungi.

The magnitude of the influx values obtained in this study indicates that soil fungi do not have unusually high uptake rates across the plasmalemma. Shaw and Bell (1989) observed values of about 100 nmol g^{-1} h^{-1} for caesium uptake by excised wheat roots with an external Cs concentration of 5 μM. A conversion to influx per unit surface area would yield a value much higher than those determined for any of the fungi investigated here. Jones *et al.* (1991) recorded short-term caesium uptake by excised roots of *Agrostis capillaris* of 600 –1000 pg Cs mg^{-1} per 15 min (equivalent to about 5–30 nmol g^{-1} h^{-1}), depending on external potassium concentration. For comparison, uptake of the major nutrient ions, potassium, sodium and chloride, across the plasmalemma of single-celled giant algae proceeds at a rate in the order of 10 000 nmol m^{-2} h^{-1} (MacRobbie, 1962, 1964), and rubidium influxes in stomatal guard cells (where Rb is used as a potassium analogue) have even higher values (Clint and MacRobbie, 1984).

11.6.2 ACCUMULATION AND IMMOBILISATION OF CAESIUM BY SOIL FUNGI

A study was made of the immobilisation capacity of fungi when exposed to solutions containing 5 μM CsCl labelled with [137]Cs. Rate of loss of caesium from "loaded" tissue was determined by measuring the radioactivity lost to successive aliquots of non-radioactive medium during known time intervals (Dighton et al., 1991).

Three common soil fungi were used in the study: Trichoderma viride, Phoma sp. and Cladosporium sp. These were isolated from soil removed from an upland pasture site in Cumbria. Accumulation of ^{137}Cs and its subsequent efflux were both found to be species-dependent. Thus, Cladosporium had the highest initial caesium content (after a 72 h loading period), but also the highest value of efflux. It also exhibited the shortest half-time for caesium efflux at the end of the washout period, t_{max} (95 min). All three species had retained large percentages of their original activity at t_{max}, and the half-times for tracer loss were all very long. It was estimated that it would take some 37 h of washout to reduce the caesium content of Cladosporium to 3 % of its original content after loading, and that Trichoderma and Phoma would require at least 308 h and 146 h respectively to bring about a similar loss.

Thus loss of radiocaesium after loading proceeds at a slower rate than the rate of original accumulation for all the species studied. This suggests that much of the Cs taken up is bound within the tissue, making it less available for exchange with the outside medium. The inference is that these fungi are potential Cs accumulators in the soil ecosystem.

Thus the influx (i.e. rate of uptake) of caesium into fungal hyphae has been shown to be modest, and inadequate to explain the high radiocaesium activities associated with fungal fruiting bodies collected from field sites in the wake of the Chernobyl accident. It would seem that a more likely explanation for these findings lies in the large capacity for accumulation shown at least by those fungal species we have studied in the laboratory. This, together with the recalcitrance with which caesium is subsequently lost from the hyphae after accumulation, means that these organisms are likely to be of great importance in the maintenance of large pools of radiocaesium in contaminated soils, and in cycling of that radiocaesium in terrestrial ecosystems. The compartmentation of caesium within fungal tissue is of great interest and clearly warrants further study.

11.6.3 THE EFFECT OF MYCORRHIZAL INFECTION ON CAESIUM INFLUX BY PLANTS

Short-term uptake of ^{137}Cs by heather (Calluna vulgaris) plants was measured in the laboratory using radio-labelled liquid media and plants grown from shoot cuttings. Uptake by mycorrhizal plants was compared with uptake by non-mycorrhizal plants, and the effects of different external Cs and K concentrations on Cs influx was also determined (Clint and Dighton, 1992).

Plants were pretreated for 24 h by immersing the cleaned roots in media containing either 5 μM or 500 μM KCl. Influx was then measured by transferring the plants to similar solutions containing also CsCl at 5 μM or 500 μM labelled with ^{137}Cs at 1000 Bq ml^{-1} for known time intervals, For each set of influxes there were thus four experimental treatments, as follows:

- 5 μM CsCl, 5 μM KCl ("low Cs, low K", LL)
- 5 μM CsCl, 500 μM KCl ("low Cs, high K", LH)
- 500 μM CsCl, 5 μM KCl ("high Cs, low K", HL)
- 500 μM CsCl, 500 μM KCl ("high Cs, high K", HH)

Mycorrhizal and non-mycorrhizal plants were used in each experimental treatment $(+, -)$.

A time course for the increase in tissue Cs content was prepared by plotting caesium content against influx period for each plant over a 3 h uptake period. The value of the influx (rate of uptake), obtained from the slope of the curve, decreased with time as the tissue content rose. In all cases, and at any given time during the 3 h uptake period, the Cs content of the mycorrhizal plants was lower than that of the non-mycorrhizal plants. Table 11.7 lists the Cs contents of plants from the various treatments towards the end of the 3 h uptake period. The non-mycorrhizal plants had between 1.7 and 3.7 times the Cs content of mycorrhizal plants. External potassium concentration had no consistent effect on uptake, whilst increasing external Cs concentration markedly increased the Cs content of the plants.

In addition, the root:shoot ratio of Cs content was determined for each plant at the end of the uptake period, as shown in Table 11.8. It can be seen

Table 11.7. Uptake of caesium from liquid media by young heather plants

Treatment[a]	Cs content after 150 min (nmol Cs g^{-1} fresh wt)
LL –	20.5 ± 3.4
LL +	5.5 ± 0.5
LH –	12.4 ± 2.1
LH +	4.6 ± 0.7
HL –	723 ± 110
HL +	430 ± 25
HH –	978 ± 84
HH +	440 ± 42

LL = low [Cs], low [K]; LH = low [Cs], high [K]; HL = high [Cs], low [K]; HH = high [Cs], high [K]. + = mycorrhizal plants; – = non-mycorrhizal plants. Values are mean and standard error, $n = 6$.

Table 11.8. Shoot:root ratios of accumulated
Cs in heather plants after incubation in liquid
media

Treatment[a]	$\dfrac{\text{Cs in shoot g}^{-1}\text{ fresh wt}}{\text{Cs in root g}^{-1}\text{ fresh wt}}$
LL –	0.02 ± 0.008
LL +	0.11 ± 0.018
LH –	0.06 ± 0.025
LH +	0.07 ± 0.017
HL –	0.05 ± 0.017
HL +	0.15 ± 0.019
HH –	0.07 ± 0.016
HH +	0.44 ± 0.035

LL = low [Cs], low [K]; LH = low [Cs], high [K];
HL = high [Cs], low [K]; HH = high [Cs], high
[K]. + = mycorrhizal plants; – = non-mycorrhizal
plants. Values are mean and standard error, $n = 6$.

that the ratio is generally higher in mycorrhizal than in non-mycorrhizal
plants. The inference is that in mycorrhizal plants, despite their lower overall
accumulation during the first 3 h, a higher proportion of the Cs taken up is
located in the shoot. In plants with high external Cs (HL, HH), a higher
absolute amount, as well as proportion, of Cs is found in shoots of
mycorrhizal plants when compared with non-mycorrhizal plants. The
implication is that mycorrhizal infection can enhance the proportion of
accumulated radiocaesium which is transported to the shoot of the plant, at
least in heather. This finding is of obvious significance in the investigation of
radiocaesium cycling and potential entry into the food chain (Clint and
Dighton, 1992).

Although these results were easily repeatable in our laboratory, it is
important to acknowledge that differences exist between the conditions
employed in these experimental studies and the conditions under which
heather plants in the field are contaminated with environmental radioactivity.
In particular, the chemical Cs concentration in the "High Cs" treatment used
here greatly exceeds the concentration that would be found in even the most
contaminated field sites in the UK.

11.6.4 THE EFFECTS OF MYCORRHIZAL INFECTION ON
LONG-TERM CAESIUM ACCUMULATION BY PLANTS

Caesium accumulation by plants was studied using radiolabelled soil in which
mycorrhizal and non-mycorrhizal plants were grown over a period of several
months (Clint, G., Dighton, J. and Monster, A. unpublished).

A single soil type (a brown earth) was mixed one part soil to three parts sharp sand on a weight basis to provide a solid medium of low nutrient status. This was sterilised by microwave treatment before use. The plant species included sheep's fescue (*Festuca ovina*) and white clover (*Trifolium repens*) which form endomycorrhizal associations, and Scots pine (*Pinus sylvestris*) which forms ectomycorrhizal associations. Pine seedlings were grown-up in perlite and peat, together with a mycorrhizal inoculum where appropriate. The other species were initially sown on sharp sand into which a standard mycorrhizal inoculum was incorporated where appropriate.

During the experiment itself, plants were grown in pots containing a central core or "exclusion zone" of 30 g soil labelled with ^{137}Cs at about 100 Bq g^{-1}. The remainder of the soil in the pot was not radiolabelled. The exclusion zones were made from dissected plastic tubes covered with meshes of two different sizes: coarse and fine. The coarse mesh (1 mm) allowed both plant roots and fungal hyphae to penetrate into the tube, whilst the fine mesh (0.002 mm) allowed only fungal hyphae to penetrate. In this way the authors hoped to be able to distinguish between plant accumulation via plant roots plus extramatrical fungal hyphae and plant accumulation via fungal hyphae only. Plants were grown for approximately 3 months and then harvested.

11.6.4.1 Radiocaesium accumulation by fescue

In fescue root biomass was much higher than shoot biomass as given by the low shoot:root weight ratios. Despite this, the shoots had higher radiocaesium concentrations than the roots, and, as a consequence, the shoot:root radiocaesium ratios were high in all treatments. In pots containing coarse-meshed exclusion zones the uptake of radiocaesium was more extensive and here the shoot:root ratio of radiocaesium was higher in the mycorrhizal plants, suggesting that there had been enhanced root:shoot transport of radiocaesium in this treatment.

11.6.4.2 Radiocaesium accumulation by clover

In clover, the mycorrhizal plants, in agreement with earlier work, took up less radiocaesium than the non-mycorrhizal plants. As with the fescue, clover plants grown with coarse-meshed exclusion zones showed greater radiocaesium uptake than plants grown within fine-meshed zones. This may indicate that two separate uptake pathways exist: one via the roots themselves and another via the fungal hyphae.

11.6.4.3 Radiocaesium accumulation by pine

Pine is a tree species which forms ectomycorrhizal associations, and since it is possible to culture a number of individual ectomycorrhizal fungi in the laboratory, the authors were able to inoculate separate batches of pine seedlings with three different fungal species (*Suillus luteus*, *Paxillus involutus* and *Cenococcum graniforme*), instead of a single crude mycorrhizal inoculum. The concentrations of radiocaesium found in the plant parts were much lower than those seen in either fescue or clover, and no radiocaesium at all was detectable in the shoots of the non-mycorrhizal plants (see Table 11.9). This may reflect differences in physiology between tree species and other pasture plants. Interestingly, there appeared to be differential effects on root-to-shoot transport of radiocaesium, depending on the fungal species involved in the mycorrhiza. Thus, with *Suillus luteus* no radiocaesium appeared to be translocated, whereas with both *Paxillus involutus* and *Cenococcun graniforme* some shoot activity was detectable. This may relate to differences in caesium-binding capacity or ionic regulation in general between the three fungal species.

Thus the effects of mycorrhizal infection on caesium uptake by plants depend upon the plant species, the type of mycorrhizal association formed, and the fungal species involved in the association. In some cases, in both short-term uptake experiments and in long-term accumulation studies, it was found that a mycorrhizal association reduced the levels of caesium taken up by the plant (Table 11.9). This perhaps surprising result was easily repeatable in the laboratory with the species used, but clearly the generality of this finding must be investigated. An explanation for it may lie in the role of extramatrical fungal hyphae as accumulators and immobilises of caesium, thereby acting in a protective role regarding the availability of caesium for uptake by plants. Conversely, the possession of a mycorrhizal infection often resulted in the enhancement of translocation of radiocaesium from the root to the shoot of the plant, which may have important radiological implications in terrestrial ecosystems.

Table 11.9. Radiocaesium uptake by pine seedlings

	Fresh weight (g)		^{137}Cs content (Bq g^{-1})	
Sample[a]	Root	Shoot	Root	Shoot
F PI	1.52 ± 0.37	1.43 ± 0.42	0.094 ± 0.028	0.049 ± 0.030
C PI	1.51 ± 0.27	1.67 ± 0.35	0	0.024 ± 0.024
F CG	1.30 ± 0.26	1.23 ± 0.33	0.076 ± 0.076	0.044 ± 0.044
C CG	1.23 ± 0.23	1.42 ± 0.20	0.110 ± 0.110	0.041 ± 0.026
F SL	1.57 ± 0.28	1.42 ± 0.26	0.096 ± 0.043	0
C SL	1.49 ± 0.20	1.46 ± 0.28	0.119 ± 0.069	0
F	0.76 ± 0.08	0.48 ± 0.12	0	0
C	0.86 ± 0.08	0.68 ± 0.13	0.048 ± 0.048	0

[a] F = fine-meshed tube; C = coarse-meshed tube; Other letters indicate mycorrhizal infection with the following fungal species: PI = *Paxillus involutus*; CG *Cenococcum graniforme*; SL = *Suillus luteus*. Values shown are mean and standard error, $n = 5$.

REFERENCES

Anderson, J. M. and Ineson, P. (1982) A soil microcosm system and its application to measurements of respiration and nutrient leaching. *Soil Biology and Biochemistry*, **14**, 415–416.

Bradley, R., Burt A. J. and Read, D. J. (1982) The biology of mycorrhiza in the Ericaceae. VIII. The role of mycorrhizal infection in heavy metal resistance. *New Phytologist*, **91**, 197–209.

Bunzle, K. and Kracke, W. (1984) Distribution of ^{210}Pb, ^{210}Po, stable lead and fallout ^{137}Cs in soil plants and moorland sheep of a heath. *The Science of the Total Environment*, **39**, 23–28.

Byrne, A. R. (1988) Radioactivity in fungi in Slovenia, Yugoslavia, following the Chernobyl accident. *Journal of Environmental Radioactivity*, **6**, 177–183.

Clint. G. M. and Dighton, J. (1992) Uptake and accumulation of radiocaesium by mycorrhizal and non-mycorrhizal heather plants. *New Phytologist*, **121**, 555–561.

Clint, G. M. and MacRobbie, E. A. C. (1984) Effects of fusicoccin in "isolated" guard cells of *Commelina communis*. *Journal of Experimental Botany*, **35**, 180–192.

Clint, G. M., Dighton, J. and Rees, S. (1991) Influx of radiocaesium into fungal hyphae in a range of basidiomycetes. *Mycological Research*, **95**, 1047–1051.

Clint. G. M., Harrison. A. F. and Howard, D. M. (1992) Rates of leaching of ^{137}Cs and potassium from different plant litters. *Journal of Environmental Radioactivity*, **16**, 65–76.

Denny, H. J. (1986) Zinc tolerance and ectomycorrhizal betula. PhD Thesis, University of Birmingham.

Dighton, J. (1986) Mycorrhizas. In: Jones, K. and Lea, P. J. (Eds) *Applied and Environmental Microbiology*, pp. 161–180. University of Lancaster, Lancaster.

Dighton, J. and Horrill, A. D. (1988) Radiocaesium accumulation in the mycorrhizal fungi *Lactarius rufus* and *Inocyle longicystis*, in upland Britain, following the Chernobyl accident. *Transactions of the British Mycological Society*, **91**, 335–337.

Dighton, J., Clint, G. M. and Poskitt, J. (1991) Uptake and accumulation of radio-caesium by upland grassland soil fungi: a potential pool of Cs immobilisation. *Mycological Research*, **95**, 1052–1056.

Elstner, E., Fink, R., Holl, W., Lengfelder, E. and Ziegler, H. (1987) Natural and Chernobyl-caused radioactivity in mushrooms, mosses and soil samples of defined biotops in S. W. Bavaria. *Oecologia*, **73**, 553–558.

Harley, J. L. (1969) *The Biology of Mycorrhiza*. Plant Science Monographs. Leonard Hill, London.

Harley, J. L. and Smith, S. E. (1983) *Mycorrhizal Symbiosis*. Academic Press, London.

Harrison, A. F., Clint, G. M., Jones, H. E., Poskitt, J. M., Howard, B. J., Howard, D. M., Beresford, N. A. and Dighton, J. (1990) Distribution and recycling of radio-caesium in heather-dominated ecosystems. Report to the Ministry of Agriculture, Fisheries and Food, London, Project N601.

Haselwandter, K. (1987) Accumulation of the radioactive nuclide ^{137}Cs in fruitbodies of basidiomycetes. *Health Physics*, **34**, 713–715.

Haselwandter, K., Berreck. M. and Brunner, P. (1988) Fungi as bioindicators of radio-caesium contamination: pre- and post-Chernobyl activities. *Transactions of the British Mycological Society*, **90**, 171–174.

Horrill, A. D., Kennedy, V. H. and Harwood, T. R. (1990) The concentrations of Chernobyl derived radionuclides in species characteristic of natural and semi-natural ecosystems. In: Desmet, G., Nassimbini, P. and Belli, M. (Eds) *Transfer of Radionuclides in Natural and Semi-natural Environments*. Elsevier, London.

Horrill, A. D. and Lindley, D. K. (1990) Monitoring method based on land classi-fication for assessing the distribution of environmental contamination. In: *Environmental Contamination Following a Major Accident*. IAEA, Vienna.

Hove, K., Pedersen, O., Garmo, T. H., Soleheim Hansen, H. and Staaland H. (1990) Fungi: a major source of radiocaesium contamination of grazing ruminants in Norway. *Health Physics*, **59**, 189–192.

Jones, H. E., Harrison, A. F., Poskitt, J. M., Roberts, J. D. and Clint, G. M. (1991) The effects of potassium nutrition on ^{137}Cs uptake in two upland species. *Journal of Environmental Radioactivity*, **14**, 279–294.

Lowe, V. P. W. and Horrill, A. D. (1991) Caesium concentration factors in wild herbivores and the fox (*Vulpes vulpe* L.), *Environmental Pollution*, **70**, 93–107.

MacRobbie, E. A. C. (1962) Ionic relations of *Nitella translucens*. *J. Gen. Physiol.*, **45**, 861–878.

MacRobbie, E. A. C. (1964) Factors affecting the fluxes of potassium and chloride ions in *Nitella translucens*. *Journal of General Physiology*, **47**, 859–877.

Oolbekkink, G. Y. and Kuyper, T. W. (1989) Radioactive caesium from Chernobyl in fungi. *The Mycologist*, **3**, 3–6.

Pierson, D. H., Crookes, R. N. and Fisher, E. M. R. (1960) *Radioactive Fallout in Air and Rain*. AERE-R 3358. HMSO London.

Read, D. J., (1986) Non-nutritional effects of mycorrhizal infection. In: Gianinazzi-Pearson, V. and Gianinazzi, S. (Eds) *Physiological and Genetical Aspects of Mycorrhizae*, pp. 169–175. INRA, Paris.

Shaw, G. and Bell, J. N. B. (1989) The kinetics of caesium absorption by roots of winter wheat and the possible consequences for the derivation of soil-to-plant transfer factors for radiocaesium. *Journal of Environmental Radioactivity*, **10**, 213–231.

Smith, F. B. and Clarke, M. J. (1989) *The Transport and Deposition of Airborne Debris from the Chernobyl Nuclear Power accident with Special Emphasis on the consequences to the United Kingdom*. Scientific Paper No. 42. Meteorological Office, HMSO, London.

Witkamp, M. and Barzansky. B. (1968) Microbial immobilisation of ^{137}Cs in forest litter. *Oikos*, **19**, 392–395.

Witkamp, M. and Frank, M. L. (1970) Effects of temperature, rainfall and fauna on transfer of ^{137}Cs, K, Mg and mass in consumer-decomposer microcosms. *Ecology*, **51**, 466–474.

12 The Importance of Nickel for Plant Growth in Ultramafic (Serpentine) Soils

JOHN PROCTOR
University of Stirling, UK

ALAN J. M. BAKER
University of Sheffield, UK

ABSTRACT

Ultramafic soils normally have 500–10 000 $\mu g\,g^{-1}$ total nickel (cf. 5–500 $\mu g\,g^{-1}$ in non-ultramafic soils). Plant-available quantities of nickel are much less, but there is no universally agreed method of assessing its availability which is likely to vary greatly between sites and even on a microscale within sites. Temporal variations in nickel availability almost certainly occur. Nickel is very toxic to many plants grown in culture solutions but its toxicity can be ameliorated by Ca^{2+} ions and ameliorated or exacerbated by Mg^{2+} ions. Plants on ultramafic soils usually have elevated ($>50\,\mu g\,g^{-1}$ foliar dry matter) nickel concentrations and some of them, "hyperaccumulator" species, have very high ($\geq 1000\,\mu g\,g^{-1}$) foliar nickel. The biochemistry of nickel accumulation is beginning to be understood although many aspects of the phenomenon remain enigmatic. There are substantial differences in innate nickel tolerance between plant species, and within-species tolerance in ultramafic races has been demonstrated. Bioassays for soil nickel have been developed and one of the best known is that for oats (*Avena sativa* L.), which show characteristic symptoms of nickel toxicity. The use of this bioassay has shown a wide variation in nickel toxicity between sites. Recently, nickel has been shown to be a truly essential general micronutrient for plants although it is required in exceedingly small quantities and its ecological importance in this respect remains to be evaluated. It is concluded that the importance of nickel in ultramafic soils ranges from negligible in some sites to considerable in others and that it may be involved in many possible interactions with other soil chemical and physical factors.

12.1 INTRODUCTION

It is implicit in many discussions of the ecology of ultramafic (serpentine) soils that chromium, cobalt and particularly nickel have an important and widespread causal role. This idea apparently dates from West (1912) who suggested

Toxic Metals in Soil–Plant Systems. Edited by S. M. Ross
© 1994 John Wiley & Sons Ltd

that chromium was possibly a toxic element on the barren areas of the Keen of Hamar ultramafic site, Shetland, Scotland. The most influential early paper about the importance of chromium, cobalt and nickel was that by Robinson *et al.* (1935) who studied ultramafic soils from Cuba, Puerto Rico and the USA and sought to apply an all-embracing explanation of their infertility. They indicated the toxic nature of chromium, cobalt and nickel in culture solutions and pointed out that large total quantities of the elements were the only factors common to the different soils that they investigated. They concluded that "the presence of comparatively large quantities of chromium and nickel, and perhaps cobalt, are the dominant causes of infertility in serpentine soils in which the physical conditions are favourable for plant growth". A critical evaluation of their work shows, however, that this conclusion was hardly justified from their data since they admitted that "in 11 of the 15 soils examined Loew's theory that an excess of magnesia over lime is responsible for the toxicity to plants may be taken as a satisfactory explanation of the observed toxicity", and of the four remaining, "the quantities of both lime and magnesia are so low that they may be limiting factors of plant growth". Moreover, the quantities of chromium, cobalt and nickel in the plants and extractable from the soils they analysed are low compared with results obtained elsewhere by later workers.

The aim of this chapter is to provide a critical assessment of the role only of nickel in ultramafic soils since the importance of chromium or cobalt is much less likely and the evidence featuring these elements has often been based on analyses of contaminated samples (Proctor and Nagy, 1992). Some recent papers have tended to play down the importance of nickel (e.g. Carter *et al.*, 1987; Kruckeberg, 1992) whilst others have continued to assert its influence (e.g. Jaffré, 1980; Robertson, 1992; Mizuno and Nosaka, 1992). The authors shall attempt to reconcile these different viewpoints and explain how the importance of nickel is likely to vary from area to area, over relatively short distances, even on a microscale, within single sites, and to vary temporally.

12.2   SOIL NICKEL

Ultramafic soils normally have from 500 to 10 000 $\mu g\,g^{-1}$ total nickel (Table 12.1), values which are much greater than those of 5–500 $\mu g\,g^{-1}$ which are normally found in non-ultramafic soils (Swaine, 1955). Of the total nickel, however, it is only a small proportion that is likely to be available to plants. In the examples given in Table 12.1, dilute acid-soluble nickel concentrations range from 11 to 601 $\mu g\,g^{-1}$, ammonium-acetate-exchangeable nickel concentrations range from 2 to 259 $\mu g\,g^{-1}$; and soil-solution nickel quantities have ranged from <0.1 to 2.4 mg litre^{-1}.

A number of attempts have been made to determine the most useful chemical extractant for "available" nickel from ultramafic soils. Shewry and

Table 12.1. The concentrations of total nickel, acetic-acid soluble nickel, exchangeable nickel, and soil solution nickel in a range of ultramafic soils (—: no data)

	Ni_{total}	$Ni_{acetic\ acid}$	$Ni_{exchangeable}$	$Ni_{soil\ solution}$
England				
Lizard Peninsula[a]	2350	11	2	—
Scotland				
Keen of Hamar[b]	9300	178	9.3	0.13
Keen of Hamar[c]	—	601	119	2.4
Meikle Kilrannoch[d]	1250	—	14.5	0.67
Rhum[e]	2090	—	1.7	0.07
Whitecairns[f]	—	114–289	26–61	—
	—	49–403	22–49	—
Guatemala				
Lake Izabal[g]	—	—	259	—
Italy[h]				
Impruneta	2500	88	—	0.17
Pian de Verra	1180	70	—	1.10
Newfoundland[i]	2700	—	10.2	—
New Caledonia[j]	4300	—	55	—
	10 400	—	8.7	—
Indonesia				
Sulawesi[k]	4050	—	40	—
Zimbabwe[l]				
Kingston Hill	6800	290	53	0.58
Noro	6600	310	41	—
Tipperary Claims	7100	351	56	0.67
Umvuma	3500	120	21	—
Sabah[m]				
Gunung Silam (280 m)	1200	25	13	—
Gunung Silam (790 m)	1500	50	12	—

[a] Mean of analyses for "Rock Heath" samples L4 and L5 of Slingsby and Brown (1977).
[b] Analyses for "debris" samples. Total values are means for samples U2, U5 and U6 of Slingsby and Brown (1977). Acetic acid and exchangeable values are the means of mean values for the debris Vegetation Groups VI–IX of Carter et al. (1987). Soil solution concentrations are means for debris samples from Proctor et al. (1981).
[c] Analysis for a soil sample under sedge–grass–heath Group V vegetation (Carter et al. 1987).
[d] Total value is the mean of two samples in Proctor and Woodell (1971); exchangeable values are from Proctor et al. (1991); soil solution values are from Proctor et al. (1981).
[e] Values are the means for a range of skeletal soils from Looney and Proctor (1989a).
[f] Values from Hunter and Vergnano (1952). Upper row is "Basin soils", Lower row is "Hill slope soils".
[g] Halstead (1968). Values for 0–10 cm deep sample.
[h] Vergnano Gambi (1992).
[i] Roberts (1992). Data for a mineral soil profile (0–7 cm deep).
[j] Jaffré (1980). Upper row is the means of two sets of samples of "sol brun eutrophe hypermagnésien". Lower row is from one set of samples of "sol ferralitique érodé".
[k] Parry (1985).
[l] Total, acetic-acid extractable, and exchangeable data are from Proctor et al. (1980). Soil solution data are from Proctor et al. (1981).
[m] Total and acetic-acid extractable data are from S. Nortcliff (unpublished). Exchangeable data are from Proctor et al. (1988).

Peterson (1976) made detailed investigations on the Green Hill and Keen of Hamar soils from Scotland but after assessing the data from the 10 extractants they used, they concluded that all were unsatisfactory. They suggested that nickel concentrations in plants gave the best measures of availability—a surprising conclusion since they showed wide interspecific variations in tissue nickel concentrations. In a study of a range of British soils, Slingsby and Brown (1977), compared bioassay (*Avena sativa* cv. Asta) shoot nickel concentrations with those extracted by acetic acid, ammonium nitrate and disodium EDTA. They concluded that none of these three extractants was universally suitable. The case for using soil solution concentrations as a measure of nickel availability was made by Proctor *et al.* (1981) but it must be admitted that the problem of the most useful extractant remains unresolved.

The proportions of total to more labile forms of soil nickel will depend on the degree to which it is complexed by various organic and inorganic substances and the extent to which soils are leached. Jenne (1968) showed how the availability of nickel is likely to be determined by soil pH and redox potential. Nickel is likely to be at non-toxic concentrations in soil solutions with a pH exceeding 6.0 and in well-aerated soils. There are a number of reports of bioassays that show the amelioration of nickel toxicity with increasing soil pH (Crooke, 1956; Halstead, 1968). Mizuno and Nosaka (1992) have shown a clear negative correlation ($r = 0.88$, $n = 66$) between pH (H_2O) and exchangeable nickel concentrations in a range of Japanese ultramafic soils. Since some of the controlling factors of nickel concentrations vary according to climate or plant exudates, it follows that the nickel concentrations themselves are subject to temporal variation. It is possible that soil solution nickel concentrations are enhanced in drying soils (Proctor and Nagy, 1992) or, more likely, that they increase on soil wetting (in Jenne's 1968 model) as Robertson (1992) has envisaged in Zimbabwe. On the other hand, very low labile nickel concentrations were found in the highly leached ultramafic soil of the Isle of Rhum (Looney and Proctor, 1989*a*). Local changes in soil pH caused by root exudates (Marschner, 1991) are likely to influence labile soil nickel concentrations, as is the release of nickel from the decomposition of litter from hyperaccumulator species (Schlegel *et al.*, 1991).

12.3 EFFECTS OF NICKEL ON PLANT GROWTH IN CULTURE SOLUTIONS

Robertson and Meakin (1980) showed that nickel concentrations of 0.032 mg litre^{-1} (6×10^{-4} mM) were toxic to roots of the tree species *Brachystegia spiciformis* in single-salt solutions of nickel sulphate. They showed that at 0.032 mg litre^{-1} nickel that cell division was entirely prevented and that 0.063 mg litre^{-1} nickel changed the properties of the cell membrane and inhibited protoplast expansion. This work has given an insight into the mechanisms of toxicity but it exaggerates the likelihood of natural nickel toxicity since when other ions occur in solutions with nickel they usually reduce its toxicity. Calcium has long been known to ameliorate the toxic effects of nickel. Magnesium – the dominant cation in many ultramafic soils – also can ameliorate nickel toxicity (Proctor and McGowan, 1976) although Johnston and Proctor (1981) showed that in their simulated soil solutions and for *Festuca rubra* at least, magnesium may exacerbate nickel toxicity. Robertson (1985) showed how calcium and magnesium can interact in ameliorating nickel toxicity in *Zea mays*. In all these instances of amelioration or exacerbation of nickel toxicity the mechanisms are unknown but it is clear that any experimental attempt using culture media to assess possible nickel toxicity should involve carefully designed culture solutions that will simulate soil solutions as far as possible. Johnston and Proctor (1981) did this for the Meikle Kilrannoch site, and, although their work can be criticised in the light of current knowledge of the influence on nickel of the Fe-EDTA used to supply iron in their solutions, their conclusions seem valid that nickel has a toxic influence, though subordinate to that of magnesium, at the Meikle Kilrannoch site (Proctor and Nagy, 1992). Other simulated soil-solution culture experiments for the Scottish ultramafic sites at Coyles of Muick, Green Hill, Hill of Towanreef, Keen of Hamar and Rhum have failed to show an effect of nickel (Proctor, J. unpublished; Looney and Proctor, 1989*b*).

12.4 NICKEL CONCENTRATIONS IN ULTRAMAFIC VEGETATION

A summary of shoot or foliar nickel concentrations from a range of ultramafic sites is given in Table 12.2. In all the samples except those from Zimbabwe, most species have less than 100 μg g^{-1} foliar or shoot nickel. The sampling of the material has usually been selective, often with the detection of "hyperaccumulator" (i.e. those with > 1000 μg g^{-1} foliar nickel) species in mind, and this has influenced the pattern of the data in Table 12.2. The least biased samples are probably those from New Caledonia and Sabah which show peaks in the 0–10 μg g^{-1} or 10–50 μg g^{-1} categories. Brooks (1987) and others have argued that there is a gap between the highest nickel concentration in non-accumulator species and the lowest (defined as 1000 μg g^{-1}) in hyperaccumulators. This is not supported by the data in Table 12.2 nor by later work by Brooks *et al.* (1992) in Brazil, and may be a misinterpretation of the data resulting from biased sampling associated with searching for hyperaccumulators. The hyperaccumulator species are in evidence in all the sites in Table 12.2 except Mount Silam in Sabah. Even there, one individual of *Shorea tenuiramulosa* (the mean Ni concentration of three individuals is 650 μg g^{-1}) had 1000 μg g^{-1} nickel.

Table 12.2. The percentage of analysed species in a range of shoot or foliar nickel concentrations from several ultramafic areas

	Nickel ($\mu g\,g^{-1}$ dry matter)									
	0–10	10–50	50–100	100–250	250–500	500–1000	1000–2500	2500–5000	5000–10 000	>10 000
New Caledonia (n = 512)[a]	8.2	42.9	16.0	15.2	6.2	5.6	2.5	0.9	0.9	0.9
Sabah[b]										
330 m (n = 42)	9.5	71.4	9.5	9.5	—	—	—	—	—	—
610 m (n = 52)	25.0	61.5	7.7	1.9	1.9	1.9	—	—	—	—
790 m (n = 19)	42.1	52.6	—	5.3	—	—	—	—	—	—
870 m (n = 30)	63.3	36.7	—	—	—	—	—	—	—	—
Zimbabwe										
(n = 20)[c]	—	30.0	10.0	25.0	10.0	15.0	5.0	—	—	5.0
(n = 5)[d]	—	20.0	20.0	40.0	20.0	—	—	—	—	—
California (n = 58)[e]	19.0	37.9	24.1	5.2	3.4	—	—	—	5.2	1.7
Italy[f]										
Impruneta (n = 11)	—	63.6	9.1	18.2	—	—	—	—	9.1	—
W. Alps (n = 17)	5.9	5.9	11.8	29.4	—	17.6	11.8	—	17.6	—
Newfoundland (n = 29)[g]	—	31.0	24.1	3.4	3.4	24.1	13.8	—	—	—
England and Scotland (n = 24)[h]	—	20.8	37.5	33.3	4.2	—	4.2	—	—	—

[a] Jaffré (1980).
[b] Proctor et al. (1989) and J. Proctor (unpublished).
[c] Brooks and Yang (1984) (data for endemic species only).
[d] Proctor et al. (1980) (mean values for several sites).
[e] Kruckeberg (1992).
[f] Vergnano Gambi (1992).
[g] Roberts (1992).
[h] Proctor (1992) and A. J. M. Baker (unpublished).

The phenomenon of hyperaccumulation has been well discussed elsewhere (e.g. Brooks, 1987) and the present discussion will be confined to a few points about it. First, it always accounts for only a small proportion of individual species in any ultramafic site and is apparently absent from many sites. Secondly, by itself it does not demonstrate an importance of nickel toxicity for the vegetation in general. Thirdly, its ecological benefit remains little known, although work by Boyd and Martens (1992) has demonstrated convincingly that it protects against herbivory in the Californian hyperaccumulator *Streptanthus polygaloides*. Fourthly, the litter from accumulator species can cause locally high soil nickel concentrations which apparently elicit nickel-tolerant strains of soil microorganisms (Schlegel et al., 1991). Finally, recent studies on some nickel hyperaccumulators from the Philippines and Indonesia (Baker et al., 1992) and with the nickel and zinc hyperaccumulator, *Thlaspi caerulescens* (Brassicaceae) from north-west Europe (Baker, A. J. M., Reeves, R. P. and Lloyd-Thomas, D. unpublished) have shown that hyperaccumulation can be a property of a plant even when it is growing on "normal" (low metal) soils.

Foliar and shoot analyses must always be of limited value in assessing the likely influence of soil nickel, however. For example, it is clear that many species retain much of the nickel they take up in their root systems (Proctor and Nagy (1992) showed this for oats (*Avena sativa*); Menezes de Sequeira and Pinto da Silva (1992) give a number of examples among native species on Portuguese ultramafic soils). Analysis of leaves of different ages has rarely been undertaken but it is known that chemical composition varies greatly with leaf age. O. Vergnano Gambi (personal communication) observed transient nickel accumulation in some species from the Val d'Ayas, in northern Italy. On Gunung Silam, Sabah, Proctor et al. (1989) found a substantially increased nickel concentration in litterfall compared with fresh leaves and interpreted this as a possible nickel excretion mechanism.

12.5 NICKEL TOLERANCE

Large interspecific differences in constitutional nickel tolerance of crop plants have been known since the work of Scharrer and Schropp (1933). Of the plants they tested, peas (*Pisum sativum*) were the most tolerant. Of four cereals investigated, the following order of tolerance was found: oats < rye < wheat < barley. Large differences between species of cultivated plants in response to nickeliferous serpentine soils were observed by Hunter and Vergnano (1952). They observed, for example, that beet (*Beta vulgaris*) was very sensitive to nickel toxicity. Of more ecological interest is the possibility of differing degrees of nickel tolerance selected in response to high soil nickel concentrations in ultramafic populations of wide-ranging species. Such were demonstrated by Proctor (1971a) in all the ultramafic *Agrostis vinealis* races

Table 12.3. Summary of the results of bioassay experiments made on oats (*Avena sativa* or *A. fatua**) plants grown in ultramafic soils

Soil source		Nickel toxicity symptoms?	Foliar nickel ($\mu g\,g^{-1}$ dry matter)		Source
			Plants with toxicity symptoms	Plants without toxicity symptoms	
Australia		Yes	91	—	Williams (1967)
England					
Lizard Peninsula	(a)	No		20	Proctor (1971b)
	(b)	No		12–23	Slingsby and Brown (1977)
Guatemala					
Lake Izabal		Yes	—	—	Halstead (1968)
Italy					
Monti Rognosi		No	—	4–15	Vergnano Gambi (1992)
Impruneta		No	—	15–20	
Japan					
Many sites		Yes	—	—	Mizuno and Nosaka (1992)
New Caledonia		Yes	160	—	Jaffré (1980)
New Zealand					
Anita Bay		No	—	—	Lee (1992)
Black Ridge		Yes	53	—	
Dun Mountain		No	—	38	
Dun Mountain (mine tailings)		Yes	72	—	
Livingstone Mountains		No	—	82	
North Cape		No	—	50	
Red Hills		No	—	12	
Red Mountain		No	—	52	

Location		Symptoms			Reference
Scotland					
Green Hill		No	26	—	Slingsby and Brown (1977)
		No	—	—	Proctor and Cottam (1982)
Hill of Towanreef		No	—	—	Proctor and Cottam (1982)
Keen of Hamar (skeletal soil)	(a)	No	—	—	Spence and Millar (1963)
	(b)	No	15	—	Proctor (1971b)
	(c)	No	19–21	—	Slingsby and Brown (1977)
	(d)	No	—	—	Proctor and Cottam (1982)
Keen of Hamar (soligenous mire)		Yes	—	—	Carter et al. (1987)
Meikle Kilrannoch		No	20	—	Proctor (1971b)
Whitecairns	(a)	Yes	—	24–199[a]	Hunter and Vergnano (1952)
	(b)	No	41[b]	—	
Sweden					
Kittelfjäll		No	—	—	Proctor and Cottam (1982)
United States, California					
Jasper Ridge*		No	—	—	J. Proctor (unpublished)
Clear Creek*		No	—	—	
Zimbabwe					
Great Dyke	(a)	Yes	—	32	Hunter (1954)
Great Dyke	(b)	Yes	—	340	Soane and Saunder (1959)
Kingston Hill		Yes	—	—	Proctor et al. (1980)
Noro		No	—	—	
Selukwe		No	—	—	
Tipperary Claims		Yes	—	—	
Umvuma		Yes	—	—	

[a] Extreme values for a range of analyses of above-ground plant parts of oats of different ages grown on nickeliferous soils. The values given above are for old leaf blades in plants showing moderate nickel toxicity symptoms ($24\ \mu g\ g^{-1}$) 93 days after sowing, and for old leaf blades in plants showing severe toxicity symptoms ($199\ \mu g\ g^{-1}$) 111 days after sowing.

[b] Plants on soil fertilised with major nutrients and calcium carbonate.

investigated. If views on the specificity of metal tolerance are accepted, then such tolerance implies a selection pressure from soil toxicity which is difficult to reconcile with the negative evidence for nickel toxicity from *Avena sativa* bioassays (discussed later) at the same sites. However, nickel tolerance was the least specific of the metal tolerances investigated by Wilkins (1957) and the possibility of non-specific tolerance is now widely accepted (Baker, 1987; Schat and Ten Bookum, 1992). Bannister (1976) showed that magnesium and nickel tolerance were highly correlated and it seems possible, because the ionic radii of magnesium and nickel are very similar, that many of the nickel tolerances observed by Proctor (1971*a*) were non-specific tolerances related to elevated soil magnesium concentrations. It must be admitted that non-magnesium-tolerant *Agrostis vinealis* from the low-magnesium Rhum site was nickel tolerant however (Proctor, 1971*a*). Later work by Bannister and Woodman (1992) has shown how much the results of tolerance tests depend on experimental conditions and particularly the growth of control plants.

12.6 BIOASSAY EXPERIMENTS

Nickel toxicity has been reported in a number of crop species (e.g. Hunter and Vergnano, 1952; Mizuno and Nosaka, 1992). The most extensive work has involved oats, usually *Avena sativa*. Hunter and Vergnano (1952) observed that oats (cv. Victory) showed characteristic symptoms (interveinal necrosis and white stripes) in relation to toxicity and this led to the use of oats (of a range of cultivars) in many subsequent bioassays (Table 12.3). Their work, which showed nickel toxicity symptoms in the field, was carried out in an acid soil derived from ultramafic drift at Whitecairns in Scotland. Nickel toxicity symptoms have been observed in oats grown in many ultramafic soils but this feature is by no means universal. Remarkably, in Britain, where several ultramafics were bioassayed (Spence and Millar, 1963; Proctor, 1971*b*; Slingsby and Brown, 1977; Proctor and Cottam, 1982), no more nickel toxicity symptoms were observed until Carter *et al.* (1987) reported a test with oats on the Keen of Hamar, Shetland. The circumstances of this test were remarkable and instructive. The soils of the impressively barren areas of the Keen were assayed many times and showed no hint of nickel toxicity. During a systematic survey of the whole of the Keen it was observed that one small area of about 400 m^2 with relatively acid and organic soils was particularly nickel-rich. Soils from this area produced nickel toxicity symptoms in oats in the glasshouse but in the field they supported closed mire vegetation of widespread species. The soil assayed by Hunter and Vergnano (1952) was also acid and organic, and both this and the Keen of Hamar mire soil just described were different from the less acid mineral soils assayed from the other Scottish sites. This observation agrees with the idea of Jenne (1968) on the availability of nickel increasing in acid wet situations although there remains a conflict with the observations of Halstead *et al.* (1969) that the "remarkable effectiveness of organic matter in preventing Ni toxicity is undoubtedly due to its ability to form stable compounds with Ni and thus to prevent its uptake by plants". But the Keen of Hamar observation demands the question of how, if high soil nickel concentrations are associated with closed vegetation of widespread species, can soil nickel ever be a cause of ultramafic barrenness and rare species?

There is no simple relationship between foliar nickel concentrations and nickel toxicity. The highest values of foliar nickel (91–340 μg g^{-1} dry matter) in Table 12.2 are associated with nickel toxicity symptoms. However, the symptoms were observed with foliar nickel as low as 24 μg g^{-1} (for old leaf blades in plants 93 days after sowing) (Hunter and Vergnano, 1952), and not observed in samples with up to 82 μg g^{-1} foliar nickel (Lee, 1992). It has been shown that most of the nickel in oat plants is likely to be retained by the root systems where it can exceed concentrations of those in the leaves of hyper-accumulators (Proctor and Nagy, 1992). Hence foliar concentrations and toxicity symptoms are likely to depend on several plant processes as well as soil supply. Mizuno and Nosaka (1992) have shown a highly positive correlation ($r = 0.94$, $n = 20$) between foliar concentrations in oats and exchangeable soil nickel. High exchangeable soil nickel concentrations are associated with nickel toxicity symptoms in oats (e.g. 259 μg g^{-1} soil nickel in a Guatemalan soil (Halstead, 1968); 119 μg g^{-1} soil nickel in a soil from the Keen of Hamar, Shetland (Carter *et al.*, 1987). However, exchangeable soil nickel concen-trations as low as 9 μg g^{-1} have produced toxicity symptoms in oats (Hunter, 1954) whilst those up to 41 μg g^{-1} (Proctor *et al.*, 1980) have not. The lack of correlations between toxicity symptoms and exchangeable soil nickel is not surprising in view of the influence, discussed earlier, of many factors on nickel toxicity and the uncertainty that exchangeable quantities are an adequate measure of plant-available nickel.

The bioassay work provides some of the best evidence for plant nickel toxicity in ultramafic soils and includes non-organic skeletal soils of near neutral pH, e.g. in Zimbabwe (Proctor *et al.*, 1980) as well as more acid organic soils such as those discussed earlier. However, oat bioassays are of limited value ecologically. Clearly, from the sample of the high-nickel soil on the Keen of Hamar, the response of oats may differ greatly from that of native species. Moreover, as at Meikle Kilrannoch, Scotland, where a critical exper-iment revealed nickel toxicity to non-ultramafic *Festuca rubra*, this toxicity was subordinate to the toxic effect of magnesium, and a bioassay with oats (Proctor, 1971*b*) had revealed only magnesium toxicity. Further problems concern the interpretation of the precise symptoms of nickel toxicity in oats and how they might be effected by other conditions.

12.7 BENEFICIAL EFFECTS OF NICKEL

Nickel has been accepted for some time as an essential element for animals (Nielsen and Sauberlich, 1970). More recently its beneficial effects for plants have been generally recognised. Dixon *et al.* (1975) discovered that nickel was a component of the enzyme urease. It was shown that with urea as the sole nitrogen supply that nickel was essential for normal growth. Later Brown *et al.* (1987*a*) proved that nickel was necessary for optimal cereal (barley, oat and wheat) growth even with ammonium as the nitrogen source. Brown *et al.* (1987*b*) have presented convincing evidence that nickel is likely to be a truly essential micronutrient generally for plants, and found that cereals were inviable when it was carefully excluded from them.

The nickel requirements of the crop species tested are all very low (tissue concentrations of $\leqslant 0.1\ \mu\text{g}\,\text{g}^{-1}$ dry matter) and it is unlikely to limit their growth in normal agricultural situations. There are as yet no critical experiments on the quantification of nickel requirements of ultramafic plants. If they have a higher requirement for nickel than the crops it is possible that the emphasis might shift from putative toxic effects of nickel to putative micronutrient effects for some plants in ultramafic soils. There are some indications already that nickel may have a positive effect on ultramafic plants in addition to the herbivory prevention in the hyperaccumulator referred to earlier (Boyd and Martens, 1992). Gabbrielli *et al.* (1989) have shown that roof-surface phosphatase from *Alyssum bertolonii* is stimulated by Ni^{2+} concentrations between 1 and 10 μM. Vergnano Gambi *et al.* (1992) have shown that nickel at 0.25 mM has a marked stimulatory effect on the germination of *A. bertolonii*.

12.8 CONCLUSIONS

The toxicity of nickel is by no means a universal feature of ultramafic soils. The influence of nickel varies between sites and even within sites, both spatially – microscale – and temporally. At some sites, e.g. the barren areas of the Keen of Hamar in Shetland, nickel toxicity has never been shown to be important; at others, e.g. some sites in Zimbabwe, there is much better evidence for nickel toxicity. At no site, however, is there adequate evidence that nickel toxicity is, above all other factors, of clear overriding importance for plant growth. There are many instances of complete vegetation cover on highly nickeliferous soils. In addition to the Keen of Hamar example discussed earlier, Proctor and Craig (1978) have described well-developed riverine forest on a nickel-rich soil in Zimbabwe. Nickel toxicity to non-tolerant species depends on the accompanying ions; and magnesium, usually the most abundant basic cation, affects nickel toxicity in several ways. By occupying a high proportion of cation exchange sites, magnesium raises the pH of ultramafic soils which is inimical to nickel toxicity. Moreover, magnesium ions are able to ameliorate or exacerbate nickel toxicity. Future advances in our understanding of the importance of nickel for plant growth in ultramafic soils depend on critical experiments and on an appreciation that its effect may vary from negligible to being of increasing importance in a series of possible interactions with a range of chemical and physical factors.

REFERENCES

Baker, A. J. M. (1987). Metal tolerance. *New Phytologist*, **106**, 93–111.
Baker, A. J. M., Proctor, J., van Balgooy, M. M. J. and Reeves, R. D. (1992) Hyperaccumulation of nickel by the ultramafic flora of Palawan, Republic of the Philippines. In: Baker, A. J. M., Proctor, J. and Reeves, R. D. (Eds) *The Vegetation of Ultramafic (Serpentine) Soils*, pp. 291–304, Intercept Ltd., Andover, Hants.
Bannister, P. (1976) *Introduction to Physiological Plant Ecology*. Blackwell Scientific, Oxford.
Bannister, P. and Woodman, B. (1992) The influence of tolerance indices and growth on metal tolerance of pasture legumes and serpentine plants. In: Proctor, J., Baker, A. J. M. and Reeves, R. D. (Eds) *Proceedings of the First International Conference on Serpentine Ecology*, University of California, Davis, 19–22 June 1991, pp. 353–366. Intercept Ltd., Andover, Hants.
Boyd, R. S. and Martens, S. N. (1992). The raisin d'être for metal hyperaccumulation by plants. In: Baker, A. J. M., Proctor, J. and Reeves, R. D. (Eds) *The Vegetation of Ultramafic (Serpentine) Soils*, pp. 279–289. Intercept Ltd., Andover, Hants.
Brooks, R. R. (1987) *Serpentine and its Vegetation. a Multidisciplinary Approach*. Dioscorides Press, Portland, Oregon.
Brooks, R. R. and Yang, X. H. (1984). Elemental levels and relationships in the endemic serpentine flora of the Great Dyke, Zimbabwe and their significance as controlling factors for the flora. *Taxon*, **33**, 392–399.
Brooks, R. R., Reeves, R. D. and Baker, A. J. M. (1992) The vegetation of Goiás State, Brazil. In: Baker, A. J. M., Proctor, J. and Reeves, R. D. (Eds) *The Vegetation of Ultramafic (Serpentine) Soils*, pp. 67–81. Intercept Ltd., Andover, Hants.
Brown, P. H., Welch, R. M. and Cary, E. E. and Chekai, R. T. (1987*a*). Beneficial effects of nickel on plant growth. *Journal of Plant Nutrition*, **10**, 2125–2135.
Brown, P. H., Welch, R. M. and Cary, E. E. (1987*b*). Nickel: a micronutrient essential for higher plants. *Plant Physiology* **85**, 801–803.
Carter, S. P., Proctor, J. and Slingsby, D. R. (1987) Ecological studies on the Keen of Hamar serpentine, Shetland. *Journal of Ecology*, **75**, 21–42.
Crooke, W. M. (1956). The effect of soil reaction on uptake of nickel from a serpentine soil. *Soil Science*, **81**, 269–276.
Dixon, N. E., Gazzola, C., Blakely, R. L. and Zerner, B. (1975) Jack bean urease (EC 3.51.5). A metalloenzyme. A simple biological role for nickel? *Journal of the American Chemical Society*, **97**, 4131–4233.
Gabbrielli, R., Grossi. L. and Vergnano Gambi, O. (1989). The effects of nickel, calcium and magnesium on the acid phosphatase activity of two *Alyssum* species. *New Phytologist*, **111**, 631–636.
Halstead, R. L. (1968) Effect of different amendments on yield and composition of oats grown on a soil derived from serpentine material. *Canadian Journal of Soil Science*, **48**, 301–305.
Halstead, R. L., Finn, B. J. and McLean, A. J. (1969) Extractability of nickel added to soils and its concentration in plants. *Canadian Journal of Soil Science*, **49**, 335–342.
Hunter, J. G. (1954) Nickel toxicity in a Southern Rhodesian soil. *South African Journal of Science*, **51**, 133–135.

Hunter, J. G. and Vergnano, O. (1952) Nickel toxicity in plants. *Annals of Applied Biology*, **39**, 279–284.

Jaffré, T. (1980) *Etude Ecologique du Peuplement Végétal des Sols Dérivés de Roches Ultrabasiques en Nouvelle Calédonie*. ORSTOM, Paris.

Jenne, E. A. (1968) Controls on Mn, Fe, Co, Ni, Cu, and Zn concentrations in soils and water: the significant role of hydrous Mn and Fe oxides. *Advances in Chemistry Series*, **73**, 337–387.

Johnston, W. R. and Proctor, J. (1981) Growth of serpentine and non-serpentine races of *Festuca rubra* in solutions simulating the chemical conditions in a toxic serpentine soil. *Journal of Ecology*, **69**, 855–869.

Kruckeberg, A. R. (1992) Plant life of western North American ultramafics. In: Roberts, B. A. and Proctor, J. (Eds). *The Ecology of Areas with Serpentinized Rocks: A World View*, pp 31–73. Kluwer Academic, Netherlands.

Lee, W. G. (1992) New Zealand ultramafics. In: B. A. Roberts and J. Proctor (Eds) *The Ecology of Areas with Serpentinized Rocks. A World View*, pp. 375–418. Kluwer Academic, Dordrecht.

Looney, J. H. H. and Proctor, J. (1989*a*). The vegetation of ultrabasic soils on the Isle of Rhum I. Physical environment, plant associations and soil chemistry. *Transactions of the Botanical Society of Edinburgh*, **45**, 351–364.

Looney, J. H. H. and Proctor, J. (1989*b*). The vegetation of ultrabasic soils on the Isle of Rhum II. The causes of the debris. *Transactions of the Botanical Society of Edinburgh*, **45**, 351–364.

Marschner, H. (1991) Plant–soil relationship: acquisition of mineral nutrients by roots from soils. In: Porter, J. R. and Lawlor, D. W. (eds) *Plant Growth: Interactions with Nutrition and Environment*, pp. 125–160. Cambridge University Press, Cambridge.

Menezes de Sequeira, E. and Pinto da Silva, A. R. (1992) Ecology of serpentinized areas of north-east Portugal. In: Roberts, B. A. and Proctor, J. (Eds) *The Ecology of Areas with Serpentinized Rocks. A World View*, pp. 169–197. Kluwer Academic, Dordrecht.

Mizuno, N. and Nosaka, S. (1992). The distribution and extent of serpentinized areas in Japan. In: Roberts, B. A. and Proctor, J. (Eds) *The Ecology of Areas with Serpentinized Rocks. A World View*, pp. 271–311. Kluwer Academic, Dordrecht.

Nielsen, F. H. and Sauberlich, H. E. (1970) Evidence of a possible requirement for nickel by the chick. *Proceedings of the Society of Experimental Biology and Medicine*, **134**, 845–849.

Parry, D. E. (1985) *Ultramafic soils of the humid tropics with particular reference to Indonesia*. Unpublished report of Hunting Technical Services Ltd., U.K.

Proctor, J. (1971*a*) The plant ecology of serpentine III. The influence of a high Mg/Ca ratio and high nickel and chromium levels in some British and Swedish serpentine soils. *Journal of Ecology*, **59**, 827–842.

Proctor, J. (1971*b*) The plant ecology of serpentine II. Plant response to serpentine soils. *Journal of Ecology*, **59**, 397–410.

Proctor, J. (1992) Chemical and ecological studies on the vegetation of ultramafic sites in Britain. In: Roberts, B. A. and Proctor, J. (Eds) *The Ecology of Areas with Serpentinized Rocks. A World View*, pp. 135–167. Kluwer Academic, Dordrecht.

Proctor, J. and Cottam, D. A. (1982) Growth of oats, beet and rape in four contrasting serpentine soils. *Transactions of the Botanical Society of Edinburgh*, **44**, 19–25.

Proctor, J. and Craig, G. C. (1978) The occurrence of woodland and riverine forest on the serpentine of the Great Dyke. *Kirkia*, **11**, 129–132.

Proctor, J. and McGowan, I. D. (1976) Influence of magnesium on nickel toxicity. *Nature, London*, **134**, 260.

Proctor, J. and Nagy, L. (1992) Ultramafic rocks and their vegetation: an overview. In: Baker, A. J. M., Proctor, J. and Reeves, R. D. (Eds) *The Vegetation of Ultramafic (Serpentine) Soils*, pp. 469–494. Intercept Ltd., Andover, Hants.

Proctor, J. and Woodell, S. R. J. (1971) The plant ecology of serpentine. I. Serpentine vegetation of England and Scotland. *Journal of Ecology*, **59**, 375–395.

Proctor, J., Burrow, J. and Craig, G. C. (1980) Plant and soil chemical analyses from a range of Zimbabwean serpentine sites. *Kirkia*, **12**, 127–139.

Proctor, J., Johnston, W. R., Cottam, D. A. and Wilson, A. B. (1981) Field capacity water extracts from serpentine soils. *Nature, London*, **294**, 245–246.

Proctor, J., Lee, Y. F., Langley, A. M., Munro, W. R. C. and Nelson, T. N. (1988) Ecological studies on Gunung Silam, a small ultrabasic mountain in Sabah I. Environment, forest structure and floristics. *Journal of Ecology*, **76**, 320–340.

Proctor, J., Phillips, C., Duff, G. K., Heaney, A. and Robertson, F. M. (1989) Ecological studies on Gunung Silam, A small ultrabasic mountain in Sabah II. Some forest processes. *Journal of Ecology*, **77**, 317–331.

Proctor, J., Bartlem, K., Carter, S. P., Dare, D. A., Jarvis, S.B. and Slingsby, D. R. (1991) Vegetation and soils of the Meikle Kilrannoch ultramafic sites. *Botanical Journal of Scotland*, **46**, 47–64.

Roberts, B. A. (1992) Ecology of serpentinized areas, Newfoundland, Canada. In: Roberts, B. A. and Proctor, J. (Eds) *The Ecology of Areas with Serpentinized Rocks: A World View*, pp. 75–113. Kluwer Academic, Netherlands.

Robertson, A. I. (1985) The poisoning of roots of *Zea mays* by nickel ions, and the protection afforded by magnesium and calcium. *New Phytologist*, **100**, 173–189.

Robertson, A. I. (1992) The relation of nickel toxicity to certain physiological aspects of serpentine ecology: some facts and a wild hypothesis. In: Baker, A. J. M., Proctor, J. and Reeves, R. D. (Eds) *The Vegetation of Ultramafic (Serpentine) Soils*, pp. 331–336. Intercept Ltd., Andover, Hants.

Robertson, A. I. and Meakin, M. E. R. (1980) The effect of nickel on cell division and growth of *Brachystegia spiciformis* seedlings. *Kirkia*, **12**, 115–126.

Robinson, W. D., Edgington, G., and Byers, H. G. (1935) *Chemical Studies of Infertile Soils Derived from Rocks High in Magnesium and Generally High in Chromium and Nickel*. Technical Bulletin of the United States Department of Agriculture No. 471.

Scharrer, K. and Schropp, W. (1933) Sand-und wasserkulturversuche mit nickel und kobalt. *Zeitschrift für Pflanzenernährung und Düngung*, **31**, 94–113.

Schat, H. and Ten Bookum, W. M. (1992) Metal-specificity of metal tolerance syndromes in higher plants. In: Baker, A. J. M., Proctor, J. and Reeves, R. D. (Eds) *The Vegetation of Ultramafic (Serpentine) Soils*, pp. 337–352. Intercept Ltd., Andover, Hants.

Schlegel, H. G., Cosson, J.-P. and Baker, A. J. M. (1991) Nickel-hyperaccumulating plants provide a niche for nickel-resistant bacteria. *Botanica Acta*, **104**, 18–25.

Shewry, P. R. and Peterson, P. J. (1976) Distribution of chromium and nickel in plants and soil from serpentine and other sites. *Journal of Ecology*, **64**, 195–212.

Slingsby, D. R. and Brown, D. H. (1977) Nickel in British serpentine soils. *Journal of Ecology*, **65**, 597–618.

Soane, B. D. and Saunder, D. M. (1959) Nickel and chromium toxicity of serpentine soils in Southern Rhodesia. *Soil Science*, **88**, 322–330.

Spence, D. H. N. and Millar, E. A. (1963) An experimental study of the infertility of Shetland serpentine soil. *Journal of Ecology*, **51**, 333–343.

Swaine, D. J. (1955) *The Trace-Element Content of Soils*. Technical Communication of the Commonwealth Bureau of Soils No. 48.

Vergnano Gambi, O. (1992) The distribution and ecology of the vegetation of ultramafic soils in Italy. In: Roberts, B. A. and Proctor, J. (Eds) *The Ecology of*

Areas with Serpentinized Rocks. A World View, pp. 217–247. Kluwer Academic, Dordrecht.

Vergnano Gambi, O., Gabbrielli, R. and Pandolfini, T. (1992) Some aspects of the metabolism of *Alyssum bertolonii* Desv. In: Baker, A. J. M., Proctor, J. and Reeves, R. D. (Eds) *The Vegetation of Ultramafic (Serpentine) Soils*, pp. 319–329. Intercept, Andover.

West, W. (1912) Notes on the flora of Shetland, with some ecological observations. *Journal of Botany, London*, **50**, 265–275, 297–306.

Wilkins, D. A. (1957) A technique for the measurement of lead tolerance in plants. *Nature, London*, **180**, 37–38.

Williams, P. C. (1967) Nickel, iron and manganese in the metabolism of the oat plant. *Nature, London*, **180**, 37–38.

13 A Metal Budget for a Monsoonal Wetland in Northern Australia

C. MAX FINLAYSON

International Waterfowl and Wetland Research Bureau, Slimbridge, UK;
Alligator Rivers Region Research Institute, Jabiru, Australia

13.1 INTRODUCTION

The monsoonally influenced areas of the Northern Territory, Australia contain some of the most valuable saline and freshwater wetland habitats in Australia (Finlayson *et al.*, 1988; Finlayson and Von Oertzen, 1992). Foremost in conservation value amongst the freshwater wetlands are the seasonally inundated floodplains and swamps of the Alligator Rivers Region (Finlayson, 1988; Finlayson *et al.*, 1988, 1991). These wetlands have attracted considerable interest over the last two decades, partly as a consequence of increased environmental concern following extensive exploration for minerals and the development of two uranium mines in the region. The catchment of Magela Creek has been at the centre of much of this mineral exploration and development activity.

Magela Creek is a seasonally flowing tributary of the East Alligator River and is situated about 250 km east of Darwin, in the Alligator Rivers Region (ARR), on the western edge of Arnhem Land (Fig. 13.1). The ARR is taken to comprise the catchments of the East Alligator, West Alligator and South Alligator Rivers. Interest in the region, and in the Magela Creek catchment in particular, increased after the discovery of two large uranium deposits— Ranger and Jabiluka—in the catchment. The deposit at Ranger is currently being mined and milled. Government approval for these activities was only forthcoming after an environmental enquiry (Fox *et al.*, 1977) that, amongst other issues, considered the potential effects of trace and heavy metals on the biota of the Magela Creek ecosystem.

The ARR is also an extremely important conservation area, containing a vast assemblage of plants and animals, some of them endemic and some rare or even threatened elsewhere in northern Australia. In order to promote the conservation of these species, Kakadu National Park was established, the first stage being declared in 1979, and now covering an area of approximately two million hectares. The park has received international acclaim under the

Toxic Metals in Soil–Plant Systems. Edited by S. M. Ross
© 1994 The Commonwealth of Australia. Published in 1994 by John Wiley & Sons Ltd

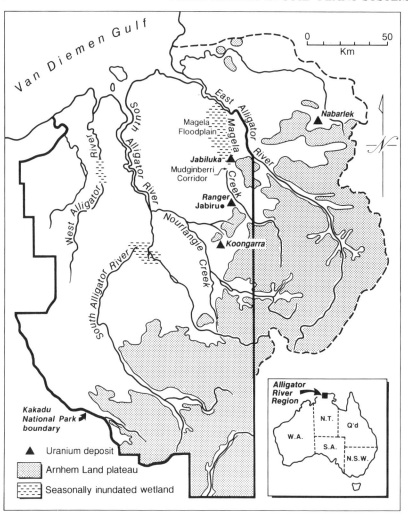

Figure 13.1. Map of the Alligator Rivers Region in the Northern Territory, Australia. Adapted from Finlayson *et al.* (1990*b*)

UNESCO World Heritage Convention and the so-called "Ramsar" Convention on Wetlands of International Importance. The natural heritage recognised by these international conservation conventions includes the seasonally inundated wetlands of the Magela Creek floodplain. The uranium deposits at Ranger and Jabiluka are located within the park boundaries, but are covered by mining leases that are excised from the park. Thus, they are not technically part of the park. (A third excised lease is also located in the catchment of the nearby Nourlangie Creek; see Figure 13.1). The administrative arrangements under which this complex situation is managed are briefly described in Finlayson (1991 a).

When mining commenced at the Ranger mine site it was anticipated that future mine management procedures may need to include release of excess water to Magela Creek and thence into the internationally acclaimed seasonally inundated wetlands. Needless to say, the potential release of water from Ranger to these wetlands provoked much contention. As a result of this contention, the tasks of collecting baseline environmental information on the Magela ecosystem and of determining the fate of potential waterborne pollutants were initiated. Components of this information are used in this case study of metal cycling in a wetland ecosystem. As detailed physiological and chemical analyses are either totally lacking, or are of a preliminary nature, the analysis of the metal budget for the floodplain is restricted to presenting a broad analysis of the major inflows and outflows, the key vegetation components and the sediment. Information on chemical turnover (cycling) of metals on the floodplain was obtained from analyses of water, sediment and plant material collected as a part of the environmental baseline investigations reported in a number of different publications, and summarised in the Annual Research Summaries of the Australian Government-financed Alligator Rivers Region Research Institute.

In this analysis of the metal budget on the Magela floodplain, vegetation and sediment analyses from 1983–1984 are compared to water analyses taken during 1982–1983. As the purpose of this chapter is to present a general overview of metal loads in various components of this wetland ecosystem, such an approach is considered adequate. At the same time it is recognised that more accuracy would result if all the metal analyses were from samples collected over the same time period.

The vegetation, water and sediments were considered to be the most important components of the ecosystem in terms of material entering the floodplain and being stored and/or cycled, or even recycled. The analysis does not include the groundwater, mainly because adequate data are not available. The extremely populous fauna on the floodplain will also affect the turnover of nutrients and trace metals on the floodplain. However, there is no information on the importance of these organisms in the turnover of heavy metals in this wetland ecosystem. Further population and feeding ecology analyses are required before the influence of the many animals on the nutrient and metal budgets can be determined. The analysis that is presented is not, therefore, totally comprehensive.

Far more detailed research is required before the links and interactions between the different components of the ecosystem involved in turnover of metals can be elucidated. The metal budget presented in this study is only the starting point in understanding the metal turnover processes. As an example, the suspended particulate matter in the creek water has a high adsorption capacity and is considered by Hart and Beckett (1986) to be potentially important in the transport of heavy metals around the ecosystem. However, the biological availability of these metals has not been determined.

Prior to considering the importance of the water, vegetation and sediment in the chemical turnover (or cycling) processes on the Magela floodplain, the main physical and biological features of the ecosystem are briefly described.

13.2 DESCRIPTION OF THE MAGELA FLOODPLAIN

13.2.1 PHYSICAL CHARACTERISTICS

Magela Creek is one of many streams that originate in the deeply dissected Arnhem Land Plateau, which dominates the physical relief of the eastern portion of the ARR. The plateau edge or escarpment rises 200–300 m above the undulating lowland plains that characterise much of the ARR. Deep and rugged gorges that contain seasonally flowing creeks dissect the plateau. Waterfalls and rapids are common features of these creeks. A generalised description of Magela Creek presents features that are also, in part, found along other creeks that originate in these gorges.

In general terms, Magela Creek consists of five distinct sections (Finlayson et al., 1990a): rocky and sandy channels that dissect the plateau and plummet over the edge of the plateau; narrow, braided and tree-lined sandbed channels that extend across the lowlands; a series of billabongs and connecting channels (the Mudginberri Corridor) that occur immediately prior to the floodplain; a seasonally inundated black-clay floodplain with a few permanent billabongs, but no distinct channel; and a single channel that discharges into the East Alligator River. Most of the environmental investigations have concentrated on those sections of Magela Creek immediately downstream of the Ranger uranium mine, that is, the Mudginberri Corridor and the black-clay flood-plain. Collectively (but not accurately), these two sections are generally referred to as the Magela floodplain.

The areal extent of the floodplain is about 225 km. The plain floods during the monsoonal wet season; usually a three-month period of heavy rainfall between November and April. Rainfall averages 1560 mm. This rainfall is extremely variable and unpredictable. For most of the remainder of the year (the dry season) there is very little rainfall. The relationship between flow rate and water depth in Magela Creek and rainfall in the catchment is shown in Figure 13.2.

Except for a few permanent billabongs and swamps, the plain dries out over the dry season. The billabongs, which are in fact more accurately referred to as lagoons, vary from remnants of deep channels on the black-clay plain, depressions in flow channels, and lagoons on small feeder streams (Hart and McGregor, 1982; Walker et al., 1984). The latter two types may, under extremely low rainfall conditions, dry out during the dry season.

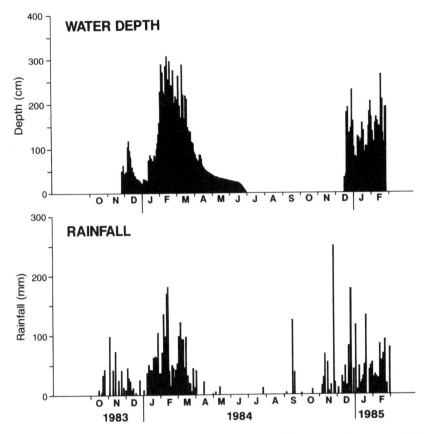

Figure 13.2. Water depth in Magela Creek and rainfall in the catchment at Jabiru airport, October 1983—February 1985. Adapted from Finlayson (1991*b*)

The monsoonal climate has two distinct seasons (see Figure 13.2). The wet season is characterised by thunderstorms, tropical cyclones and rain depressions. The dry season is characterised by south-easterly trade winds which rarely bring any rain. At the commencement of the wet season there is a period of intermittent storms that saturate the soil, followed by more consistent rains which cause the creeks to fill and flow and then to flood. As the rainfall increases, continuous flow occurs in the braided channels of the creek and eventually enters the Mudginberri Corridor before spilling out across the floodplain. The creek generally begins to flow continuously about a month after the start of the wet season. The water flow across the floodplain consists of a series of floods superimposed on a base flow. Except for the flood periods, water moves slowly over the floodplain, partly due to the low gradient and partly due to the impedance caused by the wetland vegetation (Vardavas,

1988). At the end of the wet season the creeks stop flowing and the water level falls. The drying phase can take many months before the floodplain once again begins to resemble a parched plain and the creeks and channels dry out or are reduced to a series of isolated lagoons.

13.2.2 BIOLOGICAL CHARACTERISTICS

The biota of the wetlands is extremely dynamic and diverse (Finlayson *et al.*, 1990*a*) and productive (Finlayson, 1988; Finlayson *et al.*, 1988). Two of the most apparent biological features are the large numbers of waterbirds and the diverse floristic composition and dense foliage cover of the floodplain during the wet season. A summary of the biological features of the floodplain is presented in Finlayson *et al.* (1990*a*). These features of the Magela and other wetlands in the ARR were recognised as being of international importance under the Ramsar Convention.

The few persistent swamps on the floodplains in the ARR are important refugia for many bird species (Morton *et al.*, 1990*a,b*) with some 40 000 wandering whistling-ducks (*Dendrocygna arcuata*), 70 000 plumed whistling-ducks (*Dendrocygna eytoni*), 30 000 radjah shelduck (*Tadorna radjah*), 50 000 Pacific black duck (*Anas superciliosa*), 50 000 grey teal (*Anas gibberifrons*) and 1.6×10^6 magpie geese (*Anseranas semipalmata*). Annual peak numbers of magpie geese on the Magela were estimated at 500 000 during the late dry season, but many of them disperse to other wetlands during the wet season. The Magela is one of the more important dry season refugia with some 200 000 waterbirds commonly being present, decreasing to about 50 000 during the wet season (Morton *et al.*, 1990*a,b*).

The floristic composition and vegetation foliage cover on the floodplain differs seasonally (Finlayson *et al.*, 1989). The abundance and diversity of the wet season vegetation contrasts with the sparse cover of the few dry season herbs, grasses and sedges. This is especially obvious on the open grass and sedge-dominated parts of the floodplain. During the wet season the dominant grass and sedge species are interspersed with a myriad of flowering waterlilies and herbs. These species also fringe the open water of the deeper billabongs and extend into the areas shaded by the *Melaleuca* trees that fringe the flood-plain. This fringe zone, which takes in the edge of the seasonally inundated floodplain and therefore includes the extensive *Melaleuca* forests and wood-lands, contains 158 of the 222 plant species known from these wetlands, with 94 species occurring on the seasonally inundated floodplain, excluding the permanent billabongs and swamps. The fringe zone and seasonally inundated habitats contain a large number of annual plant species; that is, species that complete their life cycle in one growing season and have a carry-over of seeds to the next growing season.

The annual plant growth habitat usually predominates in wetlands that experience alternate periods of wetting and drying (Mitchell and Rogers,

1985). As the germination and establishment requirements of annual wetland plant species are dependent on the rainfall and flooding pattern, there is significant year-to-year floristic change on the floodplain. These changes have been reported by Finlayson *et al.* (1989) in an analysis of the major vegetation communities on the floodplain, and later examined in a preliminary series of seedbank analyses (Finlayson *et al.*, 1990*b*). Despite this variation it has still been possible to identify broad vegetation communities that are dominated by one or two plant species (Finlayson *et al.*, 1989). Of the ten plant communities identified, four covered about 80% of the floodplain: *Melaleuca* open forest and woodland 44% (8680 ha); *Oryza* grassland 12% (2730 ha); *Hymenachne* grassland 9% (1930 ha); and *Pseudoraphis* grassland 14% (3050 ha). It is these communities, dominated by *Melaleuca* species, *Oryza meridionalis*, *Hymenachne acutigluma* and *Pseudoraphis spinescens* respectively, that are considered in the metal cycling budget. Of the grasses *O. meridionalis* is an annual species, and *H. acutigluma* and *P. spinescens* are perennial. The *Melaleuca* trees are evergreen with substantial litterfall throughout the year (Finlayson, 1988).

13.3 METAL BUDGET

Under the operational conditions established for the Ranger uranium mine, direct releases of stored water to Magela Creek were envisaged (Fox *et al.*, 1977) and later accepted as a necessary option in the water management strategy for the mine (Johnston, 1991). Therefore, unless alternate water disposal methods are adopted, contaminants from the Ranger mine site could be transported to Magela Creek, and possibly to the floodplain, by surface water. As part of a comprehensive programme to develop a management strategy for any such releases and to determine their most likely effects on the creek and floodplain biota, extensive chemical sampling was undertaken. These analyses are used to assess the relative importance of the water, vegetation and sediment in a metal budget for this ecosystem.

The water that flows down the creek at the start of each wet season is typically acidic, high in conductivity and contains high concentrations of sulphate (Walker and Tyler, 1985; Brown *et al.*, 1985). The source of these materials is groundwater brought to the surface by raised watertables and swept out of the pools and channels that remain in the creeks during the dry season by the advancing flood water. On entering the billabongs in the Mudginberri Corridor, the water quality, which has deteriorated over the dry season (Brown *et al.*, 1985), becomes worse. The water quality changes are well described in the above publications; an analysis of the loads of selected metals in the major inflow and outflow water from the floodplain, based on the work of Hart *et al.* (1987), is presented in this study.

The ultimate fate of metals contained in the water will depend on their physico-chemical form. Dissolved material may travel straight through the system and be discharged directly to the East Alligator River, or they may precipitate out. However, it is more likely that part, at least, will be taken up by the aquatic plants and sediments in the floodplain and perhaps recycled during subsequent growing periods. Particulate materials are more likely to become trapped on the floodplain, either by settling out because the reduced water flow on the floodplain cannot maintain the particulate matter in suspension, or by physical entrapment within the extensive areas of aquatic plants. These deposited materials may also be remobilised and transported further within the system. A conceptual model of the major processes involved in biogeochemical cycling on the Magela floodplain was developed by Hart *et al.* (1987) and is presented in Figure 13.3. The main components in this model are the water, vegetation and sediment. The contribution of these components to the budget and cycling processes on the Magela floodplain are separately considered below.

13.3.1 WATER

Hart *et al.* (1987) estimated the loads of particulate matter, major ions (sodium, potassium, calcium and magnesium) and trace metals (manganese, copper, lead, zinc and uranium) transported by Magela Creek and contained in the rainfall that fell directly on the floodplain. These data were then combined with the estimated loads of materials discharged from the floodplain

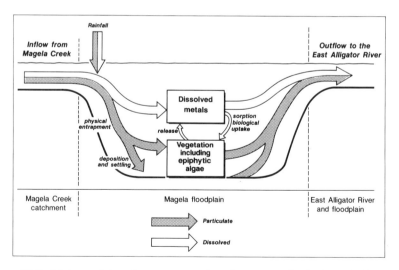

Figure 13.3. Major metal cycling processes thought to occur in the Magela floodplain ecosystem. Adapted from Hart *et al.* (1987)

to estimate the net deposition or removal of each substance from the floodplain.

A summary of the estimated annual loads of major ions and trace metals added to the catchment from rainfall and transported from the catchment to the floodplain along Magela Creek to a sampling site in the Mudginberri Corridor is given in Table 13.1. Overall, the annual loads estimated in this study are low in comparison with other tropical rivers, as is the annual denudation rate of 39 kg ha^{-1} over the whole catchment (see comparative data in Hart *et al.*, 1987). Most of the sodium and potassium transported from the catchment by Magela Creek comes from rainfall, whereas the catchment appears to contribute additional calcium and magnesium. These data point to the fact that there is very little interaction between the rainfall and the catchment.

The annual input–output budget for the Magela floodplain calculated by Hart *et al.* (1987) is given in Table 13.2. This is based on the annual input to Mudginberri Corridor and floodplain from surface runoff, the contribution to this area from annual rainfall, and the load discharged annually from the floodplain to the East Alligator River. These values were calculated from analyses on samples collected during the wet season of approximately three months duration. The net difference between the total input and output is also given, with net deposition shown as positive values and net release shown as negative values.

Information for this analysis came from sampling conducted in Magela Creek; however, about half of the water that enters the floodplain comes from sources other than Magela Creek. The quality of the water from these sources

Table 13.1. Amount of material added annually to the Magela Creek catchment and that transported from the catchment to the Mudginberri Corridor section of the Magela floodplain. Error estimates are given in brackets. See Hart *et al.* (1987) for sampling and analytical details and the estimation of errors

	Amount added by rainfall	Amount transported in creek water
Sodium (t)	110 (98)	150 (42)
Potassium (t)	43 (51)	54 (22)
Calcium (t)	14 (34)	45 (71)
Magnesium (t)	14 (24)	60 (21)
Iron (t)	—^{a}	89 (105)
Manganese (t)	1 (1)	2 (1)
Copper (kg)	290 (160)	150 (280)
Lead (kg)	215 (210)	30 (16)
Zinc (kg)	3000 (3000)	120 (115)
Uranium (kg)	60 (30)	30 (110)

^{a} Missing value.

Table 13.2. Calculation of material loads entering and transported from the Magela floodplain during the 1982–1983 wet season. Error estimates are given in brackets. See Hart *et al.* (1987) for sampling methods and the estimation of errors

	Material load of			
	Runoff	Rainfall	Output	Net load
Sodium (t)	370 (90)	33 (13)	630 (150)	−220 (180)
Potassium (t)	120 (50)	13 (16)	200 (55)	−60 (75)
Calcium (t)	120 (150)	4 (11)	310 (72)	−180 (170)
Magnesium (t)	170 (46)	4 (7)	390 (110)	−220 (120)
Iron (t)	230 (230)	*a*	680 (190)	−450 (300)
Manganese (t)	6 (3)	1 (1)	2 (1)	+4 (3)
Copper (kg)	410 (610)	89 (50)	65 (35)	+440 (610)
Lead (kg)	83 (35)	67 (65)	22 (12)	+130 (75)
Zinc (kg)	250 (250)	910 (930)	87 (47)	+1100 (970)
Uranium (kg)	42 (240)	18 (9)	11 (6)	+48 (240)

a Missing value.

was assumed, on the basis of limited data (Hart *et al.*, 1987), to be the same as that in Magela Creek. Thus, the total annual input loads given in Table 13.2 were determined by direct measurements from Magela Creek and extrapolation to the other sources.

As the uncertainty in the values given in Table 13.2 is high, it is only possible to discuss general trends. Broadly, the floodplain appears to be a net source of major ions (sodium, potassium, calcium and magnesium) and possibly a net sink for trace metals (copper, lead, zinc and uranium). As there is also a net loss of suspended materials from the inflowing water to the floodplain it is not unexpected that trace metals will also be deposited on the floodplain (Hart *et al.*, 1987), particularly as the total fraction of these metals will be influenced by the fate of particulate matter in the water.

13.3.2 VEGETATION

The macrophytic vegetation on the floodplain has the potential to absorb nutrients and trace metals from either the sediment or the water. In terms of chemical cycling, therefore, such plants have the potential to transfer nutrients and metals between the sediment and the water.

An estimate of the weights of selected metals absorbed by the grasses, *P. spinescens*, *H. acutigluma* and *O. meridionalis* on the Magela floodplain, and later added to the detrital and/or debris "pool" on the floodplain, can be calculated from above-ground biomass weight and chemical analysis data (Finlayson, *et al.*, 1987; Finlayson, 1991*b*). Plant sampling was conducted every four weeks throughout the period October 1983 to February 1985, with five replicate samples being collected on each site (Finlayson, 1991*b*). The chemical analyses were performed on acid-digested samples by the methods given in Finlayson (1991*b*). Below-ground biomass was not quantitatively assessed due to severe sampling difficulties in the wet season when the sites were covered by 1–2 m of flood water. Thus, it is not possible to include this component of the plant biomass in the load calculations. It is beyond doubt, however, that this material will play a role in the metal/nutrient cycling processes.

The total amount of selected metals contained in each grass species at the time of maximum (above-ground) biomass weight has been calculated using the available weight and chemical data (Table 13.3). The values in Table 13.3 are calculated from metal concentrations at the time of maximum above-ground biomass and extrapolated to the entire area covered by each of the grass species.

As the three grasses attained their maximum biomass weight at slightly different times during the wet season (Table 13.3 and Figure 13.2), and the metal concentrations in the individual species were not necessarily at a maximum at these times (Finlayson, 1991b), this comparison has limited value. However, it does demonstrate that the amount of each metal contained in these three species (Table 13.3) far exceeds the amount contained in the water flowing into the floodplain over the wet season (Table 13.2). As water samples from the vicinity of the plant sampling sites were not collected, it is not possible to directly compare the loads of metals in the plants and in the water on the floodplain. The comparison with the inflowing water is simplistic as it does not take into account the chemical interactions and transfer of

Table 13.3. Total amounts of selected metals contained in three grass species at the time of maximum biomass of each species on the floodplain

	Pseudoraphis spinescens	*Hymenachne acutigluma*	*Orzya meridionalis*
Sample date	15 May 1984	19 June 1984	9 April 1984
Sodium (t)	130	40	30
Potassium (t)	815	600	165
Calcium (t)	150	70	25
Magnesium (t)	80	30	20
Iron (t)	150	120	30
Manganese (t)	120	5	4
Copper (kg)	1020	680	140
Lead (kg)	20	15	20
Zinc (kg)	1940	680	280
Uranium (kg)	40	5	2

metals between the water and the floodplain sediment. However, as the plants are likely to have absorbed the metals from the sediment, and the total load of metal contained in their tissue far exceeds that in the inflowing water, it is possible to conclude that the inflowing water does not greatly contribute to the metal budget of the plants.

The values in Table 13.3 represent the total amount of each metal contained in each of the three grass species. They do not represent the amount that could be cycled back into the water by the senescing and decaying plant material. This is particularly so for the perennial *P. spinescens* and *H. acutigluma* species which do not completely senesce. Furthermore, some species have several growth and senescence phases over the year and the concentration of metals contained in the plant tissues varies (Finlayson, 1991*b*). A better indication of the role of the vegetation in the turnover of metals in this ecosystem can be derived by considering the amount of material that potentially enters the detritus–debris pool on the floodplain. For the metals absorbed, or even adsorbed, by the plants they must be released or returned to the sediment and water if they are to potentially become available to other organisms. This is most likely to occur during periods of senescence when substantial amounts of plant matter are added to the detritus–debris pool on the floodplain. This can occur at different times of the year, but is most apparent during and after the wet season when the multitude of aquatic/wetland plants grow, reproduce and senesce (Finlayson *et al.*, 1988).

The periods of biomass decline of the three major grass species are assumed to represent the loss of plant material and absorbed metallic substances to the detritus–debris pool on the floodplain (see Finlayson, 1991*b*). By subtracting the amount of each metal contained in the plant biomass at the end of the period of biomass decline from the amount at the start of this period, an estimate of the load of these substances in the detritus–debris pool can be obtained. Direct estimates of the amount of detrital material contained in the water are not available, except for the indirect and partial estimate available from particulate material analyses. Direct sampling of detrital material in such a dynamic, large and diverse ecosystem raises immense logistical problems (see, e.g. Hart *et al.*, 1987; Finlayson, 1991*b*).

The *Melaleuca* woodlands and forests also contribute a large amount of litter to the detritus and debris pool on the floodplain (Finlayson, 1988) which is potentially available for turnover between the plants, water and sediment. The total annual loads of selected metals added to this pool from the leaf litter can be calculated from the leaf litter values of Finlayson (1988) and Finlayson *et al.* (1994) and chemical analyses of this leaf litter (Finlayson, unpublished data).

For the purposes of this exercise the actual fate of these substances in the senescing and decomposing plant material has not been considered, although it is recognised that this is an important aspect of chemical turnover (that is, cycling and even recycling). The results of the above-described analyses are given in Table 13.4. The values in Table 13.4 represent the total calculated chemical content of the detritus–debris assumed to originate from each of the four dominant plant species.

Differences between the loads potentially turned over by the grasses and trees are very obvious from the data presented in Table 13.4. The *Melaleuca* litter contained substantially more magnesium, calcium and sodium than did the grass detritus–debris. In contrast, *H. acutigluma* contributed substantially more potassium, and *P. spinescens* more iron, copper, zinc and uranium to the detritus-debris pool than did the other species.

Whilst the plant species described above are dominant over large areas of the floodplain, it must be remembered that there are another 220 plant species that also accumulate nutrient and non-nutrient metals and can also potentially contribute to the chemical loads contained in the detritus–debris pool on the floodplain. Therefore, the loads given in Table 13.4 for the plant material cannot be taken as total values to this component of the metal budget. The extent to which the contribution of the vegetation to the budget is under-estimated by only considering a small number of dominant plant species is not known. However, it could be substantial.

The fate of the metals contained in the decaying plant material will be dependent on many factors, including the actual rate of decomposition. These rates have not been specifically determined, although they can be estimated from the biomass sampling reported by Finlayson (1991*b*) which showed periods of biomass decline for each species at some stage during the wet season. Without going into a complete description of the growth patterns of these species, it is evident from these data that the plants senesce and decompose quickly under the warm conditions that exist on the floodplain during the wet season. For example, 82% of the weight of *P. spinescens* (amounting to about 1.4 kg m^2) was lost or senesced over a 12 week period and

Table 13.4. Chemical content of wetland grass detritus–debris and *Melaleuca* litter deposited on the Magela floodplain over a one year period

	Pseudoraphis spinescens	Hymenachne acutigluma	Orzya meridionalis	Melaleuca species
Sodium (t)	30	15	30	60
Potassium (t)	250	1040	150	110
Calcium (t)	70	60	20	630
Magnesium (t)	80	90	20	150
Iron (t)	110	80	30	2
Manganese (t)	10	10	5	1
Copper (kg)	880	410	80	60
Lead (kg)	5	5	10	20
Zinc (kg)	1400	1200	240	390
Uranium (kg)	115	2	1	20

all the *O. meridionalis* (approximately 0.5 kg m^2) over an 8 week period. The relative roles of grazing animals and microorganisms in the breakdown of this plant material is not known.

The *Melaleuca* leaf litter decomposes at a much slower rate, especially during the dry season, with 82% of the dry matter taking seven months to decompose (Finlayson *et al.*, 1984*a*). Decomposition of the leaf litter is much faster during the wet season, although the amount of litter that accumulates at any location is greatly affected by water flows across the floodplain (Finlayson *et al.*, 1994). Movement of leaf litter across the floodplain by water is most apparent early in the wet season. During the wet season when the forests are flooded, very little litter actually accumulates under the trees. The fate of this material, mainly leaves (Finlayson, 1988; Finlayson *et al.*, 1994), is not known. As with the grasses, a complete analysis of the decomposition processes remains to be done and it is not possible to determine how differing litter accumulation and transport patterns and rates of plant tissue decomposition affect the cycling of metals in this ecosystem.

Periodically during the dry season, sections of the floodplain are burnt. This is not a regular event, but at times the fires can burn large areas of dry grass/sedgeland and even destroy large *Melaleuca* trees, as reported by Williams (1984). The influence of burning on the accumulation and cycling of metals has also not been investigated.

15.3.3   SEDIMENTS

The physico-chemical conditions on the floodplain during the wet season will have an important bearing on the availability, or otherwise, of metals to the plants, either rooted in the sediment or free-floating in the water. As this compartmental comparison is being restricted to the loads of metals without considering their specific availability, it is not necessary to dwell on these many and complicated interrelated factors. Rather, a straightforward analysis of the total metal load in the sediments of the Magela floodplain is presented. However, even this simple approach is not without inherent difficulties; for example, the sediments are neither uniform in composition nor evenly distributed across the floodplain.

To overcome some of these difficulties, sediment samples were collected from each of the four plant communities considered in the vegetation analysis above, and which cover approximately 80% of the floodplain area. These samples were acid digested and chemical analyses done by standard techniques (see Finlayson, 1991*b*). The results of these analyses were used to calculate the total metal loads in the four communities being considered in the vegetation analysis section (Table 13.5). These values could also be extrapolated to the 20% of the floodplain not dominated by these four plant communities to represent the situation over the entire floodplain. However, this does not alter

Table 13.5. Chemical content of the upper 10 cm of floodplain sediment in three different grass communities and in Melaleuca forests

	Pseudoraphis spinescens	Hymenachne acutigluma	Orzya meridionalis	Melaleuca species
Sodium (t)	2	2	5	10
Potassium (t)	10	25	30	35
Calcium (t)	5	5	10	10
Magnesium (t)	10	10	20	20
Iron (t)	45	50	140	225
Manganese (kg)	180	200	450	580
Copper (kg)	105	40	40	295
Lead (kg)	20	30	30	370
Zinc (kg)	50	80	125	225
Uranium (kg)	10	5	5	40

the basic comparison, summarised in Table 13.6, between the metal loads in the water, the vegetation and the sediments.

The uppermost 10 cm of sediment was chosen as the sample depth as it contains 60–90% of the root/rhizome biomass of the aquatic grasses on the floodplain (unpublished data). Thus, this approximates to the depth of sediment that contains metals that could be available to these plants. *Melaleuca* trees have roots that extend much deeper than this, although the extent of root penetration in these trees was not determined. A greater depth, say 100 cm, may give a better estimate of the amount of metals available to these plants. However, in terms of the comparison of metal loads, all this will indicate is that the sediments contain even more metal than is indicated in Table 13.6.

Table 13.6. Metal content of the plant detritus–debris pool, the sediment and the creekwater entering the floodplain

	Detritus–debris pool	Sediment	Creekwater
Sodium (t)	135	20×10^6	370
Potassium (t)	1550	100×10^6	120
Calcium (t)	780	30×10^6	120
Magnesium (t)	340	60×10^6	170
Iron (t)	220	460×10^6	230
Manganese (t)	30	1410×10^6	5
Copper (kg)	1430	480×10^6	410
Lead (kg)	40	450×10^6	80
Zinc (kg)	2230	480×10^6	250
Uranium (kg)	140	60×10^6	40

13.4 DISCUSSION

The combined results of the metal analyses for the water, vegetation and sediment components on the floodplain are presented in Table 13.6. The sediment component very clearly dominates the floodplain metal budget. Even allowing for the contribution made to the plant detritus–debris pool by other plant species this situation is unlikely to be altered. However, this comparison does not contain any information on the relative importance of the water and sediment in the cycling of metals on the floodplain. It only demonstrates that the greatest proportions of metals in this wetland ecosystem are contained within the sediments. The proportion of the metals that are potentially available to the plants and hence could be cycled through this pathway is, however, unknown. Further chemical analyses are necessary to determine this. The comparison contained in Table 13.6 simply demonstrates the basic mass loading differences between the vegetation, water and sediment components of the Magela wetland ecosystem.

The availability of metals to the plants will be affected by the physico-chemical conditions that prevail in the water, particularly at the sediment–water interface. As shown by chemical analyses of water from the Magela (Hart *et al.*, 1982; Walker and Tyler, 1985; Brown *et al.*, 1985), these conditions vary throughout the year and, based on analyses elsewhere in northern Australia (Finlayson *et al.*, 1980; Finlayson and Gillies, 1982), also vary over the course of a single day. Further data from the Magela ecosystem are required before the availability of individual metals to the different plant species can be assessed.

Except for sodium, iron and lead, the plant detritus–debris pool contains greater metal loads than that contained in the creekwater that enters the floodplain (Table 13.6). The immediate conclusion from the broad analysis, summarised in Table 13.6, is that the Magela catchment only contributes a small amount of the total metal load contained in the vegetation debris and detritus on the floodplain during the wet season. This corresponds with the conclusions reported in Hart *et al.* (1987) that the denudation rate in the catchment is relatively low for a tropical stream and that there is very little interaction between the rainfall and the highly weathered catchment.

Hart *et al.* (1982) reported that only small amounts of particulate matter are transported in Magela Creek, but that it contains a large proportion of the iron, manganese, copper and zinc. Furthermore, the particulate matter does, at times, contain higher concentrations of heavy metals than floodplain sediments (Hart and Beckett, 1986). As particulate matter has a high adsorption capacity it could be an important mechanism for transport of heavy metals around the Magela ecosystem.

The amount of organic matter in the water will also be an important factor in cycling and transport of heavy metals on the floodplain. Suspended particulate matter in slow-moving water on the floodplain was found by Hart and Beckett (1986) to be comprised of 30–50% organic material. This was assumed to be primarily algal biomass. The role of the algal biomass, including the large quantities of epiphytic and filamentous algae that develop on the submerged plants, in the metal transport and cycling processes is not known. The fate of the large quantities of organic matter that originate from the senescing macrophytes and decomposing litter on the floodplain is also not known.

Thus, available information on the Magela Creek ecosystem enables a broad metal budget to be calculated, but further chemical and even physiological studies will be necessary before the intricate nature of the metal cycling processes will be known. Whilst there are many gaps in this analysis of the metal budget, the available information far exceeds that available for many wetland and even lake ecosystems, particularly in northern Australia.

Finlayson et al. (1984b) determined the annual loading of a number of nutrients and heavy metals to Lake Moondarra, an artificial lake in north-western Queensland. The copper and zinc loads (59 and 182 kg respectively) were less than those calculated for the Magela floodplain, whereas for lead (74 kg) the load was similar. However, the source of these metals (and other materials) was a sewage flow that, at the time of the investigations, constituted the only continuous inflow of water into the lake; the river only flowed after occasional monsoonal storms. The large amounts of aquatic vegetation, both submerged and free-floating species, in the lake at this time, were assumed capable of accumulating a large proportion of the metal contained in the sewage (Finlayson et al., 1984b). However, since this investigation much of the aquatic vegetation has died and the sewage has been diverted.

The Lake Moondarra situation only provides part of the necessary data for a comparison to the metal budget for the Magela floodplain, but the vastly different climatic and hydrological conditions subtract even further from the value of the comparison. It does, however, demonstrate that the vegetation can accumulate a large component of the heavy metal load contained in water inputs to a wetland ecosystem. Thus, as with the Magela floodplain, this vegetation could play a major role in metal turnover in the ecosystem.

ACKNOWLEDGEMENTS

The information used in this chapter is derived from scientific papers and reports published by staff and consultants at the Alligator Rivers Region Research Institute in Jabiru, Australia. Many people have been involved in this programme and the author thanks all of those who supported his own research activities when an employee of the Institute. Bruce Bailey and Ian Cowie, in particular, made an inestimable contribution to the plant sampling projects.

REFERENCES

Brown, T. E., Morely, A. W. and Koontz, D. V. (1985) The limnology of a naturally acidic tropical water system in Australia. II. Dry season characteristics. *Verhandlung Internationale Vereinigung fur Theoretische und Angewandte Limnologie.*, **22**, 2131–2135.

Finlayson, C. M. (1988) Productivity and nutrient dynamics of seasonally inundated floodplains in the Northern Territory. In: Wade-Marshall, D. and Loveday, P. (Eds) *North Australia: Progress and Prospects, Vol. 2, Floodplain Research*, pp. 58–83. ANU Press, Darwin.

Finlayson, C. M. (1991a). Plant ecology and management of an internationally important wetland in monsoonal Australia. In: Kusler, J. A. and Day, S. (Eds) *Proceedings of an International Symposium on Wetlands and River Corridor Management*, pp. 90–98. Charleston, South Carolina.

Finlayson, C. M. (1991b). Primary production and major nutrients in three grass species on a tropical floodplain in northern Australia. *Aquatic Botany*, **41**, 263–280.

Finlayson, C. M. and Gillies, J. E. (1982) Biological and physiochemical characteristics of the Ross River Dam, Townsville. *Australian Journal of Marine and Freshwater Research*, **33**, 811–827.

Finlayson, C. M. and Von Oertzen, I. A. M. L. (1992) Wetlands of Australia: Northern (tropical) Australia. In: Whigham, D. F., Dykyova, D. and Heijny, S. (Eds) *Wetlands of the World*. Vol. 1, pp. 195–243. Kluwer, Dordrecht.

Finlayson, C. M., Farrell, T. P. and Griffiths, D. J. (1980) Studies of the hydrobiology of a tropical lake in north-western Queensland. II. Seasonal changes in thermal and dissolved oxygen characteristics. *Australian Journal of Marine and Freshwater Research*, **31**, 589–596.

Finlayson, C. M., Bailey, B. J. and Cowie, I. D. (1984a) Recycling of nutrients, heavy metals and radionuclides in paperbark swamps. Summary in Alligator Rivers Region Research Institute Research Report 1983–84, pp. 18–20. AGPS, Canberra.

Finlayson, C. M., Farrell, T. P. and Griffiths, D. J. (1984b). Studies of the hydrobiology of a tropical lake in north-western Queensland. II. Growth, chemical composition and potential for harvesting of the aquatic vegetation. *Australian Journal of Marine and Freshwater Research*, **35**, 525–536.

Finlayson, C. M., Bailey, B. J. and Cowie, I. D. (1987) Cycling of nutrients and heavy metals by floodplain vegetation. Summary in Alligator Rivers Region Research Institute Annual Research Summary 1986–87, pp. 113–117. AGPS, Canberra.

Finlayson, C. M., Bailey, B. J., Freeland, W. J. and Fleming, M. (1988) Wetlands of the Northern Territory. In: McComb, A. J. and Lake, P. S. (Eds) *The Conservation of Australian Wetlands*, pp. 103–106. Surrey Beatty & Sons, Sydney.

Finlayson, C. M., Bailey, B. J. and Cowie, I. D. (1989) Macrophytic vegetation of the Magela floodplain, northern Australia. Office of the Supervising Scientist Research Report No. 5.

Finlayson, C. M., Bailey, B. J. and Cowie, I. D. (1990a). Characteristics of a seasonally flooded freshwater system in monsoonal Australia. In: Whigham, D. J., Goode, D. F. and Kvet, J. (Eds) *Wetland Ecology and Management—Case Studies*, pp. 141–162. Kluwer Academic, Dordrecht.

Finlayson, C. M., Cowie, I. D. and Bailey, B. J. (1990b) Sediment seedbanks in grassland on the Magela Creek floodplain, northern Australia. *Aquatic Botany*, **38**, 163–176.

Finlayson, C. M., Wilson, B. and Cowie, I. D. (1991) Management of freshwater monsoonal wetlands: conservation threats and issues. In: Donohue, R. and Phillips, W. (Eds) *Proceedings of a Wetland Workshop*, pp. 109–117. ANPWS, Canberra.

Finlayson, C. M., Cowie, I. D. and Bailey, B. J. (1993) Litterfall in a *Melaleuca* forest on a seasonally inundated floodplain in tropical northern Australia. *Wetland Ecology and Management* **2**, 177–188.

Fox, R. W., Kelleher, G. G. and Kerr, C. B. (1977) *Ranger Uranium Environmental Inquiry, Second Report*. AGPS Canberra.

Johnston, A. (1991) Water management in Alligator Rivers Region: A research view. In: Hynes, R. V. (Coordinator) *Proceedings of the 29th Congress of the Australian Society of Limnology*, pp. 10–34. Jabiru, N.T. AGPS, Canberra.

Hart, B. T. and Beckett, R. (1986) The composition of suspended particulate matter from the Magela Creek system, northern Australia. *Environmental Technology Letters*, **7**, 613–624.

Hart, B. T. and McGregor, R. J. (1982) Water quality characteristics of eight billabongs in the Magela Creek catchment. Office of the Supervising Scientist Research Report No. 2.

Hart, B. T., Davies, S. H. R. and Thomas, P. A. (1982) Transport of iron, manganese, cadmium, copper and zinc by Magela Creek, Northern Territory, Australia. *Water Research*, **16**, 605–612.

Hart, B. T., Ottaway, E. M. and Noller, B. N. (1987) Magela Creek system, northern Australia. II. Material budget for the floodplain. *Australian Journal of Marine and Freshwater Research*, **38**, 861–876.

Mitchell, D. S. and Rogers, K. H. (1985) Seasonality/aseasonality of aquatic macrophytes in southern hemisphere inland waters. *Hydrobiologia*, **125**, 137–150.

Morton, S. R., Brennan, K. G. and Armstrong, M. D. (1990*a*) Distribution and abundance of magpie geese, *Anseranas semipalmata*, in the Alligator Rivers Region, Northern Territory. *Australian Journal of Ecology*, **15**, 307–320.

Morton, S. R., Brennan, K. G. and Armstrong, M. D. (1990) Distribution and abundance of ducks in the Alligator Rivers Region, Northern Territory. *Australian Wildlife Research*, **17**, 573–590.

Vardavas, I.M. (1988) A water budget for the tropical Magela floodplain. *Ecological Modelling*, **46**, 165–194.

Walker, T. D. and Tyler, P. A. (1985) Tropical Australia, a dynamic limnological environment. *Verhandlungen Internationale Vereinigung fur Theoretische und Angewandte Limnologie*, **22**, 1727–1734.

Walker, T. D., Waterhouse, J. and Tyler, P. A. (1984) Thermal stratification and the distribution of dissolved oxygen in billabongs of the Alligator Rivers Region, Northern Territory. Office of the Supervising Scientist Open File Record 28.

Williams, A. R. (1984) Changes in *Melaleuca* forest density on the Magela floodplain, Northern Territory, between 1950 and 1975. *Australian Journal of Ecology*, **9**, 199–202.

Index